Geospatial Technology for Earth Observation

Deren Li • Jie Shan • Jianya Gong
Editors

Geospatial Technology for Earth Observation

Springer

Editors
Deren Li
State Key Laboratory of Information Engineering in Surveying, Mapping and Remote Sensing
Wuhan University
129 Luoyu Road
430079 Wuhan
China, People's Republic

Jie Shan
School of Civil Engineering
Purdue University
550 Stadium Mall Drive
West Lafayette, IN 47907-2051
USA
jshan@ecn.purdue.edu

Jianya Gong
State Key Laboratory of Information Engineering in Surveying, Mapping and Remote Sensing
Wuhan University
129 Luoyu Road
430079 Wuhan
China, People's Republic

ISBN 978-1-4419-0049-4 e-ISBN 978-1-4419-0050-0
DOI 10.1007/978-1-4419-0050-0
Springer New York Dordrecht Heidelberg London

Library of Congress Control Number: 2009933773

Printed on acid-free paper

Springer is part of Springer Science+Business Media (www.springer.com)

Table of Contents

About the Editors

Deren Li, Wuhan University

He is the chair of the Academic Committees of Wuhan University and the State Key Laboratory of Information Engineering in Surveying, Mapping and Remote Sensing, academician of the Chinese Academy of Sciences and the Chinese Academy of Engineering, and holds an honorary doctorate from ETH, Zurich. He received the Samuel G. Gamble Award at the ISPRS XXI Congress in 2008, and numerous academic honors.

Jie Shan, Purdue University

He is an associate professor in the School of Civil Engineering, and the recipient of multiple academic awards, among which are the Talbert Abrams Grand Award, the ESRI Award for Best Scientific Paper in GIS (First Place), and the Excellence in GIS Award. He is an associate editor for IEEE Geoscience and Remote Sensing and the assistant editor for Photogrammetric Engineering and Remote Sensing.

Jianya Gong, Wuhan University

He is a professor and the director of the State Key Laboratory of Information Engineering in Surveying, Mapping and Remote Sensing. He has been a visiting professor at the University of Massachusetts, Boston; Hong Kong Polytechnic University; George Mason University. Prof. Gong has served multiple times as the ISPRS Commission Secretary and Working Group Chair since 1988.

Acknowledgement

As the editors of this volume, we are most grateful to each and every one of the 63 chapter authors for their commitment and collaboration during the preparation of this book. We greatly appreciate their willingness to share their recent work and their patience in iteratively revising the manuscripts. Without such tireless and persistent effort, this book would not have been published.

We also wish to extend our most sincere thanks to the following reviewers, who helped us improve the book in spite of their busy schedules and other commitments: Peggy Agouris (George Mason University), Larry Biehl (Purdue University), Chuqun Chen (South China Sea Institute of Oceanology, Chinese Academy of Sciences), Peijun Du (China University of Mining and Technology), Henri Eisenbeiss (Swiss Federal Institute of Technology, Zurich), Clive Fraser (University of Melbourne), Ejaz Hussain (Purdue University), Xiuping Jia (University College, The University of New South Wales, Australian Defence Force Academy), John Kerekes (Rochester Institute of Technology), Richard Onchaga (International Institute for GeoInformation Science and Earth Observation, ITC), Toni Schenk (The Ohio State University), Charles Toth (The Ohio State University), Chaowei Yang (George Mason University), and Zhengwei Yang (National Agricultural Statistics Service, United States Department of Agriculture). Besides, we would also like to thank David Shan, who carefully read a number of chapters, edited the language, and formatted the texts for accuracy and consistency.

Our deepest thanks go to our families, whose understanding and support allowed us to spend endless nights and numerous weekends with the authors, the reviewers, and the publisher in the past one and an half years to work on this book project.

Finally, applying our novice Microsoft Word skills, the book is a product of self-formatting and "typesetting," and errors and inconsistencies in the book remain to be ours.

Deren Li (Wuhan)
Jie Shan (West Lafayette)
Jianya Gong (Wuhan)

Preface

Earth observation is currently a critical issue of worldwide interest. It is used to map the Earth, monitor the environment, manage our limited resources, and predict, record, and respond to global change and disasters. All space programs around the world have a significant component of Earth observation to accomplish the above objectives. For such scientific explorations, geospatial technology plays a core and fundamental role throughout the entire data acquisition, processing, interpretation, and information extraction process. This rapid progression has driven geospatial technology into a transition period. On one hand, sensing technology provides abundant data with increasingly higher radiometric, spatial, spectral, and temporal resolutions. On the other hand, in terms of data processing and interpretation techniques, various new developments are being introduced to geospatial technology from a number of related disciplines, including machine vision, pattern recognition, computational science, and applied mathematics. Driven by such new developments, the solutions to classical mapping-related tasks are moving towards higher level automation and resolution. In the meantime, a new, non-classical task is consequently emerging, which is meant to serve the public and decision-makers with all types of geospatial products, including data, information, and knowledge.

The book covers a wide range of topics in Earth observation, which can be grouped into seven parts with a total of 19 chapters.

Part I pertains to system, platform, sensor and sensor integration. The discussion varies from space system (Chapter 1, Li), to aerial platform, in particular UAV (Chapter 2, Yan *et al.*) and mobile vehicles (Chapter 3, Li *et al.*), all of which utilize integrated sensors for Earth observation data collection at a different scale for a variety of applications, such as large scale topographic mapping and road surface evaluation. Chapter 1 also presents an overview on geospatial information service as an extension to the conventional Earth observation tasks.

Part II, consisting of Chapter 4 (Crespi *et al.*) and Chapter 5 (Yuan), deals with the geometric aspects of data processing in Earth observation. Both chapters discuss the principles of the physical model and the generalized or replace model, with the former more focused on satellite images while the latter is focused on both aerial and satellite images. Both chapters provide empirical examples with evaluation and comparison of the two types of models. Moreover, Chapter 4 provides numerical solutions to assure reliable results in sensor orientation.

Part III concentrates on some physical aspects of remote sensing. Chapter 6 (Liang *et al.*) presents algorithms for mapping all types of land surface radiative fluxes from MODIS data. The developed algorithms are evaluated by using well-distributed global measurements over various land cover types at different elevations. Such a study is useful in driving high-resolution land ecosystem/carbon cycle and hydrological models. Chapter 7 (Gens) starts with a concise overview on surface reflective and emissive properties, electromagnetic spectrum, energy-matter interaction, and passive and active remote sensing. It then presents a wide range of vivid applications of remote sensing techniques, including urban areas, vegetation, biodiversity and coral reefs, geology and soils, water bodies, and cryosphere, some of which are further detailed in subsequent chapters.

Part IV focuses on algorithms for remote sensing image classification and analysis. Chapter 8 (Waske *et al.*) is a review of classification methods, varying from classical and advanced individual classifiers to their combined use (multi classifiers). Such methods are used for the classification of SAR, multispectral, and hyperspectral images. The classification of multiple data sources such as optical images and LiDAR are also presented. Chapter 9 (Zhang and Zhong) discusses topics related to hyperspectral image analysis: feature selection, clustering (unsupervised classification), and spectral unmixing. After a concise review of popular methods, the authors present a framework based on artificial immune systems as an alternative solution: the clonal selection approach is used for feature selection, while an artificial immune classifier is developed for hyperspectral images. The unmixing problem is formulated as a kernel-based least squares estimation. Chapter 10 (Han *et al.*) addresses the effect of resolution on image classification. Versions of Landsat TM and SPOT images under different aggregation rates are used for analysis and evaluation.

Part V is composed of Chapters 11 and 12. The former (Chapter 11, Trinder and Sowmya) discusses automated feature extraction in photogrammetry from the machine learning point of view. A prototype tool is discussed and evaluated for road network extraction from remote sensing images. As for building extraction, the chapter presents a solution based on level set segmentation and Dempster-Shafer fusion of multiple data sources. Chapter 12 (Lisini *et al.*) presents a framework for image-based feature extraction for geospatial analysis. Results of road and building extraction from SAR, LiDAR, and optical images are presented; and such a framework is also used for automated SAR image registration and change detection as an alternative approach to the pixel-based method.

Part VI extends the content of the book to web-based geospatial service. Chapter 13 (Gong *et al.*) outlines the concepts and details the contents, concerns, standards, technologies, and design of a geospatial service web, a prototype for which is reported. Chapter 14 (Zhu *et al.*) discusses the selection and optimization of web services in terms of response time. The developed algorithm is able to deal with uncertainty in service response time, which makes the remotely sensed image processing service chain more reliable.

Part VII consists of five chapters that primarily demonstrate various applications of the geospatial technology. Chapter 15 (Konecny) reports the challenges and start-of-the-art of cadastral mapping using integrated Earth observation techniques, ranging from satellite, aerial, and terrestrial surveying to documentation using amateur handled cameras and GIS. Chapter 16 (Zhou) seeks an efficient and practical method to determine the spatio-temporal pattern of environmental change caused by land cover change by integrating multi-temporal and multi-scale remotely sensed data from various sources. It quantifies spatial pattern for analyzing land cover changes through calculating the class-level landscape metrics of change trajectories. Water remote sensing is discussed in Chapter 17 (Chen and Yu). Starting with the effects of water body on electromagnetic radiation and its optical properties, this chapter discusses *in situ* and remote sensing measurements and atmospheric correction methods to acquire reliable and precise measurements. A variety of applications are also addressed, including oil spill detection, water depth, water temperature, and total suspended sediment concentration, colored dissolved organic matter concentration, chlorophyll-a concentration, primary productivity estimation, and red tide detection. Case studies are presented to demonstrate the performance of atmospheric correction and costal water environment management. Chapter 18 (Shan *et al.*) reviews the status of flood mapping in the USA and reports a case study in the state of Indiana. Temporal Landsat TM images are used to determine flood extent, while damages are estimated by jointly using crop and road GIS layers. A web-mapping service is used for public access and information dissemination. Chapter 19 (Zhang *et al.*) presents an overview of the role of Earth observation technology in the establishment of e-government and decision-making in China. The applications of such technology and developed systems in the recent Wenchuan earthquake (May 2008, China) are highlighted through slide identification, hazards assessment, and loss estimation. In addition, three intensive prototype systems and their associate studies are reported: the Flood Disaster Monitoring and Evaluation System for managing the recent floods and their mitigation; the Water Resources and Desertification Monitoring Analysis System, which provides related information services for the management and decision-making of government; the West Development Information Service System that provides important social and economic development and resource information for governments.

As reflected from the above summary, geospatial technology has played multiple roles in Earth observation, ranging from data acquisition, processing, analysis, information extraction, service, and decision-making. It became a key and indispensible core component in Earth observation for sciences, economics, and homeland security. It is unrealistic for one book of this size to cover all aspects of this technology, however, we attempted to select and organize the materials in such a way that the essential, fundamental methods and recent, promising developments in this and other related disciplines are covered. We also attempted to promote geospatial technology by including representative and influential applications and case studies that have broad impact.

The book is primarily targeted to graduate students and professionals. It can be used as a reference to augment courses in remote sensing, geographic information systems, photogrammetry, and natural resources management. The resources provided in the book, such as reference to literatures, free and shareware, data vendors or sources, and case studies are useful for both researchers and practitioners as well as administrators. We also expect the book to be introductory material to more advanced topics and future developments and applications. Finally, we welcome corrections, comments, and suggestions that would benefit the audience and improve our future endeavors.

Deren Li (Wuhan)
Jie Shan (West Lafayette)
Jianya Gong (Wuhan)

Chapter 1

AN OVERVIEW OF EARTH OBSERVATION AND GEOSPATIAL INFORMATION SERVICE

Deren Li

1.1 The Space-Air-Ground Borne Integrated Earth Observation System

A great progress in Earth sciences in the past century is Earth Observation (EO) from space and storing, managing, and distributing remotely sensed data via networked geographic information systems (GIS). Various topographic maps, thematic maps, ortho-images, and other products can be obtained from remotely sensed data by air-borne and space-borne remote sensing sensors within the electromagnetic spectrum, including visible, infrared, and microwave bands. Weather forecasts, natural resource exploitations, environmental monitoring, crop yield estimations, land use, land cover change, and prediction and prevention of natural hazards, such as sandstorms, droughts, floods, volcanoes, and earthquakes, can be carried out with Earth observation technology. New Earth observation technology improves the living standards of human beings and the development of social economy. It has contributed greatly to the sustainable development and research of Earth science.

With the invention of airplanes by Wright Brothers in 1903, it was possible to take aerial photographs for various applications. Though the advent of the first aerial camera was in World War I, modern earth observation did not start until 1930s, when a large number of aerial photographs were collected and interpreted. After the launch of the first man-made satellite in 1957, the protocol for Earth observation was established. For the first time, human beings could observe the planet they have lived on for several millions of years. A variety of remote sensing platforms (Table 1.1), including space-borne, near space, air-borne, and ground-based, allow observation from several thousands of meters to 40 thousands kilometers above the Earth. Remote sensing technology is experiencing unprecedented changes, as shown by the following 5 features:

1. Innovative sensors continue to be developed. While traditional frame cameras (B/W , color, color-infrared, and ultraviolet) are still widely used in remote sensing, new types of sensors, such as panoramic cameras,

D. Li et al. (eds.), *Geospatial Technology for Earth Observation*,
DOI 10.1007/978-1-4419-0050-0_1,

infrared radiometers, multispectral scanners, CCD linear scanners, microwave radiometers/scatterometers, laser altimeters and Lidar also come into use. These sensors acquire EO data in various bands of electromagnetic spectra, and recorded as hardcopy or softcopy for later use.

Table 1.1 Remote sensing platforms

Platforms	Altitudes	Applications
Geostationary satellites	36,000-40,000 km	EO for a given location
EO satellites (Polar orbit)	500～1,000 km	EO for a given period
Space shuttles Or Space Lab	240～350 km	EO, space experiments
Near space Boat	20-100 km	EO for a given Location
Tele-detection unmanned aircraft	100m～100 km	inventory
Super-altitude jets	10,000～12,000 m	military reconnaissance
Middle-low altitude airplane	500～8,000m	aerial photogrammetry
Aeroboat	500～3,000 m	military reconnaissance
Helicopter	100～3,000 m	photogrammetry
Tele-detection airplanes	Less than 500 m	photogrammetry
Glider	50～500 m	photogrammetry
Balloons	Less than 800 m	inventory
Ropeways	10～40 m	heritages
Cranes	5～50 m	ground-truthing
Mobile mapping vans	0～30 m	ground-truthing

2. Multi-scale image pyramids are created, providing EO data from coarse to fine resolution. Spatial resolution of satellite sensors ranges from 1.1 km per pixel for NOAA AVHRR, 250 m per pixel for MODIS, 79 m per pixel for Landsat MSS, 30 m pixel for Landsat TM, 20~10 m per pixel for SPOT1-2, 18 m per pixel for JERS-1, 5.8 m per pixel for IRS-1C, and 1 m, 0.61 m and 0.4 m per pixel for IKONOS, Quickbird and GeoEye-1. In addition, airborne and ground-based sensors can acquire images at a spatial resolutution of centimeters. Multi-temporal remote sensing data can be acquired over a given area. For instance, METEOSAT meteoroligical satellite revisits every 30 minutes, while NOAA AVHRR revists twice a day. The temporal resolutions of other commonly used satellites are: 3 days for ERS-1, 26 days for SPOT, and 3 days for IKONOS in 1 meter resolution mode and 1.5 days in 1.5 resolution mode. Through multiple satellite constellations, the revisit interval can be reduced to a few hours.

3. Hyper-spectral remote sensing remains an active reseach and potential application field. Two trends could be observed in the past decade: one was to make use of a wider range of spectrum, and the other was concerned with the design of narrower bandwidths. For instance, the ETM+ (Enhanced Thematic Mapper) on board of Landsat 7 has a high resolution panchromatic band in addition to 7 multispectral bands used in the past sensors with wide field-of-view for oceanographic applications were developed. Presently, imaging spectrometers with spectral resolution less than 5 nm have been used, such as EO-1 from the USA.
4. As active microwave sensors, the space- and air-borne SAR systems are very important because of their all-weather and all-time capabilities. The phase of the microwave return can be measured very accurately and thus travel path changes in wavelength of very small fractions can be detected. Interferometric SAR (InSAR), Differential InSAR (D-InSAR) and Permanent Scatters InSAR (PS-InSAR) are used for creating digital elevation models (DEMs) which determine terrain deformation and extract displacement information due to target motion with high accuracy.
5. Multi-band and multi-polarization SAR as well as their integration can provide more and more information for atmosphere, oceanography and coastal zones, water resources and floods, forestry and agriculture, land resources and environment, hazards, and other applications.

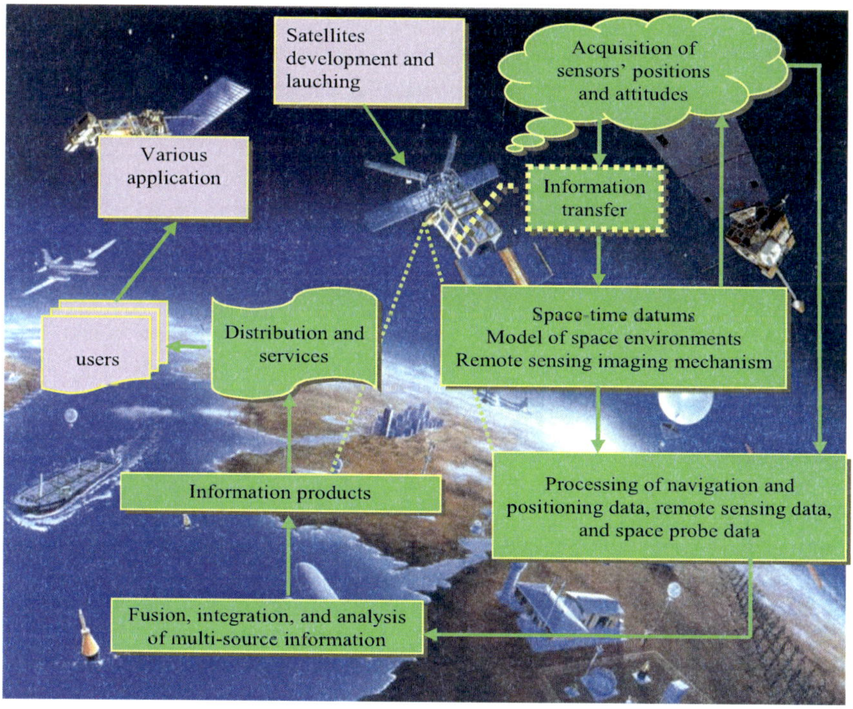

Fig. 1.1 The cycle of EO data acquisition, processing, and applications

In summary, data acquisition from space is now featured with multi-platforms, multi-sensors, and multi-scale earth observation capabilities with hyper spectral resolution, high spatial resolution, and high temporal resolution. In the future, multi-scale images will be acquired from space, near space, aerial, and ground platforms, using sensors covering a wide spectrum of ultraviolet, visible, infrared, microwave, laser radar, and terahertz with high spatial, spectral, and temporal resolutions. With the help of modern communication technology, a space-air-ground-integrated EO sensor web can be formed with the capability of all-day and all-weather-operation. Figure 1.1 describes the cycle of EO data acquisition, processing, information extraction and applications, which are the main contents of this book.

1.2 Grid Technology and Geospatial Data-Information-Knowledge -Service Transformation

1.2.1 Background

Although much progress has been made in geospatial data acquisition and applications, the utilization of acquired spatial data is still at a low level, partially due to the fact that they are acquired and managed by independent public sectors, which provide individual services to particular users. The importance of Spatial Data Infrastructure (SDI) information sharing across different organizations has been recognized by both government agencies and private sectors. However, there is a lack of effective tools to mine potentially rich information from them.

Geo-services aim at providing basic services to those who need them the most. These end users may be distributed geographically over different public sectors, and, in some cases, they may be distributed worldwide, such as the member countries of the Committee on Earth Observation Satellites (CEOS). Vögele *et al.* (2003) summarized the following 4 types of basic geo-services in the context of GeoShare Project carried out by the United Kingdom, Germany, Norway, and the Netherlands.

Web Map Services (WMS): Web Map Services produce map renderings of geo-referenced data. It is important to note that a "map" in WMS is simply a visual representation of geo-data that does not contain the data itself. A map may be rendered by WMS in raster image format or as vector graphics in the Scalable Graphics and Computer Graphics Metadata File formats. The Web Map Services Specification is approved by Open GIS Consortium (OGC).

Web Feature Services (WFS): Web Feature Services provide an assessment to geospatial data stored in a geo-database through HTTP requests. The results of such requests are returned in GML format, which is a XML-based geospatial vector data exchange format. Like WMS, WFS implements an approved OGC specification.

Geodatabases: Geodatabases contain geodata for meaningful WFS and WMS integration. Currently available geodatabases overcome the separation of geometric and attribute data, and the object-relational schema is much closer to OGC's concept of geographic features.

The intelligent category services (ICS): Intelligent Category Services play an important and central role in the GeoShare Network. It is used to manage the metadata descriptions of services, datasets, and data collections in the network. The ICS will thus provide a detailed inventory of all registered resources, and queries to ICS are the basis for the retrieval and selection of task-specific data and services.

In addition to web-services, there are many tools for LBS (Location based Services) and MLS (Mobile Location Services) through mobile communication networks. Figure 1.2 shows the LBS configuration and application interfaces.

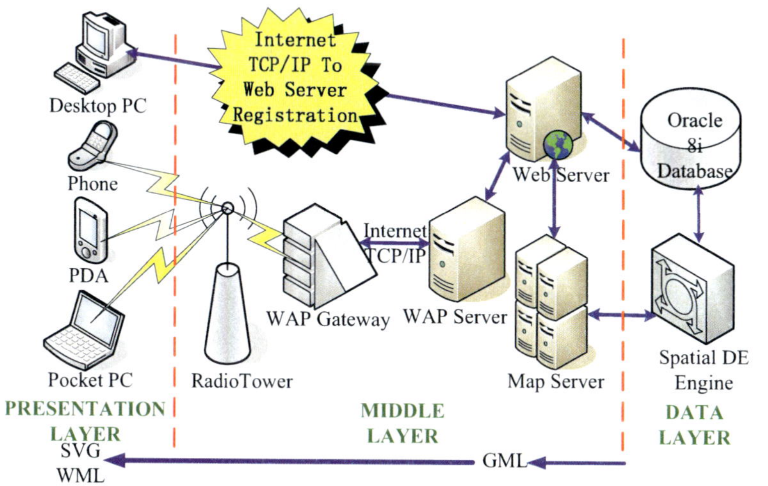

Fig. 1.2 System configuration of location based services

At least, the following problems need to be solved in order to meet the basic pre-requisites of geo-services previously mentioned:

1. The mechanism and approaches underlying intelligent services provided, including the modeling of intelligent services work flow, tasks allocation, intelligent search, and system structure of intelligent services platform
2. The standards of spatial information services and interoperability
3. The semantic model of spatial information services, which includes ontology based semantic model and spatial information service model
4. The implementation of spatial information services on different terminals, such as notebook computers, mobile phones, PDAs, TV sets, telephones, desktop computers, etc.

The central component in such Geo-Services is the integration of spatial information systems with computer networks. The information grid has been the mainstream technique in the third generation Internet. Spatial information grid aims at solving the problems listed above by the integration of grid technology and spatial information technology. This is the main trend of spatial information for the next 20 years. The GEOSS 10 Year Implementation Plan approved at the Ministerial Summit (Brussels, 2005), attended by more than 50 countries and organized by GEO, is a good example of this.

1.2.2 Grid technology: The Third Surge of the Internet

Grid technology has emerged as an important new field, distinguishing itself from conventional distributed computing by its focus on large-scale resource sharing, innovative applications, and, in some cases, performance orientation (Foster and Kesselman 1988).

Three hierarchies can be defined in grid: computing grid, information grid, and knowledge grid. Computing grid is the fundamental layer, providing infrastructure to higher levels, namely, information grid and knowledge grid. Information grid will provide end users with an intelligent processing platform for removing isolated information islands, so that users can acquire, process, and distribute information easily. Knowledge grid will provide end users an intelligent processing platform for removing isolated knowledge islands for users to acquire, process, and distribute knowledge easily.

Current research on grid technology mainly focuses on grid computing, information grid, and grid services. Grid computing is a virtual and high-performing computational environment by connecting various computers (including computer groups), databases, mass storages, and peripheral equipment, which are distributed geographically. Its applications include distributed computing, high throughput computing, collaborative engineering, and database browsing. It can be defined as a "seamless integration and collaborated computing environment" across a wide region.

Information grid provides an intelligent data processing and information extraction platform to end users based on currently available network infrastructure, protocols, the Internet, and databases. Its objective is to establish a new generation Internet-based information processing platform and software infrastructure, which is built upon operating sysmtem and the Internet. Within such a platform, information processing is distributive, collaborative, and intelligent. Users can access all information by only one interface. The ultimate goal of the information grid is to provide services on demand. The system structure, information representation, meta-information, information connectivity and consistency, and information security are hot research topics in current information grid.

Due to its potential innovative applications, research on grid technology is increasing. Industries, government agencies, and academic institutions all pay attention to it. The international organization OGSA (Open Grid Service Architecture) is now making service protocols and standards. Also, an independent, non-profit organization GCF (Global Grid Forum) is established for related research efforts.

In summary, grid is a new technology based on the Internet, which merges high speed Internet, computers, large-scale databases, sensors, and remote equipment into one computing environment. It provides researchers and common users with more resources, functionalities, and services. People use the Internet for receiving and sending emails and browsing the web, while grid gives people more powerful functions for sharing resources of computation and storage, data, information, knowledge, expertise, etc. Grid computing is a new type of software structure, which organizes numerous low cost storage modules and servers into a virtual computation environment, where resources are used effectively and transparently. Any node in the grid can share services, such as data storage, computation, databases, and location based services. Presently, grid computing is applied to many fields, such as simulation, medicine, geosciences, biological science, and military, and has achieved much better performance than conventional distributed technology.

1.2.3 Generalized spatial information grid

It is believed that 80% of natural and social phenomena are closely related to spatial location. With the rapid development of information technology, communication technology, air- and space-borne remote sensing, and satellite navigation positioning system in the 1970s, a new interdisciplinary field called geo-spatial information science (i.e. geomatics or geoinformatics) came into being, with GNSS, GIS, and remote sensing (RS) as its core subjects.

According to the ISO's definition, geomatics is "a discipline related to spatial data acquiring, measuring, analyzing, storing, managing, displaying, and applying"; it belongs to information science and technology.

Figure 1.2 shows the data flow of air- and space-borne spatial information systems. It is obvious that these systems can answer such questions as "what kind of changes have taken place with what objects, when, and why" (4 W's questions). However, the state-of-the-art of spatial information systems can not meet the requirements of seamless integration and collaborative computing that require grid technology, so it can not answer the above 4 W's questions in real-time. Main problems are as follows.

The flow of data processing, information extraction, knowledge discovery, and application services are serial and separate. Data are acquired over point or area support. None of the current data acquisition systems has the capability of on-board processing or real-time computing. Information processing is usually carried out in a single computer with man-machine interaction. Data mining and knowledge discovery are still in the early stages. Application services are isolated due to a lack of intelligent services in networked environments.

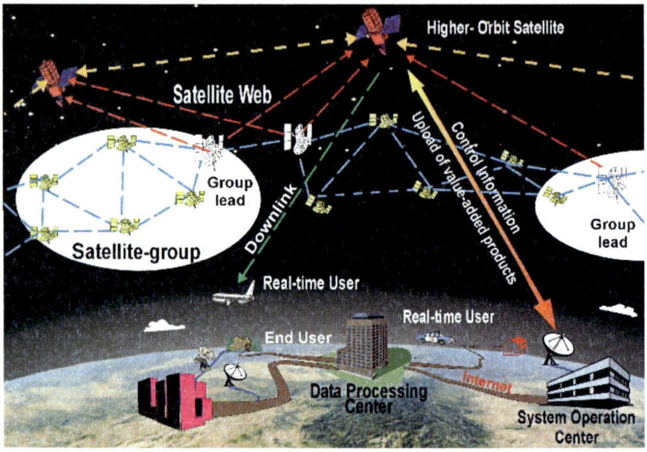

Fig. 1.3 An intelligent aerospace-borne spatial information system (from Zhou and Menas,2002)

Sharing and utilizing satellite resources are at a very low level. It is still very difficult to undertake intelligent data processing, information extraction, and knowledge discovery, both in theoretical research and algorithmic developments, leading to a massive amount of data awaiting information extraction. Earth observation systems are usually isolated, falling short of being multi-platform, multi-sensor, multi-resolution, and multi-temporal integrated systems. The current systems are not application-oriented grid systems with real-time satellite deployment, data integration, automatic information extraction, and intelligent services, as simple integration of geospatial information systems and communication systems is not enough.

The integration of grid technology with geospatial information systems remains a major problem. A solution to this problem lies in the generalized information grid (Li 2008).

In the context of grid technology and geospatial information system, the generalized spatial information grid is a kind of real-time or near real-time spatial information system supported by an information grid. It runs in an environment of an information grid with an integrated mode of data acquisition (from space-borne, air-borne, and ground-based sensors), information processing, knowledge discovery, and intelligent geo-services.

To establish such a grid, we need to do the following:

1. All-weather and all-day data acquisition by space-borne, air-borne, and ground based sensors
2. Seamless links from sensors to services by integrating information grid formed by satellite communication networks and ground-based wire/wireless computer communication networks
3. Implementation of quantitative, automatic, intelligent, and real-time grid computing in the generalized information grid, extracting information and knowledge from data

4. Intelligent services to any end-users, providing the most useful information to users who need it the most with the fastest speed

To meet the above needs, a generalized spatial information grid should be composed of the following four parts:

1.2.3.1 Smart Sensor Web

Current sensors are usually task-specified instruments for data acquisition. The raw data acquired are transferred to ground receiving stations and processed by specific software systems, with retrievals made available to end users. Such a workflow is composed of separated processing units, with man-machine interactions. Real-time processing would be impossible in such a workflow.

An intelligent sensor web at low cost and catering for large and small scale applications can be created in the near future, backed by the developments in sensor technology, computer hardware and software, and communication technology. Such a web will be operable wherever data acquisition and information services are needed. Gross (1999) said "in the next century, planet earth will don an electronic skin. It will use the Internet as a scaffold to support and transmit its sensations. This skin is already being stitched together. It consists of millions of embedded electronic measuring devices: thermostats, pressure gauges, pollution detectors, cameras, microphones, glucose sensors, EKGs, and electroencephalographs. These will probe and monitor cities and endangered species, the atmosphere, our ships, highways and fleets of trucks, our conversations, our bodies--even our dreams." Tao (2003) proposed a conceptual framework for such a web, indicating that it should be interoperable, intelligent, dynamic, and measurable.

An intelligent sensor web in the context of a generalized spatial information grid should have the following features: (1) a network composed of contact or non-contact sensors that are pervasive everywhere and possess functionalities of data acquisition and communication; (2) the capability of on-board data processing to provide user-required data in real time; and (3) an intelligent sensor web embedded into a global information grid to provide intelligent services to various users according to their specific requests..

1.2.3.2 An Intelligent Data-Information-Knowledge Transforming System Based on Grid Computing

The volume of data acquired by such a multi-sensor web is massive in the range of TBs to PBs. To pre-process and post-process such data, we need to extract both semantic and non-semantic information and to intelligently mine user-required knowledge from them. Li and Wang (2007) summarize the author's work in spatial data mining in the past 10 years.

A typical problem in the current EO system is "rich in data but lacking in information and knowledge". Grid computing brings a new prospect for solving this problem. However, the following listed issues must be considered in the context of grid computing:

- Determining the locations and attitudes of sensors under a unified spatial-temporal datum automatically and in real time
- Mathematical modeling for inverting physical and geometrical properties from data by both contact and non-contact sensors integrally
- Real-time updating and fusion of multi-source spatial data
- Intelligent feature extraction and object classification using grid computing for multi-sensor and multi-platform remotely sensed imagery
- Spatial data mining and knowledge discovery from huge volume of spatial data and attributes data

We need to make breakthroughs in spatial-temporal datum, remote sensing imaging mechanisms, pattern recognition, computer vision, and data mining in order to solve the aforementioned issues, in order to rcalizc unified inversion of physical and geometrical properties and to process data quantitatively, intelligently and automatically.

1.2.3.3 A New Generation of GIS Suitable for Grid Computing

It is well known that GIS is a spatial information system that can acquire, store, manage, analyze, describe, display, and distribute geo-referenced data. Data acquired by the intelligent sensor web described above will be the data source of GIS after they are pre-processed, with information extracted. Knowledge discovered from them will form the knowledge base of intelligent GIS.

As an information system for data storage, representation, analysis, and applications, GIS has evolved from single and isolated to networked system, such as Web GIS and Mobile GIS. The next generation of GIS should be Grid GIS which is an open and distributed geo-spatial information services system with federal databases and interoperability which runs on a grid computing environment.

1.2.3.4 Intelligent Geo-Service Agents

The concept of geo-services has been introduced in Section 1.2.1 above. An intelligent Geo-service agent would provide the end-users with the most necessary information in the shortest time and at the lowest cost. The central component in such Geo-Services is the integration of earth observation satellites, image processing system, and spatial information service system with computer grid via wire- and wireless communication networks. To set up such agent we need to solve the intelligent geo-service mechanism and modeling, related standards for service and interoperability, ontology based semantic model for geo-service, and the realization of geo-service at different terminals.

As one practitioner of geo-services, the Europe Space Agency (ESA) has, since 2000, setup the Global Monitoring of Environment and Security Plan (GMES) in order to promote sustainable development. A typical example of the plan was the earthquake rescue in Algeria in 2003. By using SPOT images before and after the earthquake, ESA delineated the destroyed areas quickly and estimated the population of people which suffered from the earthquake, providing accurate information for earthquake rescue. JAXA (Japan Aerospace Agency) established rapid response earth observation system. Japan Fishery Information Center (JAFIC) provides fishery information to vessels in the sea near Japan at near real-time. They have set up a decision support system for guiding ocean fishing by monitoring the change of oceanic environment, which influences the clustering of fish groups, so as to better locate them. Near real-time applications with MODIS/AIRS using the EOS direct broadcast are widely used in the USA for monitoring water quality, fire, air quality and polar winds by NOAA, NASA, and related agencies.

1.3 Towards a New Era for Geo-Information

1.3.1 Why a New Era of Geo-Information ?

In recent years, GIS applications have become increasingly widened as the requirements for geo-spatial information services grow. Since the concept of GIS was first proposed by Dr. Roger Tomlinson in the 1960s, it has gone through a long process of developments and has acquired remarkable achievements over the last four decades. Its application fields include mapping sciences, land use, resources management, environmental monitoring, transportation, urban planning, precision farming, digital earth, and so on.

Google Map started changing the way we see the world. The current and forthcoming Google Earth, Virtual Earth, NGI and Web 2.0, grid technology, new earth observation satellites and smart sensor web make geographic information so popular that not only professional users, but all Internet users can do a variety of work on a uniform spatial information service platform. The openness of geo-spatial information service will greatly boost its applications and popularity.

Along with the prevalence of Internet and the emergence of Web 2.0 technology, the release, distribution and publishing of geo-spatial information supported by new technologies are growing at a fast rate. Michael Goodchild proposed his idea of "Citizens as Voluntary Sensors" in order to describe the cooperative production of geo-spatial information and the transmission and sharing of geo-spatial knowledge in a more persuasive way. He outlined the prospect of the whole world equipped with facilities that are capable of uploading what they consider as important and effective sources of geo-spatial information (Goodchild 2007).

Web 2.0, as a human-oriented network, brings new opportunities for geo-spatial information services and it turns passive users into initiative and creative ones. Web 2.0 provides a variety of services with characteristics including experience, communication, variation, creativity and relation. For geo-spatial information services, visual services are the basis of experience (such as Google Earth, Virtual Earth, etc.); measurable services ensure variation and creativity, and minable services allow for relation among professional applications. This is a great revolution for our traditional geo-spatial information services based on digital maps. Web 2.0 has the potential to transform traditional measurement-by-specification to measurement-on-demand by uploading the external orientation elements of Digital Measurement Images (DMIs) and corresponding measurement software kit via the Internet. Measurement-on-demand means when digital image pairs and their exterior orientation parameters and related software uploaded to the Internet on Web 2.0, any end users can measure objects as they like. This function allows public users to be data and information providers, and geo-spatial data are changing from outdated to active participatory services through the integration of smart sensor web and Web GIS. Spatial Data Infrastructure with these kinds of geo-spatial information services provides a better solution to users from all ways of life than before.

Google intends to take photos for all streets in the world and develop a street scenes web site, as shown in Figure 1.4. Thus, all Internet and 3G network users have access to functions on a uniform spatial information service platform. With such a platform, geo-spatial data integrated with statistical data, population data, and social and economic data, offer comprehensive services of measurement-on-demand for e-government services, e-commerce, public security and transportation industries, and can answer the so-called "4W questions", i.e., when, where, what object, and what change. These herald that a new geospatial information era is coming.

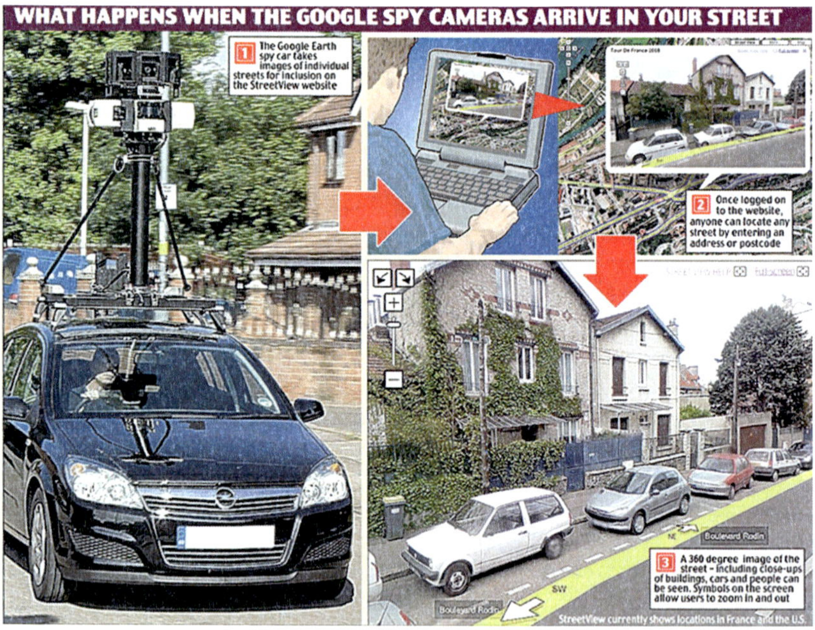

Fig. 1.4 Google intends to take photo for all streets in the world and develop a street scene site (UK Daily Mail on July 11, 2008)

1.3.2 Main Features of the New Geo-information Era

Compared with the traditional geo-information systems that are based on electronic maps, main features of new geo-information systems can be summarized as follows:

1.3.2.1 Data Providers: From Professionals to All Public Users

In the new geo-information era, data providers are not only professionals but also the general public. A great number of users require fundamental information which is concerned with professional and individual applications, such as electric power facilities, municipal facilities, security facilities, transportation information, location-based information, and so on. Unfortunately, these kinds of information can not be discovered from traditional 4D (DEM, DOQ, DLG, and DRG) products directly. Traditional 4D products can not satisfy the needs of integrity, richness, accuracy and reality of geo-spatial information. However, in new geo-information era, such information can be acquired from DMI (Digital Measurable Images) which are released on the Internet according to specific requirements. As illustrated in Figure 1.5, a user is issuing rental advertisement through the ImageCity website platform, and other users can decide whether to bid for renting or not by browsing the ImageCity website as well. Thus, richer and more comprehensive information services including maps, images, and POI can be provided by the new geo-information systems, which can also support on-line access through distributed systems and interoperability, ensuring active participation.

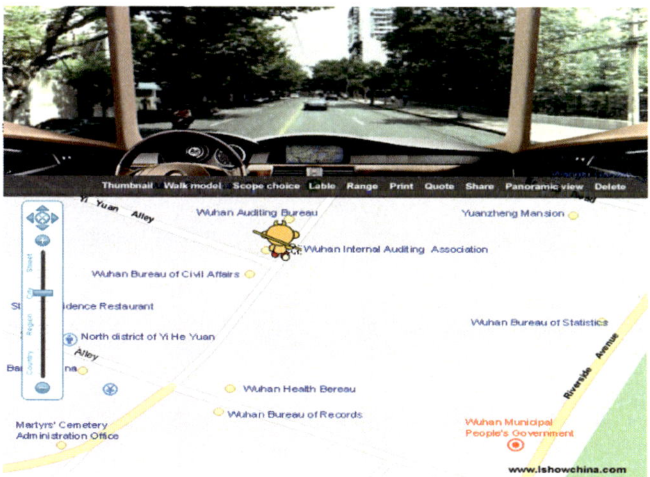

Fig. 1.5 A user issues rental advertisement through the ImageCity website platform, which is helpful for other users bidding for it.

1.3.2.2 The Users: Data and Information Providers as Well

In the traditional geo-information era, there existed a clear boundary between data provides and data users. However, in the new geo-information era, this boundary has been blurred, as data updating and maintenance can be fulfilled by data providers and data users. For example, data users can also upload or annotate new geospatial information in addition to traditional downloading.

Web 2.0 is a revolution and it advocates openness, anticipation, and sharing and creativity, which can satisfy individual needs and Web 2.0 turns passive users into initiative and creative users. According to characteristics of Web 2.0 technology, not only professionals but also public users play the role of data providers in the new geo-information era. Provided data and services are transferred from fixed updating at regular intervals to more popular updating forms in the new geo-information era, i.e., from static updating to dynamic updating.

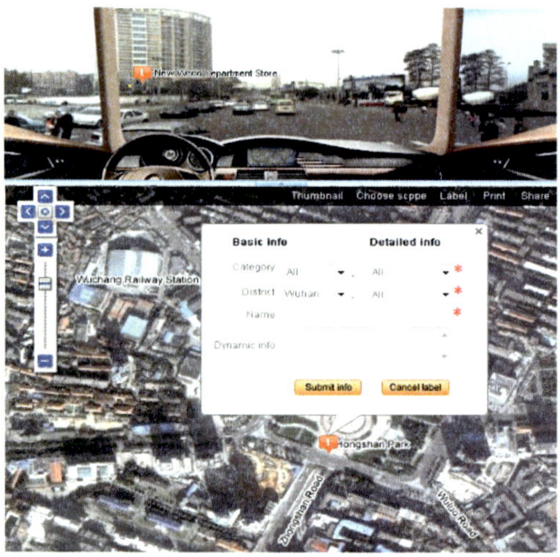

Fig. 1.6 The annotation function of ImageCity website

Take the ImageCity platform developed by the author and his team as an example. ImageCity offers an annotation function which actualizes communication among map-makers, users and intelligent services. With the convenience of Web 2.0 technology, business users only need to finish registration and payment on this website to annotate their company and brands on image maps. Moreover, registered business users can offer what they think is important to consumers on the website regularly and dynamically. Generally, the information contents include business websites, telephone numbers, scene images and text information etc, as shown in Figure 1.6. At the same time, the annotation function gives access to registered public users to finish free publishing and sharing of location-based images and text information.

A lot of POIs consisting of graphics, images, audios and videos are added. Take ImageCity Wuhan as an example. The whole data volume is about 2TB, with more than 300,000 POIs added, which provides geo-spatial information services for individuals, business, industry and government users.

1.3.2.3 Geospatial Data: From Outdated to Live Through Smart Sensor Web

In the new geo-information era, information services are provided on the basis of data sources acquired through space-borne, airborne, and ground-based, non-contact and contact sensors. Data and information transmission operates in the next generation of the Internet based on Web 2.0 and 3G mobile networks, enabling live information services. Geo-spatial information services based on specifications, as in the traditional mode, are not adequate, because data provided from spatial databases are static. In the new geo-information era, the adoption of sensor web technology provides access to services according to specific needs for professional and general users in multimedia and dynamic service environments. For example, for an ITS (Intelligent Transportation System) based on sensor web, mobile sensors on vehicles, fixed sensors at road intersections, captured video data, data in monitoring centers, road condition inspection data and data for emergence are helpful to improve transportation, reduce traffic delay, traffic accidents, and decrease gas consumption to a great degree. Furthermore, these kind of sensors can be installed on 3G phones, which have been developed in Wuhan Urban Grid Management and Service System by using cameras as data collection tools.

As mentioned previously, all aerial and space sensors can be integrated together to build a big and smart sensor web. In this way, the accomplishment of real-time data updating, information extraction and services will come true.

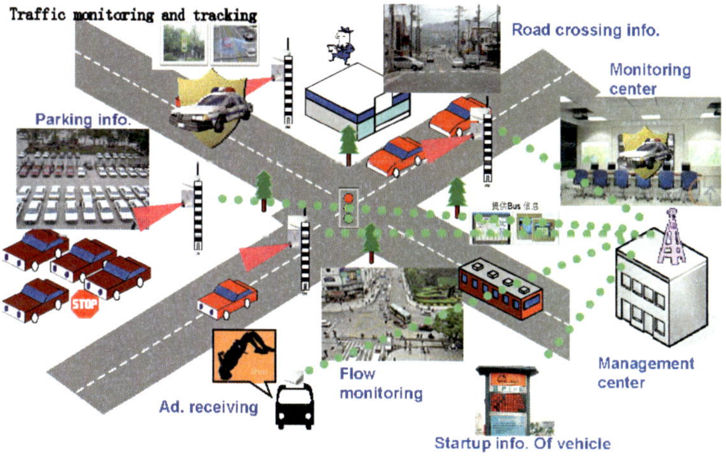

Fig. 1.7 Application system in ITS based on sensor web, in which traffic jam can be decreased by about 20%. Traffic delay can be decreased by about 10%~25%. Traffic accident can be reduced by about 50%~80%. Gas consumption can be reduces by about 30%.

Figure 1.8 illustrates the reference architecture for an interoperable sensor web. In this architecture, different kinds of data and services can register through registration center and build catalog services. For an end user, there are three ways to implement data mining and knowledge discovering based on the sensor web.

The first way is based on sensor web model. When an end user sends a request on the client side, Decision Support System transfers the received request to a service chain and searches for corresponding registered services in catalog services through the workflow sequence. Afterwards, registered services acquire information of interest and return it to the end user.

The second way is based on direct feedback of the sensor web, which is applied when the corresponding registered service in catalog services can not be found through the workflow sequence. In this case, new sensor web feedback is further searched. If the required service can be found, it will be returned to the end user and will be registered at the registration center.

The third way is based on retrieval of digital products, which is also applied when the corresponding registered service in catalog services can not be found through workflow sequence. In this case, required data is further searched through a sensor web node instead of the sensor web feedback.

Recently, OGC has released many standards dealing with smart sensing web such as Sensor Model Language (SensorML), Sensor Alert Service, Sensor Observation Service, Sensor Planning Service, Observations and Measurement, Transdure Markup Language etc.

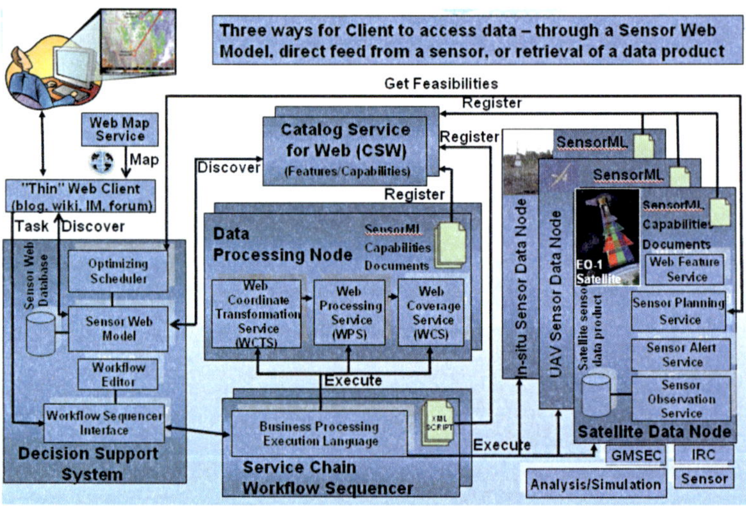

Fig. 1.8 Reference Architecture for an interoperable sensor Web

Figure 1.9 shows conceptual system architecture of future Grid GIS. In this open architecture, the registration of vector data, image data, DEM data and professional data and information is completed by the operating and maintenance personnel of different types of data through registration services.

Service providers of different types of software platforms register their typical functions in the registration center. The registered functions can then be employed for search, integration, sharing of data, information and knowledge through service discovery and service chain integration.

Different types of end users also need to be registered as legal users through user security authentication and after that the registered data and information based on grid sharing protocol are available for them.

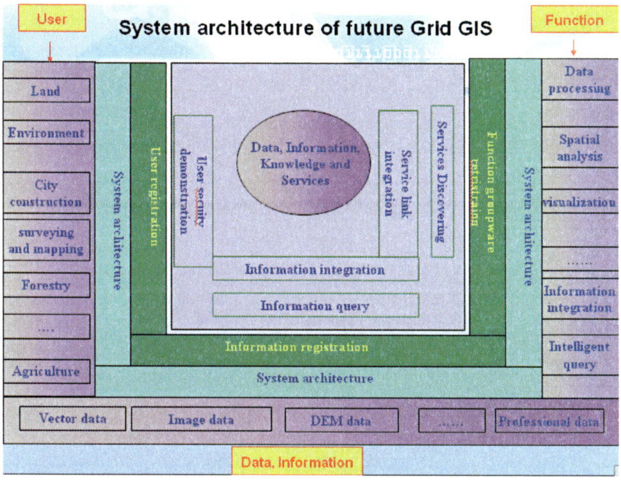

Fig. 1.9 The system architecture of the future Grid GIS

Figure 1.10 provides a description about the three-layered registration center. The first layer is data registration, the second is software or function registration, and the last is user registration.

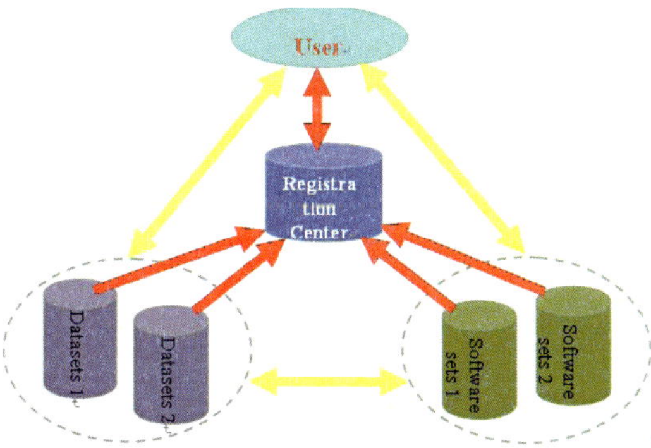

Fig. 1.10 Geo-spatial information processing and service model of the new generation

What follows is the procedure of heterogeneous and distributed geo-spatial information services, as illustrated in Figures 11-15.

Step 1: a service provider registers one function (e.g. Binarization Algorithm and its service type is transfer service) on the internet. The annotation of this service aims for binarization of the image and registers supplementary information including relative information of this service provider (e.g. name, phone number, email, department, address etc.)

Step 2: another provider registers another function (e.g. change detection algorithm and its service type is image analyzing) on the Internet. The annotation of this service aims for change detection between two different images. In the same way, relative information of this service provider is also registered as supplementary information.

Step 3: an end user logs in successfully and queries the service needed at a registration center.

Step 4: the abstract service chain is established through the project requirements and the existing functions. For instance, a service request, such as a flooding area analyzing model, can be established as the combination of the following services.

Step 5: the abstract service chain is mapped into the executed service chain.

Step 6: results computed according to service chain are shown, from which the flooding area is calculated.

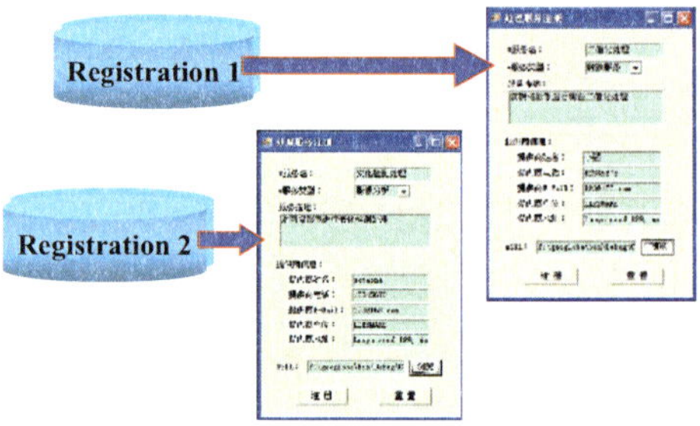

Fig. 1.11 Registration of geo-spatial information services

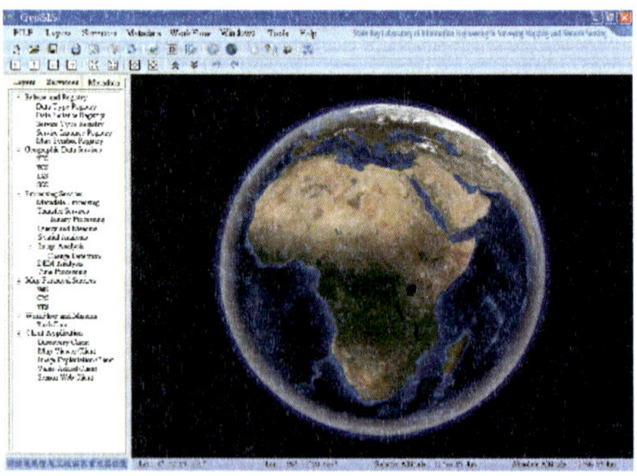

Fig. 1.12 Inquire of geo-spatial information services

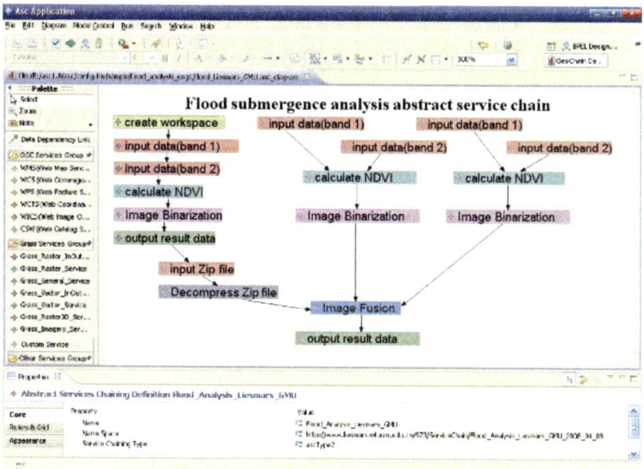

Fig. 1.13 Service chain setup

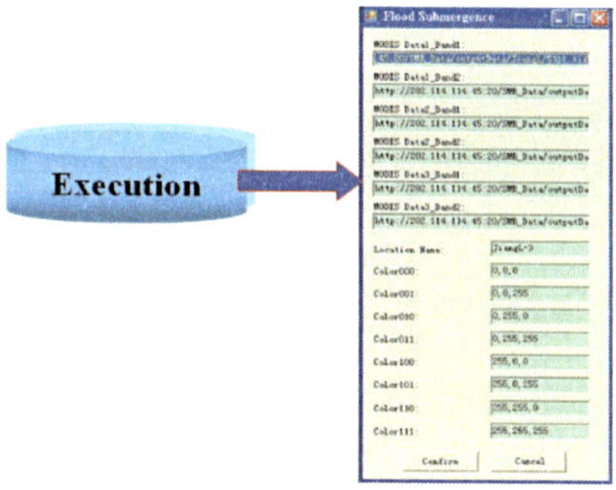

Fig. 1.14 Execution of geo-spatial information services

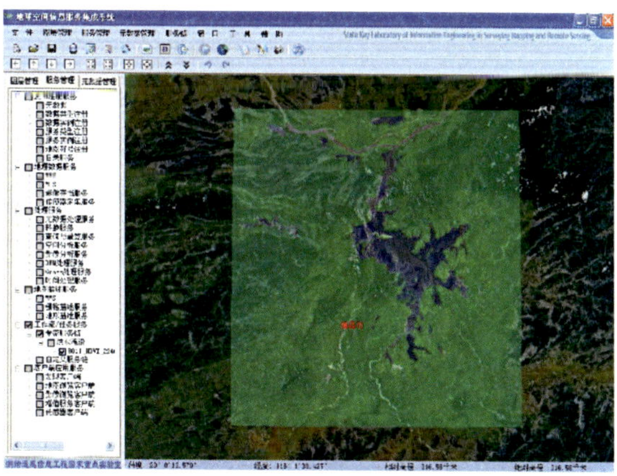

Fig. 1.15 Results of heterogeneous distributed geo-spatial information services

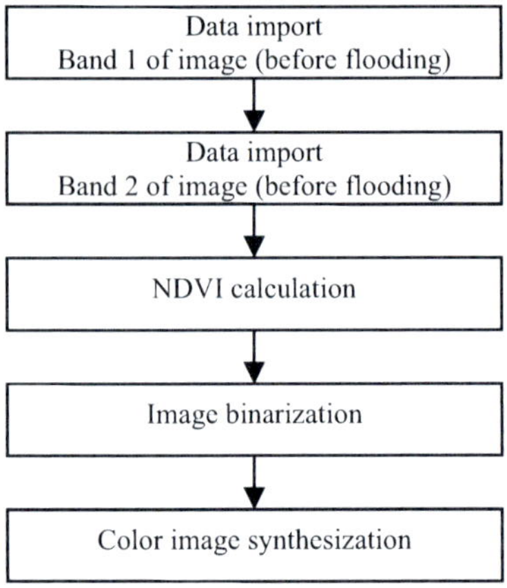

Fig. 1.16 An example of the establishment of an abstract service chain

1.3.2.4 From Measurement by Specification to Measurement on Demand

In the new geo-information era, geo-spatial data in service are DMI instead of simple image maps. The so-called DMI are digital stereo images appended with six exterior orientation elements acquired by Mobile Measurement Systems (MMS). With DMI on the internet accompanied with measuring software kits, measurement of a special object at centimeter level precision is available. Undoubtedly, the overlapping of DMI with GIS data makes the representation of geographic objects more comprehensive and vivid and facilitates visible, searchable, measurable and minable functions.

In Figure 1.17, the overlapping of a DMI and an orthoimage of Beijing Bird's Nest, which can support measurement on demand, is illustrated.

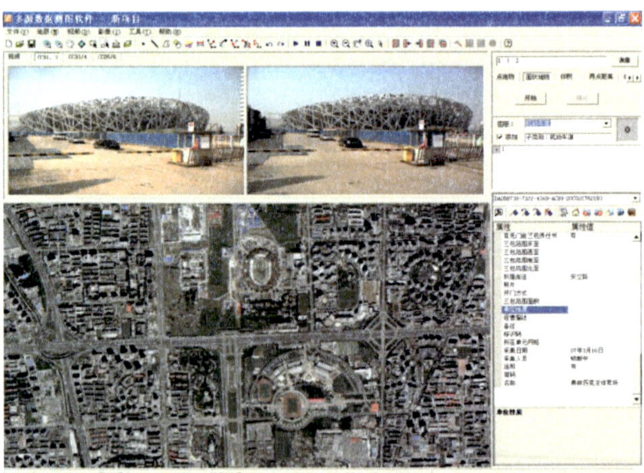

Fig. 1.17 A DMI and an Ortho image of Beijing Bird's Nest

1.3.2.5 From Data-Driving to Application-Driving: Service Oriented Architecture (SOA)

In the new geo-information era, service-oriented, spatial information sharing facilitates services with different granularity, data updating that is dynamic and continuous, integration and interoperability between CORBA, DCOM, and EJB in the standard protocol. This is elaborated below.

Service-Oriented Architecture (SOA) is a software system structure that achieves interoperability by packaging the program units, which can accomplish a given task. It originates from distributed computing model, promoted by OMG (Object Management Group) and IONA and is widely accepted as the standard.

Service-Oriented Integration (SOI) integrates the traditional objects with highly flexible Web Services. SOI provides an abstract interface. Systems are able to communicate with one another by this interface, instead of using low-layer protocols and self-defined programming interfaces to prescribe communication with other systems. The system only needs to appear in a form of service, select the interactive system, make simple discovery and bind with this service at runtime or in design.

Fig. 1.18 An example of service-oriented architecture

There is a more intuitive example, as shown in Figure 18. Service-oriented spatial information sharing based on SOA provides data and tools for active measurement-on-demand service for users instead of just information, such as width of roads, position of toll gates, and so on, which is a characteristic of data-oriented spatial information sharing.

1.3.3 Problems and Challenges in the New Geo-information Era

As discussed above, the new geo-information era provides rich geospatial information. However, this brings about problems and challenges as summarized below.

1.3.3.1 Out-of-order Issues in Geospatial Data Collection and Information Proliferation

The out-of-order issues in data collection mean inconsistency among data format, data content, temporal-spatial characteristics, etc., which needs to be solved through data interoperability and standardization of various types of original data.

Information proliferation is also a hot issue in the new era, as conflicts surface between the limited storage space and the information contents. Limited bandwidth transfer rate is not compatible with mass information transmission due to unprecedented volume of data uploading and downloading in case of emergency. How to distinguish useful and useless data efficiently and how to realize rapid publishing and sharing of geo-spatial information by using the grid technology are major issues.

1.3.3.2 Quality Issues in Geo-Information Updating

Traditionally, data uploading and updating are implemented based on standards by professionals, while, in the new geo-information era, public users are also data providers. The participation of end users is conducive to data updating in real time, but how to guarantee updating quality is a problem, which may be resolved by some effective web standards and online data cleaning tools.

1.3.3.3 Security Issues in Geographic Information Services

Security issues are also a problem, which should not be neglected. For instance, high-resolution images, semantic information, orientation and attribute information about important departments and locations involve data security issues. How to set an automatic security filter on the Internet needs to be solved.

1.3.3.4 Privacy Issues in Sharing Geo-Information

How to protect individual privacy in the web environment has attracted more and more attention nowadays. In the new geo-information era, the spatial information providers release services in the network, the related spatial data would be at high precision.

When high-resolution data are concerned with individual privacy, their protection becomes a pressing problem: how to find the balance between protecting personal privacy and sharing spatial information, when the high-precision data would come down to personal privacy. Currently, the consensus among many countries is that privacy needs to be protected although divergences exist among different policies. For example, in America, street view websites have been complained that they infringe individual privacy in some aspects and have been warned by Privacy International. Therefore, legislative work is necessary. Automatic detecting and removing of individual privacy related objects from high resolution image data is one of the most important tasks for open geo-services system.

1.3.3.5 Property Issues in Sharing Geo-Information

Although existing policies, laws and rules in surveying and mapping still play important role in monitoring and managing of geo-spatial information, they are not so effective in the new geo-information era.

There are many pressing problems, such as how to specify the ownership of property rights, the contents of property rights and the property location in geographic information sharing, and to satisfy the privacy requirement. These require new laws and rules.

1.3.4 Strategies

The coming of new geo-information era represents developments and advancement of information technology. Some strategies in terms of standard, planning, law, technology and application are essential for the opportunities and challenges discussed above.

1.3.4.1 Standards

New issues in new geo-information era need us to standardize rules and laws in the first instance and then to accomplish geo-spatial information services better.

Nowadays, conflicts between mounting earth observation data and pressing requirements from every walk of life lead to an embarrassing situation where there are more data but little information. In new geo-information era, geospatial information services based on DMI represents new direction in spatial data services, and its integration with grid services, interpretation services and telecommunication services provides solution to automation, intelligence, and popularization of geo-spatial information.

1.3.4.2 Planning

In the new geo-information era, such rules as product secrecy protection and product quality control requirements need to be standardized. One typical example is to blur the vehicle identification number or human face. Furthermore, what needs to be standardized includes definition, content and relationship with other basic geographic information products. Also, technical specifications, technical requirements, testing methods, testing rules, distributing format, and secrecy requirements of new products should be considered.

1.3.4.3 Laws

Such issues as right of privacy, property right and security brought by the new geo-information era need to be resolved by regulations and decrees on the legal level to restrict the behavior of the public. It is important that no infringement to privacy be conducted.

1.3.4.4 Technology

According to the characteristics of the new geo--information era, developments and promotion of high-tech with independent property rights on data obtaining, information processing, knowledge discovery and intelligent services are indispensable for providing technical support to the construction of the generalized spatial information grid. At the same time, it is necessary to offer solutions to information security and removal of personal privacy information automatically and technically.

1.3.4.5 Applications

Applications in different industries are encouraged to improve and enrich the annotations to geo-information. It is necessary to make the contents and forms of the new generation of geo-information as plentiful, vivid, and abundant as possible to satisfy various needs. These can be applied in many fields, such as digital cities management, intelligent transport, police, roads, railways, and so on. Accordingly, the experiences on data collection, processing, quality control, standardization, and applications will be accumulated correspondingly.

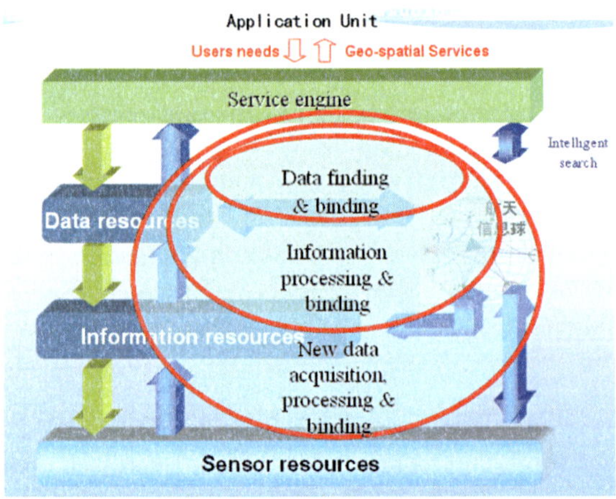

Fig. 1.19 The task-oriented service mechanism

1.4 Summary

In this chapter, the background to the new era of geo-information is analyzed at first. Main features of the new geo-information era are then compared with those of the traditional era. Problems and challenges faced by the new geo-information era are discussed, followed by an outline of strategies including standards, planning, laws, technology, and applications.

The new Geo-Information era offers unprecedented opportunities for us. We need to develop better tools for storage, management, on-board processing, distribution, and service of geo-spatial information in the environment of Grid computing and the next generation of Internet. The goal of the new geo-information era is to realize one-data collection-many-applications and measurement-on-demand. We will explore new strategies for meeting the challenges.

In prospect, new geo-information era will see the geospatial information industry chain grow to prosperity. Geo-spatial information sharing will bring tremendous economic benefits to the whole world. The ultimate benefit of the new geo-information era is to realize geo-information for all.

References

M.F. Goodchild (2007): Citizens as sensors, the world of volunteered geography, GeoJournal, 2007, **69**, 211-221

N. Gross (1999): The earth will don an electronic skin. Buviness Week, Aug.30. Available from m:http://www.bwsiness/week/com/1999/99-35/63644024,htm

I. Foster, C. Kesselman (1988): The Grid: Blueprint for a New Computing Infrastructure.1988,

D.R. Li (2008): On generalised and specialised spatial information grids: are geo-services ready? International Journal of Digital Earth, Taylor & Francis, 2008(4): 315-325

D.R. Li and SL Wang (2007): Spatial data mining and knowledge discovery, ISPRS Book Series, Volume 5: Advances in Spatio-Temporal Analysis, pp.171-192, Taylor &Francis, 2007

V. Tao, 2003. The Smart Sensor Web: A Revolutionary Leap in Earth Observation Systems, http://www.geoplace.com, GEO World.

T. Vögele, R. Spittel, U. Visser, S. Hübner (2003): GeoShare -building a transnational geodata infrastructure for the North Sea region. Geodaten- und Geodienste- Infrastrukturen - von der Forschung zur praktischen Anwendung, IFGI Press, Münster, Germany,18,1–15. http://www.gi-tage.de/archive/2003/downloads/gitage2003/tagungsband/voegele.pdf

GQ Zhou and K. Menas.(2002):.Future Intelligent Earth Observing Satellites. Pecora 15/Land Satellite Information IV/ISPRS Commission I/FIEOS 2002 Conference Proceeding.

Chapter 2

A UAV REMOTE SENSING SYSTEM: DESIGN AND TESTS

Lei Yan, Zhiyang Gou, Yini Duan

2.1 Introduction

Airborne remote sensing refers to reconnaissance techniques using aircraft, balloon, or other platforms. Comparing to satellite remote sensing, it has four main advantages: relative low cost, flexibility in the frequency and time of data acquisition, and the ability to record spatial details finer than current satellite technology can (Mei *et al.* 2001).

With the development of sensors and communication techniques, high resolution airborne remote sensing has been successfully applied to large-scale topographic mapping and surveying for detailed ground information. Although conventional airborne remote sensing has some drawbacks, such as altitude, endurance, attitude control, all-weather operations, and monitoring of the dynamics, it is still an important technique of studying and exploring the Earth's resources and environment.

Airborne remote sensing can be categorized to manned aerial vehicle remote sensing and unmanned aerial vehicle (UAV) remote sensing according to the platform. The name UAV covers all vehicles which are flying in the air with no person onboard with the capability of controlling the aircraft (Eisenbeiss and Zhand 2004). Thanks to GPS and communication technology, UAVs can be remotely controlled or flown autonomously based on pre-programmed flight plans or more complex dynamic automation systems.

Over the past years, with the rapid development of micro-electronics, communications, and materials and propulsion systems, research on UAV has made an obvious progress. The number of UAV systems is growing fast, and most of them are for military use or part of a development project (Blyenburgh 2008). In this international activity, 49 countries around the world are involved (Europe: 23; Asia: 14; South-America: 4; North-America: 3; Australia: 2; Africa: 2) (Everaerts 2008). Some typical UAV are the Darkstar (Lockheed Martin and Boeing, USA), Predator (General's Atomic Energy, USA), Heron (Aircraft Industries, Israel), and Alenia Mirach (Italy). Several UAV models are shown in Fig. 2.1 (www.livingroom.org.au/uavblog). Take the most well-known UAV "Global Hawk"

D. Li et al. (eds.), *Geospatial Technology for Earth Observation*,
DOI 10.1007/978-1-4419-0050-0_2,

as an example. It flies at an altitude of more than 10,000 m with a fuselage of 13.5m long and 4.62m high, a wingspan of 35.4m, maximum take-off weight of 11,622kg, maximum payload of 885kg, cruise speed of 635km/h, practical ceiling of 20,500m. So far, it is the largest UAV that can implement a high-altitude, long-distance and long-time continuous reconnaissance mission http://www.livingroom.org.au/uavblog/archives/1_global_hawk.jpg. It uses a variety of cameras including electro optical and infrared and has capabilities of transmitting pictures in real time to bases.

a b c

Fig. 2.1 Examples of unmanned aerial vehicles: (a) Predator; (b) Cypher; (c) Global Hawk. The images are from http://www.livingroom.org.au/uavblog/archives/Predator-UAV_nh_100201.jpg

UAV remote sensing system is based on UAV which has both the common characteristics of aerial remote sensing and its own unique features. Compared with manned aerial vehicles, remote sensing systems with the platform of UAV can work all-day and all-weather and perform flight tasks in high-risk areas. Moreover, UAVs are able to operate rather close to the object and acquire images with few centimeter resolution (Eisenbeiss and Zhand 2006), providing sufficient detail.

Due to the cost of the mission, the need for rapid response or the fact that observations need to be carried out in an environment that may be harmful or dangerous to an aircrew, scientific interest in this type of platforms is growing, and a number of experiences have already been reported (Berni *et al.* 2009). Many remote sensing applications have benefited from the use of UAVs. A remarkable example is the adoption of remote sensing using UAVs in archaeology (Çabuk *et al.* 2007, Eisenbeiss and Zhand 2006). UAVs have also been used successfully in vegetation monitoring (Herwitz *et al.* 2004, Rango *et al.* 2006, Berni *et al.* 2009), and in Japan these systems are considered to be an integral part of farm equipment. With the ability to get close to the object, they were also used for road condition assessment (Zhang 2008). Rapid response imaging using UAVs has received a lot of attention as well (Everaerts 2008). This has been demonstrated in road accident simulations (Haarbrink and Koers 2006) and in many cases of forest fire monitoring (Réstas 2006, Martínez-de Dios *et al.* 2006). UAVs have also been proposed as platforms to monitor volcanoes (Puri *et al.* 2007, Doherty 2004). Some UAV systems have also used Lidar and SAR (Wang *et al.* 2009, Edrich 2006).

Besides the UAV with a fixed wing, there are many other kinds of UAVs, some of which even were used in remote sensing before the manned aircraft. In some sense, the kite is the simplest and most primitive model of the UAV. M. Arthur Batut took the first aerial photograph using a kite. It was taken over Labruguiere, France in the late 1880s.

A tethered balloon (Çabuk *et al.* 2007, Vierling *et al.* 2006) or blimp is also a simple UAV. Fig. 2.2 shows a modern blimp (en.wikipedia.org/wiki/Blimp). It is easily controlled (especially its altitude), but of course quite unstable if the wind speeds increase. The balloon can be adapted to the size and mass of the instruments that need to be carried. Tethered balloons have been widely used for remote sensing purposes for over a century. In 1858, Gaspard Felix Tournachon (later known as "Nadar") captured the first recorded aerial photograph from a balloon tethered in Paris, in which the houses can be seen clearly (Robert 1975). Recently, true color and infrared aerial photographs taken from balloons or kites have been used in photography-based studies of periglacial features, vegetation growth, and soil properties (e.g. Boike and Yoshikawa 2003, Buerkert *et al.* 1996, Friedli *et al.* 1998, Gerard *et al.* 1997, Chen and Lee 2006).

Fig. 2.2 A modern blimp

Fig. 2.3 An unmanned powered paraglider (Thamm and Judex 2006)

Fig. 2.4 Yamaha R-MAX

Unmanned powered paragliders (Fig. 2.3) are interesting because they need very little ground support, use proven technology, carry substantial loads, and are low-cost compared to other low-altitude systems(Everaerts 2008). They have been flown successfully in remote areas that cannot be addressed economically with conventional survey equipment (Thamm and Judex 2006).

Another kind of UAV is the unmanned helicopters which come in many types and sizes. Some companies already use helicopters with/without GPS navigation for the production of aerial imagery of single buildings and cities or for documentation of industrial constructions. Zischinsky *et al.* (2000) used images taken from a modelhelicopter partly for the generation of a 3D-model of a historical mill. Fig.2.4 (www.livingroom.org.au/uavblog) shows the model of RMAX which came out in the year 1997 and was equipped with an azimuth and a differential GPS sensor system 3 years later. The RMAX UAV system from Yamaha was successfully used as a ground truth measurement system (Hongoh 2001). In Japan, hundreds are used in agriculture as platforms to plough, sow, spray, etc (Everaerts 2008).

UAV is playing a more and more important role in land resource surveying, city planning, environmental protection, pollution monitoring, disaster monitoring, and other applications. This chapter introduces the design and tests of an airborne remote sensing system based on UAV. In the following sections, a real sample of UAV remote sensing system will be introduced.

2.2 UAV Remote Sensing System

The architecture of the proposed UAV remote sensing system is shown in Fig. 2.5. One or more unmanned airplanes may serve as the airborne remote sensing platform. Even though the UAV can fly automatically under the control of the preset program, it is always controlled and monitored from the Ground Control Station, so it needs reliable communication links to and from the aircraft. The Ground Control Station provides a working space for a pilot, navigator, instrument operator and usually a mission commander. The sensors onboard are controlled by the airborne control system to capture images of the working area. After basic and real-time processing on board, the data can be downloaded to the Ground Control Station for on-site processing or forwarded to a processing center. The data processing center or ground receiving station will then further process, archive, manage and distribute the data for expert users. Advanced processing may involve information extraction and application of RS images. This way the business operation of the UAV remote sensing system is formed and a complete set of protocols and standards can be realized.

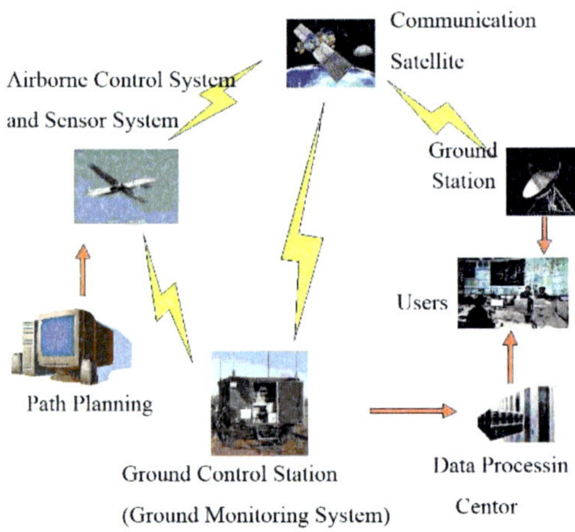

Fig. 2.5 General view of the architecture of UAV remote sensing system

In general, the four core components of a UAV remote sensing system include remote sensing airborne control system, remote sensing ground monitoring system, multimode airborne digital camera system, and automatic path-planning system. They are respectively introduced in the following subsections.

2.2.1 Remote Sensing Airborne Control System

The unmanned airplane implements automatic judgment and performance. Once a malfunction occurs in any component or module of the UAV remote sensing airborne control system, the quality of the results will be affected. Therefore, the design and development of a stable and reliable remote sensing airborne control system is the primary task for the implementation of unmanned airplanes for remote sensing.

The remote sensing airborne control system carries out positioning, camera auto-triggering, and data storage and transmission between the camera and the unmanned airplane platform, etc. It can implement fast update and reliable connection between airplane and remote sensing sensors. It can also load different kinds of remote sensing sensors which meet the requirements of interfaces.

2.2.1.1 Composition

According to the task demands, the control system is constituted of three modules: the photographing control module, data processing and transmission module, and power supply management module, which are depicted by Fig. 2.6.

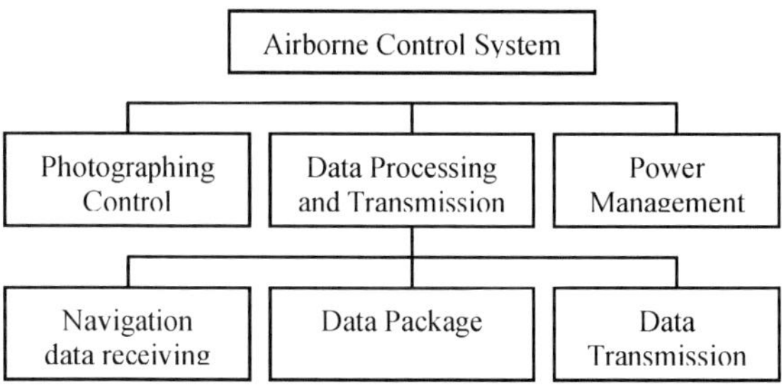

Fig. 2.6 Composition of remote sensing airborne control system

1) Photographing Control Module

This module is the core of the whole airborne control system. Since the survey tasks lay much emphasis on data accuracy, different control strategies should be adopted according to topographies and weathers. In this module, two schemes are available. The first scheme is to photograph by triggering the camera shutter in fixed time interval. The second scheme is to photograph at fixed positions. When the airplane enters the target regions according to the position information from the unmanned airplane, it will automatically take pictures. Because the airflow disturbance has much influence on the flight route of the unmanned airplane, some places will be redundantly photographed and there exist some gaps or incompletely photographed regions if the first scheme is adopted in bad weather. In this condition, the airplane must accurately photograph at selected positions. The flowchart of photographing control procedure is illustrated in Fig. 2.7.

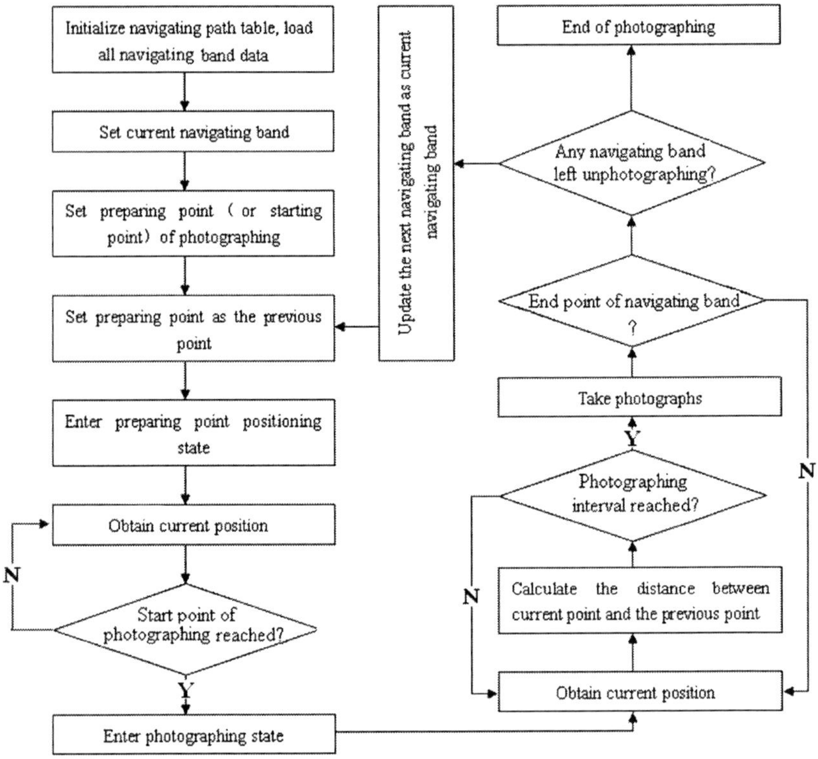

Fig. 2.7 Flowchart of UAV photographing control

2) Data Processing and Transmission Module

After the scene is photographed, the raw images are stored in the airborne storage and sub-sampled to be transmitted to the control system. The data processing and transmitting module packages the information such as position, attitude, and time and image data, and then sends them to the ground.

3) Power Supply Management Module

The unmanned airplane contains a 27V DC power supply, while in general the power supply of industrial compute is 5V or 12V. Thus voltage must be converted before being used. The power supply management module transforms voltage, provides stable power output, and controls the power supply of the industrial computer system.

2.2.1.2 Software Design

This subsystem's main control software is a multi-thread program including three main threads: the position and attitude data acquisition thread, image data acquisition thread, and photographing control thread.

The position and attitude data acquisition thread receives data of the plane by the low speed RS422 interface, and these data will be transmitted to the photographing control program for further processing.

The image data acquisition thread receives the snapshots transmitted from the camera system. The communication mode adopts C/S structure, which is based on TCP/IP protocol, implementing computer interconnection with the camera system being the client and the control system being the server. The camera system cannot work until the interconnection with the server is set up.

The photographing control thread controls the process of photographing by analyzing the current position of the airplane or the time according to different scheme.

2.2.1.3 Hardware Design

The hardware of the system includes a photographing control board and a power control board. The control program manipulates circuit board by setting the state of the parallel interface. The photo control board controls the camera exposal by triggering the camera shutter with relays, and the power control board controls the power switch of the camera with relays. In order to guarantee the proper working of the switch in a high-vibration condition, aero electric relays instead of common ones are adopted. The parallel interface has a certain voltage output during hardware initialization after the industrial computer starts, but the system requires that the parallel interface should be set after the start of the control program. So a time delay circuit may be added to the hardware circuits to keep the relay off during the starts of the industrial computer. Fig. 2.8 shows the photos of two kinds of circuit boards (a and b) and a photo control device (c).

a b c

Fig. 2.8 Examples of UAV Hardware Components: (a)Photo Control Board; (b)Power Control Board; (c)Photo Control Device

2.2.2 Remote Sensing Ground Monitoring System

The ground monitoring system is developed to provide the pilot, navigator, instrument operator, and mission commander with real-time downloaded image data and navigating data and report sensor working state. With the help of the ground transmission channel of the unmanned airplane, the ground monitoring system displays working states of the on-board devices during the flight as often as possible to ensure the data collection quality.

2.2.2.1 Composition

The ground monitoring system is constituted of four modules: the data acquisition module, data processing module, data display/storage module, and data downloading module, which are illustrated by Fig. 2.9.

During the flight, the ground monitoring system receives and sends the data to the data processing module. Then the valid data is transmitted to the data display/storage module for real-time monitoring and storing of the data for ex-post analysis.

The remote sensing data downloading module downloads the raw image data to computers. The module empties the storage space of the on-board devices and implements consecutive flight without refitting the devices. The four modules of the ground monitoring system are illustrated by Fig. 2.9.

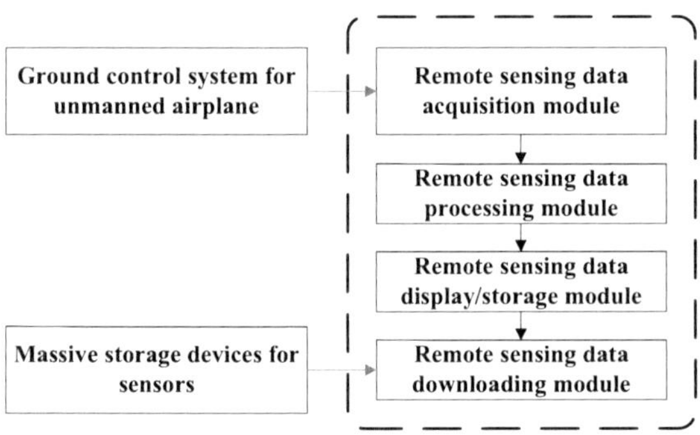

Fig. 2.9 Modules of the ground monitoring system

1) Data Acquisition Module

The function of the remote sensing acquisition module is to receive and send the data to the navigation data including position (longitude, latitude and height), altitude (pitching angle, rolling angle, yaw angle), and the real-time image data to the processing module. The air-to-ground transmission channel of the unmanned airplane packages the remote sensing data, and every data package has a specified mark. By detecting the marks, the data acquisition module sends the specified data to the remote sensing processing module for further processing.

2) Data Processing Module

The function of the remote sensing data processing module is to unpack data. The remote sensing airborne control system firstly marks the navigating path data and image data before transmitting them, then re-packs the data according to the specifications of the navigating task and finally transmits the data to the ground through the air-to-ground transmission channel. The remote sensing data processing module acquires the validated data after removing the navigating task marks. The data is directly sent to the remote sensing data display/storage module.

3) Data Display/Storage Module

The remote sensing data display/storage module displays image data together with the navigating path data corresponding to the image. Simultaneously, this module records the remote sensing data in the database for post playback and analysis, which provides an objective evaluation for the photographing quality. This module is the key part for the ground monitoring system.

4) Data Downloading Module

The remote sensing downloading module is relatively independent. After the unmanned airplane lands and the cabin is opened without changing the load, all the remote sensing data photographed during the flight are downloaded to ground computers. The unmanned airplane can then instantly implement the next task.

2.2.2.2 Software Design

The software of the ground monitoring system is implemented by the following three programs: data acquisition program, data processing program, data display/storage program. The data acquisition program is developed based on pipeline communication. It works in the following procedure: initialize pipeline, set up connection with pipeline server, read data into pipeline, recognize package symbols and send data to the data processing program. The data processing program is in fact an unpacking program which searches marks of the navigating data and image data. The data are then sent to the remote sensing data display/storage module. The communication between data display/storage module and data processing module adopts C/S structure based on TCP/IP.

2.2.3 Multi-Mode Digital Camera System

UAVs are used to carry off-the-shelf light-weight instruments such as consumer digital cameras (Haarbrink and Koers 2006, Shortis *et a.l* 2006), miniature RADAR, passive microwave radiometers, and LiDAR (Vierling 2006, Sugiura *et al.* 2005, Sugiura *et al.* 2007, Martínez-de Dios 2006, Archer *et al.* 2004).

In order to realize the combination of high-resolution, wide view field, multi-spectra and stereo imaging, a set of multi-mode airborne digital cameras (MADC) have been developed. Limited by the actual load of the UAV, we use a small efficient digital camera which has a 4k×4k area array CCD and a focal length of 80mm. The ground resolutions corresponding to different flight altitudes are shown in Table 2.1.

Table 2.1 Ground resolution at different altitudes (focal length=80mm)

Altitude (m)	500	1000	1500	2000	3000	4000	5000	6000	7000
Ground spacing (cm)	5.6	11.2	16.9	22.5	33.8	45.0	56.2	67.5	78.8

Here three cameras are installed on a simple supporting rack. Relative position is designed differently for different modes, such as wide vision field, multi-spectrums. Synchronized triggering of the shutters for the three cameras is performed by the control system. At the same time, the synchronizing pulse is sent to the input interface of the position and orientation system (POS) to sign the time of the photographing moment. After the exposure, image data are transmitted to an industrial computer through and stored in the high speed hard drive. By changing or adjusting the positions and orientations of the three cameras on the simple camera support rack, the following two modes can be composed.

2.2.3.1 High-Efficiency Wide Angle Airborne Digital Imaging Mode

This is one trend of improving the efficiency of airborne photography. This mode aims at implementing high efficiency of digital airborne photogrammetry. This system improves the spatial and geometrical stability of airborne photographing images with the technical innovation based on wide-angle imaging technology, which mainly involves large (wide) area array. The images acquired are actually from three different cameras with certain overlap between each other. The final wide angle field image with larger area is formed from those three images by matching, correcting and splicing. The installation of the three cameras for wide field imaging is shown in Fig. 2.10. The real configuration (Fig. 2.10 b) also has a low resolution color camera.

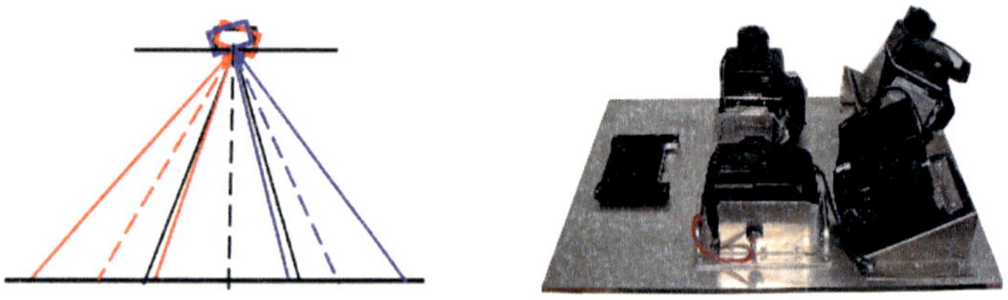

a b

Fig. 2.10 Wide field of view mode: (a) imaging mode; (b) photograph of the installed cameras

2.2.3.2 Multi Spectrum Remote Sensing Data Acquisition Mode

This digital airborne photographing system shown above can also service as a multi spectrual remote sensing sensor system after reset the geometric relationship of the three cameras. When three cameras photograph the same scene simultaneously, different camera catches different spectral information with a filter.

2.2.4 Automatic Path-Planning System

Aerial photogrammetry based on the airplane platform has a history of 70 to 80 years after the Wright brothers invented the airplane. However, the aerial photography planning (especially the path planning) has long been depending on the planners' manual operations. This method involves much work and can not easily accommodate airborne route change. This traditional planning strategy cannot satisfy the needs in UAV mapping. During the unmanned flight, the mission controllers on the ground have to conduct real-time monitoring on the state of aircrafts and sensors. Thus, a computerized system specially designed for path planning is needed to design flight path automatically, upload the data to the controlling system of unmanned aircrafts and sensors before flight, receive the data from the aircraft in flight, and display on a digital map. This ensures that the mission controllers on the ground can learn about the state of spacecrafts and sensors and control them properly.

Airborne route planning software will provide necessary interface for users to read a vector/raster map or Digital Orthogonal Map (DOM) used for path planning. The data must meet the requirement of geometric precision, and must be transformed to a common geographic reference. The software can enable the technicians to plan the target area interactively and get the optimum airborne route that meets the requirements for remote sensing and airborne surveying. In addition, the technicians can calculate and get several parameters such as the width of strip, degree of overlapping of the image, internal of exposure time, etc. In the end, the related digital aerial file is formed and can be uploaded to the control systems of unmanned aircrafts and sensors to control the aircrafts to fly in the scheduled path and the sensors to get data as designed.

The path-planning system developed in this book includes the following functions: map/image input, flying area planning, and path-planning.

(1) map/image input

It is designed to read vector/raster map or image with geographical coordinates in the format of SHP, GeoTIFF, etc. The input data can be arranged as different layers.

(2) flying area planning

It is an interactive interface by which users can define and modify the flying area on the map or image.

(3) path-planning

It provides the user with visual and non-visual path-planning. The visual path-planning allows the user to plan the path under the background of a map or image which can directly show the relationship between the path and the geographic element. With the non-visual path-planning mode, user can only do it with flying area information without the map and image background.

Fig. 2.11 shows an example of the path-planning system application. This example uses digital map as background. The area with semitransparent mask is the working area and the line indicates the air route. The yellow dot, green dot and red dot represent the preparation point, start point and end point of every air route, respectively.

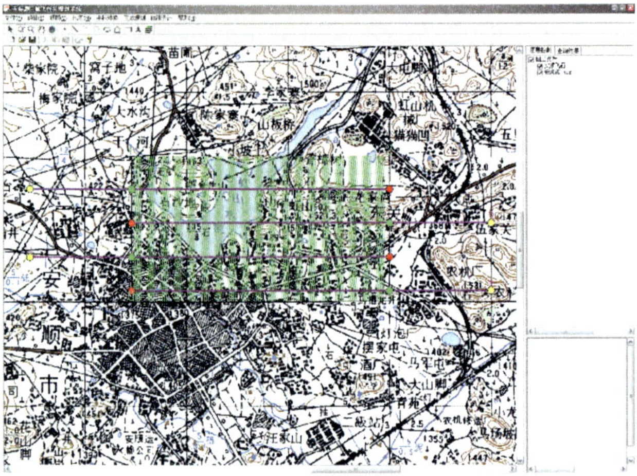

Fig. 2.11 The main interface of the path-planning system

2.3 Experiment of UAV Remote Sensing System

China's first high-end and multi-functional remote sensing based on unmanned aerial vehicle (UAV) made its successful test flight on August 8, 2005 in Anshun City, southwest of China's Guizhou Province. The remote sensing system, jointly developed by Peking University and Guizhou Aviation Industry (Group) Co Ltd, adopts intelligent and high-definition data retrieving technologies. This experiment had three flights (see Table 2.2). The first flights aimed at testing the imaging device on-board. The second time was carried out according to the requirement of photogrammetry, which tested the whole UAV remote sensing system. And, the last one tested the ability of real-time compression and decompression and validated some improvements.

What is more notable is that the resolution has reached 2.25cm at low altitude, with clear identification of heads and tails of ducklings on the farm (Fig. 2.12). Fig. 2.13 shows the air route designed for remotes sensing experimental flight of the UAV. The lines are preset air routes and the dots are the start and end of a line. The area in the box is the working area. An image auto-spliced from images photographed by the high-efficiency wide angle airborne digital imaging mode of MADC is shown by Fig. 2.14.

Table 2.2 General results of experimental flights in August, 2005

	First Flight	Second Flight	Third flight
Date and time	10:00 , Aug. 8	15:30 , Aug. 18	9:30 , Aug. 24
area	An airport	Anshun City	An airport
weather	cloudy	sunny	cloudy
Relative altitude	200m	1000m	200m，400m
Ground resolution	0.02m	0.10m	0.02m，0.04m
flight velocity	40m/s	40m/s	40m/s
Lens focus	80mm	80mm	80mm
shutter	1/1000s	1/1000s	1/500s
Aperture value	2.8	2.8	2.8
Photograph mode	fixed time interval	fixed position	fixed position
longitudinal overlap	0	54.79%	0
lateral overlap	0	30%	0
Number of images	33	158	48

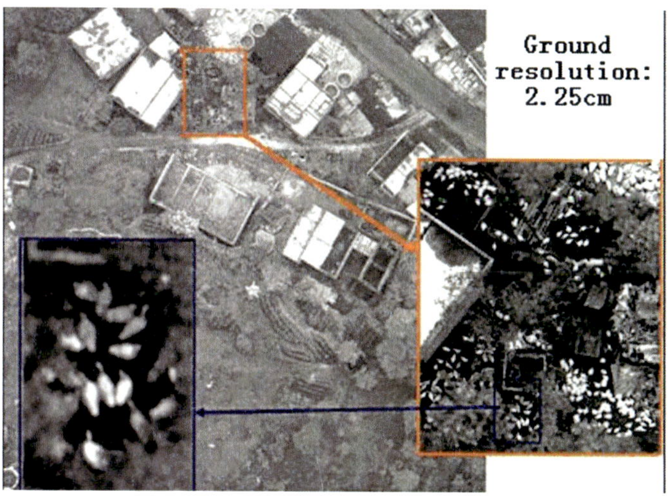

Fig. 2.12 Enlarged images of the duck farm

Fig. 2.13 The air route designed for remotes sensing experimental flight of the unmanned aircraft in Anshun, Guizhou Province in 2005

Fig. 2.14 Image auto-spliced by images taken by three parallel high-resolution digital cameras

Fig. 2.15 shows four images acquired at an altitude of 0.2 km by the first experimental flight. Fig. 2.16 gives a part of the real-time mosaic of the snapshots. Fig. 2.17 is an image spliced by images photographed at an altitude of 1km on August 18, 2005.

Generally, this experiment was carried out successfully. In the field of unmanned airborne remote sensing, it firstly adopted integrated, intelligent, and high-resolution spatial data acquisition and other important technologies. The application capability has been qualified, especially in the aspects of reliability, flying altitude, stabilization, navigating precision, operation costs and image acquisition. The experimental flights successfully validate airborne control, ground monitoring, image acquisition and path-planning of the UAV remote sensing system. Even though, the results have shown that UAV is able to be a complement of conventional airborne remote sensing and satellite remote sensing, it still has some problems to be solved, which are listed as follows:

- Complete the path planning system to implement professional task flight of unmanned airplane;
- Study the movement compensation of images to solve the bottleneck problem in high-speed ultra low situation;
- Complete information downloading and compressing/decompressing

functions to provide the basis for the real-time application;

- Minimize the size and weight of digital cameras and remote sensing systems to solve the bottleneck problem of limited payload;
- Develop a complete set of software systems for the automatic processing of UAV remote sensing.

Fig. 2.15 Four images acquired by the UAV system in Anshun City, China

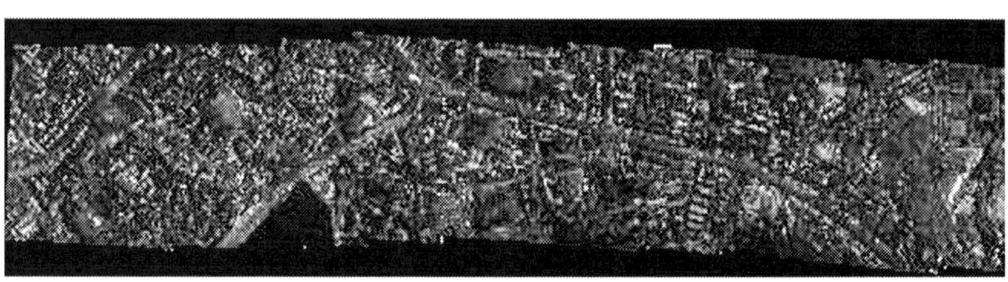

Fig. 2.16 The real-time mosaic of the snapshots

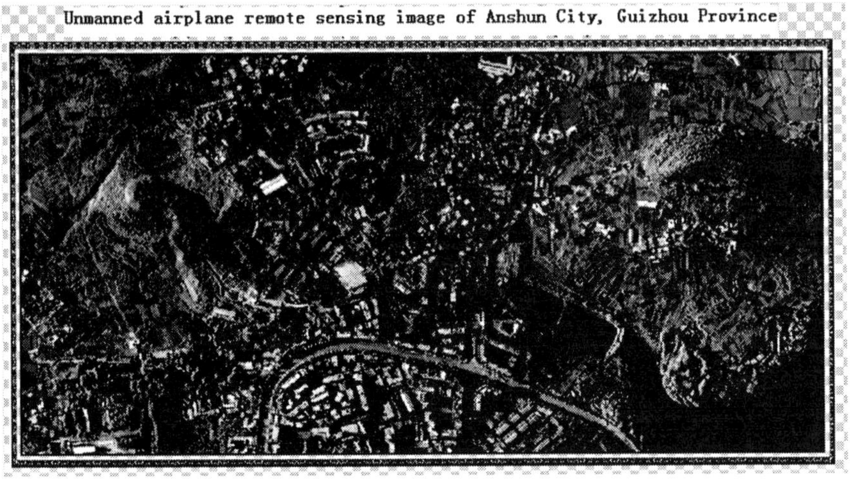

Fig. 2.17 Image spliced by images photographed at an altitude of 1km

2.4 Conclusions

Since UAV has a limited loading capacity, the load must be reasonably designed to reduce the volume, weight, and power consumption while reaching the required performance index. Considering the current development of UAV, UAV remote sensing system should be equipped with one sensor in the early stage. However, the system should have the ability to change according to different sensors quickly (such as optical digital camera, infrared digital camera and SAR). Multiple sensors can be equipped only within the range of the loading capacity of the system. UAV usually has a high precision GPS/INS navigation and positioning device and needs real-time position and altitude data of the station. In order to reduce the weight and power consumption, the positioning data of the remote sensor also shares the GPS/INS data of the UAV. Thus, there is no need to install positioning and altitude measuring equipments any more.

The number of UAV systems used in remote sensing and mapping has been increasing fast in the past years. Many of them are still under research phase, and there are few systems that offer complete solutions to a user. UAVs have given remote sensing a new appeal for scientists and will certainly become the preferred platform for development of remote sensing instruments and applications in the near future when the necessary regulations and techniques are solved.

UAVs are unlikely to replace more conventional remote sensing platforms, but they do offer advantages as means of supplementing conventional field data, especially by providing data that can be coordinated with broad-scale imagery from aircraft and satellite platforms.

Acknowledgments

The authors would like to thank Academician Qingxi Tong, Prof. Shaowen Yang, Qiming Qin, Bing Zhang, Qiming Zeng, Associate Prof. Jian Jiao, Hongying Zhao for their UAV RS research work and contribution to this chapter.

References

Archer F *et al* (2004) Introduction, overview, and status of the Microwave Autonomous Copter System (MACS). Proceedings of *IGARSS 200*, Anchorage, Alaska, USA.

Berni JAJ, Zarco-Tejada PJ, Suárez L, Fereres E(2009) Thermal and Narrowband Multispectral Remote Sensing for Vegetation Monitoring From an Unmanned Aerial Vehicle. TRANSACTIONS ON GEOSCIENCE AND REMOTE SENSING, 47:722-738

Blyenburgh PV (2008) UAS Unmanned Aircraft Systems. The Global Perspective 2007/2008. www.uvs-info.com

Boike J, Yoshikawa K (2003). Mapping of periglacial geomorphology using kite/balloon aerial photography. Permafrost and Periglacial Processes, 14:81−85

Buerkert A, Mahler F, Marschner H (1996). Soil productivity management and plant growth in the Sahel: Potential of an aerial monitoring technique. Plant and Soil, 180:29−38

Çabuk A, Deveci A, Ergincan F (2007) Improving Heritage Documentation. GIM International 21, 9, September 2007

Chen XX, Lee V (2006) Spectral mixture analyses of hyperspectral data acquired using a tethered balloon. Remote Sensing of Environment 103(3)338-350

Doherty P (2004) Advanced Research with Autonomous Unmanned Aerial Vehicles. Proceedings of the 9th International Conference on Principles of Knowledge Representation and Reasoning

Damià Vericat , James Brasington, Joe Wheaton, Michaela Cowie (2008) Accuracy assessment of aerial photographs acquired using lighter-than-air blimps: Low-cost tools for mapping river corridors. River Research and Applications. Published online in Wiley InterScience www.interscience.wiley.com DOI: 10.1002/rra.1198

Edrich M (2006) Ultra-lightweight synthetic aperture radar based on a 35 GHz FMCW sensor concept and online raw data transmission, Proc. Inst. Elect. Eng.—Radar Sonar Navig., 153:129–134

Eisenbeiss H, Zhand L (2006) Comparison of DSMs generated from mini UAV imagery and terrestrial laserscanner in a cultural heritage application. The International Archives of Photogrammetry, Remote Sensing and Spatial Information Sciences, Dresden, Germany, Vol. XXXVI part 5

Everaerts J (2008) The Use of Unmanned Aerial Vehicles for Remote Sensing and Mapping, Proceeding of ISPRS, The International Archives of the Photogrammetry, Remote Sensing and Spatial Information Sciences. Vol. XXXVII. Part B1. Beijing 1187-1191

Friedli B, Tobias S, Fritsch M (1998) Quality assessment of restored soils: Combination of classical soil science methods with ground penetrating radar and near infrared aerial photography. Soil and Tillage Research 46: 103−115

Gerard B, Buerkert A, Hiernaux P, Marschner H (1997) Nondestructive measurement of plant growth and nitrogen status of pearl millet with low-altitude aerial photography. Soil Science and Plant Nutrition 43: 993−998

Hongoh D, Kajiwara K, Honda Y (2001) Developing Ground Truth Measurement System using RC Helicopter and BRDF Model in Forest Area. 22nd Asian Conference on Remote Sensing.

Henri Eisenbeiss (2004) A MINI UNMANNED AERIAL VEHICLE (UAV): SYSTEM OVERVIEW AND IMAGE ACQUISITION. International Workshop on "PROCESSING AND VISUALIZATION USING HIGH-RESOLUTION IMAGERY" 18-20 November, Pitsanulok, Thailand.

Herwitz SR *et al.* (2004) Imaging from an unmanned aerial vehicle: agricultural surveillance and decision support. Computers and Electronics in Agriculture 44: 49-61.

Haarbrink RB, Koers E (2006) Helicopter UAV for Photogrammetry and Rapid Response. The International Archives of Photogrammetry, Remote Sensing and Spatial Information Sciences, Antwerp, Belgium, Vol. XXXVI-1/W44.

James BC (2002) Introduction to remote sensing. The Guilford Press, New York

Martinez Rubio J *et al* (2005) IMAP3D: Low cost photogrammetry for cultural heritage. *Proceedings of CIPA2005 XX International Symposium*, Torino, Italy.

Martínez-de Dios JR *et al*, (2006) Experimental results of automatic fire detection and monitoring with UAVs. Forest Ecology and Management

Mei AX, Peng WL,Qin QM, Liu HP(2001) Introduction to Remote Sensing. Higher Education Press, Beijing

Puri A *et al.* (2007) Statistical Profile Generation for Traffic Monitoring Using Real-time UAV based Video Data. Proceedings of Mediterranean Conference on Control & Automation 2007, Athens, Greece.,

Rango A *et al.* (2006) Using unmanned vehicles for rangelands: Current applications and future potentials. Environmental Practise, 8 (3): 159-169.

Réstas A (2006) The regulation Unmanned Aerial Vehicle of the Szendro Fire Department supporting fighting against forest fires 1st in the world! Forest Ecology and Management, 234S (2006) S233.air vehicles. International Journal of Systems Science 37 (6):351-360.

Robert G R(1975). Manual of remote sensing. American Society of photogrammetry,vol.1..

Shortis MR *et al.* (2006) Stability of zoom and fixed lenses used with digital SLR cameras. *The International Archives of Photogrammetry, Remote Sensing and Spatial Information Sciences*, Dresden, Germany, Vol. XXXVI part 5.

Sugiura R *et al.* (2005) Remote-sensing technology for Vegetation Monitoring using an Unmanned Helicopter. Biosystems Engineering 90(4):369-379.

Sugiura R. *et al,* (2007). Correction of Low-altitude Thermal Images applied to estimating Soil Water Status. Biosystems Engineering 96(3):301-313.

Thamm HP, Judex M (2006) The "Low cost drone" – An interesting tool for process monitoring in a high spatial and temporal resolution. The International Archives of Photogrammetry, Remote Sensing and Spatial Information Sciences, Enschede, The Netherlands, Vol. XXXVI part 7.

Vierling LA *et al.* (2006) The Short Wave Aerostat-Mounted Imager (SWAMI): A novel platform for acquiring remotelysensed data from a tethered balloon. Remote Sensing of Environment 103: 255-264.

Wang WQ, Peng QC, Cai JY (2009) Waveform-Diversity-Based Millimeter-Wave UAV SAR Remote Sensing. TRANSACTIONS ON GEOSCIENCE AND REMOTE SENSING 47(3):691-700

White LP (1977) Aerial Photograph and Remote Sensing for Soil Survey. Clarendon Press, Oxford,UK

Zhang CS (2008) AN UAV-BASED PHOTOGRAMMETRIC MAPPING SYSTEM FOR ROAD CONDITION ASSESSMENT.The International Archives of the Photogrammetry, Remote Sensing and Spatial Information Sciences. Vol. XXXVII. Part B5.627-632

Zischinsky Th, Dorfner L, Rottensteiner F (2000) Application of a new Model Helicopter System in Architectural Photogrammetry. Amsterdam. IAPRS Vol. XXXIII. B5/2.

Chapter 3

DESIGN AND APPLICATIONS OF AN INTEGRATED MULTI-SENSOR MOBILE SYSTEM FOR ROAD SURFACE CONDITION DETECTION

Qingquan Li, Yong Liu, Qingzhou Mao

3.1 Introduction

Landborne Multi-Sensor Integration System (LMSIS) has great advantages in data acquisition, i.e., accurate and comprehensive acquisition, rapid and automatic data processing. Moreover, its application has been expanded to vehicle navigation, road survey and design, traffic surveillance, and so on.

In the past few decades, with the development of information technology and automatic controlling and high precision micro-measurement technology, IMSMS has advanced greatly in the types and properties of sensors and has been put into service in road surface condition detection. For instance, the Road Surface Profilometer, a product of the Dynatest Company in Denmark, is a portable road surface detection system designed to provide advanced, automatic pavement roughness measurement in high quality for engineers and construction managers (Dynatest 2009). This Top-of-the-line Profilometer involves up to 21 lasers, accelerometers, Inertial Motion Sensor, GPS receivers, etc. Moreover, ARAN—Automatic Road Analyzer (Roadware 2008) of RoadWare Company in Canada is an advanced platform available for collecting pavement condition and asset data. Its sensors include Distance Measuring Instrument (DMI), GPS receivers, laser SDP, laser XVP, etc. Those products are mainly used to measure road roughness, rutting, landslide, detect cracks, distress, and so on. In addition, the Hawkeye2000 series (Arrb 2008) of ARRB company in Australia, Digital Highway Data Vehicle (Waylink 2008) (DHDV) of WayLink company in USA and Road Assessment Vehicle (WDM 2008) (RAV) of WDM company in UK have similar functions in road surface detection. All those international products apply the latest and most sophisticated technology, but they are expensive for many consumers in developing countries, such as China. A low-cost integrated multi-sensor mobile system needs to be developed.

Starting from the 1990s, some research institutes and universities in China, such as Research Institute of Highway Ministry of Communications (RIHMC), Wuhan University, Nanjing University of Science and Technology (NJUST), etc.,

D. Li et al. (eds.), *Geospatial Technology for Earth Observation*,
DOI 10.1007/978-1-4419-0050-0_3,

have begun to develop IMSMS which aims at road surface detection. The CiCS of RIHMC, an experimental system, has been developed to measure roughness, detect pavement distress and grab front images while the JG-1 of NJUST, can measure roughness, rutting and detect crack at the vehicle speed of 70Km/h. Later in 2000s, supported by National Natural Science Foundation, Enterprise Innovation Fund of the Ministry of Science and Technology and National 985 Project, State Key Laboratory for Information Engineering in Surveying, Mapping and Remote Sensing in Wuhan University developed an IMSMS —SmartV. However, most researchers have only focused on one function of IMSMS (Aris *et al.* 2006, Chen *et al.* 2006, Randeniya *et al.* 2008, Kelvin *et al.* 2002, Li and Yuan 2002, Li and Liu 2007, Nagai *et al.* 2004, Sreenivas *et al.* 2006). Few studies examine the whole system. Given this, this paper sets out to examine the functions of the whole system of SmartV and its application comprehensively.

3.2 SmartV System Overview

3.2.1 Components of SmartV

The hardware of SmartV refers to all the sensors and the vehicle which are used as the sensor carrier. Besides the vehicle, the sensors are divided into two categories:

- **Vehicle Sensors:** the sensors assembled in the vehicle by the automobile manufacturer. The outputs of these sensors are useful for some customer sensors. For example, pulse signal of vehicle odometer and vehicle reversing signal can be used as inputs of GPS/DR.
- **Customer Sensors:** the sensors equipped by consumers according to the requirements of some special application. For example, positioning sensors can locate vehicles, and laser range finders can measure distance.

The software of SmartV comprises onboard grabbing software and indoor post-processing software.

- **Onboard data acquisition software:** the software which is installed on the workstation of the SmartV's vehicle, and is mainly used to acquire in real-time and store massive data. The performance requirements of this software are automatic, real-time processing and steadily running.
- **Indoor data post-processing software:** the software which is used to post-process the acquired data and operated by the indoor workers. This software needs friendly user interface, performance of managing the massive data, and diverse report forms.

3.2.2 Major Functions

SmartV has various functions in application. In road surface detection, SmartV can be divided into the following subsystems, as shown in Figure 3.1.

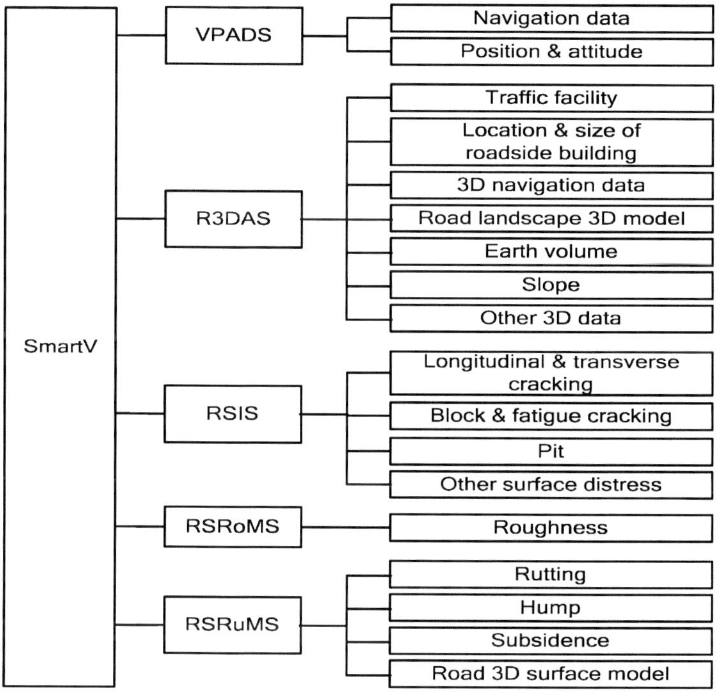

Fig. 3.1 Functions of SmartV in road surface detection

The functions of each subsystem are as follows:

1) The sensor of *Vehicle Positioning and Attitude Determination Subsystem/System (VPADS)* assisted by the positioning technology of road linear reference system (RLRS) is dead reckoning GPS receiver. This system can provide Spatial Reference Coordinate System (SRCS, vehicle current WGS coordinate and attitude) and Reference Time System (RTS, synthesized GPS time) for other subsystems.

2) The sensors of *Road 3D Data Acquisition Subsystem/System (R3DAS)* are multiple industrial CCD cameras and multiple laser scanners. Geometric size of objects in the public field of view of stereo cameras and object properties can be obtained. On the other hand, 3D point cloud of both sides of the scanned road can be obtained by registering different profile scans together, and needed road information can be analyzed.

3) The sensor of *Road Surface Imaging Subsystem/System (RSIS)* is a high resolution linear CCD camera. This system can grab high definition road surface

images by scanning and then the surface distress can be analyzed through the images.

4) The sensors of *Road Surface Roughness Measurement Subsystem/System (RSRoMS)* are high precision laser range finders and accelerometers, and this system can measure roughness of the road surface.

5) The sensors used in *Road Surface Rutting Measurement Subsystem/System (RSRuMS)* are linearly structured light and CCD camera. Thus, high resolution and precision 3D road surface models can be obtained through this system by triangulation technology, and rutting of road surface can also be obtained.

3.2.3 Applications

According to the above functions, SmartV can be widely applied in the following areas:

- **Navigation:** The trajectory of vehicles can be used to create digital maps which are the fundamental database for navigation. Property data of navigation, such as building, gas station and traffic light, can be obtained by front CCD cameras.
- **Road facilities management:** The data of road facilities, such as facility location, road planting, roadside slope and road fence etc., can be collected.
- **Road survey and design:** The road 3D data can be used to analyze earth volume calculation and evaluation of construction workload of road maintenance or reconstruction.
- **Road surface condition detection.**

3.3 System Design

For sequential data merging and information extracting, the output data of sensors should be effectively obtained. In order to achieve this, the following two criteria of the system must be achieved:

1) The system should satisfy the condition of the installation, power supply of the sensors and the condition of sensor data acquisition, transmission, and storage. Firstly, the sensors must be installed in the limited space of the vehicle. Secondly, enormous demand on power for sensors, accessory equipments, and workstations need extra power supply since the power supply of the vehicle can only meet its own demands.
2) The system should satisfy the condition of time synchronization and coordinate unification among the sensors. In order to query and locate certain data from massive information, and analyze the correlation among them, the data must be stamped by SRCS and RTS. It involves two aspects: firstly, how to generate the SRCS and RTS; secondly, how to unify the data with them.

The following sections will discuss how to achieve the two criteria from hardware architecture (sensors controlling and synchronization) and software architecture (coordinate transformation).

3.3.1 Sensors Controlling

Sensors controlled in SmartV are shown in Figure 3.2. According to the figure, extra power supply, SRCS and RTS supply for other subsystems comprehensively. For one thing, there are two schemes for extra power supply, i.e., (1) gasoline generator which works when driving, (2) commercial power interface which is used when parking. So enough power supply is provided in whatever condition the vehicle is. What is more, gasoline can be saved and work environment in the vehicle can be improved. SRCS and RTS are provided by VPADS. SRCS is based on the location of the GPS receiver, which is a unique coordinator in WGS. RTS benefits from the atomic clock which is mounted on the GPS satellite and provides a high precision time benchmark.

In addition, there is an independent controller for each subsystem. The major functions are discussed as follows:

- It can provide the power supply for the sensors and workstations of its system (except for high power equipments). There are many kinds of sensors which use different voltage and the controller can transform voltage for them.
- It can provide trigger pulse signal by time interval or distance interval. This signal can be used to grab the image at every 10 meters when driving or at every 2 seconds when parking.
- It can provide high precision time output at high frequency for data acquisition.

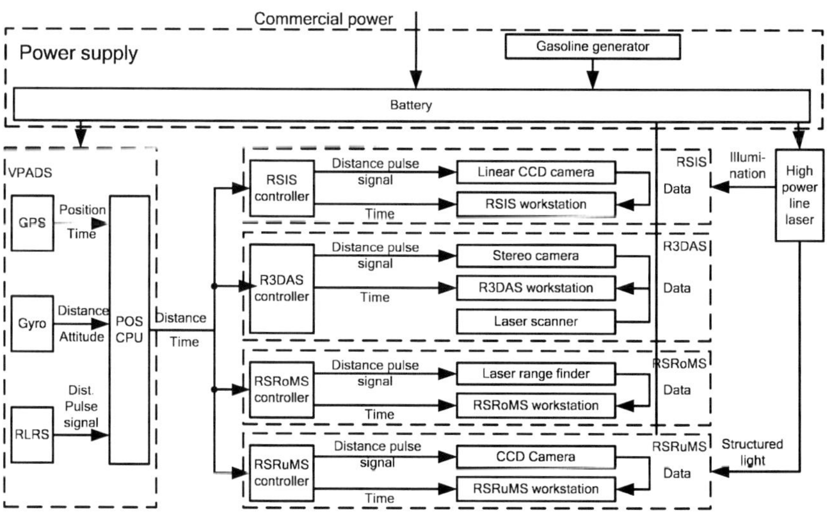

Fig. 3.2 Control of sensors in SmartV

3.3.2 Time Benchmark and Sensor Synchronization

Although GPS can provide high precision time in UTC at a high level accuracy of 50 nanoseconds, it can not be used directly for data acquisition workstation. The following two reasons account for this:

1) GPS signal is not guaranteed on vehicles. Especially when the vehicle is in the tunnel or under a shelter, the GPS receiver in the controller can not obtain a GPS signal. So time precision can not be guaranteed.
2) Raw GPS refresh rate is too low. In a high-speed data acquisition system, such a refresh rate is too low to stamp time and location for each data frame.

To solve these problems, a timer has been embedded into the controller and it has two functions as follows:

1) The controller in each subsystem has the function of timing. When the signal is lost, the timer starts to time based on the latest received GPS time; and when the signal is recaptured again, the timer calibrates the current time, which can prevent error cumulating. Time error of the timer, using a high stable crystal oscillator as the timing unit, is 1 millisecond in 12 hours.
2) Time can be subdivided. The timer takes GPS satellite as time resource and provides high frequency time output with the identical high precision as raw GPS time, e.g., at a rate of 100Hz.

The controller also provides the function of synchronous controlling. This is achieved by outputting multiple trigger pulse signals, such as the signal for stereo cameras synchronous exposing.

3.3.3 Coordinate Transformation and Unification

SRCS of SmartV is established on GPS with twin antenna and gyroscope and is a world geodetic system. The location of road surface customarily uses road linear reference (i.e., pile numbers and odometer offset). Thus, RLRS must be transformed into world geodetic system at first. The following Figure 3.3 shows the reference systems used in this transformation.

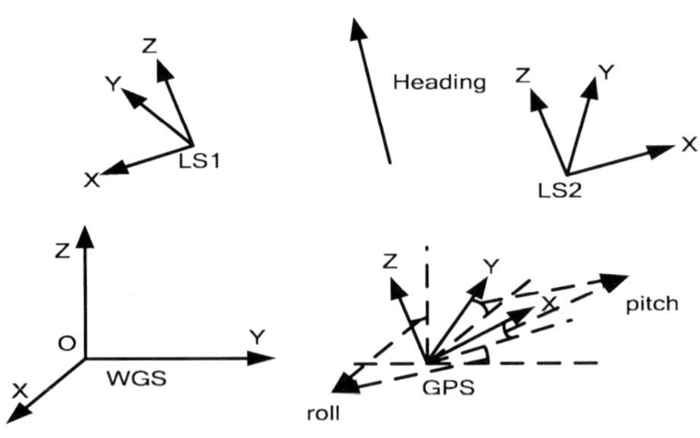

Figure 3.3 WGS84 and road linear reference system

In Figure 3.3, LS1-XYZ and LS2-XYZ are road linear reference systems. After transformation, GPS combined with RLRS form the SRCS. But the data from other subsystems is located in the local coordination system fixed on vehicle or sensor. So this requires another transformation from the local coordinate system into SRCS. Moreover, the transformation between SRCS and local coordinate systems are rotation and translation.

3.4 Multi-Source Data Acquisition and Processing

3.4.1 Vehicle Positioning and Attitude Determination

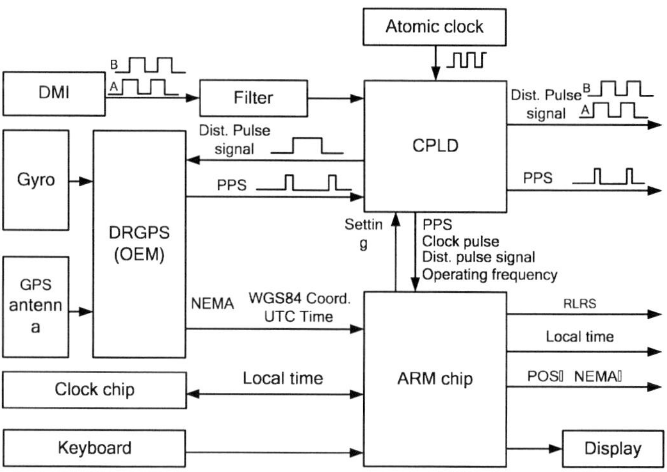

Fig. 3.4 Principle diagram of VPADS

VPADS is composed of GPS receiver, i.e., an absolute positioning technology, and gyroscope, i.e., a relative positioning sensor. It combines the advantages of the two technologies. So vehicle location can be accurately estimated. The following Figure 3.4 shows the principle diagram of VPADS. First, dead reckoning GPS (DRGPS) module takes GPS signal, gyroscope output and DMI signal as input. Next, it exports GPS data as NEMA format to ARM single chip after merging them together by serial port, along with pulse-per-second (PPS) signal to CPLD chip. Then, CPLD receives the ARM's setting parameters. Finally, it merges DMI signal, atomic clock signal and PPS together to export the following signals:

1) Distance pulse signal used as dead reckoning signal to DRGPS
2) Distance pulse signal to ARM chip to estimate distance
3) Shaped signal of AB phase signal of DMI to other controllers

4) Shaped signal of PPS to other controllers and ARM chip used as time correcting signal
5) Millisecond signal and operating frequency signal to ARM chip

In this way, ARM chip actually works as a CPU of VPADS, by receiving outside control command, transmitting data, merging different data and uploading position, and time data to workstation and other controllers.

3.4.2 Road 3D Data Acquisition and Processing

Road 3D data include two types of data: stereo image database and 3D point cloud. On the one hand, stereo image is grabbed from a stereo camera mounted in the front of the vehicle by an image grab board and imaging road scene. Consequently, geometric size of viewable objects can be measured by photogrammetry and properties of them can also be obtained. Also, they can simply serve as landscape images. Stereo cameras expose synchronously by the distance signal from the R3DAS controller. On the other hand, 3D point cloud is obtained through a laser scanner at a high scanning frequency. By these means, the laser scanner continuously exports scanned data which are stamped with high precision time in workstation. In this process, time is crucial for registering every data frame. The following Figure 3.5 describes the processing of road 3D data.

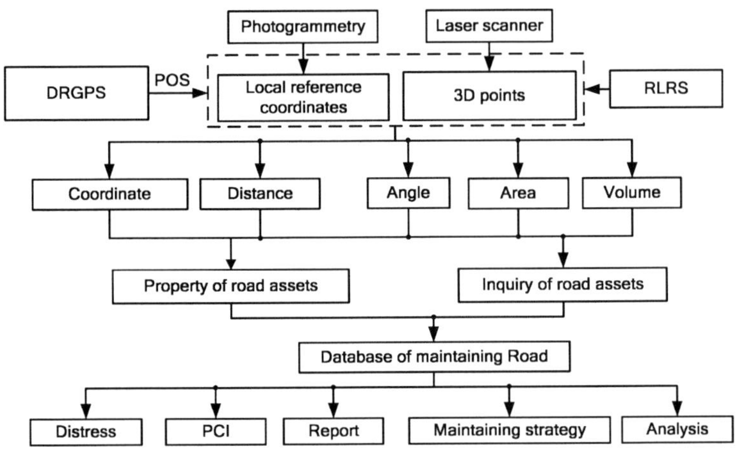

Fig. 3.5 Processing of road 3D data

3.4.3 Laser Illumination, Road Surface Imaging and Crack Detection

The linear CCD camera is used as an image sensor which works through scanning imaging. Figure 3.6 is the flow chart of imaging and image processing.

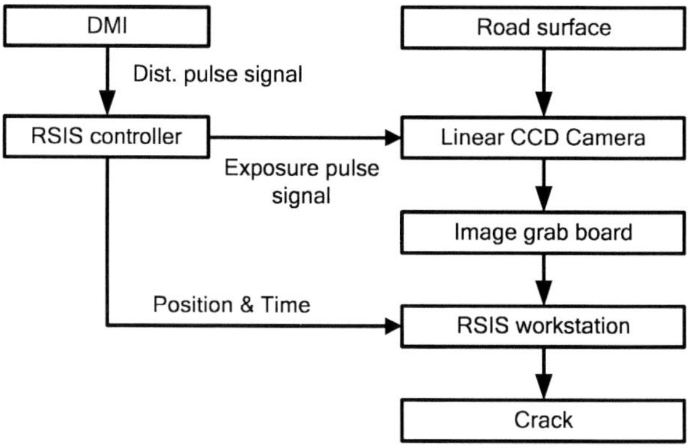

Fig. 3.6 Road surface imaging and image processing

From the figure, we can see that the RSIS controller receives the distance signal of DMI and exports an exposure pulse signal to the linear CCD camera and a high power liner laser is used as illuminating light. Due to the high line rate of the linear CCD camera, the exposure time is very short. Moreover, in natural conditions, illumination is changed significantly. For instance, in shadow or in a cloudy day the illumination is much lower than in a sunny day. So, high illumination is needed. Because of the high illumination that the laser can provide, the high power line laser can be used to provide enough illumination in any environment. Moreover, it is also served as structured light for RSRuMS.

Crack detection can be achieved from road surface image. Figure 3.7 shows the processing flow.

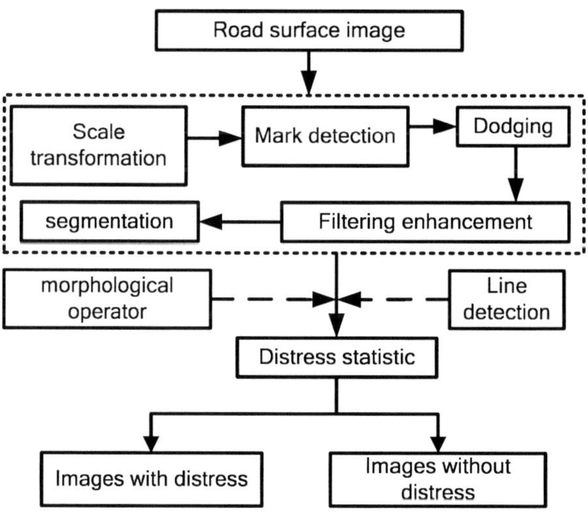

Fig. 3.7 Crack detection processing flow

The following steps illustrate the process in the figure. The first step is ***Scale transform***. That is, multi-level pyramid images are created for improving processing efficiency and low-pass filter is used to keep the texture edge. The next step is ***Road surface mark detection.*** This involves detect marks, such as lane line, zebra crossing or guiding signs. As a result, the detection region narrows between two lane lines. Then, ***Dodging Processing,*** based on projection compensation, can effectively remove the uneven illumination in the image. Consequently, ***Filtering enhancement*** can reduce noise and enhance the distress texture by morphological filtering and linear enhancement filtering. Finally, in ***Image segmentation***, the distress feature is segmented by differential statistics in neighborhood. After all these steps, statistics are used to detect the planar features and calculate the degree of road distress.

3.4.4 Road Surface Roughness Measurement

Laser range finder method with inertia compensation is a longitudinal section measurement method. First, vertical distance between the vehicle and road surface is obtained by a high precision laser range finder sensor. Next, the vertical bumpiness caused by vehicle driving is measured with accelerometer. Then, the position can be located accurately with DMI. Finally, all the data is processed and analyzed, and therefore, the international roughness index (IRI) can be obtained. The following Figure 3.8 shows the principle diagram of road surface roughness measurement.

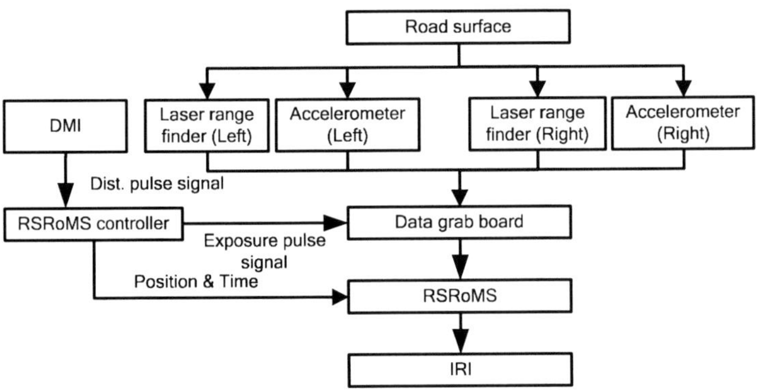

Fig. 3.8 Principle diagram of road surface roughness measurement

Laser range finder and accelerometer are rigidly connected together, and mounted on the vehicle. The vertical distance from the road surface to laser range finder cannot exceed the work distance of the laser range finder. Besides, the controller of RSRoMS sends a pulse signal to the data grab board. So the data acquisition, transmission and processing can be done synchronously. At the same time, DMI provides the horizontal translation for calculating IRI.

3.4.5 Road Surface Rutting Measurement

As shown in Figure 3.9, the following principle diagram illustrates the process of road surface rutting measurement.

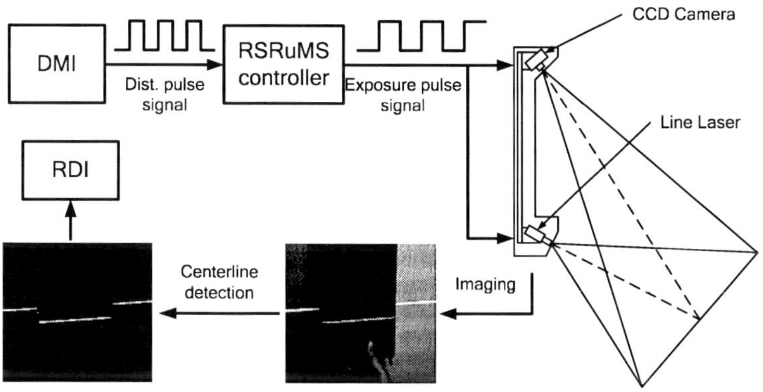

Fig. 3.9 Principle diagram of road surface rutting measurement

Figure 3.9 shows that rutting measurement is based on the technology of structured light triangulation. Line laser is used as structured light. Planar CCD camera with a high frame rate can image transverse sections of road surface which are illuminated by laser.

Furthermore, Figure 3.10 shows the steps of calculating rutting and distortion index (RDI). First, centerline of the laser line in the image is extracted and outliers are removed. Next, world coordinate of the centerline is interpolated by a look-up table which has been built as a Delaunay triangle. Then, a key point is set at every 20 points and all key points are connected as line segments one-by-one. For each line segment, if there is a key point above it, this line will be deleted. Then there is a line segment that all key points are below it, and this line segment is a fitting line to the road transverse section profile. Finally, from the first point of the centerline to the last one, the distance between the point and the line is calculated and the maximal distance is the current rutting.

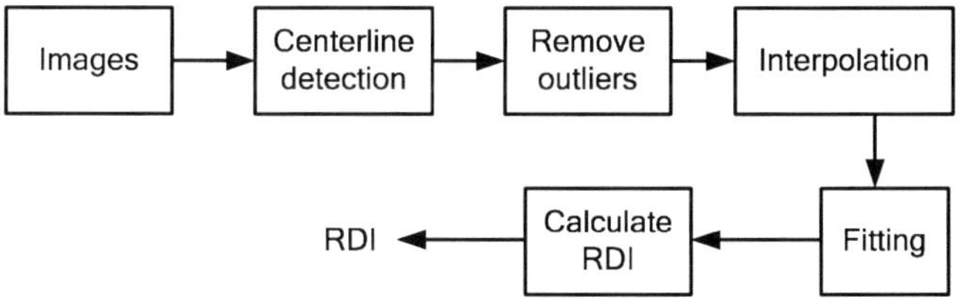

Fig. 3.10 The steps of calculating rutting and distortion index

If all transverse section profiles are registered together, 3D road surface model can be obtained.

3.5 Experimental System and Applications

3.5.1 Experiments System

As shown in the following Figure 3.11, Wuhan University has developed an experimental IMSMS –SmartV. The experimental system has been put into tests and applications. As a result, accurate measurement results have been obtained. For example, some calibration station tests in crack detection, roughness and rutting measurement. The results are showed in Tables 3.1-3.3. Table 3.2 compares the values with standard IRI; the correlation coefficient is 0.9932, which is above the 0.98 demand. Table 3.3 compares the values with manual measurement results; the correlation coefficient is 0.9814, which is above the 0.95 demand.

Fig. 3.11 SmartV of Wuhan University

Table 3.1 Contents of crack detection test for RSIS

Contents	Achieve goal	Result
Crack detection	100% for crack (crack width bigger than 2mm) Width error ≤ 1mm	100%
Raveling detection	Area error ≤ 5%	3.5%
	Auto detection rate (2mm width crack) ≥ 80%	85%
	Auto detection rate ≥ 80%	85%
Other distress detection	Area error ≤ 5%	4.0%
	Auto detection rate ≥ 80%	80%
DMI error	≤ 0.1%	0.1%
Repetition	Cv ≤ 5%	2.0%

Table 3.2 Roughness test for RSRoMS

Contents	Road No.				
	1	2	3	4	5
Average	1.53	3.23	2.01	4.86	6.43
Standard deviation	0.070	0.073	0.080	0.172	0.199
variation coefficient	4.58%	2.28%	4.01%	3.54%	3.10%
Standard IRI	1.57	2.59	1.88	3.74	5.13

Table 3.3 Rutting test for RSRuMS

Contents	Road No.						
	1	2	3	4	5	6	7
Average	30.6	27.6	24.6	17.5	13.5	11.8	3.6
Standard deviation	0.297	0.164	0.245	0.396	0.789	0.122	0.114
variation coefficient	0.97%	0.59%	1.00%	2.26%	5.83%	1.04%	3.20%
Manual measurement	31.2	25.4	22.6	17.7	14.1	11.9	6.2

3.5.2 Applications

Presently, there are several IMSMS for road surface condition detection that have equipped some highway engineering Co., Ltd. and have been successfully put into use. This system can be used to investigate the Pavement Surface Condition Index (PCI), the Riding Quality Index (RQI), and the Rutting Depth Index (RDI) of the highway. Furthermore, the report form can satisfy the standard of the CPMS pavement management system of Ministry of Transport of the People's Republic of China.

For example, in 2008 October, the system had been used to detect the road surface conditions of the Fu-Ning highway. This highway, connecting from Fuding city to Ningde city of Fujian province, China, was opened for traffic in 2003. Furthermore, its whole length is 141 Km and it is a bituminous pavement, but mixed with some concrete pavement. The following reports the investigation results.

a) Evaluation of overall distress

PCI, which can be calculated from Pavement Distress Ratio (DR), is often used as evaluation index for road surface distress. Moreover, there are five scales range from Excellent, Good, Average, Fair to Poor. Readers who are interested in this field can refer to the highway performance assessment standards (China Communication Press, 2008) for details about the evaluation system.

As a result, Table 3.4 is the overall distress evaluation, and Table 3.5 and Table 3.6 are respectively the distress of bituminous pavement and concrete pavement.

From these tables, we can draw a conclusion that this highway is still in good condition, and we can also investigate the damage level of each distress in detail.

Table 3.4 Overall distress evaluation

Evaluation Object	DR (%)	PCI	Evaluation
Up-Direction	0.69	88.2	Good
Down-Direction	0.57	88.8	Good
Overall	0.63	88.5	Good

Table 3.5 Distress of bituminous pavement

Distress Types	(Unit)	Up-Direction	Down-Direction
Alligator Cracking	(m×m)	920.1	348.3
Block Cracking	(m×m)	0.0	1.2
Longitudinal Cracking	(m)	686.5	519.8
Transverse Cracking	(m)	1043.9	1543.8
Pot Holes	(m×m)	269.2	77.9
Ravelling	(m×m)	161.6	583.1
Subsidence	(m×m)	0.0	0.0
Rutting	(m)	0.0	0.0
Upheaval	(m×m)	0.0	0.0
Bleeding	(m×m)	0.0	0.0
Repairing	(m×m)	1084.8	1339.4

b) Evaluation of riding quality Index

RQI is generally defined as an expression of irregularities in the pavement surface, and can be calculated from International Roughness Index (IRI). Table 3.7 is the overall RQI evaluation.

Table 3.7 shows that the riding quality is satisfactory, i.e. the drivers or passengers traveling on this highway must feel comfortable.

c) Evaluation of rutting depth index

Table 3.8 shows the overall RDI evaluation which RDI is calculated from Rutting Depth (RD). We can see that the distress of rutting is a bit serious.

Table 3.6 Distress of concrete pavement

Distress Types	(Unit)	Up-Direction	Down-Direction
Slab Damage	(m×m)	1090.0	375.5
Cracking	m	336.9	280.8
Corner Damage of Slab	(m×m)	97.1	71.1
Faulting of Slab Ends	m	0.0	0.0
Pumping	m	0.0	0.0
Edge Failure	m	20.4	8.4
Jointing Material Damage	(m×m)	14.8	13.9
Hole	(m×m)	6.5	5.5
Arching	(m×m)	0.0	0.0
Surface Angularity	(m×m)	7.9	0.0
Repairing	(m×m)	76.3	94.7

Table 3.7 Overall RQI evaluation

Evaluation Object	IRI (Unit: m/km)	RQI	Evaluation
Up-Direction	2.54	88.05	Good
Down-Direction	2.41	88.95	Good
Overall	2.47	88.50	Good

Table 3.8 Overall RDI evaluation

Evaluation Object	RD (Unit: mm)	RDI	Evaluation
Up-Direction	10.53	78.94	Average
Down-Direction	10.70	78.60	Average
Overall	10.61	78.77	Average

Besides of the overall evaluation, detailed evaluation can be also obtained, e.g. evaluating the highway per kilometer or hectometer. Similarly, taking the Fu-Ning highway as example, the distress evaluation per Km is shown as Table 3.9.

Table 3.9 Distress evaluation per kilometer

Evaluation per Kilometer (Unit)	Excellent	Good	Average	Fair	Poor
Up-Direction (Km)	105.1	29.7	7.0	2.1	0.2
Down-Direction (Km)	99.7	35.9	7.0	1.6	0.0

Table 3.6 shows the road length that lies in each evaluation scale, and most of the road is good condition. Moreover, PCI can be obtained at each pile number and this is shown in Figure 3.12.

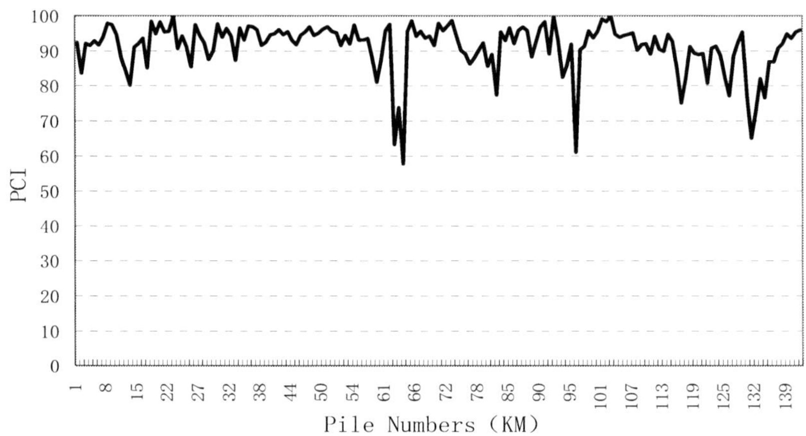

Fig. 3.12 PCI curve per kilometer

From this figure, we know each part of the highway's PCI. For example, the PCI between 63Km and 66Km is lower than the others. That is to say the condition of this part is not good, and the owner should determine whether it needs to be repaired or not.

3.6. Conclusions

To summarize, this paper reports the significant findings of an ongoing research, Integrated Multi-Sensor Mobile System in the application of road surface detection. The SmartV, composed of many subsystems, has made great contributions in multi-source data acquisition, storing, and processing. Experiments and applications have proved that the measurement results are accurate and reliable. Consequently, SmartV would also contribute to great social and economic benefits for the industries.

With the development of sensors, IMSMS should be advanced in the future. Therefore, further studies need to improve the performance of the system. More research can be carried out to meet new challenges and problems. Firstly, further exploration is required to examine data fusion and information extraction. How to achieve multi-source data fusion and high-level information extraction remains to be a challenging work. Further research can be carried out in these areas. Secondly, future study needs to consider cost reduction since new sensors which IMSMS are mainly based on are costly before widely used. Thirdly, IMSMS should update continuously with the latest development of technology and sensor. Therefore, researchers must keep pace with the development of new technology, familiarize themselves with the performance of new sensors, and master how to use those new sensors.

References

Arrb, 2008. www.arrb.com.au/index.php?option=com_content&task=view&id=64&Itemid=52 (Acc. 2008-09-25)

Yunfang Chen, Ye Zetian, Xie Caixiang, Shi Bo,Wang Guibin, 2006. 3D Data Collection and Modeling by Vehicle-borne CCD and Laser Scanner Aided by IMU and GPS, Science of Surveying and Mapping, 2006.9, Vol.31 No.5, P91~92.

China Communication Press, 2008. Highway Performance Assessment Standards

Dynatest, 2009. www.dynatest.com/functional-rsp.php?tab=functional (Acc. 2009-06-13)

Deren Li, Yuan Xiuxiao, 2002. Error Processing and Reliability Theory [M]. Wuhan: Wuhan University Press, 2002.7.

Qingquan Li, Liu Xianglong, 2007. An Algorithm to Image-based Pavement Cracks Geometry Features Extraction [J]. Science paper Online. 2007.7, 2(7):P517-522.

M. Nagai, R. Shibasaki, H. Kumagai, S. Mizukami, D. Manandhar, H. Zhao, 2004. Construction of Digital Surface Model by Multi-Sensor Integration from an Unmanned Helicopter, Proceedings of the ISPRS Working Group V/6, Volume XXXVI, Part 5/W1.

Aris Polychronopoulos, Nikos Floudas, Angelos, Dirk Bank, Bas van den Broek, 2006. Data Fusion in Multi-sensor Platforms for Wide-area Perception [C]. Intelligent Vehicles Symposium, 2006.6, Tokyo, Japan.

D.I.B. Randeniya, M. Gunaratne, S. Sarkar, A. Nazef, 2008. Calibration of Inertial and Vision Systems as a Prelude to Multi-sensor Fusion [J]. Transportation Research Part C, 16 (2008), P255~274

Roadware, 2008. www.roadware.com/products_services/aran/ (Acc. 2008-09-25)

Sreenivas R. Sukumar, Sijie Yu, David L. Page, Andreas F. Koschan, Mongi A. Abidi, 2006. Multi-sensor Integration for Unmanned Terrain Modeling. In Proc. SPIE Unmanned Systems Technology VIII, Vol.6230, Orlando, FL, P65~74, 2006.4.

Kelvin C.P. Wang and Weiguo Gong, 2002. Automated Pavement Distress Survey: A Review and A New Direction, 2002 Pavement Evaluation Conference, 2002, Roanoke, Virginia.

Waylink, 2008. www.waylink.com/ (Acc. 2008-09-25)

WDM, 2008. www.wdm.co.uk/product.htm (Acc. 2008-09-25)

Chapter 4

HIGH RESOLUTION SATELLITE IMAGE ORIENTATION MODELS

Mattia Crespi, Francesca Fratarcangeli, Francesca Giannone, Francesca Pieralice

4.1 Introduction

A few years ago high resolution satellite imagery became available to a limited number of government and defense agencies that managed such imagery with highly sophisticated software and hardware tools. Such images became available to civil users in 1999 with the launch of Ikonos, the first civil satellite offering a spatial resolution of 1 m. Since then other high resolution satellites have been launched, among which there are EROS-A (1.8 m), QuickBird (0.61 m), Orbview-3 (1 m), EROS-B (0.7 m), Worldview-1 (0.5 m) and GeoEye-1 (0.41 m), with many others being planned to launch in the near future. High resolution satellite imagery is now available in different formats and processing levels at an affordable price. The diverse types of sensors and their growing availability are revolutionizing the role of satellite imagery in a number of applications, ranging from intelligency to insurance, media, marketing, agriculture, utilities, urban planning, forestry, environmental monitoring, transportation, real estate etc. As a possible alternative to aerial imagery, high resolution satellite imagery has also impact in cartographic applications, such as in orthophoto production, especially for areas where the organization of photogrammetric surveying may be critical.

Moreover, an increasing demand for large scale mapping and terrain modelling exists so that almost all the satellites have along-track stereo acquisition capability. Many new satellites dedicated to stereo viewing, for example Cartosat-1 (2.5 m), have been launched. This compensates the limited capacity of very high resolution satellites for three-dimensional point determination and enables the generation of Digital Elevation Models (DEMs) and Digital Surface Models (DSMs), and also for 3D feature extraction (e.g. for city modelling).

However, the possibility of using high resolution satellite images for cartography depends on several factors: mapping specifications, sensor characteristics (geometric and radiometric resolution and quality), types of products made available by the companies managing the satellites, quality of the software used to produce the cartographic products, and quality of the final results.

D. Li et al. (eds.), *Geospatial Technology for Earth Observation*,
DOI 10.1007/978-1-4419-0050-0_4,

One of the primary barriers to a wider adaptation and utilization of satellite imagery is the sensor model being able to provide a high level geometric correction through image orientation. The distortion sources can be related to two general categories: the acquisition system, which includes the platform orientation and movement, and the imaging sensor optical-geometric characteristics; the atmosphere refraction, causing a remarkable deviation from the collinearity hypothesis (Noerdlinger 1999).

Two different types of orientation models are usually adopted: the physical sensor models (also called rigorous models) and the generalized sensor models. In the first ones, based on a standard photogrammetric approach where the image and the ground coordinates are linked through the collinearity equations and the involved parameters have a physical meaning. Besides, they require knowledge on the specific satellite and orbit characteristics. On the contrary, the generalized models are usually based on the Rational Polynomial Functions (RPFs), which link image and terrain coordinates by the Rational Polynomial Coefficients (RPCs) and do not need the knowledge about the sensor and acquisition features. The RPCs can be calculated by the final users via a Least Squares (LS) estimation directly from Ground Control Points (GCPs), or proprietarily generated by the sensor managing companies based on their own physical sensor models and distributed to users through imagery metadata. Nevertheless, the first strategy (also called terrain-dependent) is not recommended if a reliable and accurate orientation is required. In the second strategy, they can be generated according to a terrain-independent scenario, using known physical sensor. In order to avoid instability due to high correlations among the coefficients, two different methods can be used: Tichonov regularization or an innovative method based on Singular Value Decomposition (SVD) and QR decomposition, estimating only the strictly required coefficients.

This chapter will discuss many features of the sensor models, both for single images and stereopairs. Specifically, discussions will be focused on the rigorous model for the orientation of the basic image (level 1A) (Sect. 4.2, 4.3, 4.6) and of the image projected to a specific object surface (usually an expanded ellipsoid derived from the WGS84) (level 1B) (Sect. 4.4). The RPC model for the orientation of single and stereopairs images is discussed in Sections 4.7, 4.8. Section 4.9 addresses the methods for accuracy assessment, while Section 4.10 presents application examples of different sensor modelling.

4.2 Rigorous Models

The rigorous model is based on a standard photogrammetric approach, i.e., the collinearity equations describing the physical-geometrical image acquisition. It must consider that an image from a pushbroom sensor is formed by many (from thousands to tens of thousands) individual lines, each acquired with proper position (projection center) and attitude values. All the acquisition positions are related by the orbital dynamics. Therefore, the rigorous model is based on the reconstruction of the orbital segment during image acquisition through the knowledge of the acquisition mode, sensor parameters, satellite position and attitude parameters. The approximate values of these parameters can be computed by using the information contained in the image metadata file, delivered with each image. These approximate values are then corrected by a LS estimation process based on a suitable number of GCPs. Moreover, in order to relate the images to the ground coordinates, expressed in an Earth Centered - Earth Fixed (ECEF) reference frame, a translation and a set of rotation matrices depending on orbital parameters (Keplerian elements) and sensor attitude have to be used. These matrices include those needed to shift between sensor, platform, orbital and Earth Centered Inertial (ECI) coordinate systems, while the transformation between ECI and ECEF coordinate systems must take into account precession, nutation, polar motion and Earth rotation matrices (Kaula 1966).

4.2.1 Coordinate Systems

In order to introduce the collinearity equations, the definitions of some coordinate systems are needed (Westin 1990):

Image system (I): is a 2-dimensional system describing a pixel position in an image. The origin is in the upper left corner, and the pixel position is defined by its row (*J*) and column (*I*). The column numbers increases toward the right and row numbers increases downwards (Fig. 4.1a).

Sensor system (S): the origin is at the perspective center (whose orbital motion may be described as if it were the satellite center of mass), the *z*-axis is directed from the perspective center to pixel array, the *x*-axis is approximately tangent to the orbit directed as the satellite motion (see paragraph 4.2.4, 4.2.5), and *y*-axis forms a right-handed Cartesian system. Note that *y*-axis is approximately parallel to the pixel array. The principal point is the orthogonal projection of the perspective center onto the pixel array (Fig. 4.1b).

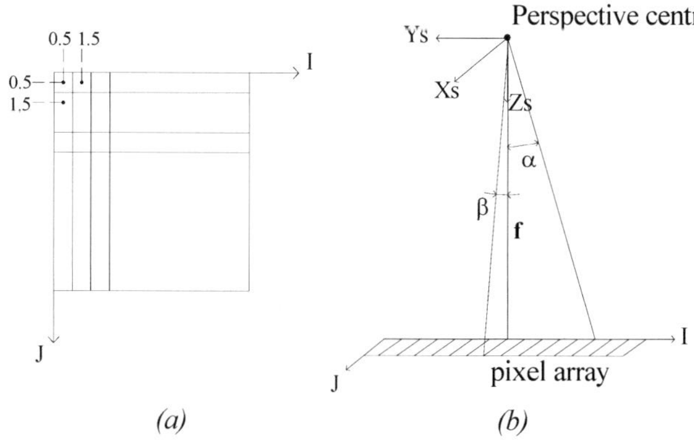

Fig. 4.1 Image system (a) and sensor system (b)

Body system (B): it is aligned to the flight system (see below) when the angle roll (φ), pitch (θ) and yaw (ψ) are zero.

Flight system (F): the origin is at the perspective center, the *X*-axis is tangent to the orbit along the satellite motion, the *Z*-axis is in the orbital plane towards the Earth center of mass and the *Y*-axis completes the right-handed coordinate system.

Orbital system (O): the *Xo-Yo* plane coincides to the orbital plane, which is defined by right ascension of the ascending node (Ω) and by the orbit inclination (i). *X*-axis is along the nodal line, *Y*-axis and *Z*-axis complete the right-handed coordinate system. With the hypothesis of Keplerian orbit, *Z*-coordinate of satellite in the orbital system is zero.

Earth Centered Inertial system - ECI (I): the origin is at the Earth center of mass, the *X*-axis points to vernal equinox (epoch J2000 - 1 January 2000, hours 12 UT), the *Z*-axis points to celestial north pole (epoch J2000) and the *Y*-axis completes the right-handed coordinate system (Teunissen and Kleusberg 1998, Hofmann et al. 2008).

Earth-Centered Earth-Fixed system — ECEF (E): the origin is at the Earth center of mass, the *X*-axis is the intersection of equatorial plane and the plane of reference meridian (epoch 1984.0), the *Z*-axis is the mean rotational axis (epoch 1984.0) and the *Y*-axis completes the right-handed coordinate system (Teunissen and Kleusberg 1998, Hofmann et al. 2008).

Geodetic Local system (L): the origin is a chosen point on the ellipsoid (here the WGS84 is used) the *N*-axis is tangent to the local meridian, *E*-axis is tangent to the local parallel and *h*-axis (elevation axis) is along the ellipsoid normal.

4.2.2 Orbital Parameters

The satellite orbit can be described using the well-known Keplerian elements (Fig. 4.2). According to Keplerian laws, a satellite (considered as a material point), under the effect of a gravitational field generated by a mass concentrated in a point, moves in a plane describing an elliptic orbit. The satellite position at each generic epoch *T* is represented by seven parameters (Kaula 1966, Westin 1990).

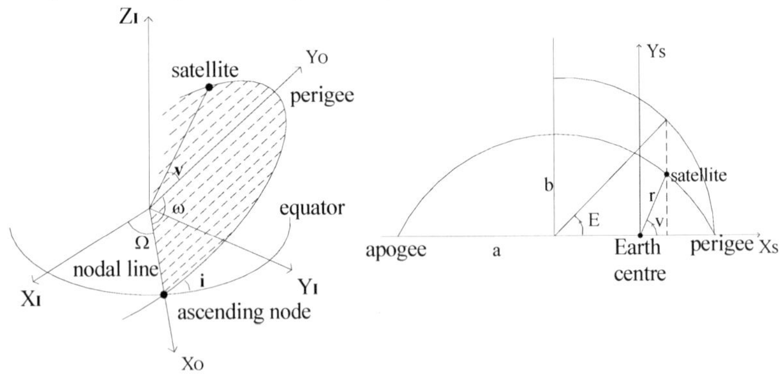

Fig. 4.2 Keplerian parameters

semi-major axis (a): it is semi-major axis of the orbital ellipse.

orbit inclination (i): it is the angle (positive if counter clockwise) between the orbital plane and the equatorial plane. By convention it is between 0 and π.

right ascension of the ascending node ($\boldsymbol{\Omega}$): it is the angle (positive if counter clockwise observed from the North Pole) at the center of the Earth from the vernal equinox to the ascending node.

eccentricity (e): it is the eccentricity of the orbital ellipse.

true anomaly (ν): it is the angle measured in the center of the ellipse between the perigee and the position of the satellite at generic epoch T defined to be 0 at perigee.

argument of the perigee ($\boldsymbol{\omega}$): it is the angle between the nodal line (intersection between the orbital plane and the equatorial plane) and the semi-major axis, measured in the orbital plane from the ascending node to the perigee.

time of the perigee passage (T_p): it is the time referring to the epoch when the satellite is nearest to the Earth.

The approximate values of these parameters can be computed based on the ephemeris information in the metadata file released together with the image.

In some cases (e.g. Ikonos, QuickBird Standard OrthoReady, Cartosat-1), the metadata file released by the sensor managing Companies do not include the ephemeris file (spacecraft position and velocity every few seconds) but only one satellite mean position that is described by two angles:

azimuth ($\boldsymbol{\alpha}$): it specifies the satellite position relative to the area that is collected on Earth, it is measured clockwise from the North.

elevation (e): it is the angle from the horizon up to the satellite.

Nevertheless, some main features of the orbit (a, i) are always known, so that the approximate values of the other Keplerian elements may still need to be computed.

4.2.3 Attitude Angles

To define the sensor during the acquisition it is necessary to know its attitude described by the roll (φ), pitch (ϑ) and yaw (ψ) angles, respectively referred as X, Y, Z axes of the flight system. The approximate values of these angles are calculated with the metadata file information. The corrections to these approximate values may be modeled by second order polynomials. Although there is not any physical meaning in doing this, good results seem to support this choice (Westin 1990)

$$\begin{cases} \varphi = \tilde{\varphi} + a_0 + a_1\tau + a_2\tau^2 \\ \vartheta = \tilde{\vartheta} + b_0 + b_1\tau + b_2\tau^2 \\ \psi = \tilde{\psi} + c_0 + c_1\tau + c_2\tau^2 \end{cases} \tag{4.1}$$

where τ is the time, in seconds, such as $\tau = J_S \cdot t$ where t is the time needed to scan a row on the ground and J_s is the row of the pixel. The nine coefficients (a_i, b_i, c_i) are unknown and need to be estimated with the LS adjustment.

In some case (e.g. Ikonos) the satellite can acquire images at a stated scan azimuth (the exact angle from the starting point of the scan and it is measured clockwise from North) and following a stated scan direction ("forward" or "reverse"); a reverse scan is generally from North to South, whereas a forward scan being from South to North.

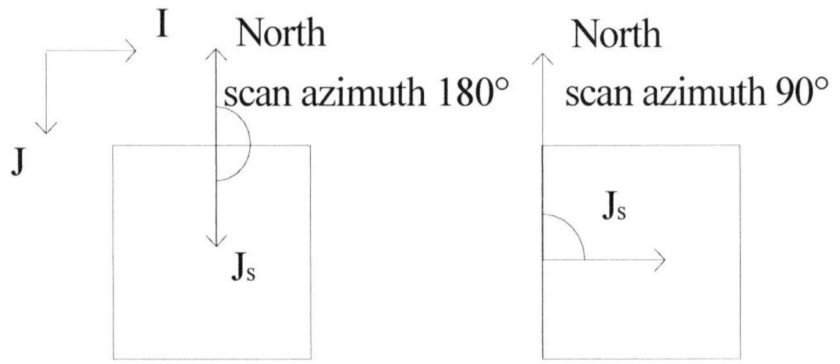

Fig. 4.3 Example of *Js*-axis direction

If the scan azimuth is 180° and the scan direction is "reverse", the image is collected from North to South, or if scan azimuth is 90° and the scan direction is "forward", the image is collected from West to East. In the former case, the J-axis of image system is directed to the scanning direction, whereas in the latter the scanning direction is perpendicular to the J-axis of the image system (Fig. 4.3). So the position of a generic point p (I, J) has to be projected on scanning direction (J_s-axis) with the following relation:

$$J_S = -J \cdot \cos\beta + I \cdot \sin\beta \tag{4.2}$$

where β is the scan azimuth and I, J are the image coordinates.

4.2.4 Coordinate System Transformations

The global rotation matrix from the sensor system to the ECI can be expressed through three rotations (Westin 1990)

$$R_{SI} = R_{FI} \cdot R_{BF} \cdot R_{SB} \tag{4.3}$$

which can be detailed recalling the standard from the matrices representing the rotations around the axes of the right-handed Cartesian coordinate system:

$$R_X(\alpha)=\begin{bmatrix}1 & 0 & 0\\ 0 & \cos\alpha & sen\alpha\\ 0 & -sen\alpha & \cos\alpha\end{bmatrix} \quad R_Y(\beta)=\begin{bmatrix}\cos\beta & 0 & -sen\beta\\ 0 & 1 & 0\\ sen\beta & 0 & \cos\beta\end{bmatrix} \quad R_Z(\gamma)=\begin{bmatrix}\cos\gamma & sen\gamma & 0\\ -sen\gamma & \cos\gamma & 0\\ 0 & 0 & 1\end{bmatrix} \tag{4.4}$$

(rotations α, β, γ are positive if they are counter clockwise when seen from the positive semi axes).

Inertial-Orbital matrix (R_{OI}): this allows the passage from inertial geocentric system (ECI) to the orbital one. It is a function of right ascension of ascending node (Ω) and of orbit inclination (i)

$$R_{OI} = R_X(\mathrm{i}) \cdot R_Z(\Omega) \tag{4.5}$$

Inertial-Flight matrix (R_{FI}): this allows the passage from the inertial geocentric system (ECI) to the flight one. It is a function of Keplerian orbital parameters and varies with the time inside each scene (for each image row J)

$$R_{FI} = R_X(-\frac{\pi}{2}) \cdot R_Z(\frac{\pi}{2}) \cdot R_Z(U) \cdot R_X(\mathrm{i}) \cdot R_Z(\Omega) \tag{4.6}$$

where i is the inclination, Ω the right ascension of the ascending node, $U = \omega + v$ with ω argument of the perigee and v true anomaly

Flight-Body matrix (R_{BF}): it allows the passage from the orbital system to the body one through the attitude angles (φ, ϑ, ψ), which depends on time (for each pixel row)

$$R_{BF} = R_Z(\psi) \cdot R_Y(\vartheta) \cdot R_X(\varphi) \tag{4.7}$$

Body-Sensor matrix (R_{SB}): it allows the passage from the body to the sensor system. This matrix considers the deviation of the parallelism between axes $(X,Y,Z)_S$ and $(X,Y,Z)_B$ and it is considered constant during a scene for each sensor. Its elements may be provided in the metadata files.

The rotation matrix for the transformation from ECI system to ECEF system (R_{EI}) can be subdivided into four sequential steps, considering the motions of the Earth in space: precession, the secular change in the orientation of the Earth's rotation axis and the vernal equinox (described by the matrix R_P); nutation, the periodic and short-term variation of the equator and the vernal equinox (described by the matrix R_N); polar motion, the coordinates of the rotation axis relative to the IERS Reference Pole (described by the matrix R_M); and Earth's rotation about its axis (described by the Sideral Time through the matrix R_S) (Montenbruck Gill 2001).

$$R_{EI} = R_M \cdot R_S \cdot R_N \cdot R_P \tag{4.8}$$

The product of R_{EI} and R_{SI}^T matrices allows the passage from sensor S to ECEF system, with the final rotation matrix being:

$$R_{ES} = R_{EI} \cdot R^{T}_{SI} = R_Z(K) \cdot R_Y(P) \cdot R_X(W) \tag{4.9}$$

where the angles (K, P, W) define the satellite attitude at the epoch of the acquisition of image row J with respect to the ECEF system.

4.2.5 Interior Orientation and Self-Calibration Parameters

The interior orientation parameters describe the intrinsic geometric features of the sensor. Moreover, self-calibration parameters are used to correct the geometric errors in the CCD linear array and the optical system. Note that, referring to the linear array design, one CCD line can be made of ns segments being N_{pi} is the number of pixels contained in i-th segment. The modeling of the geometric errors is easy if carried out in the sensor system (S). For the sake of simplicity, here we consider models under the assumption of $ns = 1$. More details may be found in (Poli 2005).

4.2.5.1 CCD Linear Array Geometric Errors

The geometric errors that may occur in CCD linear array sensors are briefly described and modeled hereafter:

1. The change of the pixel size. It has the effect to change the image scale (Fig. 4.4).

If (p_x, p_y) are the pixel dimensions and (dp_x, dp_y) a change of the pixel size in x direction and in y direction respectively, the errors (dx_p, dy_p) result:

$$dx_p = dp_x \qquad dy_p = y \cdot \frac{dp_y}{p_y} \tag{4.10}$$

The error dy_p may be also due to the focal length variation and the radial distortion. Note that the first order terms in 4.10, 4.13 are highly correlated. Therefore, it is not possible to estimate both the pixel dimension variation together the focal length variation.

Fig. 4.4 Effect of pixel size change in y_S direction "from (Poli 2005)"

2. The shifts and rotations of the CCD segments in the focal plane. These errors are described and modelled as follows:

- shifts in x- and y-directions (Figs. 4.5a, 4.5b): constant quantity dx_c and dy_c, respectively

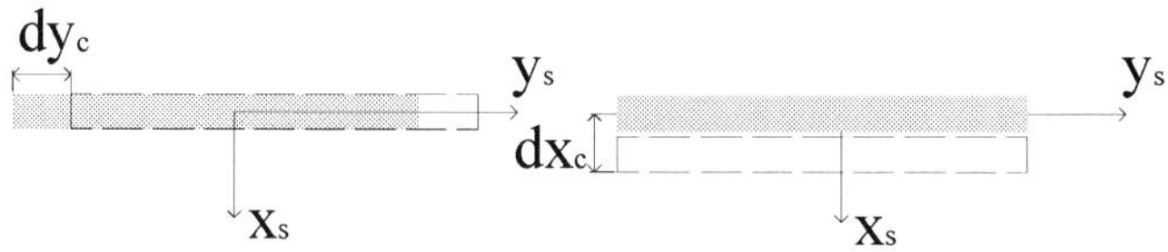

Fig. 4.5 Shift of CCD segment in yS direction (a) and shift of CCD segment in xS direction (b) "from (Poli 2005)"

- effect of horizontal rotation θ in the CCD plane. The rotation produces the error dy_θ in y_S direction and dx_θ in x_S direction (Fig. 4.6), but only the latter has to be considered since θ is small:

$$dx_\theta = y \cdot \sin\theta \tag{4.11}$$

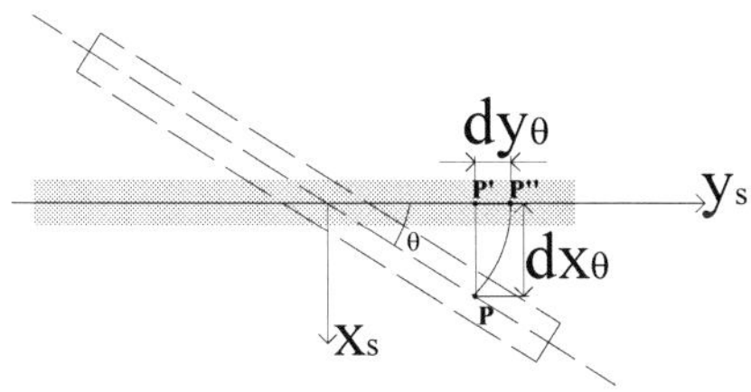

Fig. 4.6 Effects of rotation of CCD segment in the focal plane "from (Poli 2005)"

3. The line bending in the focal plane. The straight CCD line is supposed to be deformed into an arc if the size of the bending is described by the central angle δ that subtends the arc described by the deformed line and the central angle δ' is related to the generic pixel position, the error results Fig. 4.7,

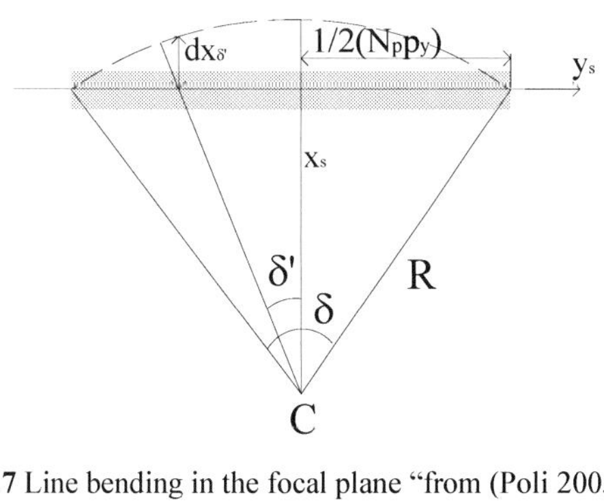

Fig. 4.7 Line bending in the focal plane "from (Poli 2005)"

where $R = \frac{N_p \cdot p_y}{2sen\frac{\delta}{2}}$ and the bending is supposed in the plane defined by $x_S > 0$.

$$dx_{\delta'} = R\left(\cos\delta' - \cos\frac{\delta}{2}\right) \quad (4.12)$$

If the CCD line is bending in the plane where $x_S < 0$, equation (4.12) is valid with the opposite sign.

4.2.5.2 Optical System Errors

The possible errors that may occur in an optical systems have been deeply investigated in close range, airborne and satellite photogrammetry (Brown 1971, Beyer 1992, Jacobsen 1998).

1. The displacement of the lens principal point. This error is modelled with constant shifts Δx_p, Δy_p applied to the principal point coordinates (x_p, y_p) in x and y directions and is totally correlated with the shift of the CCD linear array in the focal plane.
2. The change of the focal length f. The effect of this error Δf in x and y directions is modeled as

$$\begin{aligned} dx_f &= -\frac{\Delta f}{f} \cdot \overline{x}_p \\ dy_f &= -\frac{\Delta f}{f} \cdot \overline{y}_p \end{aligned} \quad (4.13)$$

where $\overline{x}_p = x - x_p$ and $\overline{y}_p = y - y_p$.

3. The symmetric lens distortion is described by the coefficients k_1 and k_2 and modelled as

$$\begin{aligned} dx_r &= \left(k_1 \cdot r^2 + k_2 \cdot r^4\right) \cdot \overline{x}_p \\ dy_r &= \left(k_1 \cdot r^2 + k_2 \cdot r^4\right) \cdot \overline{y}_p \end{aligned} \quad (4.14)$$

where $r^2 = \overline{x}_p^2 + \overline{y}_p^2$. For pushbroom sensors with CCD linear array only dy_r may be significant and $r^2 \cong \overline{y}_p^2$.

4. The decentering lens distortion is modeled as

$$dx_d = p_1 \cdot (r^2 + 2\overline{x}_p^2) + 2p_2 \cdot \overline{x}_p \overline{y}_p$$
$$dy_d = 2p_1 \cdot \overline{x}_p \overline{y}_p + p_2 \cdot r^2 + 2\overline{y}_p^2 \tag{4.15}$$

Again, only dy_d may be significant.

4.2.5.3 Final Consideration about the Self-Calibration Parameters

Accounting for the described general models for the CCD linear array geometric errors and for the optical system errors and considering their significance and correlations, only a few self-calibration parameters need to be introduced. Moreover, these parameters can be conveniently chosen to directly model self-calibration with respect to the image coordinates (I, J). In detail, the following self-calibration parameters are worth to be set up:

1. The position of principal point (I_0, J_0), accounting also for the shift of the CCD linear array in the focal plane (as described in Fig. 4.5).
2. The rotation of the CCD linear array in the focal plane (k) (as described in Fig. 4.6).
3. The change of the focal length (δf), accounting also for the scale variation and the isotropic change of pixel dimension.
4. The symmetric lens distortion up to the third order in y direction only (d_2)

Another possible parameters is δ representing the line bending, however, it will not be considered in the following rigorous model implementation.

4.3 Rigorous Model for Original Images (Level 1A)

As mentioned before, a rigorous model is based on the collinearity equations and describes the imagery acquisition both from the geometrical and physical (and stochastic) points of view. It is now possible to write the collinearity equations relating to the position of a point in the image space to the corresponding point in the object space, according to a central projection. In our case, the collinearity equations may be conveniently expressed in the Earth Centered Inertial (ECI) system starting from the relationship (Fig. 4.8)

$$\begin{bmatrix} X_t \\ Y_t \\ Z_t \end{bmatrix} = \begin{bmatrix} X_S \\ Y_S \\ Z_S \end{bmatrix} + d \begin{bmatrix} u_x \\ u_y \\ u_z \end{bmatrix} \quad \Rightarrow \quad \begin{bmatrix} u_x \\ u_y \\ u_z \end{bmatrix} = \frac{1}{d} \begin{bmatrix} X_t - X_S \\ Y_t - Y_S \\ Z_t - Z_S \end{bmatrix} \tag{4.16}$$

where $(X_t, Y_t, Z_t)_I$ are the ECI coordinates of the ground point, $(X_s, Y_s, Z_s)_I$ are the ECI coordinates of the perspective center, $(u_x, u_y, u_z)_I$ are the components in the

ECI system of the unit vector $\hat{u}_{SI}$ directed from satellite to the ground point, and d is the distance from the perspective center to the ground point.

$$d = \sqrt{(X_t - X_S)^2 + (Y_t - Y_S)^2 + (Z_t - Z_S)^2} \tag{4.17}$$

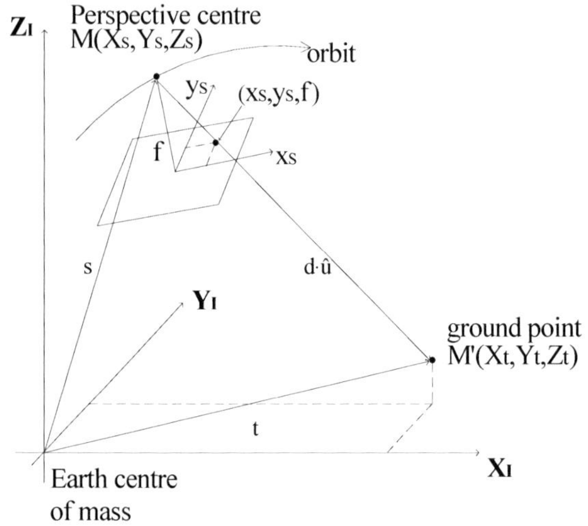

Fig. 4.8 Central projection mode

Therefore, introducing the sensor system the collinearity equations read:

$$\frac{1}{d_s}\begin{bmatrix} x_S \\ y_S \\ f \end{bmatrix} = R_{SI}\begin{bmatrix} u_x \\ u_y \\ u_z \end{bmatrix}_I \quad \Rightarrow \quad \begin{bmatrix} x_S \\ y_S \\ f \end{bmatrix} = \frac{d_s}{d} R_{SI}\begin{bmatrix} X_t - X_S \\ Y_t - Y_S \\ Z_t - Z_S \end{bmatrix}_I \tag{4.18}$$

where f is the focal distance, R_{SI} is the rotation matrix from ECI to sensor system and d_s is the perspective center to image point distance

$$d_s = \sqrt{x_S^2 + y_S^2 + f^2} \tag{4.19}$$

Finally, the standard form of two collinearity equations for each ground point is obtained by dividing the first two equations of (4.18) by the third one:

$$\begin{cases} \dfrac{x_S}{f} = \dfrac{R_{SI,11}|X_t - X_S| + R_{SI,12}|Y_t - Y_S| + R_{SI,13}|Z_t - Z_S|}{R_{SI,31}|X_t - X_S| + R_{SI,32}|Y_t - Y_S| + R_{SI,33}|Z_t - Z_S|} \\ \dfrac{y_S}{f} = \dfrac{R_{SI,21}|X_t - X_S| + R_{SI,22}|Y_t - Y_S| + R_{SI,23}|Z_t - Z_S|}{R_{SI,31}|X_t - X_S| + R_{SI,32}|Y_t - Y_S| + R_{SI,33}|Z_t - Z_S|} \end{cases} \tag{4.20}$$

With simple geometric consideration (Fig. 4.1b) the collinearity equations can be written as a function of image coordinates, interior orientation and self-calibration parameters previously set up, Keplerian orbital and attitude parameters.

$$\begin{cases} \dfrac{x_S}{f} = \tan\beta = \dfrac{d_{pix}}{f}\left[J - \mathrm{int}(J) - 0.5 - J_0 - k(I - I_0)\right] \\ \dfrac{y_S}{f} = \tan\alpha = -\dfrac{d_{pix}}{f}\left\{(I - I_0) + d_2(I - I_0)^3 + k\left[J - \mathrm{int}(J) - 0.5 - J_0\right]\right\} \end{cases} \tag{4.21}$$

Therefore, equating (4.20) and (4.21) the relationship between the image and the ground coordinates is found:

$$\begin{aligned} F_1 &= \frac{d_{pix}}{f}\left[J - \mathrm{int}(J) - 0.5 - J_0 - k(I - I_0)\right] + \\ &- \frac{R_{SI,11}\left|X_t - X_S\right| + R_{SI,12}\left|Y_t - Y_S\right| + R_{SI,13}\left|Z_t - Z_S\right|}{R_{SI,31}\left|X_t - X_S\right| + R_{SI,32}\left|Y_t - Y_S\right| + R_{SI,33}\left|Z_t - Z_S\right|} = 0 \\ F_2 &= -\frac{d_{pix}}{f}\left\{(I - I_0) + d_2(I - I_0)^3 + k\left[J - \mathrm{int}(J) - 0.5 - J_0\right]\right\} + \\ &- \frac{R_{SI,21}\left|X_t - X_S\right| + R_{SI,22}\left|Y_t - Y_S\right| + R_{SI,23}\left|Z_t - Z_S\right|}{R_{SI,31}\left|X_t - X_S\right| + R_{SI,32}\left|Y_t - Y_S\right| + R_{SI,33}\left|Z_t - Z_S\right|} = 0 \end{aligned} \tag{4.22}$$

The collinearity equations depend on the parameters described in the paragraphs 4.2.2, 4.2.3 and 4.2.5. In some cases the image metadata file is not supplied with the satellite position and velocity at every interval time; thereby the orbit is reconstructed using the procedure described in paragraph 4.4.1 (Crespi et al. 2008b).

As mentioned, the approximate values for all parameters may be derived from the information contained in the metadata files, released together with the image (a, i, Ω, e, v, ω, φ, ϑ, ψ, f and I_0) or they are simply fixed to zero (a_i, b_i, c_i, J_0, k and d_2). In theory, these approximate values must be corrected by an estimation process based on a suitable number of GCPs, for which the collinearity equations are written. Nevertheless, since the orbital arc related to each image acquisition is extremely short (a few hundreds of kilometers) compared to the whole orbit length (tens of thousands). Some Keplerian parameters are not estimable at all (a, e, ω) and others (i, Ω, T_P) are extremely correlated both among themselves and with respect to the sensor attitude, interior orientation and self-calibration parameters (f, I_0, J_0, k, d_2) (Giannone 2006). The parameters estimable are (a_i, b_i, c_i, T_P, f, I_0, J_0, k, d_2). Regarding the stochastic model, the standard deviations of the image observations are set equal, since manual measurement tests carried out independently by different operators range from 1/3 to 1/2 pixel in accuracy. For the GCP coordinates standard deviations are usually set equal to the mean values obtained during their direct surveying or cartographic selection (Brovelli et al. 2008).

In this rigorous model also the refraction effect has been taken in account. Atmospheric refraction effect causes a well know bending of the optical paths due to the variation of atmospheric density (Fig. 4.9a). The purpose of atmospheric refraction modeling is the correction of the image coordinates in order to remove the refraction effect and to estimate the orientation parameters under the hypothesis of straight optical paths, which are actually modeled by collinearity equations (4.22). In fact, the ground point P is projected onto the image along a refracted path in p', but equations (4.22) model the collinearity condition along the straight path POp. Therefore, we need to compute the correction from p′ to p in order to properly adopt the model (4.22). The deviation from the collinearity assumption due to the atmospheric refraction is computed by a model duly described in (Noerdlinger 1999) (Fig. 4.9b, Table 4.1) which basically allows to calculate the displacement d due to the refraction effect and the position of P' on the ground, starting from the off-nadir angle. Therefore, a first rough orientation is performed, neglecting the refraction effect, in order to estimate the off-nadir angle under which each ground point is imaged, then the corresponding P'.

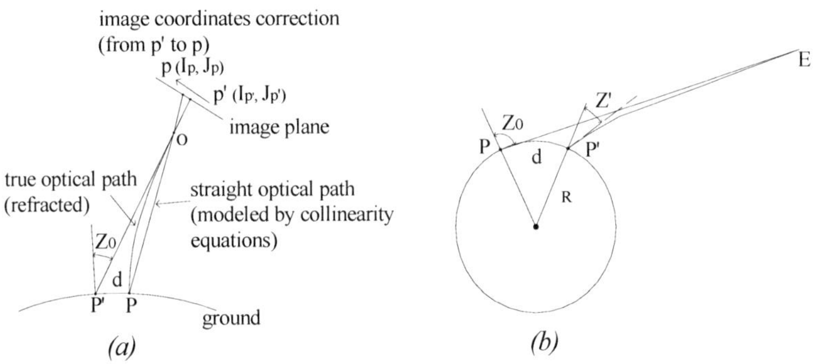

Fig. 4.9 Effect of atmospheric refraction

Moreover, through the collinearity equations, starting from P and P', the corresponding image position p and p' are computed, so that the components of the vector a suitable to remove the atmospheric refraction effect from the image coordinates can be computed by:

$$\begin{bmatrix} I_a \\ J_a \end{bmatrix} = \begin{bmatrix} I_p - I_{p'} \\ J_p - J_{p'} \end{bmatrix} \tag{4.23}$$

The vector $\vec{a}$ is applied to the coordinates collected on the image, from which a new estimation process is performed in order to refine the orientation parameters. Note that one iteration is usually enough, since the refraction is well estimated on the basis of the first rough orientation.

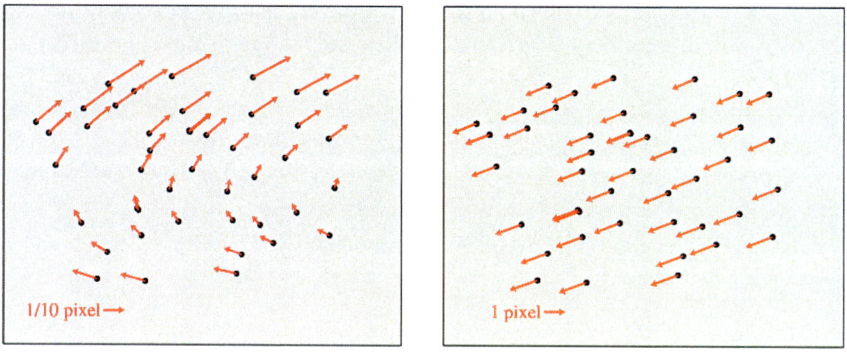

(a) (b)

Fig. 4.10 Distribution of atmospheric refraction effect, represented by vector $\vec{a}$ computed for each ground point

It is important to consider the refraction effect especially when the satellite attitude variation during the image acquisition causes a refraction effect not uniformly distributed (Fig. 4.10a). On the contrary, its impact is lower when the satellite attitude variation causes a rather uniform refraction effect (Fig. 4.10b).

Table 4.1 Displacement d over the local sphere approximating the ellipsoid due to refraction effect function of z_0

z_0 (°)	z' (°)	z_0 - z' (°)	d (m)
10	9.99710	0.0029	0.47
20	19.9939	0.0061	1.06
30	29.9904	0.0096	1.97
40	39.9860	0.0140	3.62
45	44.9834	0.0166	5.03
50	49.9802	0.0198	7.21

In order to avoid instability due to high correlations among some parameters leading to design matrix pseudo-singularity, Singular Value Decomposition (SVD) and QR decomposition are employed to evaluate the actual rank of the design matrix, to select the actually estimable parameters and finally to solve the linearized collinearity equations system in the LS sense (see Sect. 4.5). Moreover, the statistical significance of each estimable parameter is checked by a Student T-test so to avoid over-parameterization. In case of a statistically non-significant parameter, it is removed and the estimation process is repeated until all parameters are significant.

Pseudo-observation on the GCPs are also allowed, in order to account for their accuracy, which may vary depending on their source (cartography or direct surveying).

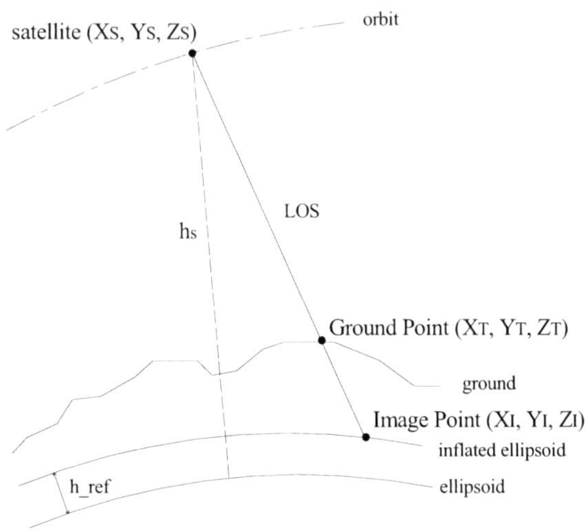

Fig. 4.11 Model geometry

4.4 Rigorous Model for Pre-Processed Images (Level 1B)

In this case it has to be noted that the images are projected onto a specific object (usually an "inflated" ellipsoid, derived from the WGS84 choosing a certain ellipsoidal height) (level 1B). The collinearity equations link points on the ground and points projected on the mentioned "inflated" ellipsoid (Pieralice 2007, see Fig. 4.11).

Each point on the ground surface corresponds to a point on "inflated" ellipsoid, identified from line of sight (LOS), i.e. the line directed from the perspective centre to the point on the ground. The collinearity condition is satisfied when $\hat{u}_{SI}$ (the unit vector directed from perspective centre to image point) coincides with $\hat{u}_{ST}$ (the unit vector directed from perspective centre to ground point), i.e., ground point and image point are lined up on LOS. The collinearity equations may be conveniently expressed in the ECEF system in vector form:

$$\hat{u}_{SI} = R \cdot \hat{u}_{ST} \tag{4.24}$$

where R is a rotation matrix. In fact, relative "small" translation of ground with respect to ellipsoid can be expressed with an infinitesimal rotation around the perspective center, because the height of satellite platform (h_S) is much more than the difference of elevation between ground surface and the "inflated" ellipsoid (Δh).

Under this infinitesimal rotation hypothesis ($\cos\varphi,\theta,\psi \cong 1$, $\sin\varphi \cong \varphi, \sin\theta \cong \theta, \sin\psi \cong \psi$) the rotation matrix R is reduced to the sum of the unit matrix and an anti-symmetric matrix.

$$R = \widetilde{R} + \delta R = I + \delta R \qquad \delta R = \begin{bmatrix} 0 & \varphi & \theta \\ -\varphi & 0 & \psi \\ -\theta & -\psi & 0 \end{bmatrix} \Rightarrow R = \begin{bmatrix} 1 & \varphi & \theta \\ -\varphi & 1 & \psi \\ -\theta & -\psi & 1 \end{bmatrix} \tag{4.25}$$

where the attitude angles are supposed to be modelled by a time-dependent function up to the second order, similar to (4.1). The (4.24) can also be expressed in the following way:

$$\begin{bmatrix} X_I - X_S \\ Y_I - Y_S \\ Z_I - Z_S \end{bmatrix} = \rho R \begin{bmatrix} X_T - X_S \\ Y_T - Y_S \\ Z_T - Z_S \end{bmatrix} \tag{4.26}$$

where

ρ is the scale factor, (ratio of perspective centre-image point distance d_{SI} and perspective centre-ground point distance d_{ST}: $\rho = d_{SI} / d_{ST}$;

X_T, Y_T, Z_T are the ground coordinates in the ECEF system;

X_I, Y_I, Z_I are the image coordinates in the ECEF system;

X_S, Y_S, Z_S are the perspective centre coordinates in the ECEF system.

Finally, note that in this case the collinearity equations on the basis of previous consideration, now reads

$$\begin{cases} F_1 = \dfrac{|X_I - X_S|}{|Z_I - Z_S|} - \dfrac{R_{SE,11}|X_T - X_S| + R_{SE,12}|Y_T - Y_S| + R_{SE,13}|Z_T - Z_S|}{R_{SE,31}|X_T - X_S| + R_{SE,32}|Y_T - Y_S| + R_{SE,33}|Z_T - Z_S|} = 0 \\ F_2 = \dfrac{|Y_I - Y_S|}{|Z_I - Z_S|} - \dfrac{R_{SE,21}|X_T - X_S| + R_{SE,22}|Y_T - Y_S| + R_{SE,23}|Z_T - Z_S|}{R_{SE,31}|X_T - X_S| + R_{SE,32}|Y_T - Y_S| + R_{SE,33}|Z_T - Z_S|} = 0 \end{cases} \tag{4.27}$$

The collinearity equations depend on the parameters described in paragraphs 4.2.2 and 4.2.3. The approximate values for all parameters may be derived from the information contained in the metadata files, released together with the image (α, β, e, i, Ω) or they are simply fixed to zero (a_i, b_i, c_i). Again, these approximate values must be corrected by an estimation process based on a suitable number of GCPs, for which the collinearity equations are written. Finally, the parameters estimable are (a_i, b_i, c_i).

It has to be noted that the image coordinates in the collinearity equations (4.27) must be expressed in the ECEF system, while on the other hand the image coordinates are obtained by point measurement on the image so that only I and J are known.

First of all, it is required to change image coordinates into cartographic coordinate through the simple equations:

$$\begin{cases} N_P = N_A - J \cdot p \\ E_P = E_A + I \cdot p \end{cases} \tag{4.28}$$

where

N_P, E_P are the North and East UTM WGS84 coordinates of a generic point P.
N_A, E_A are the North and East UTM WGS84 coordinates of upper left corner A of the image.
I, J are the coordinates of a generic point P in the image system.
p is the pixel size (in meters).

Moreover, the cartographic coordinates are converted into geographic coordinates (latitude φ and longitude λ). Finally, since the ellipsoidic height (h) of points on the image is the elevation of "inflated" ellipsoid, the geodetic coordinate (φ, λ, h) are converted in Cartesian coordinates (ECEF system).

4.4.1 Computation of Satellite Positions

In general the detailed information about the satellite position are not supplied for the level 1B images, therefore the satellite coordinates can be computed only on the basis of the angles (azimuth and elevation) that define satellite position with respect to image center (Fig. 4.12).

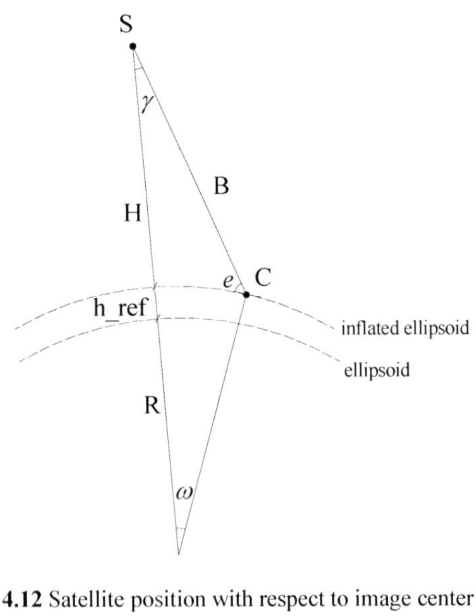

Fig. 4.12 Satellite position with respect to image center

Referring to Fig. 4.12 the off-nadir angle γ and B distance between the image center C and the satellite position S are calculated with the sine theorem.

$$\sin\gamma = \frac{(R + h\_ref)\cdot\sin(e + \pi/2)}{(R + H)} \tag{4.29}$$

$$B = \frac{(R + h\_ref)\cdot\sin\omega}{\sin\gamma} \tag{4.30}$$

where R is the radius of local sphere, H is the height of satellite, h_ref is the height of "inflated" ellipsoid in reference to the WGS84 ellipsoid, e is the elevation angle, γ is the off-nadir angle and $\omega = \pi - (\gamma + e + \pi/2)$. The satellite coordinates in a geodetic local system, whose origin is the center of image (C) are:

$$\begin{cases} X_{S(C)} = B\cdot\cos e\cdot\sin\alpha \\ Y_{S(C)} = B\cdot\cos e\cdot\cos\alpha \\ Z_{S(C)} = B\cdot\sin e \end{cases} \tag{4.31}$$

Local coordinates are transformed into ECEF coordinates, and from the unique satellite position it is possible to reconstruct the orbit segment. Satellite coordinates are converted from ECEF to ECI system (4.8), then from ECI to the orbital system with R_{OI} matrix (4.5).

Generally the inclination i is known, on the contrary the right ascension of ascending node Ω can be calculated under the hypothesis of Keplerian orbit ($Z_{SOrbit} = 0$). This equation has two solutions (Ω_1, Ω_2) that correspond to ascending and descending orbit respectively. If the satellite moves on descending orbit while acquires images, the right value of Ω is obtained with $X_{S(C)} < 0$ (abscissa in the orbital system), otherwise the solution is with $X_{S(C)} > 0$ if the satellite is ascending.

On the orbital plane the satellite position relative to image's center ($S_{(C)}$) identifies the U_C angle, where $U_C = \arctan(Y_{S(C)}/X_{S(C)})$; consequently the satellite position relative to each image row is obtained, moving the satellite forward and backward on the orbit with respect to central position:

$$U_i = U_C - (Js_C - Js_i)\cdot\Delta U \tag{4.32}$$

where J_{sc} is the scanning row of image's center, and J_{si} is a generic scanning row of the image and ΔU is the angular displacement relative to one scanning row.

$$\Delta U = \frac{ATR}{R + h\_ref} \tag{4.33}$$

The satellite position (S_i) in the orbital system is:

$$\begin{cases} X_{S(i)} = (R+H)\cdot \cos U_i \\ Y_{S(i)} = (R+H)\cdot \sin U_i \\ Z_{S(i)} = 0 \end{cases} \tag{4.34}$$

where R is the radius of local sphere and H is height of satellite platform (Fig. 4.12). Then the satellite coordinates must be converted into the ECI system and into the ECEF system. The ECEF coordinates enter in collinearity equations.

4.5 Computation Techniques

As mentioned, the design matrix is likely to be close to singularity, so that rank estimation and estimable parameter selection is mandatory. In this respect SVD and QR decomposition are quite useful tools and they will be shortly recalled here. The Singular Value Decomposition (SVD) and the QR decomposition are employed to solve the linearized collinearity equations system in the LS sense (Strang and Borre 1997, Golub and Van Loan 1993). As usual, the solution is obtained iteratively due to non-linearity of the system; the iterative procedure stops when the estimated variance of the unit weight observation stabilizes.

4.5.1 Singular Value Decomposition

The Singular Value Decomposition (SVD) is a very powerful technique to deal with sets of equations or matrices that are either singular or numerically very close to being singular. The SVD of a matrix $A \in \Re^{m \cdot n}$ (with $m \geq n$) is any factorization of the form:

$$A = UWV^T \tag{4.35}$$

where $W \in \Re^{n \cdot n}$ is a diagonal matrix with positive or zero elements (w_{ij}) that are the singular values of A; $U \in \Re^{m \cdot n}$ and $V \in \Re^{m \cdot n}$ are orthogonal matrices, whose columns (u_j, v_j) are called the left and right singular vectors. For a system of linear equations $(Ax = b)$, using the SVD we can write (Golub and Van Loan 1993):

$$\mathrm{UWV^T x = b} \tag{4.36}$$

and the LS solution x minimizes $\|Ax - b\|_2^2$. Since the orthogonal matrix preserves the norm, for any $x \in \Re^n$ we have:

$$\|Ax-b\|_2^2 = \left\|(U^T AV)(V^T x) - U^T b\right\|_2^2 = \sum_{i=1}^{r}(w_i z_i - u_i^T b)^2 + \sum_{i=r+1}^{m}(u_i^T b)^2 \quad (4.37)$$

where $z = V^T x$ and r is the rank of A. $\|Ax-b\|_2^{\;2} = \min$ holds, if

$$\sum_{i=1}^{r}(w_i z_i - u_i^T b)^2 = 0 \quad (4.38)$$

then using the SVD, the LS problem is now in form of a diagonal matrix, and finally

$$x_i = \begin{cases} x_i = \dfrac{u_i^T b}{w_i} v_i & \text{if } w_i \neq 0 \\ \text{undetermined} & \text{if } w_i = 0 \end{cases} \quad (4.39)$$

The advantage of using the SVD is that it can reliably handle the rank deficient case as well as the full rank case.

4.5.2 QR Decomposition

The QR decomposition of a matrix $A \in \Re^{m \cdot n}$ (with $m \geq n$) is given by:

$$A = QR \quad (4.40)$$

where $Q \in \Re^{m \cdot m}$ is an orthogonal matrix and $R \in \Re^{m \cdot n}$ is an upper triangular matrix. If the rank of A is equal to n, the first n columns of Q form an orthonormal basis for the Rank(A). Thus, the calculation of the QR factorization is a way to compute an orthonormal basis for a set of vectors.

The standard algorithm for the QR decomposition involves sequential evaluation of Householder transformations. An appropriate Householder matrix, applied to a given matrix, can zero all the elements, situated below a given element, in a column of the matrix. For the first column of the matrix A, an appropriate matrix H_1 is evaluated, which puts on zero all the elements below the first element in the first column of A. Similarly H_2 zeroes all elements in the second column below the second element and so on up to H_{n-1}

$$R = H_{n-1} \cdots H_1 A \quad (4.41)$$

where $Q^T = H_{n-1} \ldots H_1$, i.e., $Q = H_1 \ldots H_{n-1}$.

The generic matrix H_i zeroes all elements in the first column below the first element for a sub-matrix of A ($A_i \in \Re^{((m-i)\cdot(n-i))}$). If A is rank deficient, the QR factorization does not give a basis for the Rank(A). In this case to calculate an orthonormal basis for Rank(A), it is necessary to compute the QR decomposition of a column-permuted version of A, i.e., $AP = QR$ (Golub and Van Loan 1993)

$$Q^T AP = \begin{bmatrix} R_{11} & R_{12} \\ 0 & 0 \end{bmatrix} \begin{matrix} \rightarrow & r \\ \rightarrow & m-r \end{matrix} \qquad \begin{matrix} \downarrow & \downarrow \\ r & n-r \end{matrix} \tag{4.42}$$

where P is a permutation, r is the rank of A, R_{11} is an upper triangular and non singular matrix and Q and P are products of Householder matrices $Q = H_1 \ldots H_r$, $P = P_1 \ldots P_r$.

For understanding the role of the permutation matrix, it is necessary to define the vector $N \in \Re^m$ for a generic matrix $A \in \Re^{m \cdot n}$:

$$A_{mn} = \begin{vmatrix} a_{11} & a_{12} & \cdots & a_{1n} \\ a_{21} & a_{22} & \cdots & a_{2n} \\ \vdots & \cdots & & \vdots \\ a_{m1} & & \cdots & a_{mn} \end{vmatrix} \tag{4.43}$$

$$N_m = \begin{vmatrix} \sum_{k=1}^{m} a_{k1}^2 & \sum_{k=1}^{m} a_{k2}^2 & \cdots & \sum_{k=1}^{m} a_{kn}^2 \end{vmatrix} \tag{4.44}$$

The element of the N are the square value of norm calculated for each column of A. The permutation matrix P applied at the generic matrix A makes a matrix $A_P = AP$ such that the elements of the corresponding vector N are placed in descending order. As for the generic matrix H_i, the generic matrix P_i permutes the column of a sub-matrix of $A_i \in \Re^{((m-i)\cdot(n-i))}$; if k is the column with the maximum value of norm, the permutation matrix P_i exchanges the columns i and k.

In a system of linear equations $(Ax = b)$, if $A \in \Re^{m \cdot n}$ and has a rank r, the QR decomposition produces the factorization $AP = QR$ where R is described in the equation (4.41). As for the LS problem we have:

$$\|Ax - b\|_2^2 = \left\|(Q^T AP)(P^T x) - Q^T b\right\|_2^2 = \|R_{11}t - (c - R_{12}z)\|_2^2 + \|d\|_2^2 \tag{4.45}$$

where

$$P^T x = \begin{bmatrix} t \\ z \end{bmatrix} \begin{matrix} \rightarrow & r \\ \rightarrow & n-r \end{matrix}$$

$$Q^T b = \begin{bmatrix} c \\ d \end{bmatrix} \begin{matrix} \rightarrow & r \\ \rightarrow & m-r \end{matrix} \tag{4.46}$$

If x is a LS minimizer we have

$$x = P \begin{bmatrix} R_{11}^{-1}(c - R_{12}z) \\ z \end{bmatrix} \tag{4.47}$$

If z is a set of zeroes in this expression, we obtain the basic solution:

$$x_B = P \begin{bmatrix} R_{11}^{-1}c \\ 0 \end{bmatrix} \tag{4.48}$$

x_B has at most r non-zero components and so Ax_B involves a subset of A columns.

4.5.3 Subset Selection Using SVD and QR

For a system of linear equations $(Ax = b)$, with $A \in \Re^{m \cdot n}$ (with $m \geq n$) it is necessary to select the estimable parameters. We describe an SVD-based subset selection procedure, due to Golub, Klema and Stewart (Golub and Van Loan 1993), that proceeds as follows:

- we compute the SVD $A = UWV^T$ and use it to determine a rank estimate r
- with the QR decomposition $QR = AP$ we select an independent subset of A columns; if $R_{11}x_B - Q^T b$ with $x_B \in \Re^r$ and we set

$$t = P \begin{bmatrix} x_B \\ 0 \end{bmatrix} \tag{4.49}$$

then $A \cdot t$ is an approximate LS predictor of b that involves the first r columns of AP. The permutation matrix P is calculated so that the columns of the matrix $B_1 \in \Re^{m \cdot r}$ in $AP = [B_1, B_2]$ are "sufficiently independent"

- we predict b with the vector $A \cdot t$ where t is described in the equation (4.49), and z minimizes $\|B_1 x_B - b\|_2$.

4.6 Stereo Rigorous Model

The rigorous model developed to orientate both level 1A and level 1B single scene can be extended to manage both along-track and across-track stereopairs (Crespi et al. 2008a). In this case, it has to be noted that the orbital elements are the same for two images if they were acquired during the same orbital path (along-track stereopairs) or are different if the images are acquired during two different orbital paths at different epochs (across-track stereopairs).

Again, the approximate values of these parameters can be computed by using the information in the metadata file and have to be corrected by a least square estimation process based on a suitable number of GCPs. The estimable parameters can be selected using the procedure described in Section 4.5. In this respect, since the tie points may be conveniently considered, it is necessary to establish a procedure for the computation of their approximate ground coordinates, which have to be LS estimated together with all other parameters.

First of all, the single scenes have to be separately oriented adopting the already described rigorous model. This separated orientation has to be considered just as approximate ones; they have to be refined in a block adjustment possibly including suited tie points. In theory, the homologous rays should intersect, identifying a unique ground point for each couple of homologous points chosen over the image. Nevertheless, errors remaining in the separate orientations cause the well known parallaxes, so that homologous rays do not intersect and no ground point can be found by intersection. Therefore, it is necessary to set up a rule to compute the approximate tie point ground positions. To this aim, the choice was made to compute the positions of the two points on the homologous rays at minimal distance and then to average their coordinates. The minimum distance between the two rays is computed. The equations of two rays supposed straight can be written in parametric form

$$r_1 : \begin{vmatrix} X_1 \\ Y_1 \\ Z_1 \end{vmatrix} = \begin{vmatrix} X_{01} \\ Y_{01} \\ Z_{01} \end{vmatrix} + s \cdot \begin{vmatrix} \mathrm{a}_1 \\ \mathrm{b}_1 \\ \mathrm{c}_1 \end{vmatrix} \qquad r_2 : \begin{vmatrix} X_2 \\ Y_2 \\ Z_2 \end{vmatrix} = \begin{vmatrix} X_{02} \\ Y_{02} \\ Z_{02} \end{vmatrix} + t \cdot \begin{vmatrix} \mathrm{a}_2 \\ \mathrm{b}_2 \\ \mathrm{c}_2 \end{vmatrix} \tag{4.50}$$

where $(X_0, Y_0, Z_0)_{1,2}$ are the coordinates of perspective centers in the ECEF system for the two images, $(a, b, c)_{1,2}$ are direction cosine known from the separate orientations. The condition to identify the two points on the rays at minimal distance reads:

$$\begin{cases} \dfrac{\partial d^2(s,t)}{\partial t} = 0 \\ \dfrac{\partial d^2(s,t)}{\partial s} = 0 \end{cases} \quad \text{with } d^2(s,t) = (X_2 - X_1)^2 + (Y_2 - Y_1)^2 + (Z_2 - Z_1)^2 \tag{4.51}$$

4.7 Rational Polynomial Function with Rational Polynomial Coefficients

A few years ago high resolution satellite imagery were available to a limited number of government and defence agencies that managed such imagery with highly sophisticated software and hardware tools. High resolution satellite imagery are now available in different formats and processing levels and at an affordable price. These types of sensors and their growing availability are revolutionizing the role of satellite imagery in a number of applications ranging from intelligency to insurance, media, marketing, agriculture, utilities, urban planning, forestry, environmental monitoring, transportation, real estate etc.

One of the primary barriers to a wider adaptation and utilization of satellite imagery is the sensor model being able to provide a high level geometric correction through the image orientation. Sensor models are a key component to represent the functional relationships between the image space and the object space, and are essential for single/multi imagery orientation.

Even if the rigorous models should theoretically provide the highest accuracy, they are only available for some satellites and can be managed by some commercial available software. Moreover, in order to estimate the unknown parameters of rigorous models, users are still faced with the challenging task of recovering the exterior orientation of the sensor using a set of GCPs usually no small than 10. When no or few GCPs are available, users cannot recover the exterior orientation of the sensor and therefore are unable to perform various mapping and data collection operations. With the introduction of generalized sensor models, this situation has changed considerably. Generalized sensor models, such as the RPF (Tao and Hu 2001a), have smoothed the requirement to manage a physical sensor model. Furthermore, as the RPC implicitly provides the interior and (approximate) exterior sensor orientation, the availability of several GCPs is no longer a mandatory requirement. Consequently, the use of the RPC for photogrammetric mapping is becoming a new standard in high-resolution satellite imagery that has already been implemented in various high-resolution sensors, such as Ikonos, QuickBird and WorldView.

4.7.1 RPC Usage and Orientation Refinement

As mentioned before, some companies (for example DigitalGlobe for QuickBird and WorldView and Space Imaging for Ikonos, India Space Research Organization for Cartosat-1) usually supply the RPCs, as part of the image metadata to enable image orientation via RPF.

The RPF relate object point coordinates (latitude, longitude and height) to pixel coordinates (I, J), as a physical sensor models, but in the form of ratios of polynomial expressions:

$$I = \frac{P_1(\varphi, \lambda, h)}{P_2(\varphi, \lambda, h)} \quad J = \frac{P_3(\varphi, \lambda, h)}{P_4(\varphi, \lambda, h)} \tag{4.52}$$

where φ, λ are the geographic coordinates, h is the height above the WGS84 ellipsoid and (I, J) are the image coordinates. The order of these four polynomials is usually limited to 3 so that each polynomial takes the generic form:

$$P_n = \sum_{i=0}^{m_1} \sum_{j=0}^{m_2} \sum_{k=0}^{m_3} t_{ijk} \varphi^i \lambda^j h^k \tag{4.53}$$

with $0 \le m_1 \le 3; 0 \le m_2 \le 3; 0 \le m_3 \le 3$ and $m_1 + m_2 + m_3 \le 3$, where t_{ijk} are the RPC.

The number of RPC depends obviously on the polynomial order: if the equations (4.52) are written with third order polynomials, the maximum number of coefficients is 80 (20 for each polynomial). Actually, the total number of RPC is reduced to 78, because the two equations can be divided for the zero order terms of the denominators.

The great power of these equations is the independence from the physical characteristic of the image acquisition (NIMA 2000). Although ground coordinates are not directly connected with the acquisition physics, it is possible taking into account the further approximated considerations (Tao and Hu 2002): ratios of the first order terms can represent distortions caused by the optical projection, while corrections such as Earth curvature, atmospheric refraction and lens distortion can be well modelled by the second-order terms; other unknown and more complex distortions with high-order components may be absorbed by the third-order terms.

The ground coordinates (φ, λ, h) in the equation (4.52) are normalized to (-1, +1) range using normalization parameters supplied in the metadata file, in order to improve the numerical precision during the computation.

The generic simple formula utilized for the normalization, is:

$$T_n = \frac{T - T_{offset}}{T_{scale}} \tag{4.54}$$

where T_n are the normalized coordinates, T_{offset}, T_{scale} are the normalization parameters available in the metadata file and T is the original ground or image coordinate (T=I, J; φ, λ, h).

Since the residual bias may be present into the RPC, the orientation can be refined on the basis of the known GPs, acting as GCPs. A possible refinement of the model (4.52) (written in normalized coordinates), allowing for bias compensation, is accomplished in a quite common way with the introduction of a simple first order polynomial in the RPF (4.55) whose parameters are estimated, provided a suitable number of GCPs is known (Hanley and Fraser 2004, Fraser and Hanley 2003).

$$I_n = A_o + I_n \cdot A_1 + J_n \cdot A_2 + \frac{P_1(\varphi_n, \lambda_n, h_n)}{P_2(\varphi_n, \lambda_n, h_n)} =$$

$$= A_o + I_n \cdot A_1 + J_n \cdot A_2 + \frac{a_0 + a_1\lambda_n + a_2\varphi_n + a_3h_n + a_4\lambda_n\varphi_n + ... + a_{17}\lambda_n^3 + a_{18}\varphi_n^3 + a_{19}h_n^3}{1 + b_1\lambda_n + b_2\varphi_n + b_3h_n + b_4\lambda_n\varphi_n + ... + b_{17}\lambda_n^3 + b_{18}\varphi_n^3 + b_{19}h_n^3} \quad (4.55)$$

$$J_n = B_o + J_n \cdot B_1 + I_n \cdot B_2 + \frac{P_3(\varphi_n, \lambda_n, h_n)}{P_4(\varphi_n, \lambda_n, h_n)} =$$

$$= B_o + J_n \cdot B_1 + I_n \cdot B_2 + \frac{c_0 + c_1\lambda_n + c_2\varphi_n + c_3h_n + c_4\lambda_n\varphi_n + ... + c_{17}\lambda_n^3 + c_{18}\varphi_n^3 + c_{19}h_n^3}{1 + d_1\lambda_n + d_2\varphi_n + d_3h_n + d_4\lambda_n\varphi_n + ... + d_{17}\lambda_n^3 + d_{18}\varphi_n^3 + d_{19}h_n^3}$$

where (I_n, J_n) are the normalized images coordinates, and P_i are third order polynomial functions of object space normalized coordinates (φ_n, λ_n, h_n); A_i and B_i terms describe image shift and drift effects in particular:

A_0, A_1, A_2, B_0, B_1, B_2 describe a complete affine transformation.
A_0, A_1, B_0, B_1 model the shift and drift.
A_0, B_0, describe a simple coordinate shift.

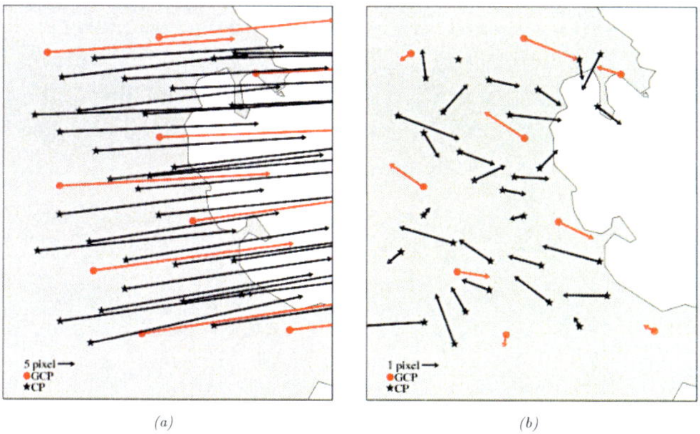

Fig. 4.13 Example of residuals adjustment with an affine transformation on a QuickBird image. Uncorrected image (a), corrected (b)

The six new coefficients (A_i, B_i) are LS estimated based on GCPs. It is noted that in theory the model is not linear, since the 2nd and 3rd terms of the right side involve both observations (I_n, J_n) and parameters (A_i, B_i). Nevertheless, usually in the right side observations I_n, J_n are considered as fixed coefficients, so that the model is treated as linear with respect to the six coefficients (A_i, B_i) (Fig. 4.13a, 4.13b).

4.7.2 RPC Generation

The RPC can be generated by terrain-dependent scenario without using physical sensor model (Tao and Hu 2001b) or according to a terrain-independent scenario, using known physical sensor model.

For the terrain-dependent scenario, which is a kind of registration of the original image to 3D geometry represented by the GCPs, the RPF model tries to approximate the complicated imaging geometry across the image scene using polynomial terms. The solution is highly dependent on the actual terrain relief, the distribution and the number of GCPs. The RPCs have to be estimated in a LS adjustment so that the number of GCPs could be very high (at least 39 if RPC up to the third order are looked for). This method is very weak and vulnerable in presence of outliers and it is likely to cause deformations far from the GCPs returning not good accuracies. Therefore, the RPFs solved by terrain-dependent approach must not be used as a replacement sensor model if high accuracy is required (Tao and Hu 2001b, Tao and Hu 2001c, Toutin et al. 2000) and will not be considered anymore hereafter.

For a terrain-independent scenario, a 2D image grid covering the full extent of the image is established and its corresponding 3D object grid with several layers (e.g., four or more layers for the third-order case) slicing the entire elevation range is generated. The horizontal coordinates (X, Y) of a point of the 3D object grid are calculated from a point (I, J) of the image grid using the physical sensor model with an a priori selected elevation Z. Then, the RPC are LS estimated with the object grid points and the image grid points. This terrain-independent computational scenario can make the RPF model a good replacement to the physical sensor models, and has been widely used to determine the RPCs.

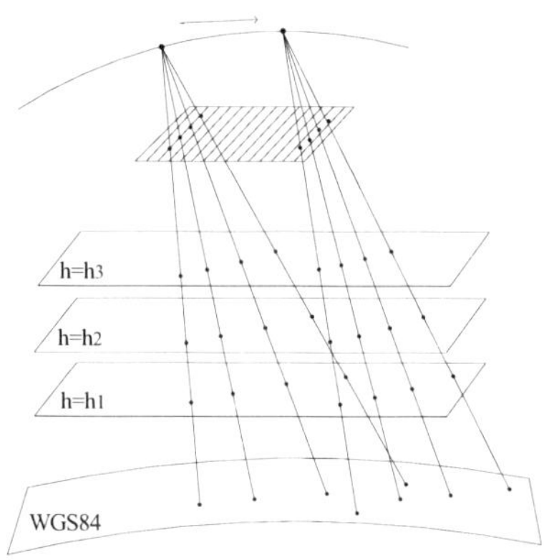

Fig. 4.14 Grid for RPC generation in the terrain-independent approach

It has to be underlined that in the usually adopted terrain-independent approach, the least square solution is often carried out through a regularization, since unknown RPCs may be highly correlated so that the design matrix is almost rank deficient (Neumaier 1998). In order to overcome the regularization requirements, an innovative algorithm for the RPC extraction, with a terrain independent approach, is analyzed. In details, at first an image discretization is made, dividing the full extent image space in a 2D grid. Then, the points of the 2D image grid are used to generate the 3D ground grid: the image was oriented and by the knowledge of the rigorous orientation sensor model, the collinearity equations were derived and used to create the 3D grid, starting from each point of the 2D grid image. In this respect, it has to be underlined that the 2D grid is actually a regular grid, whereas the 3D one is not strictly regular, due to the image attitude. Moreover, the 3D grid points were generated intersecting the straight lines modelled by the collinearity equations with surfaces (approximately ellipsoids) concentric to the WGS84 ellipsoid, placed at regular elevation steps. So, the dimension of the 3D grid is both based on the full extent of the image and the elevation range of the terrain. The grid contains several elevation layers uniformly distributed, and the points on one layer have the same elevation value (Fig. 4.14).

Note that the finest subdivision depends on the incompressible error of the rigorous model used to generate the RPCs, so that a very fine discretization is unuseful and an upper discretization limit also exists. The RPCs least squares estimation (Tao and Hu 2001c) is based on the linearization of the generic RPFs equations, which can be written as:

$$\begin{aligned}
&I_n + b_1\lambda_n I_n + b_2\varphi_n I_n + \ldots + b_{18}\varphi_n^3 I_n + b_{19}h_n^3 I_n - a_0 - a_1\lambda_n - a_2\varphi_n \ldots - a_{18}\varphi_n^3 - a_{19}h_n^3 = 0 \\
&J_n + d_1\lambda_n J_n + d_2\varphi_n J_n + \ldots + d_{18}\varphi_n^3 J_n + d_{19}h_n^3 J_n - c_0 - c_1\lambda_n - c_2\varphi_n \ldots - c_{18}\varphi_n^3 - c_{19}h_n^3 = 0
\end{aligned} \quad (4.56)$$

where a_i, b_i, c_i, d_i are the RPCs (78 coefficients for third order polynomials), (I_n, J_n) and (φ_n, λ_n, h_n) are the normalized coordinates obtained throughout the equation (4.54), with scale and offset factors computed according to:

$$\begin{aligned}
&\begin{cases} w_{offset} = \min(w_k) \\ w_{scale} = \max(w_k) - \min(w_k) \end{cases} \quad where \quad w = \varphi, \lambda, h \\
&\begin{cases} I_{offset} = J_{offset} = 1 \\ I_{scale} = n^\circ column - 1 \\ J_{scale} = n^\circ row - 1 \end{cases}
\end{aligned} \quad (4.57)$$

where k is the number of available GCPs and n° column/row are the overall columns/rows of the image; the normalization range is (0, 1).

Deeper investigations underlined that many RPCs are highly correlated. In order to avoid instability due to high correlations, leading to a pseudo-singular design matrix, Tickhonov regularization is usually used. Generally, the regularization is exploited in a Tickhonov fashion, adopting a damping factor to the diagonal of the normal matrix, in order to guarantee its non singularity. A new alternative approach is based on the Singular Value Decomposition (SVD) and QR decomposition which are employed to evaluate the actual rank of the design matrix and to select the actual estimable coefficients (Bianconi et al. 2008, Brovelli et al. 2008); again, the SVD-based subset selection procedure is due to Golub, Klema and Stewart (Strang and Borre 1997, Golub and Van Loan 1993).

4.8 Stereo Model via RPC

RPF model represents an attractive tool also for managing stereopairs orientation and possible subsequent DSM generation. Also in this case, the first problem to solve is the computation of terrain point's approximate coordinates (φ, λ, h). A straight forward procedure is based on the Direct Linear Transformation (DLT). The near-linear projection of the high resolution satellite image ensures rapid convergence of the spatial intersection from even very coarse initial values for the object point coordinates.

The DLT is not using any pre-information about image orientation. The 22 unknowns (11 for each images) for the transformation of the object coordinates to the image coordinates have to be determined with at least 6 control points. The DLT equations are:

$$\begin{aligned} I &= \frac{L_1 \cdot E + L_2 \cdot N + L_3 \cdot U + L_4}{L_9 \cdot E + L_{10} \cdot N + L_{11} \cdot U + 1} \\ J &= \frac{L_5 \cdot E + L_6 \cdot N + L_7 \cdot U + L_8}{L_9 \cdot E + L_{10} \cdot N + L_{11} \cdot U + 1} \end{aligned} \tag{4.58}$$

where (I, J) are the image coordinates, (E, N, U) are the ground coordinates respect to the Cartesian Local system centered in the center of the image and the L_i are the DLT parameters. In case of stereopairs the equations (4.58) are doubled and expressed in the following forms for every GCP:

$$\begin{cases} \left(L_9^1 E + L_{10}^1 N + L_{11}^1 U + 1\right) \cdot I^{(1)} - \left(L_1^1 E + L_2^1 N + L_3^1 U + L_4^1\right) = 0 \\ \left(L_9^1 E + L_{10}^1 N + L_{11}^1 U + 1\right) \cdot J^{(1)} - \left(L_5^1 E + L_6^1 N + L_7^1 U + L_8^1\right) = 0 \\ \left(L_9^2 E + L_{10}^2 N + L_{11}^2 U + 1\right) \cdot I^{(2)} - \left(L_1^2 E + L_2^2 N + L_3^2 U + L_4^2\right) = 0 \\ \left(L_9^2 E + L_{10}^2 N + L_{11}^2 U + 1\right) \cdot J^{(2)} - \left(L_5^2 E + L_6^2 N + L_7^2 U + L_8^2\right) = 0 \end{cases} \tag{4.59}$$

where superscripts 1 or 2 are related to the first and the second image respectively.

4.9 Accuracy Assessment of High Resolution Satellite Imagery Orientation by Leave-One-Out Method

4.9.1 Hold-Out Validation

Currently, the most used method to assess spatial accuracy of oriented high resolution satellite image is the Hold-Out Validation (HOV), also known as test sample estimation. According to it, the data set (known ground points) is partitioned in two subsets: the first one used to determine the orientation-orthorectification model (GCPs) and the second to validate the model itself (check points or CPs). The only restriction on such selection is to have both sets sufficiently well-distributed on the whole image; apart from this consideration, the selection should be random. Once the model is trained, accuracy is usually evaluated as Root Mean Squared Error (RMSE) of residuals between imagery derived coordinates with respect to CPs coordinates, independently determined and used as reference.

This method has the advantage of being simple and easy to compute, but it also has some drawbacks, as it is generally not reliable and it is not applicable when a low number of ground points is available. First of all, once the two sets are selected, accuracy estimate is not reliable since it is strictly dependent on the points used as CPs; if outliers or poor quality points are included in the CPs set, accuracy estimate is biased. In addition, when a low number of ground points is available, almost all of them are used as GCPs and very few CPs remain, so that RMSE may be computed on a poor (not significant) sample. In these cases, accuracy assessment with the usual procedure is essentially lost. In addition, this method displays a low efficiency, making a poor use of the available information, as a large part of it must be collected and used only for validation purpose.

4.9.2 Leave One Out Cross Validation

In order to overcome the drawbacks of HOV a possible alternative procedure to perform accuracy assessments of orthorectified image is the Leave-One-Out cross-validation (LOOCV) method. The LOOCV is a statistical estimation technique currently applied in different fields such as machine learning (Elisseeff and Pontil 2002), bioinformatics (Simon et al. 2003) and generally in any other field requiring an evaluation of the performance of a learning algorithm (e.g. in geostatistics). It is a special case of the *k*-fold cross-validation method (Stone 1974, Geisser 1975), which involves the partitioning of the original data set in k subsets of equal size (approximately). The model is trained k times, using each subset in turn as the test set, with the remaining subsets being the training set. The overall accuracy can be obtained averaging the accuracy values computed on each subset. The LOOCV is a *k*-fold cross-validation computed with $k=n$, where n is the size of the original data set. Each test set is therefore of size 1, which implies that the model is trained n times. Therefore, the alternative proposal consists in applying the LOOCV as an effective accuracy evaluation method for image orientation, being particularly useful when a low number of ground points is available.

This method applied involves the iterative application of the orientation model, using all the known ground points as GCPs except one, different in each iteration, used as a check point. In every iteration, the residuals between image derived coordinates and the CP coordinates (prediction error of the model on CP coordinates) are calculated. The overall spatial accuracy achievable from the oriented image may be estimated by calculating the usual RMSE or, better, a robust accuracy index like the median Absolute Deviation (mAD) of the prediction errors on all the iterations.

In this way we solve both mentioned drawbacks of the classical procedure: it is a reliable and robust method, not dependent on a particular set of CPs and on outliers, and it allows us to use each known ground point both as a GCP and as a CP, maximizing all the available ground information. Obviously, this is of particular relevancy when the ground point number is kept as low as possible due to budget and/or logistic constraints. LOOCV may obviously apply to both with a rigorous and with a RPC-based (with possible zero or first order correction) orientation model.

Some experiments were carried out to assess how well LOOCV derivable accuracy indices (mAD and RMSE) are able to represent the overall accuracy and which are their advantages with respect to the HOV RMSE. They led to the following main conclusions, pointing out that the LOOCV method with accuracy evaluated by mAD seems promising and useful for practical cases:

- the LOOCV RMSE and HOV RMSE are too sensitive to outliers and "critical" points (mainly located along the perimeter of the area covered by ground points), which may display high residuals when they act as CPs.
- HOV RMSE displays the risk to be too dependent on the geometric distribution of CPs, so that the HOV derived accuracy is likely to be not representative for the whole image when only a few CPs are available.
- the LOOCV mAD is a robust index able to filter out the effect of the high residuals; this is of particular relevancy for the "critical" points, which are not representative of the mean achievable accuracy.

Finally a simple decreasing exponential function was proposed to represent the accuracy trend versus the number of GCPs. This model may be conveniently applied to LOOCV mAD to find the minimum number of GCPs for accuracy assessment when a number of ground points is available:

$$y = a \cdot e^{bx^{-t}} \tag{4.60}$$

where y is the RMSE, x is the number of GCPs used to build the model, a and b are estimated with standard deviation (σ_a, σ_b) by LS adjustment.

The value of t is calculated iteratively, starting with $t = 1$ and estimating the two coefficients a and b. If the difference from the asymptotic value at $x = nGCP/2$ (that is $y(nGCP/2) - a$), is larger than a chosen threshold (e.g. >1 cm) the value of t is increased by a unit and the estimates of a and b have to be recomputed. This choice to constrain the slope of the function to the asymptotic value has been done in order to avoid that possible false oscillations of estimated accuracy on a few CPs can affect the estimation of the a and b parameters. The asymptotic value $a \pm \sigma_a$ enables the determination of the number of GCPs ($\tilde{n}_{GCP}$) sufficient to achieve the maximum accuracy, that is the x value corresponding to $y = a + 2\sigma_a$ (2σ upper confidence limit) rounded to the nearest upper integer (Fig. 4.15).

$$\widetilde{n}_{GCP} = \text{int}\sup t \sqrt{\frac{b}{\ln(a+2\sigma_a)-\ln a}} \tag{4.61}$$

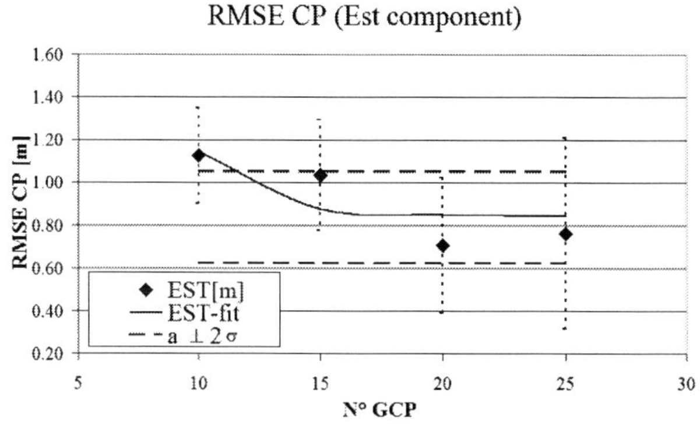

Fig. 4.15 Example of RMSE CP vs. n° GCP fit

4.10 Application of the Orientation Models

This section presents some results obtained using the previously described orientation models. The orientation models are implemented in a scientific software (SISAR - Software per Immagini Satellitari ad Alta Risoluzione) developed at the DITS (Dipartimento di Idraulica Trasporti e Strade), Area di Geodesia e Geomatica, Sapienza Università di Roma. SISAR results are compared with orientation models implemented in commercial software (OrthoEngine 10.0 PCI Geomatica, Erdas 9.0, Leica Geosystems) (Crespi et al. 2008a, Bianconi et al 2008, Crespi et al 2008b). All images of Rome (2 EROS-A, 2 QuickBird, 1 Ikonos, 1 Cartosat-1) cover areas of different dimensions; the GPs were surveyed with static or fast static procedures by a Trimble 5700 GPS receiver and their coordinates are estimated by Trimble Geomatic Office software with respect to available GPS permanent stations (MOSE at Rome Faculty of Engineering). The mean horizontal and vertical accuracies of the coordinates are between 10 and 20 cm. The two Augusta (Sicily) scenes are a stereopair, with ground points (GPs) positioned in the overlapping area. The Salerno image includes three different QuickBird standard format images coming from the same orbital segment; the particularity of this image is the latitude extension (around 48 Km).

The ground points for the Augusta and Salerno images were surveyed by geodetic quality GPS in RTK mode; the mean horizontal and vertical coordinate accuracies are between 5 and 10 cm. The point distribution on Bagnoli stereopair is not homogeneous due to the sea in the South-West area. The ground points of Bagnoli were acquired by GPS surveys using Topcon Legacy receivers in post processing procedure with respect to quite far permanent stations. The mean horizontal and vertical accuracies of the coordinates are between 10 and 20 cm. The ground points for the Castelgandolfo area were collected in RTK mode, the mean horizontal and vertical coordinate accuracies are around 15 cm. The ground coordinates for all points are expressed in the WGS84 system, while the orthometric heights were obtained applying geoid undulations from the ITALGEO95 public model.

For each image and for each software the orientation was carried out several times, varying the number of GCPs, and the related accuracies, represented by the RMSE computed over check point residuals (RMSE CP), were computed and analyzed. The RMSEs were computed both for the North and East residual components separately.

The available images for the experimentation have been acquired by several sensors. Their features are summarized in Table 4.2[1].

Table 4.2 Properties of the used data sets

Sensor	Area	GSD [m]	Off-nadir angle (°)		Scene coverage (Km x Km)	Available GPs
			start	End		
EROS A	Rome (R1)	1.80	9.1	9.4	13x10	49
	Rome (R2)	2.60	31.0	40.1	17x12	49
QuickBird	Rome (level 1A)	0.61	3.0		17x17	49
	Rome (level 1B)	0.60	2.2		17x17	24
	Augusta (*P001)	0.77	29.2		21x20	39
	Augusta (*P002)	0.75	28.2		20x19	39
	Salerno "joint"	0.67	20.0		48x19	57
Ikonos	Bagnoli-1	1.00	22.98		13x9	25
	Bagnoli-2	1.00	24.27		13x9	25
	Rome	1.00	18.7		11x10	27
Cartosat-1	Rome BandA	2.5	4.97		7.5x30	43
	Rome BandF	2.5	26.09		7.5x30	43
	Castelgandolfo BandA	2.5	12.35		30x30	25
	Castelgandolfo BandF	2.5	28.20		30x30	25

[1] EROS A Rome (R1): ITA1-e1038452;

EROS A Rome (R2): ITA1-e1090453;

QuickBird Salerno "joint" is obtained to stitch three QuickBird images in order to have a single "strip" image:

QuickBird Salerno (*P001): 05JUL17100900-P1BS-005520834030_01_P001;

QuickBird Salerno (*P002): 05JUL17100903-P1BS-005520834030_01_P002;

QuickBird Salerno (*P003): 05JUL17100906-P1BS-005520834030_01_P003;

QuickBird Augusta (*P001): 04JAN06093201-P1BS-000000130187_01_P001;

QuickBird Augusta (*P002): 04JAN06093307-P1BS-000000130187_01_P002;

QuickBird Rome (level 1A): 02JUN03100558-P1BS-000000032060_01_P001;

QuickBird Rome (level 1B):05APR28101432-P2AS-005746807010_01_P001

Ikonos Rome: po_15194;

Ikonos Bagnoli-1: po_918_pan_0000010001;

Ikonos Bagnoli-2: po_918_pan_0010000001;

4.10.1 Rigorous Model for Single Image (Level 1A)

An image EROS A of Rome (Fig. 4.16) was oriented with the two rigorous models implemented in SISAR and in OrthoEngine. Results in terms of RMSE on CPs are compared. RMSE trend are globally similar for both software and its value is comparable with the GSD, except for the North component.

4.10.2 Rigorous Model for Single Image (Level 1B)

As example for the level 1B imagery, one image was selected, acquired respectively by Ikonos (Fig. 4.17) satellites. Image represents the area of city of Rome. The image has been oriented with rigorous models implemented in SISAR and in OrthoEngine. Also in this case the RMSE trend is similar for both software and accuracy is around the GSD value.

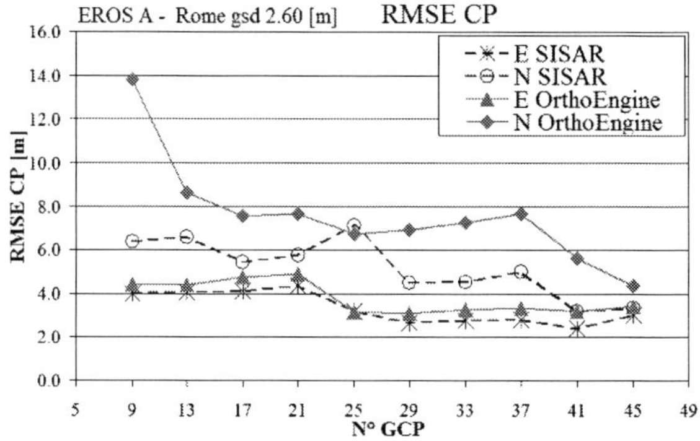

Fig. 4.16 Image accuracy vs. GCP number for EROS A (ITA1-e1090724)

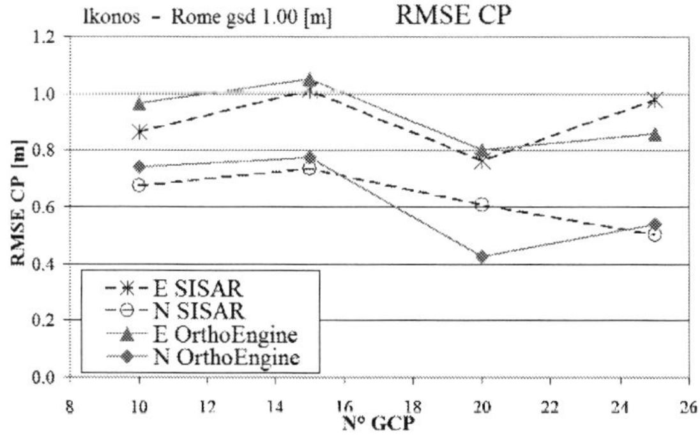

Fig. 4.17 Image accuracy vs. GCP number for Ikonos image of Rome

4.10.3 Rigorous Model for Stereopair (Level 1A)

One example of stereopair orientations using rigorous models is presented. It concerns along-track stereopair acquired by Cartosat-1 satellite (Fig. 4.18). Notice that Cartosat-1 is a satellite dedicated expressly to stereo viewing, having two CCD-line sensor cameras, looking respectively in forward direction with a nadir angle of 26° and in aft direction with a nadir angle of 5°. For the Cartosat-1 stereopair RMSE CP trend in all components is similar, except for the East component where SISAR results are less than 2.0 m, while OrthoEngine ones are slightly worse.

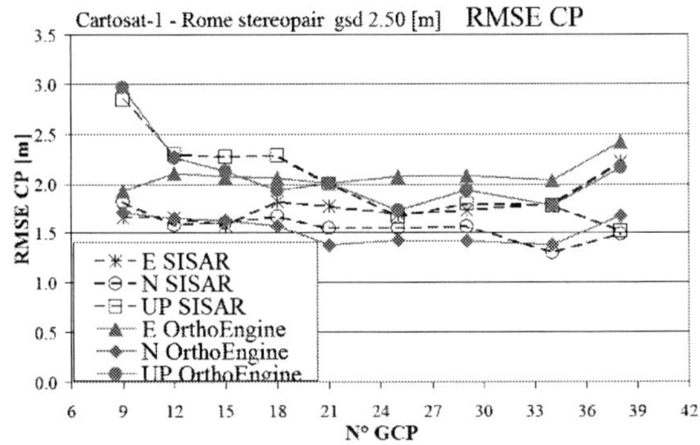

Fig. 4.18 Image accuracy vs. GCP number for Cartosat-1 stereopair of Rome

4.10.4 Usage and Generation RPC for Single Image

The next example is dedicated to the application of RPC model, supplied by the sensor managing companies, to a QuickBird image of Rome. Model was tested with three software, in fact the SISAR results are compared with the OrthoEngine and Erdas ones (Fig. 4.19).

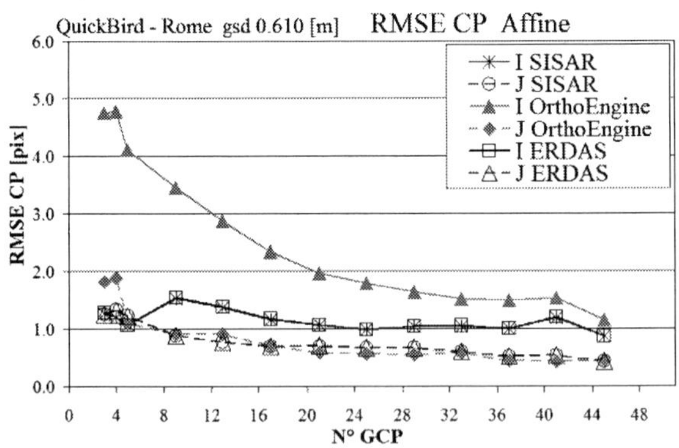

Fig. 4.19 Image accuracy vs. GCP number for QuickBird image of Rome

Moreover, one example of RPCs generation is presented, considering the QuickBird Salerno "joint" image. These example is very interesting, since RPCs for the whole strip are not available by the sensor managing companies, but they provide RPCs only for the three separated images.

In this case the results obtained from RPC generated in SISAR are compared with the results of rigorous model implemented in OrthoEngine (Fig. 4.20). The greater differences between the two model are showed in J direction, the RMSE CP trend with RPC model is more consistent than RMSE CP trend obtained with rigorous model.

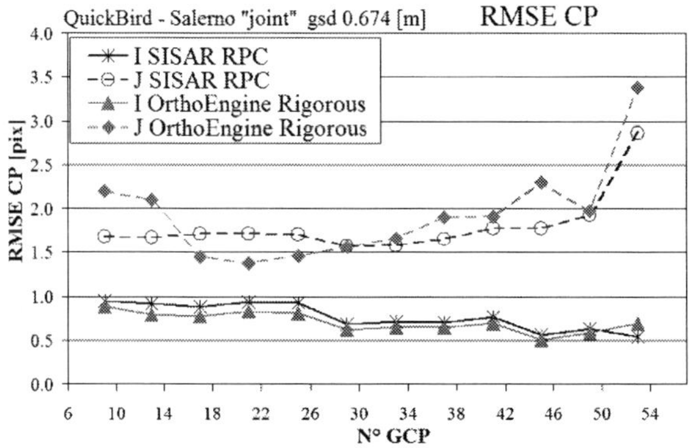

Fig. 4.20 Image accuracy vs. GCP number for QuickBird Salerno "joint"

4.10.5 Stereo Model via RPC

Finally, an example of stereopair orientation via RPC are presented. The stereopair used is acquired by Cartosat-1 satellite on the area of Castelgandolfo. Results of SISAR software are compared with the Erdas ones (Fig. 4.21).

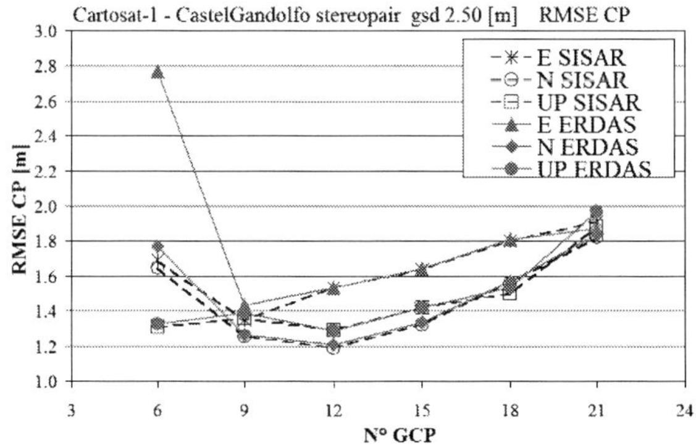

Fig. 4.21 Image accuracy vs. GCP number for Cartosat-1 stereopair of CastelGandolfo

4.10.6 Summarizing Results

In this paragraph, results on satellite images, described in the data set table (Tab. 4.2), are presented. Single images and stereopairs have been oriented with rigorous and RPC model. The next tables (Table 4.3, 4.4, 4.5) list the accuracy for the orientation tests performed with a number of GCPs, evaluated with the (4.61), suited to achieve the maximum accuracy.

Table 4.3 Results of rigorous model applied to single images

Rigorous model single image							
Sensor	Area	RMSE CP					
		SISAR			OrthoEngine		
		n° GCP	E[m]	N[m]	n° GCP	E[m]	N[m]
QuickBird	Salerno "joint"	17	0.52	0.84	21	0.72	0.65
	Rome (level 1B)	10	0.54	0.30	10	0.49	0.38
Ikonos	Rome	10	0.87	0.67	10	0.97	0.74
Eros A	Rome (R2)	17	4.11	5.45	17	4.75	7.54

Table 4.4 Results of rigorous model and RPC model applied to stereopairs

Rigorous model for stereo images									
Sensor	Area	RMSE CP							
		SISAR				OrthoEngine			
		n° GCP	E[m]	N[m]	Up[m]	n° GCP	E[m]	N[m]	Up[m]
Eros A	Rome	17	2.55	2.75	5.85	21	3.17	4.40	6.97
QuickBird	Augusta	13	0.58	0.84	1.02	21	0.78	1.07	1.33
Cartosat	Rome	15	1.58	1.60	2.27	15	2.07	1.63	2.13
Ikonos	Bagnoli	11	1.93	1.35	1.95	13	0.90	1.55	1.95
RPC model for stereo images									
Sensor	Area	RMSE CP							
		SISAR				Erdas			
		n° GCP	E[m]	N[m]	Up[m]	n° GCP	E[m]	N[m]	Up[m]
Cartosat-1	Castelgandolfo	12	1.53	1.19	1.29	12	1.53	1.21	1.29

Table 4.5 Comparison of results of RPC generated with SISAR and rigorous model ones

RPC generation							
Sensor	Area	RMSE CP					
		SISAR RPC			OrthoEngine rigorous model		
		n° GCP	I [pix]	J [pix]	n° GCP	I [pix]	J [pix]
Eros A	Rome	13	2.01	2.75	13	1.85	3.17
QuickBird	Salerno "joint"	13	0.91	1.67	17	0.78	1.45

4.10.7 Accuracy Assessment

In the following, the results of accuracy assessment by LOOCV method are presented. The results are related to the QuickBird image of Augusta. To test the proposed method a new routine performing LOOCV into the software SISAR were implemented to perform the rigorous orientation of HRSI. Moreover, the image was also oriented by OrthoEngine, manually performing LOOCV since only HOV is possible with this software. No large residuals are evidenced in this case and the LOOCV accuracies (RMSEs and mADs) have essentially the same order of magnitude of HOV RMSE (Table 4.6), but again mADs are slightly lower than HOV RMSEs and they are significantly closer to HOV RMSEs than LOOCV RMSEs.

Table 4.6 Comparison between models and accuracy indices (in pixels) for Augusta image

Accuracy index	**SISAR**			**OrthoEngine**		
LOOCV	North	East	Module	North	East	Module
RMSE	1.52	1.45	2.09	1.83	1.56	2.40
mAD	1.01	1.17	1.77	1.19	0.96	1.68
Abs max	3.36	3.64	3.78	3.62	3.54	4.59
HOV	North	East	Module	North	East	Module
RMSE	1.16	1.38	1.80	1.32	1.33	1.87

4.11 Summary

High resolution satellite imagery became available to civilian users in 1999 with the launch of Ikonos, the first civilian satellite offering a spatial resolution of 1 m. Since then, other high resolution satellites have been launched, among which are EROS-A (1.8 m), QuickBird (0.61 m), Orbview-3 (1 m), EROS-B (0.7 m), Worldview-1 (0.5 m) and GeoEye-1 (0.41 m), with many others being planned to launch in the near future.

High resolution satellite imagery is now available in different formats and processing levels at an affordable price so that they already represent a possible alternative to aerial imagery for cartographic applications and orthophoto production, especially for areas where the organization of photogrammetric surveying may be critical. Moreover, an increasing demand for terrain modeling exists so that almost all the satellites have along-track stereo acquisition capability. Many new satellites dedicated to stereo viewing, for example Cartosat-1 (2.5 m), have been launched. This enables the generation of digital elevation models, digital surface models, and 3D features, e.g. city models.

The geomatic utilizations of satellite imagery for cartographic applications and terrain modeling requires a high level geometric correction through image orientation. Some fundamental problems related to sensor models and their parameters estimation, both for single images and stereopairs, were addressed and some real applications were discussed.

Specifically, the discussions were concerned with both physical sensor models and generalized sensor models for the orientation of basic images (level 1A) and of the image projected onto a specific object surface (usually an expanded ellipsoid derived from the WGS84, level 1B). As for the rigorous models, a thorough investigation on the fundamentals of their functional model was developed and the issue of parameter estimability was concerned. A solution was proposed based on SVD and QR decompositions. RPC models were discussed not only with respect to possible refinements by zero and first order transformations, but also (and mainly) with respect to the RPCs generation, based on previously established rigorous model. Thanks to SVD and QR decompositions, it was shown that many RPCs are not estimable parameters, therefore, they are not necessary to obtain the best achievable accuracy level.

Real applications demonstrated that rigorous and RPC models both for Level 1A and Level 1B imagery can provide an orientation accuracy at the level of 1-1.5 pixels in the horizontal components, and of 1-2 pixels in the height for stereopairs (even better with Cartosat-1 and slightly worse with EROS-A).

Moreover, HOV and LOOCV methods for accuracy assessment were discussed and compared, showing that the drawbacks of the usually adopted HOV can be overcome by LOOCV. LOOCV is reliable and robust, not dependent on a particular set of CPs and on outliers, which allows the use of each known ground point both as a GCP and as a CP, maximizing all the available ground information. Obviously, this is of particular relevancy when the number of ground points is kept as low as possible due to budget and/or logistic constraints. LOOCV may obviously apply to both rigorous and RPC-based (with possible zero or first order correction) orientation models.

Finally, a simple but effective method to represent the accuracy trend versus the number of GCPs was proposed. This model may conveniently be applied to LOOCV to find the minimum number of GCPs for accuracy assessment when a number of ground points are available.

Acknowledgments

The authors would like to thank very much:

Maria A. Brovelli, Armin Gruen, Karsten Jacobsen and Eugenio Realini for many fruitful discussions;

the Leica Geosystems company who supplied the Erdas software; Valerio Caroselli (Informatica per il Territorio S.r.l., Rome, Italy), Fabrizio Filiberti (Sysdeco Italia, Rome, Italy) and Fabio Volpe (Eurimage S.p.A., Rome, Italy), who kindly supplied the EROS-A1, Ikonos and QuickBird imagery, respectively;

Laura De Vendictis, Lucia Luzietti and Augusto Mazzoni who carried out some of the ground points Global Navigation Satellite System (GNSS) surveys.

References

Beyer HA (1992) Geometric and radiometric analysis of a CCD-Camera based photogrammetric close-range system. PhD thesis, Institut fur Geodasie und Photogrammetry, Nr. 51, ETH, Zurich, May 1992

Bianconi M, Crespi M, Fratarcangeli F, Giannone F, Pieralice F (2008) A new strategy for rational polynomial coefficients generation. Proceeding EARSeL Joint Workshop Remote Sensing, New Challenges of High Resolution, Bochum (Germany) March 5-7 2008

Brovelli M.A, Crespi M, Fratarcangeli F, Giannone F, Realini E (2008) Accuracy assessment of high resolution satellite imagery orientation by leave-one-out method, ISPRS Journal of Photogrammetry and Remote Sensing, Vol.63 Issue 4 pags. 427-440

Brown DC (1971) Close-range camera calibration. Photogrammetric Engineering. Vol.37, No. 8, pp.855-866

Crespi M, Fratarcangeli F, Giannone F, Pieralice F (2008a) Orientation Of Quickbird, Ikonos and Eros A Stereopairs by an Original Rigorous Model International Calibration and Orientation Workshop. Proceeding of EuroCOW 2008, Castelldefels (Spain) January 30-February 1 2008

Crespi M, Fratarcangeli F, Giannone F, Jacobsen K, Pieralice F (2008b) Orientation of Cartosat-1 Stereo Imagery. Proceeding of EARSeL Joint Workshop Remote Sensing, New Challenges of High Resolution, Bochum (Germany) March 5-7 2008

Elisseeff A, Pontil M (2002) Leave-one-out error and stability of learning algorithms with applications. Advances in Learning Theory: Methods, Models and Applications:111-130. NATO Advanced Study Institute on Learning Theory and Practice

Fraser C S and Hanley H B (2003) Bias compensation in rational functions for Ikonos satellite imagery Photogrammetric Engineering and Remote Sensing, Vol. 69(1), pp. 53-57

Geisser S (1975) The predictive sample reuse method with applications. Journal of the American Statistical Association, Vol. 70, No. 350, pp.320-328

Giannone F (2006) A rigorous model for High Resolution Satellite Imagery Orientation. Phd Thesis of the Sapienza University of Rome. Supervisors: M. Crespi. Available: http://w3.uniroma1.it/geodgeom/personale.htm#DottoriRicerca

Golub G, Van Loan C F (1993) Matrix computation. The Johns Hopkins University Press, Baltimore and London

Hanley H B, Fraser C S (2004) Sensor orientation for high-resolution satellite imagery: further insights into bias-compensated RPC, Available: http://www.isprs.org/istanbul2004/comm1/papers/5.pdf

Hofmann Wellenhof B, Lichtenegger H, Wasle E (2008) GNSS Global Navigation Satellite System, Spinger-Verlag. ISBN: 978-3-211-73012-6

Jacobsen K (1998) Geometric calibration of space remote sensing cameras for efficient processing. IAPRS, Vol. 32, Part 1, pp. 33-43

Kaula WM (1966) Theory of Satellite Geodesy. Blaisedell Publishing Company

Montenbruck O, Gill E (2001) Satellite orbits. Springer, Berlin

Neumaier A (1998) Solving ill-conditioned and singular linear systems: a tutorial on regularization SIAM Review, Issue 3, Vol. 40 pp. 636-666

NIMA (2000) The Compendium of Controlled Extensions (CE) for the National Imagery Transmission Format (Version 2.1) NITFS technical board

Noerdlinger PD (1999) Atmospheric refraction effects in earth remote sensing. ISPRS Journal of Photogrammetry & Remote Sensing Vol. 54, pp. 360–373

Pieralice F (2007) Orthorectification of IKONOS High Resolution Satellite Imagery: definition, implementation and accuracy assessment of an original orientation model. Degree thesis of the Sapienza University of Rome. Supervisors: M. Crespi. Not published

Poli D (2005) Modelling of spaceborne linear array sensors. Diss., Technische Wissenschaften ETH Zurich, Nr. 15894, IGP Mitteilung

Simon R, Dobbin K, McShane L M (2003) Pitfalls in the Use of DNA Microarray Data for Diagnostic and Prognostic Classification. JNCI Journal of the National Cancer Institute 2003 95(1), 14-18. Oxford University Press

Stone M (1974) Cross-validatory choice and assessment of statistical predictions (with discussion). Journal of the Royal Statistical Society B, No. 36, pp.111-147

Strang G, Borre K (1997) Linear algebra, Geodesy and GPS. Wellesley-Cambridge Press, Wellesley

Tao C V, Hu Y (2001a) The rational function model-A tool for processing high resolution imagery Earth Observation Magazine, Vol. 10 (1), pp. 13-16

Tao C V, Hu Y (2001b) A comprehensive study of the rational function model for photogrammetric processing Photogrammefric Engineering & Remote Sensing, Vol. 67(12), pp. 1347-1357

Tao C V, Hu Y (2001c) Use of the rational function model for image rectification Canadian Journal of Remote Sensing, Vol. 27(6), pp. 593-602

Tao C V and Hu Y (2002) 3D reconstruction methods based on the rational function model. Photogrammetric Engineering &Remote Sensing, vol. 68(7), pp.705-714

Teunissen P.J.G, Kleusberg A (1998) GPS for Geodesy, Springer-Verlag. ISBN: 3-540-63661-7

Toutin T, Chénier R, Carbonneau Y (2000) 3D models for high resolution images: examples with Quickbird, Ikonos and EROS In Proceedings of ISPRS Commission IV Symposium, Joint International Symposium on Geospatial Theory, Processing and Applications, Ottawa, pp. 547-551

Westin T (1990) Precision rectification of SPOT imagery. Photogrammetric Engineering and Remote Sensing Vol.56, n. 2, pp. 247–253

Chapter 5

GEOMETRIC PROCESSING MODELS FOR REMOTELY SENSED IMAGERY AND THEIR ACCURACY ASSESSMENT

Xiuxiao Yuan

5.1 Introduction

For a long time, geometrical positioning through frame perspective images has been a key research issue in the field of photogrammetry and remote sensing. During the era of analogue photogrammetry, frame perspective cameras are the only means to acquire remote sensing images whether in aerial photogrammetry or in space photogrammetry, which have provided large amount of high quality images for photogrammetric applications (Li and Zheng 1992). For aerial or satellite imagery which meet the requirements for overlapping, the aerial triangular network can be reconstructed after retrieving the relative orientation parameters of the image pair. Then the object space coordinates can be calculated by implementing a bundle block adjustment. This mature technology can achieve centimeter-level accuracy and realize the goal of high-precision remote sensing georeferencing.

With the progress of space positioning technology during the recent twenty years, humans are able to obtain orbit positions and sensor attitudes during image acquisition, with the aid of the modern navigation systems such as GPS (Global Positioning System) and POS (Position and Orientation System). Combing the auxiliary data and photogrammetric observations, we can greatly reduce or even avoid the use of ground control points (GCPs), leading to a new aerial image positioning method called GPS/POS-supported aerial triangulation. This technology has proved its practicability and played an important role in topographic surveying for border areas and inaccessible regions.

With the development of sensor technology, the dominancy of frame camera has been weakened. At the mean time, linear push-broom sensor gradually attracts public's attention for its obvious advantages in aerial and space photogrammetry: i) light but steady geometric structure, which maintains stable geometric relationship between the images and corresponding objects; ii) low power consumption, help realize multispectral scanning with high temporal and spatial resolution; iii) large coverage, under long strip scanning mode the spaceborne sensor can perform continuous global observation; iv) flexible observing manner, perfect base-height ste-

D. Li et al. (eds.), *Geospatial Technology for Earth Observation*,
DOI 10.1007/978-1-4419-0050-0_5,

reo images can be easily collected in both along-track and cross-track modes; v) integrated with advanced POS, capable to execute direct georeferencing or support aerial triangulation for high precision location-fixing applications. On the other hand, due to the geometric distinction between linear push-broom and frame imagery, the traditional geometric processing models based on frame imaging are no longer applicable for modern sensors, leading to a new research focus in photogrammetry and remote sensing science.

For high precision positioning with linear push-broom imagery, image orientation is a key step, the accuracy of which has a direct influence on the subsequent production flow. Linear array imagery are acquired through line-by-line scanning with CCD sensors, each line corresponding to different scanning time and external orientation elements. Consequently, orientation adjustment methods for linear-array imagery are more complicated than traditional methods for frame image (Shao et al. 2000). Scholars have proposed several kinds of rigorous models according to different assumptions concerning platform's attitudes and positions (Kratky 1989, Westin 1990, Kornus et al. 1999, Poli 2002, Dowman and Michalis 2003, Jacobsen and Passini 2003, Toutin 2004). It should be noted that all these models are based on the classical collinearity equation so they are theoretically strict and able to achieve high georeferencing accuracy. However, these rigorous geometric models are also demanding in image geometric processing and require users having certain knowledge background in photogrammetry.

For the sake of concealing complicated parameters of the sensor and realizing real-time calculation, general sensor models which mathematically express the relation between image coordinates and corresponding object coordinates emerge. General sensor models possess important advantages, such as simplicity and less demanding computation. After the launch of the IKONOS-2 satellite, methods based on rational polynomial coefficients (RPCs) fully show their unique characteristics, such as platform-independence, excellent interpolation performance and high precision comparable to rigorous model (Yang 2000). All these elements make it widely used. Later on, more and more high-resolution imagery vendors deliver RPCs as the geometric parameters of the imagery to users. Moreover, mainstream remote sensing software systems including ERDAS, PCI and ENVI provide the RPCs support module for satellite imagery processing. Rational function models (RFMs) have become universal for high-resolution imagery processing. The relevant researches ranging from high-precision resolution of model parameters, error propagation properties to optimized schemes for adjustment are going on.

Compared with spaceborne linear array sensors, exterior orientation elements of airborne linear array sensor vary more in flight, which makes it even harder to be processed. As to the processing of aerial linear array imagery, orientation image method (Ebner et al. 1992) is commonly employed nowadays. Its processing flow can be described as follows: First, orientation images are extracted at equal time or space intervals. Second, exterior orientation elements of any linear array are interpolated from adjacent orientation images using Lagrange polynomials. Finally, combined adjustments are implemented to calculate the exact orientation elements of all the orientation images and the coordinates of the object points. Relevant researches have proved that linear push-broom imagery can achieve locational accuracy similar to that of frame images and can be applied in 4D production, i.e. digital line graphs (DLG), digital raster graphs (DRG), digital elevation models (DEM), and digital orthophoto maps (DOM).

5.2 Geometric Processing Models for Central Projection-planar Array Remote Sensing Images

5.2.1 Self-Calibration Bundle Block Adjustment with Additional Parameters

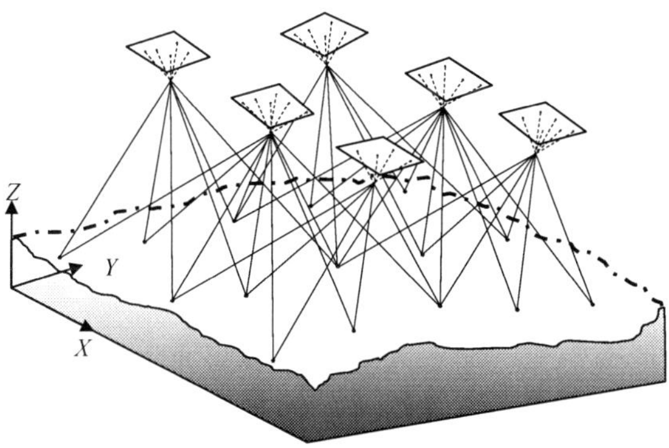

Fig. 5.1 Principle map of bundle block adjustment

Bundle block adjustment is a rigorous photogrammetric point determination method, which regards image point coordinates as observations, single space ray as an adjusted unit, and collinearity condition of central projection as basic equation. It makes homologous lays among images intersect optimally to one point and bring it to the coordinate system of GCPs by the rotation and translation of each space ray. It is a positioning method of remote sensing images, which computes ground coordinates of object points and image exterior orientation elements by image coordinates directly. This method has been widely used in digital photogrammetric systems. The basic principle of bundle block adjustment is shown in Fig. 5.1.

The theoretical basis of bundle block adjustment is the three-point collinearity equation, which represents the object point – its corresponding image point – the camera station lie on the same line. From the collinearity condition, two equations for each image point are as follows:

$$\begin{cases} x = -f\dfrac{a_1(X-X_S)+b_1(Y-Y_S)+c_1(Z-Z_S)}{a_3(X-X_S)+b_3(Y-Y_S)+c_3(Z-Z_S)} \\ y = -f\dfrac{a_2(X-X_S)+b_2(Y-Y_S)+c_2(Z-Z_S)}{a_3(X-X_S)+b_3(Y-Y_S)+c_3(Z-Z_S)} \end{cases} \tag{5.1}$$

where x, y represent image coordinates, whose origin is the principal point;

f represents the camera principal length;

X, Y, Z represent the ground coordinates of the object point;

X_S, Y_S, Z_S represent the linear elements of the exterior orientation elements of image;

$a_1, a_2, \cdots, c_3$ represent cosine values of angular elements of the exterior orientation elements of image.

A lot of factors will affect the quality of images, such as lens distortion, film distortion, atmospheric refraction, earth curvature, image digitalization and so on. Image measurement will also result in certain errors, so the observation data of image coordinates must have systematic errors. Additional parameters are used to compensate for the systematic errors present in observation data (Wang 1990). Generally, Bauer model with 3 additional parameters, Ebner model with 12 additional parameters and Brown model with 18 additional parameters are often selected and used to compensate the systematic errors of image point coordinates in bundle block adjustment (Li and Yuan 2002). In order to make following discussion convenient, take the simplest model proposed by Bauer as an example.

$$\begin{cases} \Delta x = s_1 x(x^2 + y^2 - 100) - s_3 x \\ \Delta y = s_1 y(x^2 + y^2 - 100) + s_2 x + s_3 y \end{cases} \tag{5.2}$$

where s_1, s_2, s_3 represent additional parameters. The collinearity equations with correction for systematic errors in image point coordinates are as Eq. (5.3) below, where the image point coordinates are regarded as observations, and object space coordinates, exterior orientation elements and additional parameters are regarded as unknowns.

$$\begin{cases} x = -f\dfrac{a_1(X-X_S)+b_1(Y-Y_S)+c_1(Z-Z_S)}{a_3(X-X_S)+b_3(Y-Y_S)+c_3(Z-Z_S)} + \Delta x \\ y = -f\dfrac{a_2(X-X_S)+b_2(Y-Y_S)+c_2(Z-Z_S)}{a_3(X-X_S)+b_3(Y-Y_S)+c_3(Z-Z_S)} + \Delta y \end{cases} \tag{5.3}$$

When the interior orientation elements of aerial camera are known accurately, Eq. (5.3) can be expanded by Taylor series in some neighborhood of the unknowns' approximations. If the first order derivatives are considered only, the error equation can be concluded:

$$\begin{cases} v_x = \dfrac{\partial x}{\partial X_S}\Delta X_S + \dfrac{\partial x}{\partial Y_S}\Delta Y_S + \dfrac{\partial x}{\partial Z_S}\Delta Z_S + \dfrac{\partial x}{\partial \varphi}\Delta\varphi + \dfrac{\partial x}{\partial \omega}\Delta\omega + \dfrac{\partial x}{\partial \kappa}\Delta\kappa + \\ \quad \dfrac{\partial x}{\partial X}\Delta X + \dfrac{\partial x}{\partial Y}\Delta Y + \dfrac{\partial x}{\partial Z}\Delta Z + \dfrac{\partial x}{\partial s_1}\Delta s_1 + \dfrac{\partial x}{\partial s_2}\Delta s_2 + \dfrac{\partial x}{\partial s_3}\Delta s_3 - (x - x^0) \\ v_y = \dfrac{\partial y}{\partial X_S}\Delta X_S + \dfrac{\partial y}{\partial Y_S}\Delta Y_S + \dfrac{\partial y}{\partial Z_S}\Delta Z_S + \dfrac{\partial y}{\partial \varphi}\Delta\varphi + \dfrac{\partial y}{\partial \omega}\Delta\omega + \dfrac{\partial y}{\partial \kappa}\Delta\kappa + \\ \quad \dfrac{\partial y}{\partial X}\Delta X + \dfrac{\partial y}{\partial Y}\Delta Y + \dfrac{\partial y}{\partial Z}\Delta Z + \dfrac{\partial y}{\partial s_1}\Delta s_1 + \dfrac{\partial y}{\partial s_2}\Delta s_2 + \dfrac{\partial y}{\partial s_3}\Delta s_3 - (y - y^0) \end{cases} \tag{5.4}$$

where x^0, y^0 represent the coordinate values computed by taking the approximations of exterior orientation elements and ground coordinates into Eq. (5.3).

To avoid over-parameterization, additional parameters are not regarded as free unknowns in bundle block adjustment, but regarded as virtual observations. Then, virtual observation equations can be set:

$$\begin{cases} v_{s_1} = \Delta s_1 \\ v_{s_2} = \Delta s_2 \\ v_{s_2} = \Delta s_3 \end{cases} \tag{5.5}$$

When n image points are measured, n sets of error equations in the form of Eqs. (5.4) and (5.5) can be written in matrix form as:

$$\begin{cases} \mathbf{V}_x = \mathbf{A}_x\mathbf{t} + \mathbf{B}\mathbf{x} + \mathbf{S}\mathbf{s} - \mathbf{L}_x, & \mathbf{E} \\ \mathbf{V}_S = \qquad\qquad \mathbf{E}\mathbf{s}, & \mathbf{P}_S \end{cases} \tag{5.6}$$

where $\mathbf{V}_x, \mathbf{V}_S$ represent the correction vectors of image point observations and virtual additional parameters observations;

$\mathbf{x} = [\Delta X \quad \Delta Y \quad \Delta Z]^{\mathrm{T}}$ represents the incremental vector of ground coordinates of the object point;

$\mathbf{t} = [\Delta\varphi \quad \Delta\omega \quad \Delta\kappa \quad \Delta X_S \quad \Delta Y_S \quad \Delta Z_S]^{\mathrm{T}}$ represents the incremental vector of exterior orientation elements of image;

$\mathbf{s} - [\Delta s_1 \quad \Delta s_2 \quad \Delta s_3]^{\mathrm{T}}$ represents the incremental vector of additional parameters;

$\mathbf{A}_x, \mathbf{B}, \mathbf{S}$ represent the coefficient matrices of corresponding unknowns;

$\mathbf{E}$ represents the unit matrix;

$\mathbf{L}_x = \begin{bmatrix} x - x^0 \\ y - y^0 \end{bmatrix}$ represents the residual vector of image point coordinate observations;

$\mathbf{P}_S = \frac{\sigma_0^2}{\sigma_S^2}\mathbf{E}$ represents the weight matrix of virtual observations of the additional parameters, with σ_0 being the measuring accuracy of image point coordinates, and σ_S the systematic errors of image point coordinates.

The generally highly over-determined system of equations given by Eq. (5.6) is solved via least-squares estimation. Then, the most probable values of the ground coordinates, the exterior orientation elements and the additional parameters can be obtained.

5.2.2 GPS-Supported Bundle Block Adjustment

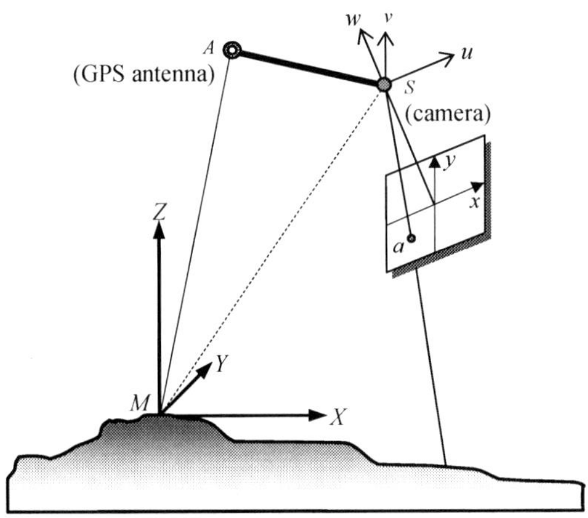

Fig. 5.2 Principle map of the aerophotogrammetry with GPS

Based on self-calibration bundle block adjustment, GPS data are combined with photogrammetric data in GPS-supported bundle block adjustment. Here, the camera station coordinates obtained via GPS at exposure instant are regarded as weighted observations. The offset vector between phase center A of airborne GPS antenna and perspective center S of aerial camera are shown in Fig. 5.2.

The object space coordinates of phase center of airborne GPS antenna and the perspective center of aerial camera is (X_A, Y_A, Z_A) and (X_S, Y_S, Z_S) in the coordinate system M-XYZ, respectively. If the three dimensional coordinate for the phase center of airborne GPS antenna is (u, v, w) (called GPS spatial offset) in the ideal camera coordinate system, then the transformation is accomplished with orientation matrix **R** consisting of three image attitude angles φ, ω, κ. The following relationship equation can be concluded:

$$\begin{bmatrix} X_A \\ Y_A \\ Z_A \end{bmatrix} = \begin{bmatrix} X_S \\ Y_S \\ Z_S \end{bmatrix} + \mathbf{R} \cdot \begin{bmatrix} u \\ v \\ w \end{bmatrix} \tag{5.7}$$

Friess (1991) found that GPS dynamic positioning based on carrier phase measurements generates drift systematic errors that have a linear relationship with flight time t when the continuous flight period is not too long. This systematic error is referred to as time-dependent GPS errors and datum inconsistencies between the GPS and the given GCP system (including localized distortion in national geodetic networks), which can be compensated for by the following model:

$$\begin{bmatrix} X_A \\ Y_A \\ Z_A \end{bmatrix} = \begin{bmatrix} X_S \\ Y_S \\ Z_S \end{bmatrix} + \mathbf{R} \cdot \begin{bmatrix} u \\ v \\ w \end{bmatrix} + \begin{bmatrix} a_X \\ a_Y \\ a_Z \end{bmatrix} + (t - t_0) \cdot \begin{bmatrix} b_X \\ b_Y \\ b_Z \end{bmatrix} \tag{5.8}$$

where t_0 represents the reference time; when the drift errors are regarded as a block invariant or strip invariant, t_0 may be the exposure time of the first image in the block or the strip;

$a_X, a_Y, a_Z, b_X, b_Y, b_Z$ represent the correction coefficients of the GPS drift systematic errors.

The rigorous geometric relationship between GPS station coordinates and the perspective center coordinates of aerial camera are shown in Eq. (5.8). To make the camera station coordinates obtained via GPS can be regarded as weighted observations in the self-calibration bundle block adjustment, the observation equations have to be linearized. Here, X_A, Y_A, Z_A are regarded as observations, the exterior orientation elements are regarded as unknowns. Considering the measuring errors of GPS spatial offset (u, v, w), Eq. (5.8) can be expanded by Taylor series in some neighborhood of the unknowns' approximations. If the first order derivatives are considered only, the error equation of GPS station coordinates can be concluded:

$$\begin{cases} v_{X_A} = \frac{\partial X_A}{\partial X_S}\Delta X_S + \frac{\partial X_A}{\partial Y_S}\Delta Y_S + \frac{\partial X_A}{\partial Z_S}\Delta Z_S + \frac{\partial X_A}{\partial \varphi}\Delta\varphi + \frac{\partial X_A}{\partial \omega}\Delta\omega + \frac{\partial X_A}{\partial \kappa}\Delta\kappa + \frac{\partial X_A}{\partial u}\Delta u + \frac{\partial X_A}{\partial v}\Delta v + \\ \quad \frac{\partial X_A}{\partial w}\Delta w + \frac{\partial X_A}{\partial a_X}\Delta a_X + \frac{\partial X_A}{\partial a_Y}\Delta a_Y + \frac{\partial X_A}{\partial a_Z}\Delta a_Z + \frac{\partial X_A}{\partial b_X}\Delta b_X + \frac{\partial X_A}{\partial b_Y}\Delta b_Y + \frac{\partial X_A}{\partial b_Z}\Delta b_Z - (X_A - X_A^0) \\ v_{Y_A} = \frac{\partial Y_A}{\partial X_S}\Delta X_S + \frac{\partial Y_A}{\partial Y_S}\Delta Y_S + \frac{\partial Y_A}{\partial Z_S}\Delta Z_S + \frac{\partial Y_A}{\partial \varphi}\Delta\varphi + \frac{\partial Y_A}{\partial \omega}\Delta\omega + \frac{\partial Y_A}{\partial \kappa}\Delta\kappa + \frac{\partial Y_A}{\partial u}\Delta u + \frac{\partial Y_A}{\partial v}\Delta v + \\ \quad \frac{\partial Y_A}{\partial w}\Delta w + \frac{\partial Y_A}{\partial a_X}\Delta a_X + \frac{\partial Y_A}{\partial a_Y}\Delta a_Y + \frac{\partial Y_A}{\partial a_Z}\Delta a_Z + \frac{\partial Y_A}{\partial b_X}\Delta b_X + \frac{\partial Y_A}{\partial b_Y}\Delta b_Y + \frac{\partial Y_A}{\partial b_Z}\Delta b_Z - (Y_A - Y_A^0) \\ v_{Z_A} = \frac{\partial Z_A}{\partial X_S}\Delta X_S + \frac{\partial Z_A}{\partial Y_S}\Delta Y_S + \frac{\partial Z_A}{\partial Z_S}\Delta Z_S + \frac{\partial Z_A}{\partial \varphi}\Delta\varphi + \frac{\partial Z_A}{\partial \omega}\Delta\omega + \frac{\partial Z_A}{\partial \kappa}\Delta\kappa + \frac{\partial Z_A}{\partial u}\Delta u + \frac{\partial Z_A}{\partial v}\Delta v + \\ \quad \frac{\partial Z_A}{\partial w}\Delta w + \frac{\partial Z_A}{\partial a_X}\Delta a_X + \frac{\partial Z_A}{\partial a_Y}\Delta a_Y + \frac{\partial Z_A}{\partial a_Z}\Delta a_Z + \frac{\partial Z_A}{\partial b_X}\Delta b_X + \frac{\partial Z_A}{\partial b_Y}\Delta b_Y + \frac{\partial Z_A}{\partial b_Z}\Delta b_Z - (Z_A - Z_A^0) \end{cases} \tag{5.9}$$

where X_A^0, Y_A^0, Z_A^0 represents the camera station coordinates calculated by taking the approximations of exterior orientation elements and GPS spatial offset into Eq. (5.8). It can be written as the matrix form,

$$\mathbf{V}_g = \mathbf{A}_g\mathbf{t} + \mathbf{R}\mathbf{r} + \mathbf{D}_g\mathbf{d}_g - \mathbf{L}_g \tag{5.10}$$

where $\mathbf{r} = [\Delta u \quad \Delta v \quad \Delta w]^{\mathrm{T}}$ represents the incremental vector of GPS offset unknowns;

$\mathbf{d}_g = [\Delta a_X \quad \Delta a_Y \quad \Delta a_Z \quad \Delta b_X \quad \Delta b_Y \quad \Delta b_Z]^{\mathrm{T}}$ represents the incremental vector of correction coefficients of GPS drift errors;

$\mathbf{A}_g, \mathbf{D}$ represent the coefficient matrices of corresponding unknowns;

$\mathbf{L}_g = \begin{bmatrix} X_A \\ Y_A \\ Z_A \end{bmatrix} - \begin{bmatrix} X_A^0 \\ Y_A^0 \\ Z_A^0 \end{bmatrix}$ represents the residual vector of the observations of the camera station coordinates obtained via GPS.

According to Eq. (5.5), the correction coefficients of GPS drift errors can be treated as virtual observations, and the error equations can be built:

$$\begin{matrix} v_{a_X} = \Delta a_X & v_{a_Y} = \Delta a_Y & v_{a_Z} = \Delta a_Z \\ v_{b_X} = \Delta b_X & v_{b_Y} = \Delta b_Y & v_{b_Z} = \Delta b_Z \end{matrix} \tag{5.11}$$

Comparing the Eq. (5.4) with Eq. (5.9), there are six same unknowns $\Delta X_S, \Delta Y_S, \Delta Z_S, \Delta\varphi, \Delta\omega, \Delta\kappa$. Therefore, two equations can be solved simultaneously. Considering Eqs. (5.5) and (5.11), the basic equation of GPS-supported bundle block adjustment is formed:

$$\begin{cases} \mathbf{V}_X = \mathbf{A}_X\mathbf{t} + \mathbf{B}\mathbf{x} + \mathbf{S}\mathbf{s} & -\mathbf{L}_X, & \mathbf{E} \\ \mathbf{V}_S = \mathbf{E}\mathbf{s} & -\mathbf{L}_S, & \mathbf{P}_S \\ \mathbf{V}_g = \mathbf{A}_g\mathbf{t} + \mathbf{R}\mathbf{r} + \mathbf{D}_g\mathbf{d}_g & -\mathbf{L}_g, & \mathbf{P}_g \\ \mathbf{V}_{d_g} = +\mathbf{E}\mathbf{d}_g, & & \mathbf{P}_{d_g} \end{cases} \tag{5.12}$$

If the corresponding weights are consisted with their respective accuracy of image point coordinate measurements and camera station coordinates obtained via GPS, then the most probable values of object coordinates can be solved using the least squares estimation. The systematic errors of the camera station coordinates obtained via GPS can be eliminated simultaneously; the accurate image exterior orientation elements can be obtained.

5.2.3 POS-Supported Aerial Triangulation

Based on the GPS-supported bundle block adjustment, POS data are combined with photogrammetric data in POS-supported bundle block adjustment. Here, the camera attitude data obtained via IMU are regarded as weighted observations. The aerophotogrammetric principle for the POS is shown in Fig. 5.3.

The IMU body coordinate system $\mathrm{I}-\mathrm{x}_I\mathrm{y}_I\mathrm{z}_I$ and camera coordinate system are not totally parallel due to limitations of equipment installation and there is a tiny direction shift $(\varphi_I, \omega_I, \kappa_I)$ between the respective axes in the two coordinate systems, which is known as boresight misalignment (Bäumker and Heimes 2002). For simplicity, the coordinate system $\mathrm{I}-\mathrm{x}_I\mathrm{y}_I\mathrm{z}_I$ can be regarded as the coordinate system $\mathrm{S}-\mathrm{uvw}$ after rotating $\varphi_I, \omega_I, \kappa_I$ around axes v, u, w sequentially.

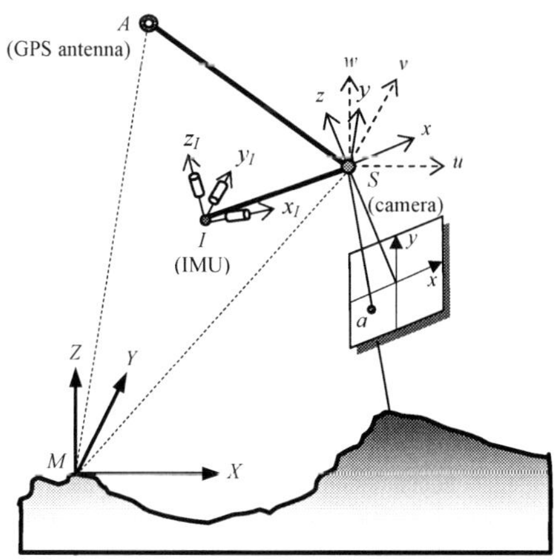

Fig. 5.3 Principle map of the aerophotogrammetry with POS

When the attitude angles of the camera determined via the IMU are $\varphi', \omega', \kappa'$, the orthogonal transformation matrix $\mathbf{R}$ consisting of image rotation angles φ, ω, κ can be represented as:

$$\mathbf{R}_{IMU} = \mathbf{R} \cdot \mathbf{R}_B^{\mathrm{T}} \tag{5.13}$$

where $\mathbf{R}_{IMU} = \mathbf{R}_{\varphi'} \mathbf{R}_{\omega'} \mathbf{R}_{\kappa'}$; $\mathbf{R}_B = \mathbf{R}_{\varphi_1} \mathbf{R}_{\omega_1} \mathbf{R}_{\kappa_1}$.

Setting $\mathbf{R} \cdot \mathbf{R}_B^{\mathrm{T}} = \begin{bmatrix} a_1 & a_2 & a_3 \\ b_1 & b_2 & b_3 \\ c_1 & c_2 & c_3 \end{bmatrix}$ gives

$$\begin{cases} \varphi' = -\mathrm{arctg}(\dfrac{a_3}{c_3}) \\ \omega' = -\arcsin(b_3) \\ \kappa' = \mathrm{arctg}(\dfrac{b_1}{b_2}) \end{cases} \tag{5.14}$$

The camera attitude angels obtained via the IMU will have bigger drift errors but still follow a linear relationship with flight time t. According to Eq. (5.8), they can be corrected by:

$$\begin{bmatrix} \varphi' \\ \omega' \\ \kappa' \end{bmatrix} = \begin{bmatrix} -\mathrm{arctg}(\dfrac{a_3}{c_3}) \\ -\arcsin(b_3) \\ \mathrm{arctg}(\dfrac{b_1}{b_2}) \end{bmatrix} + \begin{bmatrix} a_\varphi \\ a_\omega \\ a_\kappa \end{bmatrix} + (t - t_0) \cdot \begin{bmatrix} b_\varphi \\ b_\omega \\ b_\kappa \end{bmatrix} \tag{5.15}$$

where t_0 represents the reference time. When the drift errors are regarded as a block invariant or strip invariant, t_0 may be the exposure time of the first image in the block or the strip.

Eq. (5.15) represents the rigorous geometric relationship between camera attitude angles $\varphi', \omega', \kappa'$ and exterior orientation elements φ, ω, κ. Camera attitude angles obtained via IMU can be regarded as weighted observations in bundle block adjustment, i.e. $\varphi', \omega', \kappa'$ are regarded as observations, and exterior orientation elements, boresight misalignment, drift error correction coefficients are regarded as unknowns. The error equations of camera attitude angles obtained via IMU are as follows:

$$\begin{cases} v_{\varphi'} = \dfrac{\partial\varphi'}{\partial\varphi}\Delta\varphi + \dfrac{\partial\varphi'}{\partial\omega}\Delta\omega + \dfrac{\partial\varphi'}{\partial\kappa}\Delta\kappa + \dfrac{\partial\varphi'}{\partial\varphi_I}\Delta\varphi_I + \dfrac{\partial\varphi'}{\partial\omega_I}\Delta\omega_I + \dfrac{\partial\varphi'}{\partial\kappa_I}\Delta\kappa_I + \\ \quad \dfrac{\partial\varphi'}{\partial a_\varphi}\Delta a_\varphi + \dfrac{\partial\varphi'}{\partial a_\omega}\Delta a_\omega + \dfrac{\partial\varphi'}{\partial a_\kappa}\Delta a_\kappa + \dfrac{\partial\varphi'}{\partial b_\varphi}\Delta b_\varphi + \dfrac{\partial\varphi'}{\partial b_\omega}\Delta b_\omega + \dfrac{\partial\varphi'}{\partial b_\kappa}\Delta b_\kappa - (\varphi' - \varphi'^0) \\ v_{\omega'} = \dfrac{\partial\omega'}{\partial\varphi}\Delta\varphi + \dfrac{\partial\omega'}{\partial\omega}\Delta\omega + \dfrac{\partial\omega'}{\partial\kappa}\Delta\kappa + \dfrac{\partial\omega'}{\partial\varphi_I}\Delta\varphi_I + \dfrac{\partial\omega'}{\partial\omega_I}\Delta\omega_I + \dfrac{\partial\omega'}{\partial\kappa_I}\Delta\kappa_I + \\ \quad \dfrac{\partial\omega'}{\partial a_\varphi}\Delta a_\varphi + \dfrac{\partial\omega'}{\partial a_\omega}\Delta a_\omega + \dfrac{\partial\omega'}{\partial a_\kappa}\Delta a_\kappa + \dfrac{\partial\omega'}{\partial b_\varphi}\Delta b_\varphi + \dfrac{\partial\omega'}{\partial b_\omega}\Delta b_\omega + \dfrac{\partial\omega'}{\partial b_\kappa}\Delta b_\kappa - (\omega' - \omega'^0) \\ v_{\kappa'} = \dfrac{\partial\kappa'}{\partial\varphi}\Delta\varphi + \dfrac{\partial\kappa'}{\partial\omega}\Delta\omega + \dfrac{\partial\kappa'}{\partial\kappa}\Delta\kappa + \dfrac{\partial\kappa'}{\partial\varphi_I}\Delta\varphi_I + \dfrac{\partial\kappa'}{\partial\omega_I}\Delta\omega_I + \dfrac{\partial\kappa'}{\partial\kappa_I}\Delta\kappa_I + \\ \quad \dfrac{\partial\kappa'}{\partial a_\varphi}\Delta a_\varphi + \dfrac{\partial\kappa'}{\partial a_\omega}\Delta a_\omega + \dfrac{\partial\kappa'}{\partial a_\kappa}\Delta a_\kappa + \dfrac{\partial\kappa'}{\partial b_\varphi}\Delta b_\varphi + \dfrac{\partial\kappa'}{\partial b_\omega}\Delta b_\omega + \dfrac{\partial\kappa'}{\partial b_\kappa}\Delta b_\kappa - (\kappa' - \kappa'^0) \end{cases} \tag{5.16}$$

where $\varphi'^0, \omega'^0, \kappa'^0$ are the camera attitude angles computed by taking approximations of exterior orientation elements into Eq.(5.15).

Comparing Eqs. (5.4), (5.9) and (5.16), it can be seen that the same unknowns $\Delta\varphi, \Delta\omega, \Delta\kappa$ exist in three equations. Combining Eqs. (5.6), (5.11) and (5.16), the basic error equations of POS-supported bundle block adjustment are:

$$\begin{cases} \mathbf{V}_X = \mathbf{A}_X\mathbf{t} + \mathbf{B}\mathbf{x} + \mathbf{S}\mathbf{s} & & - \mathbf{L}_X, & \mathbf{E} \\ \mathbf{V}_S = & \mathbf{E}\mathbf{s} & - \mathbf{L}_S, & \mathbf{P}_S \\ \mathbf{V}_g = \mathbf{A}_g\mathbf{t} & + \mathbf{R}\mathbf{r} + \mathbf{D}_g\mathbf{d}_g & - \mathbf{L}_g, & \mathbf{P}_g \\ \mathbf{V}_{d_g} = & + \mathbf{E}\mathbf{d}_g, & & \mathbf{P}_{d_g} \\ \mathbf{V}_I = \mathbf{A}_I\mathbf{t} & + \mathbf{M}\mathbf{m} + \mathbf{D}_I\mathbf{d}_I & - \mathbf{L}_I, & \mathbf{P}_I \\ \mathbf{V}_{d_I} = & + \mathbf{E}\mathbf{d}_I, & & \mathbf{P}_{d_I} \end{cases} \tag{5.17}$$

where $\mathbf{m} = [\Delta\varphi_I \quad \Delta\omega_I \quad \Delta\kappa_I]^{\mathrm{T}}$ represents the incremental vector of the boresight misalignment;

$\mathbf{d}_I = [\Delta a_\varphi \quad \Delta a_\omega \quad \Delta a_\kappa \quad \Delta b_\varphi \quad \Delta b_\omega \quad \Delta b_\kappa]^{\mathrm{T}}$ represents the incremental vector of correction coefficients for IMU drift errors;

$\mathbf{A}_I, \mathbf{M}, \mathbf{D}_I$ represent the coefficient matrices of the corresponding unknowns;

$\mathbf{L}_I = \begin{bmatrix} \varphi' - \varphi'^0 \\ \omega' - \omega'^0 \\ \kappa' - \kappa'^0 \end{bmatrix}$ represents the residual vector of camera attitude angles obtained via IMU, $\varphi'^0, \omega'^0, \kappa'^0$ represent attitude angles computed by taking approximations of unknowns into Eq.(5.15);

$\mathbf{P}_I = \frac{\sigma_0^2}{\sigma_I^2}\mathbf{E}$ represents the weight matrix of camera attitude angles obtained via IMU, σ_I can be set as the nominal accuracy of camera attitude angles determined via IMU.

When the orientation parameters of m images taken over the adjusted block are determined via POS and n image point coordinates are measured, there will be 2n+6m error equations in the form of Eq. (5.17). If their corresponding weights are consistent with their respective accuracies in image point coordinates, the exposure station coordinates obtained via GPS, and camera attitude angles obtained via IMU, then the most probable values of object space coordinates and the exterior orientation elements of the images can be solved by the least squares estimation. Because the cumulative errors of the POS measurements are considered in error equations and the proper compensation models of POS systematic errors are imported, the correction coefficients of systematic errors are solved together in the adjustment. When the adjustment has iterative convergence, the position translation errors and linear drift errors generated by the POS, can be self-calibrated and self-eliminated, which improves the accuracy of the exterior orientation elements of the images. That not only implements direct georeferencing of aerial images, but also refines image exterior orientation elements obtained via POS.

5.3 Geometric Processing Models for Imagery Acquired by Linear Array Sensors

5.3.1 Rigorous Geometrical Models for Linear Array Imagery

5.3.1.1 Position-fixing Based on Satellite Linear Array Imagery

For geometric location-fixing based on satellite imagery, the optimal ground coordinate system would be the geocentric coordinate system (O_C-$X_CY_CZ_C$ illustrated in Fig. 5.4), which is conducive to introduce the platform's ephemeris data directly. The main function of geometric processing model is to establish the conversion formula from sensor coordinates (x, y) to the ground coordinates (X_A, Y_A, Z_A).

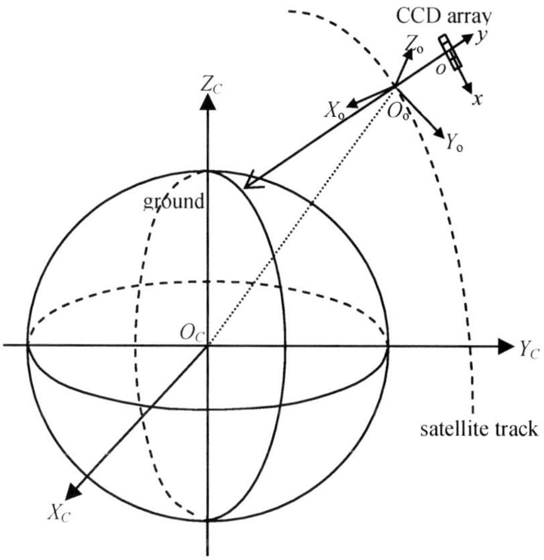

Fig. 5.4 The space relationship of relevant coordinate systems in remote sensing location

Assuming the sensor coordinates of any point in scan-line i as $(x_i,0)$, the corresponding object space coordinates as (X_A,Y_A,Z_A), a series of orthogonal transformation processes are executed to convert $(x_i,0)$ to (X_A,Y_A,Z_A). For different types of satellite imagery, the specific forms of transformation are not definitely consistent. Generally, the process can be described as:

$$\begin{bmatrix} X_A \\ Y_A \\ Z_A \end{bmatrix} = \lambda \mathbf{SRT} \begin{bmatrix} x_i \\ 0 \\ -f \end{bmatrix} + \begin{bmatrix} X_{S_i} \\ Y_{S_i} \\ Z_{S_i} \end{bmatrix} \tag{5.18}$$

where λ is the scale factor;

f is the focus length of the sensor lens;

S is the rotation matrix from satellite orbit coordinate system O_O-$X_OY_OZ_O$ to the geocentric coordinate system O_C-$X_CY_CZ_C$, which is determined by the position and velocity vector of the satellite platform;

R is the rotation matrix from satellite body coordinate system O_B-$X_BY_BZ_B$ to the satellite orbit coordinate system O_O-$X_OY_OZ_O$, which is determined by the sensor attitude angles $(pitch_i, roll_i, yaw_i)$;

T is the rotation matrix from sensor coordinate system o-xy to satellite body coordinate system O_B-$X_BY_BZ_B$;

$(X_{S_i}, Y_{S_i}, Z_{S_i})$ are the coordinates of perspective center of scan-line i in the object-space coordinate system.

From Eq. (5.18), the relationship between the sensor coordinates and its corresponding object-space coordinates can be described as:

$$\begin{cases} x_i = -f\dfrac{a_{i1}(X_A - X_{Si}) + b_{i1}(Y_A - Y_{Si}) + c_{i1}(Z_A - Z_{Si})}{a_{i3}(X_A - X_{Si}) + b_{i3}(Y_A - Y_{Si}) + c_{i3}(Z_A - Z_{Si})} \\ 0 = -f\dfrac{a_{i2}(X_A - X_{Si}) + b_{i2}(Y_A - Y_{Si}) + c_{i2}(Z_A - Z_{Si})}{a_{i3}(X_A - X_{Si}) + b_{i3}(Y_A - Y_{Si}) + c_{i3}(Z_A - Z_{Si})} \end{cases} \tag{5.19}$$

where a_{ij}, b_{ij}, c_{ij} $(j = 1,2,3)$ are direction cosine values of orthogonal transformation matrix $\mathbf{Q} = \mathbf{SRT}$.

Recently, high resolution remote sensing satellites usually carry high-performance orbit and attitude measurement system onboard, such as GPS, DORIS, stellar camera, laser gyro, and so on, which provide measurement of image orientation parameters in a certain frequency. The satellite orbit parameters determined via GPS or DORIS can reach decimeter-level accuracy, while sensor attitude angles determined by the stellar camera can reach arc-second-level accuracy. From orbit and attitude observation data mentioned above, the approximate exterior orientation elements of any scan-line i can be obtained with appropriate interpolation algorithm.

For direct georeferencing with single imagery, the elevation information is indispensable for determining the intersection point of imaging ray with ground surface (Fig. 5.5).

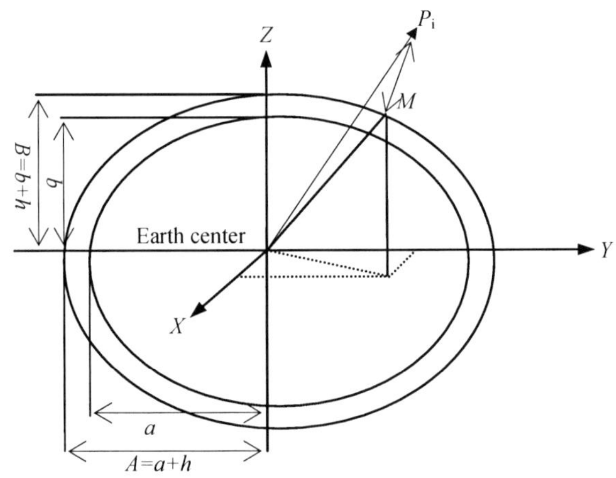

Fig. 5.5 Intersection point of imaging ray with the Earth ellipsoid

If stereo pair of imagery are available, the three dimensional coordinates of target point can be obtained with space intersection method, which realizes the direct georeferencing of linear array satellite imagery (Poli 2002).

While not considering the model error and other complicated distortion, the accuracy of direct georeferencing completely depends on the accuracy of the exterior orientation elements. It is necessary to analyze the effects of each error of the exterior orientation elements on object location. Figure 5.6 illustrates the location discrepancy caused by errors of each linear and angular elements with imagery acquired in the orbit height of 400 km and 600 km with viewing angel as 25°. The errors of the exterior orientation elements are set as 0~50 m for linear elements and 0~50 arcsecond for angular elements, respectively.

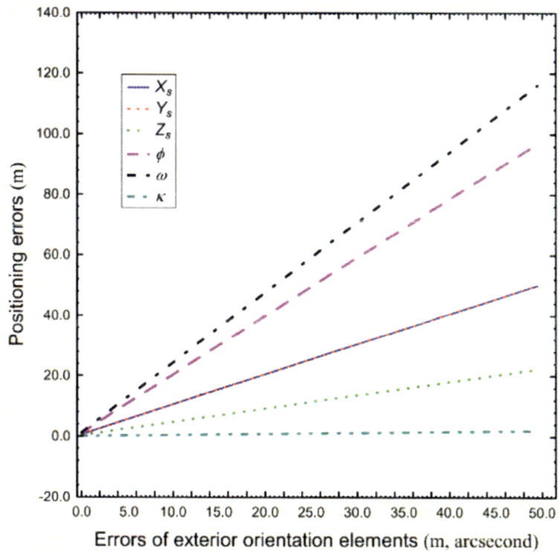

(a)

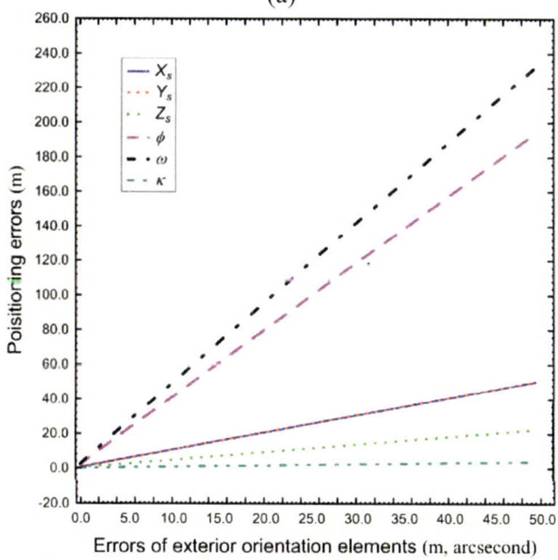

(b)

Fig. 5.6 Effect of image exterior orientation element errors on locational accuracy in the orbit height of 400 km (a) and 600 km (b)

Figure 5.6 indicates that the accuracies of angles φ and ω have the most obvious impact on location, which would increase with orbit height. Angles φ and ω are definitely the main error sources. Comparatively, the impact of linear elements Xs and Ys is much smaller and they belong to translation error which is also irrelevant to the orbit height. The impact of angular element κ is the minimum. From the analysis above, the refinement for the angular elements especially φ and ω should be emphasized.

Due to the stability of the satellite trajectory, the motion of the platform and the change of the attitude angles can be expressed by multi-order polynomial functions as (Lee et al 2000):

$$\begin{cases} X_{S_i} = m_0 + m_1 t + m_2 t^2 + \cdots\cdots + m_k t^k \\ Y_{S_i} = n_0 + n_1 t + n_2 t^2 + \cdots\cdots + n_k t^k \\ Z_{S_i} = s_0 + s_1 t + s_2 t^2 + \cdots\cdots + s_k t^k \\ \omega_i = d_0 + d_1 t + d_2 t^2 + \cdots\cdots + d_k t^k \\ \varphi_i = e_0 + e_1 t + e_2 t^2 + \cdots\cdots + e_k t^k \\ \kappa_i = f_0 + f_1 t + f_2 t^2 + \cdots\cdots + f_k t^k \end{cases} \tag{5.20}$$

where $m_0, n_0, s_0, d_0, e_0, f_0$ are the orientation elements of the centre scan line;

$m_1, n_1, s_1, d_1, e_1, f_1$ are the coefficients of the polynomials and they can be computed by the discrete ephemeris and attitude observations provided in metadata file. Experiment results show that the order of polynomial set to be 2 is appropriate (Amnon 2000).

When there exist sufficient and well distributed GCPs, the optimal orientation process is to adjust all the exterior orientation elements even including interior orientation elements simultaneously. To overcome the strong correlation between unknowns, the orientation elements should be treated as virtual observations. The basic equation can be described as:

$$\begin{cases} \mathbf{V}_x = \mathbf{B}\mathbf{x} + \mathbf{A}_1\mathbf{t}_1 + \mathbf{A}_2\mathbf{t}_2 - \mathbf{L}_x, & \mathbf{E} \\ \mathbf{V}_1 = \qquad \mathbf{E}_1\mathbf{t}_1, & \mathbf{P}_1 \\ \mathbf{V}_2 = \qquad\qquad \mathbf{E}_2\mathbf{t}_2, & \mathbf{P}_2 \end{cases} \tag{5.21}$$

where $\mathbf{V}_x, \mathbf{V}_1, \mathbf{V}_2$ are residual vectors of image coordinate observations, virtual observations of linear element coefficients and virtual observations of angular element coefficients, respectively;

$\mathbf{x} = [\Delta X_1 \quad \Delta Y_1 \quad \Delta Z_1 \quad \Delta X_2 \quad \Delta Y_2 \quad \Delta Z_2 ......]^{\mathrm{T}}$ is the incremental vector of the object-space coordinates of the object points;

$\mathbf{t}_1 = [\Delta m_0 \quad \Delta n_0 \quad \Delta s_0 \quad \Delta m_1 \quad \Delta n_1 \quad \Delta s_1 ......]^{\mathrm{T}}$ is the incremental vector of linear element coefficients;

$\mathbf{t}_2 = [\Delta d_0 \quad \Delta e_0 \quad \Delta f_0 \quad \Delta d_1 \quad \Delta e_1 \quad \Delta f_1 ......]^{\mathrm{T}}$ is the incremental vector of angular element coefficients;

$\mathbf{A}_1, \mathbf{A}_2$ are coefficient matrices of unknowns $\mathbf{t}_1, \mathbf{t}_2$ respectively;

$\mathbf{E}, \mathbf{E}_1, \mathbf{E}_2$ are unit matrices;

$\mathbf{L}_x$ is the discrepancy vector of image coordinates of GCPs;

$\mathbf{P}_1, \mathbf{P}_2$ are weight matrices of virtual observations of linear and angular element coefficients, which should be assigned according to the accuracy of observed exterior orientation elements to restrict the corrections of the systematic errors within the expected range.

For each GCP, error equations in the form of Eq. (5.19) can be established. When there are sufficient GCPs, the least-squares estimation are able to use to calculate the exterior orientation elements of the linear array imagery. However, due to the extreme narrow view field angle, there exists strong correlation among the unknowns. The column vector of the error equations present linear correlation and the state of the norm equations is seriously ill conditioned, making the solution very poor or even fail (Huang 1998). For situations in lack of GCPs, the simplified method can be adopted. As analyzed above, angular element errors are the main reasons for location errors. If the accuracy of linear elements is accurate enough, the angular elements are only refined. By this way, it not only avoids the strong correlation between linear and angular elements, but also greatly reduces the quantity requirement of the GCPs. At present, the orbit positioning accuracy of most high-resolution remote sensing satellite have exceed the meter level, which fully satisfy the requiremnt of simplified processing.

5.3.1.2 Position-fixing Based on Airborne Liner Array Imagery

Generally, a satellite runs in a stable track and spaceborne linear array sensors can easily acquire continuous high quality remotely sensed imagery. However, impacted by the strong airflow, the airborne platforms are unstable, leading to irregular variation of the sensor attitude angles. Consequently, aerial imagery collected by linear push-broom sensors always contains serious geometric distortion. To solve this problem, it is necessary to maintain a stable platform to restrain the change of exterior orientation elements or install POS to record the exterior orientation elements of each scan line.

Take three-linear array aviation digital camera Leica ADS 40 as an example. It carries PAV 30 peg-top stable platform and high precision POS. During the aerial photography, object points would be captured by three linear CCD placed in the same focal plane and imaged in three different strips. Thus the target area is three-fold covered and any two of the imagery can construct a stereo image pair. The imagery and their orientation elements are recorded in the synchronous mode.

As shown in Fig. 5.7, for a object point P, $p_A(x_A, y_A)$, $p_B(x_B, y_B)$ and $p_C(x_C, y_C)$ corresponding to different imaging time N_A, N_B and N_C are homogenous points in front, middle and back view imagery. Three looking rays E_A, E_B and E_C intersect at point P. If exterior orientation elements of front, middle and back view images are known, the three-dimensional coordinates of the object point can be calculated through multi-ray space intersection. This is the basic principle of position-fixing with three-linear array imagery. So the key issue is to obtain the accurate exterior orientation elements.

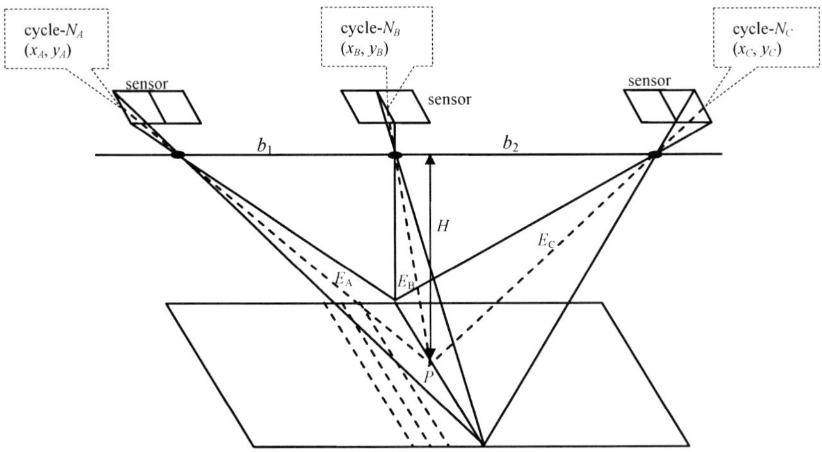

Fig. 5.7 Imaging principle of three-linear push-broom sensor (Zhao 2007)

Within a POS system, GPS can provide three-dimensional coordinates of the exposure station with high accuracy, and IMU measures three attitude angles of the sensor. GPS has the characteristics of low frequency and high accuracy while IMU being of high frequency and high accuracy. Both GPS and IMU data can be integrated to compensate for the systematic errors effectively and obtain sensor position and attitude data with high accuracy. On the other hand, for there exist spatial offset between the sensor perspective center and the phase center of GPS antenna, the boresight misalignment between the IMU body coordinate system and the image space coordinate system, and the drift errors caused by long flight, the position and attitude data obtained via POS are not the true image orientation elements defined in photogrammetry. These errors can be corrected in the adjustment process with a few of GCPs (Zhao and Li 2006).

The main idea of adjustment is to extract several linear images as orientation images at certain time intervals supported by the POS-provided image orientation elements. The exterior elements of any scan line ($X_P, Y_P, Z_P, \varphi_P, \omega_P, \kappa_P$) can be interpolated from exterior orientation elements $(X_j, Y_j, Z_j, \varphi_j, \omega_j, \kappa_j)(j \in [1, n])$ of the nearest n orientation images by Lagrange polynomials as Eq. (5.22). In the process of adjustment, for scan line p where the measured point lies, the partial coefficients in the error equation would be decomposed and redistributed to the corresponding partial coefficients of the referred orientation imagery with contribution parameter Wj. By this way, the exterior orientation elements of the orientation imagery can be calculated through minimizing interpolating errors in a least-squares adjustment.

$$
\begin{aligned}
X_P &= \sum_{j=1}^{n} X_j W_j \quad Y_P = \sum_{j=1}^{n} Y_j W_j \quad Z_P = \sum_{j=1}^{n} Z_j W_j \\
\varphi_P &= \sum_{j=1}^{n} \varphi_j W_j \quad \omega_P = \sum_{j=1}^{n} \omega_j W_j \quad \kappa_P = \sum_{j=1}^{n} \kappa_j W_j
\end{aligned}
\tag{5.22}
$$

where $W_j = \prod_{\substack{k=1 \\ k \neq j}}^{n} \frac{t - t_k}{t_j - t_k}$ can be regarded as the contribution parameter of orientation image j for scan line p.

5.3.2 General Geometric Models for Linear Array Pushbroom Remote Sensing Imagery

5.3.2.1 Rational Function Models (RFMs)

Rigorous geometric processing models based on collinearity equations are theoretically strict, but is dependent on sensor and complicated. What is more, iterative computations have to be done when coordinates are transformed from object-space to image-space. At present, RFMs are utilized much more in the positioning and orientation of high resolution remote sensing imagery. In a RFM, image pixel coordinates are expressed as the ratios of rational polynomials of ground point coordinates. For the object-to-image transformation, the defined ratios of polynomials are as follows:

$$
\begin{cases}
l = \dfrac{Num_l(P, L, H)}{Den_l(P, L, H)} \\[2ex]
s = \dfrac{Num_S(P, L, H)}{Den_S(P, L, H)}
\end{cases}
\tag{5.23}
$$

where l, s are the normalized image coordinates;

P, L, H are the normalized object coordinates (e.g. geocentric coordinates, geodetic coordinates and projection coordinates).

The definitions of the polynomials are as follows:

$$
\begin{aligned}
Num_l(P,L,H) = {} & a_1 + a_2L + a_3P + a_4H + a_5LP + a_6LH + a_7PH + a_8L^2 + a_9P^2 + \\
& a_{10}H^2 + a_{11}PLH + a_{12}L^3 + a_{13}LP^2 + a_{14}LH^2 + a_{15}L^2P + a_{16}P^3 + \\
& a_{17}PH^2 + a_{18}L^2H + a_{19}P^2H + a_{20}H^3 \\
Den_l(P,L,H) = {} & b_1 + b_2L + b_3P + b_4H + b_5LP + b_6LH + b_7PH + b_8L^2 + b_9P^2 + \\
& b_{10}H^2 + b_{11}PLH + b_{12}L^3 + b_{13}LP^2 + b_{14}LH^2 + b_{15}L^2P + b_{16}P^3 + \\
& b_{17}PH^2 + b_{18}L^2H + b_{19}P^2H + b_{20}H^3 \\
Num_S(P,L,H) = {} & c_1 + c_2L + c_3P + c_4H + c_5LP + c_6LH + c_7PH + c_8L^2 + c_9P^2 + \\
& c_{10}H^2 + c_{11}PLH + c_{12}L^3 + c_{13}LP^2 + c_{14}LH^2 + c_{15}L^2P + c_{16}P^3 + \\
& c_{17}PH^2 + c_{18}L^2H + c_{19}P^2H + c_{20}H^3 \\
Den_S(P,L,H) = {} & d_1 + d_2L + d_3P + d_4H + d_5LP + d_6LH + d_7PH + d_8L^2 + d_9P^2 + \\
& d_{10}H^2 + d_{11}PLH + d_{12}L^3 + d_{13}LP^2 + d_{14}LH^2 + d_{15}L^2P + d_{16}P^3 + \\
& d_{17}PH^2 + d_{18}L^2H + d_{19}P^2H + d_{20}H^3
\end{aligned}
$$

where $a_i, b_i, c_i, d_i (i = 1,2,\cdots,20)$ are rational polynomial coefficients (RPCs), in which b_1, d_1 are normally set to 1.

The order of polynomial functions is typically limited to 3. In a RFM, the distortions caused by the optical projection can generally be represented by the ratios of first-order terms, while corrections such as Earth curvature, atmospheric refraction, lens distortion, and so on, can be well approximated by the second-order terms. Some other unknown distortions with high-order components, such as camera vibration, can be modeled with the third-order terms.

Eq. (5.23) expresses the object-to-image coordinate transformation called forward transformation. Also a RFM could represent the inverse image-to-object transformation, i.e.

$$
\begin{cases}
P = \dfrac{Num_P(l,s,H)}{Den_P(l,s,H)} \\
L = \dfrac{Num_L(l,s,H)}{Den_L(l,s,H)}
\end{cases}
\tag{5.24}
$$

5.3.2.2 Solution to the RPCs

The RFMs are independent of sensors, and the RPCs have no significant physical meanings. With the rigorous geometric processing models unknown, the RPCs can be solved according to the actual GCPs, i.e. terrain-dependent computational scenarios. In this case, the solution is highly dependent on the actual terrain relief, the number of GCPs, and their distribution, which may cause the solution unstable even unreliable. In practice, terrain-independent computational scenarios are more preferable. Having the rigorous geometric models established, intensive and evenly distributed object grids (Fig. 5.8) are firstly generated according to the rigorous geometric processing models, and then the RPCs are solved using the object grid points as virtual GCPs through least squares estimation. Actually, the terrain-independent computational scenarios are utilizing the RFM to mathematically fit the rigorous geometric processing models and some relevant research has reported that the fitting accuracy could reach as high as ±0.01 pixels for IKONOS and QuickBird imagery (Grodecki and Dial 2001). The specific computational flow is shown in Fig. 5.9.

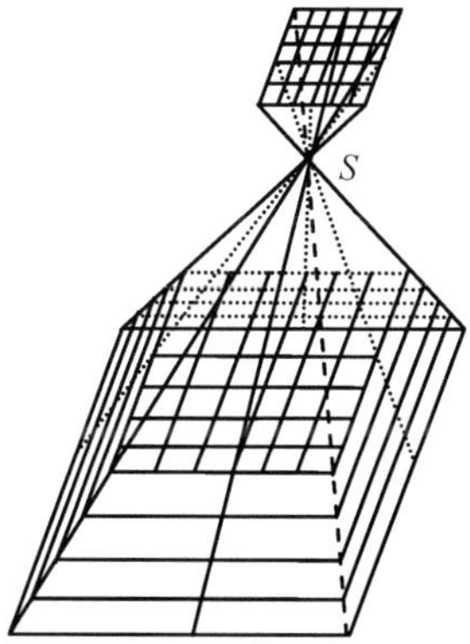

Fig. 5.8 The virtual control grids for solving the RPCs

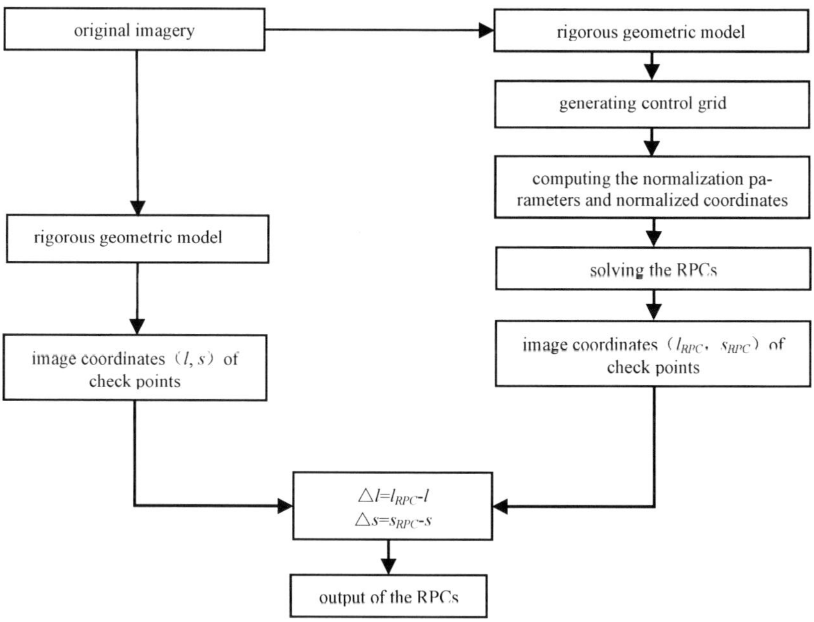

Fig. 5.9 Calculation flow of the RPCs

The division of the control grids should rely on the terrain relief of the coverage area. High accuracy solution could not be attained with the grid too sparse, while an intensive grid will result in much more computation and has a negative effect on the speed of RPCs solution. Generally, having the gird divided into 10 rows by 10 columns or 20 rows by 20 columns in planimetry and 4 or 5 layers in elevation, the solution of the RPCs could meet the accuracy requirement (Tao and Hu 2001). When generating the control grids, a check grid could be generated in the same way in order to check the solution accuracy of the RPCs.

In order to use the least squares estimation to solve the RPCs, Eq. (5.23) can then be formed as:

$$\begin{cases} v_l = \begin{bmatrix} \frac{1}{B} & \frac{L}{B} & \cdots & \frac{P^2H}{B} & \frac{H^3}{B} & -\frac{lL}{B} & -\frac{lP}{B} & \cdots & -\frac{lP^2H}{B} & -\frac{lH^3}{B} \end{bmatrix} \mathbf{J} - \frac{l}{B} \\ v_s = \begin{bmatrix} \frac{1}{D} & \frac{L}{D} & \cdots & \frac{P^2H}{D} & \frac{H^3}{D} & -\frac{sL}{D} & -\frac{sP}{D} & \cdots & -\frac{sP^2H}{D} & \frac{-sH^3}{D} \end{bmatrix} \mathbf{K} - \frac{s}{D} \end{cases} \tag{5.25}$$

where

$$B = \begin{bmatrix} 1 & L & P & H & \cdots & L^2H & P^2H & H^3 \end{bmatrix} \begin{bmatrix} 1 & b_2 & b_3 & b_4 & \cdots & b_{18} & b_{19} & b_{20} \end{bmatrix}^{\mathrm{T}};$$

$$D = \begin{bmatrix} 1 & L & P & H & \cdots & L^2H & P^2H & H^3 \end{bmatrix} \begin{bmatrix} 1 & d_2 & d_3 & d_4 & \cdots & d_{18} & d_{19} & d_{20} \end{bmatrix}^{\mathrm{T}};$$

$$\mathbf{J} = \begin{bmatrix} a_1 & a_2 \cdots a_{20} & b_2 & b_3 \cdots b_{20} \end{bmatrix}^{\mathrm{T}};$$

$$\mathbf{K} = \begin{bmatrix} c_1 & c_2 \cdots c_{20} & d_2 & d_3 \cdots d_{20} \end{bmatrix}^{\mathrm{T}}.$$

The third-order RFMs with different denominators have 78 parameters for each set, which are much more than that in the rigorous geometric processing models and inevitably result in correlation among the parameters. Hence, in order to overcome the correlation and achieve high accuracy solution, the classical least squares estimation should be improved by introducing generalized ridge estimation or employing iteration method by correcting characteristic values. After solving the RPCs, one should take the coordinates of check points into the RFMs so as to check the solution accuracy. When the RMSE of check points is in accordance with that of control points, it is indicated the solution is acceptable and reliable.

5.3.2.3 Georeferencing of Remotely Sensed Imagery Based on RFMs

Having the RFMs of both the left and the right images of a stereo pair established respectively, how to get the object-space coordinates of corresponding object point according to the image-space coordinates of conjugate image points is a RFM-based 3D reconstruction problem. Both the forward and the inverse RFMs could be utilized for 3D reconstruction, but the former is preferable in practice (Fraser and Hanley 2004), which considers the coordinates of object points as unknowns and has Eq. (5.23) linearized:

$$\begin{cases} l = (l) + \dfrac{\partial l}{\partial P}\Delta P + \dfrac{\partial l}{\partial L}\Delta L + \dfrac{\partial l}{\partial H}\Delta H \\ s = (s) + \dfrac{\partial s}{\partial P}\Delta P + \dfrac{\partial s}{\partial L}\Delta L + \dfrac{\partial s}{\partial H}\Delta H \end{cases} \tag{5.26}$$

Given a pair of conjugate image points $(l_l, s_l), (l_r, s_r)$, four error equations could be listed as Eq. (5.26) and the object-space coordinates of corresponding object point can be derived using the least squares estimation.

5.3.2.4 Refinement of the RPCs

The accuracy achieved using the RPCs provided to users with the imagery by the vendor can only be as high as that of direct georeferencing, which can not meet the mapping requirements in most situations. Therefore, it is necessary for users to refine the RPCs with GCPs. With the relevant information about the rigorous geometric processing models unavailable for general users, it is difficult to refine the RPCs directly. In practice, additional affine transformation terms are introduced into the original models in order to compensate for the systematic errors (Di et al. 2003, Fraser and Hanley 2003, Hu et al. 2004), i.e.

$$\begin{cases} l'_n = l_n + e_0 + e_1 l_n + e_2 s_n \\ s'_n = s_n + f_0 + f_1 l_n + f_2 s_n \end{cases} \tag{5.27}$$

Correction terms $\Delta l = e_0 + e_1 l_n + e_2 s_n$ and $\Delta s = f_0 + f_1 l_n + f_2 s_n$ can effectively compensate for the effects caused by positioning and orientation errors along-track and across-track, and significantly enhance the locational accuracy. For each GCP, two error equations are as follows:

$$\begin{bmatrix} v_l \\ v_s \end{bmatrix} = \begin{bmatrix} \dfrac{\partial l'_n}{\partial e_0} & \dfrac{\partial l'_n}{\partial e_1} & \dfrac{\partial l'_n}{\partial e_2} & 0 & 0 & 0 \\ 0 & 0 & 0 & \dfrac{\partial s'_n}{\partial f_0} & \dfrac{\partial s'_n}{\partial f_1} & \dfrac{\partial s'_n}{\partial f_2} \end{bmatrix} \begin{bmatrix} \Delta e_0 \\ \Delta e_1 \\ \Delta e_2 \\ \Delta f_0 \\ \Delta f_1 \\ \Delta f_2 \end{bmatrix} - \begin{bmatrix} l'_n - (l'_n) \\ s'_n - (s'_n) \end{bmatrix} \tag{5.28}$$

Three or more GCPs are needed to solve the six additional parameters for each image. And for a stereo pair of images, additional parameters and coordinates of the connection points could be solved together by introducing constraint conditions of the tie points, whereby few GCPs used as adjustment reference could be enough to compensate for the systematic errors. Presently such a method is called RFM-based bundle adjustment. Comparing to the case of single image, bundle adjustment with stereo images is stricter in theory and may achieve higher accuracy. The error equations of bundle adjustment are as follows:

$$\begin{bmatrix} v_l \\ v_S \end{bmatrix} = \begin{bmatrix} \frac{\partial l'_n}{\partial e_0} & \frac{\partial l'_n}{\partial e_1} & \frac{\partial l'_n}{\partial e_2} & 0 & 0 & 0 & \frac{\partial l'_n}{\partial X} & \frac{\partial l'_n}{\partial Y} & \frac{\partial l'_n}{\partial Z} \\ 0 & 0 & 0 & \frac{\partial s'_n}{\partial f_0} & \frac{\partial s'_n}{\partial f_1} & \frac{\partial s'_n}{\partial f_2} & \frac{\partial s'_n}{\partial X} & \frac{\partial s'_n}{\partial Y} & \frac{\partial s'_n}{\partial Z} \end{bmatrix} \begin{bmatrix} \Delta e_0 \\ \Delta e_1 \\ \Delta e_2 \\ \Delta f_0 \\ \Delta f_1 \\ \Delta f_2 \\ \Delta X \\ \Delta Y \\ \Delta Z \end{bmatrix} - \begin{bmatrix} l'_n - (l'_n) \\ s'_n - (s'_n) \end{bmatrix} \tag{5.29}$$

where $\Delta X, \Delta Y, \Delta Z$ are corrections for the object-space coordinates of image points.

5.4 Accuracy Assessment of Object Positioning in Remote Sensing

Object positioning in aerial or spaceborne photogrammetry is to determine the spatial positions of object points. As a key quality evaluation index, positioning accuracy is concerned mostly. Usually, the posterior variance after adjustment, theoretical and empirical accuracy of object-space coordinates are set as measures.

5.4.1 Root Mean Squares Error of the Unit Weight

The essence of object positioning in photogrammetry is to inverse the three-dimensional spatial coordinates of object points from 2-dimensional images. The method of least squares adjustment is adopted commonly. According to the principle of $\sum pvv = \min$, the three-dimensional spatial coordinates of object points can be solved iteratively.

After iteration convergence, posterior root mean squared error of the unit weight can be computed according to the residuals of all observations:

$$\sigma_0 = \sqrt{\mathbf{V}^{\mathrm{T}}\mathbf{P}\mathbf{V}/r} \tag{5.30}$$

where **V** represents the residual vector of observations;
P represents the weight matrix of observations;
r represents the redundancy of adjustment system.

5.4.2 Theoretical Accuracy of Object-Space Coordinates

Regarding corrections to object-space coordinates as random variables, theoretical accuracy of object-space coordinates will be analyzed in theory by computing the variance-covariance matrix **D** in least squares estimation.

$$\mathbf{D} = \sigma_0^2 \mathbf{Q} \tag{5.31}$$

is set as measure, and the theoretical accuracy formula of individual unknowns i can be written as:

$$m_i = \sigma_0 \sqrt{Q_i}, \quad (i = X, Y, Z) \tag{5.32}$$

When a photogrammetric block contains n object points, Eq. (5.33) can be used to express the overall theoretical accuracy for planimetry and for elevation determination.

$$\begin{cases} \overline{m}_{XY} = \sigma_0 \sqrt{\dfrac{1}{n} \sum_{i=1}^{n} (Q_{XX}^i + Q_{YY}^i)} \\ \overline{m}_Z = \sigma_0 \sqrt{\dfrac{1}{n} \sum_{i=1}^{n} Q_{ZZ}^i} \end{cases} \tag{5.33}$$

$Q_{XX}^i, Q_{YY}^i, Q_{ZZ}^i$ in Eqs. (5.31)~(5.33) are the diagonal elements of the i-th unknown coordinate X, Y, Z 's co-factor matrix **Q**.

5.4.3 Empirical Accuracy of Object-Space Coordinates

Comparing the adjusted coordinates with field surveying coordinates, and regarding residual errors of n check points $\Delta X_i, \Delta Y_i, \Delta Z_i$ as true errors, the empirical accuracy of object-space coordinates is the average RMSE of n check points calculated by Eq. (5.34).

$$\begin{cases} \overline{\mu}_{XY} = \sqrt{\dfrac{1}{n} \sum_{i=1}^{n} (\Delta X_i^2 + \Delta Y_i^2)} \\ \overline{\mu}_Z = \sqrt{\dfrac{1}{n} \sum_{i=1}^{n} \Delta Z_i^2} \end{cases} \tag{5.34}$$

5.5 Application Demonstration

In order to validate the validity and practicability of the theory, this section will describe some empirical studies using aerial digitalized images with POS data, three-line-array digital aerial imagery and high resolution satellite imagery (SPOT-5) for object positioning.

5.5.1 Object Positioning in Aerial Photogrammetry

5.5.1.1 GPS-Supported Aerial Triangulation

In order to discuss the feasibility of GPS precise point positioning (PPP) in dynamic aerophotogrammetry and the accuracy of GPS-supported bundle block adjustment with GPS exposure station coordinates obtained by the PPP technique, four sets of actual aerial photos taken from four experimental projects which are different in terrain and photographic scale are selected and used. The photographic scales of them are 1:2 500, 1:3 000, 1:32 000 and 1:60 000, respectively. The distributions of the test projects are shown in Fig. 5.10. The main technical parameters of the empirical images are given in Table 5.1.

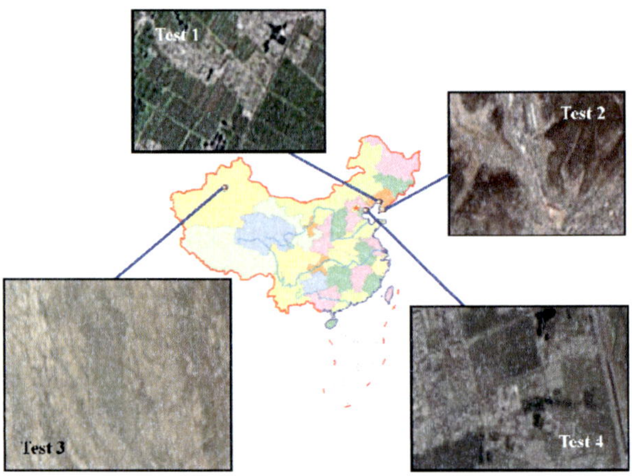

Fig. 5.10 Distributions of empirical projects

After all negatives are scanned with a resolution of 21μm, WuCAPS (Yuan 2008) and JX-4 digital photogrammetry workstation (Geo-vision 2009) are used for automatic point transfer. The corresponding image coordinates of all GCPs are measured manually in the stereoscopic mode. The root mean square error (RMSE) of all image coordinates is less than ±6.0 μm according to the results of the consecutive relative orientation with conditions for model connection and blunder elimination by WuCAPS. The observations obtained via GPS/IMU are processed together using POS data post-processing software system POSPac™ (Applanix 2008). Six elements of exterior orientation for each image can be determined after coordinate transformation, and they are the exterior orientation elements determined via POS.

Table 5.1 Details of images used in the empirical studies

specifications	Test 1	Test 2	Test 3	Test 4
date	Nov, 2004	Oct, 2005	Sep, 2005	Oct, 2005
aircraft	Yun-12	Yun-12	Yun-8	Citation II
aerial camera	Leica RC-30	Leica RC-30	Leica RC-30	Leica RC-30
flight control system	Track Air	Track Air	CCNS 4	CCNS 4
POS system	POS AV 510	POS AV 510	POS AV 510	POS AV 510
film	Kodak 2444	Kodak 2044	Kodak 2402	Kodak 2402
principal length (mm)	153.84	303.64	154.06	153.53
frame (cm^2)	23 × 23	23 × 23	23 × 23	23 × 23
photo scale	1:2 500	1:3 000	1:32 000	1:60 000
longitudinal overlap (%)	61	63	64	64
lateral overlap (%)	32	33	33	30
strips	9	10	9	4
control strips	2	2	2	0
number of photographs images	255	405	244	48
GCPs	72	155	34	29
object points	3632	6538	2957	712
area (km^2)	4 × 5	5 × 8	47 × 52	40 × 57
maximum terrain undulation (m)	38.6 (flat)	181.6 (general mountain)	729.3 (high mountain)	109.3 (upland)
airborne GPS receiver	Trimble BD 950	Trimble BD 950	Trimble BD 950	TrimbleBD 950
GPS receiver of reference station	Ashtech	Trimble 5700	Trimble 5700	Trimble 5700
sampling interval of airborne GPS receiver (second)	0.1	0.1	0.1	0.1
sampling interval of ground GPS receiver (second)	2.0	0.5	1.0	1.0
maximal range from aircraft to reference station (km)	28.988	28.459	187.592	145.400
GPS initialization (min)	5	5	5	5
antenna-camera offset (m)	0.303,-0.110,-2.029	0.303,-0.110,-2.002	-2.015,-0.030,3.102	2.034,-0.520,1.320

In order to analyze the accuracy of the actual aerial photogrammetric point determination, three different photogrammetric adjustment methods are employed in the test projects distributed in Fig. 5.10. They are divided into traditional bundle block adjustment (referred to as "TRA"), GPS-supported bundle block adjustment with one cross-strip on each end of the adjusted block (referred to as "GPS") and POS-supported bundle block adjustment with 4 full GCPs in the four corners of the adjusted block (referred to as "POS"). Moreover, the theoretical accuracy of three different methods is evaluated. The results are shown in Table 5.2.

Table 5.2 Accuracy of bundle block adjustment

images		Test 1			Test 2			Test 3			Test 4		
methods		TRA	GPS	POS	TRA	GPS	POS	TRA	GPS	POS	TRA	GPS	POS
σ_0 (μm)		5.7	6.9	6.3	4.8	6.6	5.2	6.4	6.9	6.6	5.9	6.4	6.2
# GCPs	*XY*	23	4	4	39	4	4	20	4	4	15	4	4
	Z	39	4	4	69	4	4	25	4	4	19	4	4
# check points	*XY*	49	67	67	116	151	151	14	30	30	10	25	25
	Z	33	67	67	86	150	150	9	30	30	14	25	25
max. residuals of check points (m)	*X*	0.18	0.26	-0.20	-0.20	-0.29	-0.29	-1.44	1.54	-1.36	3.13	4.02	4.22
	Y	0.14	0.23	-0.33	0.17	0.19	0.25	-0.98	-2.54	1.49	-2.73	4.68	4.59
	XY	0.198	0.232	0.212	0.205	0.328	0.338	1.738	2.637	1.648	3.385	4.893	5.117
	Z	-0.241	0.237	-0.265	-0.382	0.383	0.502	0.892	-1.666	1.593	2.054	-2.644	-2.653
empirical accuracy (m)	*X*	0.08	0.08	0.08	0.06	0.11	0.11	0.67	0.80	0.68	1.28	1.46	1.52
	Y	0.06	0.08	0.08	0.06	0.07	0.07	0.47	0.74	0.83	1.29	1.88	1.86
	XY	0.103	0.119	0.119	0.087	0.131	0.129	0.818	1.089	1.068	1.818	2.375	2.402
	Z	0.079	0.105	0.099	0.126	0.157	0.178	0.639	0.738	0.913	1.416	1.350	1.320
theoretical accuracy (m)	*X*	0.01	0.04	0.02	0.01	0.04	0.02	0.27	0.41	0.25	0.31	0.35	0.33
	Y	0.02	0.04	0.02	0.02	0.04	0.03	0.28	0.64	0.76	0.36	0.42	0.41
	XY	0.022	0.058	0.030	0.022	0.053	0.035	0.391	0.760	0.802	0.474	0.547	0.523
	Z	0.029	0.061	0.095	0.062	0.115	0.189	0.678	0.617	0.879	0.771	1.053	1.015

Note:
'TRA' denotes the self-calibration bundle block adjustment with 3 additional parameters;
'GPS' denotes the GPS-supported bundle block adjustment;
'POS' denotes the POS-supported bundle block adjustment.

In Table 5.1, images in Tests 1 and 2 can be used for 4D production at a scale of 1:500~1:2000, images in Test 3 can be used for 4D production at a scale of 1:5000~1:10000, and images in Test 4 can be used for 4D production at a scale of 1:50000, respectively. According to the existing specifications for aerophotogrammetric office operation (GB 12340-90 1991, GB/T 13990-92 1993, GB 7930-

87 1998), the empirical blocks of Test 1, Test 4, Test 2 and Test 3 can be considered as representing flat areas, uplands, hills, and high mountains, respectively. The following conclusions can be deduced by analyzing the results in Table 5.2:

1) The accuracy of photogrammetric points determined by three adjustment methods all can satisfy the requirement of 4D production at corresponding scales. However, there are 4 height check points exceeding the specifications' tolerance in the POS-supported bundle block adjustment of Test 2, which accounts 2.7% of the 150 height check points. After correction, these points can meet the requirements of photogrammetric point determination. These results show that any photogrammetric method in this section is practicable in 4D production.

2) As far as theoretical accuracy is concerned, accuracy of traditional bundle block adjustment is the highest. Comparing the accuracy of GPS-supported bundle block adjustment with that of POS-supported bundle block adjustment, the former is higher in elevation, while the latter is higher in planimetry; and the larger the scale, the greater the advantage. The reason is that the use of cross-strip can increase the number of image point observations and import corresponding measurement errors (the change of σ_0 can valid this conclusion). The scales of images for mapping are different from that of cross-strip images, and then, the errors of point transfer in control images are larger than that in the images for mapping. In photogrammetric point determination at a large scale, measurement errors are small, so the point transfer error becomes obvious.

3) As far as each block is concerned, the empirical accuracy of traditional bundle block adjustment is the highest, which is consistent with the theoretical analysis. There is no material difference between accuracy of GPS-supported bundle block adjustment with one cross-strip on each end of the adjusted block and POS-supported bundle block adjustment with 4 full GCPs in the four corners of the block. The reason is that there are some identification errors because all the check points are outstanding points and are measured manually in the stereoscopic mode. Additionally, negative distortion, scan errors and errors of exterior orientation elements obtained via POS will affect the orientation results. Although a series of systematic error compensation are incorporated in the adjustment, there are errors in photogrammetric results all the same. Therefore, the empirical accuracy is much lower than theoretical accuracy, for planimetry, especially.

4) As far as ground control is concerned, accuracy of traditional bundle block adjustment depends on the distribution and distance between ground control points. The larger the region, the more the number of ground control points. On the other hand, GPS-supported bundle block adjustment and POS-supported bundle block adjustment require four full GCPs in the four corners of the adjusted block only, regardless of the size of block. However, when the block is large, GPS-supported bundle block adjustment needs to lay one cross-strip on each end of the adjusted block, in order to ensure the photogrammetric accuracy in elevation.

5.5.1.2 Positioning by Airborne Liner Array Camera ADS 40

Test data were collected by three-linear camera ADS 40 in January 2007. The whole area is composed of 9 strips. The size of the area is about 17×12 km^2. The average elevation is 775 m and elevation variation is within 150 m. The camera principal distance is 62.7 mm and the flight height is about 2 000 m. The CCD pixel size is 6.5 μm with a ground sampling distance of 20 cm. Side overlap is about 25%. The image length is 186 000 rows, and 372 orientation images at the interval time of 8 seconds are processed. There are 9 horizontal and vertical GCPs (Fig. 5.11) evenly distributed in the area. All GCPs are obtained from GPS static net surveying with an accuracy of ±5 cm on the ground. Five tests of the three-linear array imagery were performed by Zhao and Li (2006). The results are listed in Table 5.3.

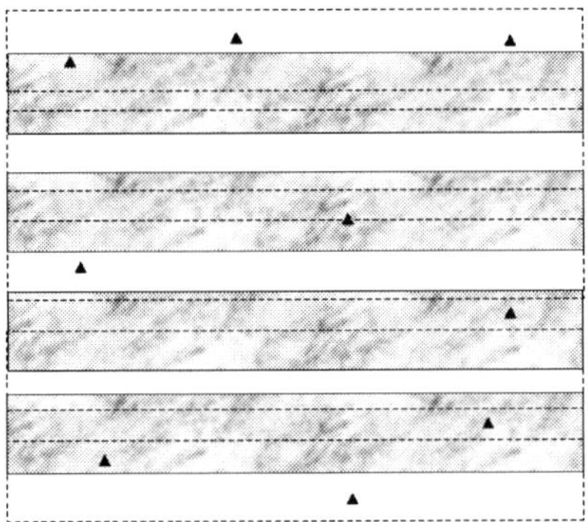

Fig. 5.11 GCP distribution of test area

Table 5.3 Location results of three-linear imagery of ADS 40

location methods	check points	maximum residuals of check points (m)			empirical accuracy (m)			
		X	Y	Z	X	Y	XY	Z
direct georeferencing	9	0.875	1.028	1.912	0.542	0.633	0.834	1.010
bundle block adjustment without GCPs	9	0.299	0.158	0.825	0.213	0.086	0.229	0.640
bundle block adjustment with one GCP	8	0.296	0.142	0.825	0.194	0.091	0.214	0.632
bundle block adjustment with four corner GCPs (three-linear array)	5	0.190	0.194	0.226	0.153	0.129	0.200	0.122
bundle block adjustment with four corner GCPs (twin-linear array)	5	0.254	0.301	0.289	0.197	0.211	0.288	0.140

The following conclusion can be drawn by analyzing the results in Table 5.3:

1) Using the exterior orientation elements of each image obtained via POS to implement multi-ray space intersection, the locational accuracy is assessed by the 9 check points according to Eq. (5.34). The accuracy is better than ±4 pixels for planimetry and about ±5 pixels for elevation, respectively. Obviously, the locational accuracy is not very good.

2) Using GPS exposure station as control condition, the bundle block adjustment was implemented without any GCPs. Both the planimetric accuracy and height accuracy is obviously improved. Comparing with direct georeferencing, the planimetric accuracy increase by 72.5%=(0.8338-0.2294)/0.8338, and the height accuracy by 36.6%=(1.0097-0.6398)/1.0097.

3) When one center GCP is selected and used as orientation point in the bundle block adjustment, the accuracy is close to that of bundle block adjustment without any GCPs.

4) When four GCPs distributed in four corners are selected and used as the orientation points in the bundle block adjustment, the planimetric accuracy is better than ±1 pixel and the height accuracy is ±0.6 pixels. It is obvious that the POS systematic errors can be compensated for effectively.

5) Considering the imaging geometry of airborne three-linear camera, both imagery of forward view and back view are used to implement the bundle block adjustment separately. Comparing to three-linear array case, the height accuracy is similar while the planimetric accuracy decreases by 44.3%=(0.2884-0.1998)/0.1998. This result illustrates that the height accuracy of the object points depends on the forward view and back view images. However, geometric stability of three-linear array is stronger than that of twin-linear array.

5.5.2 Positioning Based on High Resolution Satellite Imagery

5.5.2.1 Experiment Design

The empirical imagery are image pairs of HRG-1A level SPOT-5 across-track stereo scenes in Qianxi, China. Scene 1 was acquired on October 19th, 2004 and the viewing angle is 26.73°. Scene 2 was acquired on October 21st, 2004 and the viewing angle is -24.04°. The geometric condition of intersection is excellent with a base-height ratio of 0.9. Both scenes are only processed to enhance the radiometric quality. The size is 12 000 pixels by 12 000 pixels and the ground sampling distance is about 6 m. The terrain of the covering area is flat and the maximum height difference is within 120 m. In this area, 55 ground characteristic points (Fig. 5.12) are collected and the 3-D coordinates of them are obtained by GPS-supported block bundle adjustment with aerial images at 1:60 000 scale. The calculation was performed on the basis of WuCAPS system. The densification accuracy was ±2.5 m for planimetry and ±2.0 m for elevation. These points were used as orientation points and check points for assessing the geometric locational accuracy of the SPOT-5 scenes.

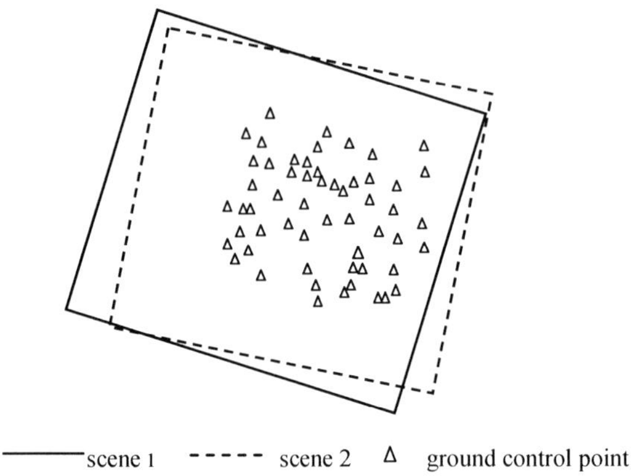

Fig. 5.12 GCP distribution of test area

Table 5.4 Accuracy of positioning using rigorous geometric model for SPOT-5 imagery

GCPs	check points	maximum residuals of check points (m)				empirical accuracy (m)			
		X	*Y*	*XY*	*Z*	*X*	*Y*	*XY*	*Z*
0	55	14.72	15.15	16.62	21.58	8.82	5.55	10.42	9.46
4	51	8.38	14.51	14.81	17.58	3.77	4.45	5.83	5.37
6	49	7.63	16.05	16.06	17.44	2.82	4.63	5.29	5.37
8	47	7.66	13.13	13.13	15.85	2.81	4.31	5.14	4.48
10	45	7.28	10.88	10.89	13.96	2.78	3.49	4.47	3.69
12	43	7.16	9.40	9.43	13.39	2.84	3.17	4.26	3.49
14	41	8.10	9.42	9.43	13.42	2.98	3.18	4.35	3.50

Table 5.5 Accuracy of positioning using rational function model for SPOT-5 imagery

GCPs	check points	maximum residuals of check points (m)				empirical accuracy (m)			
		X	*Y*	*XY*	*Z*	*X*	*Y*	*XY*	*Z*
0	55	14.20	15.21	16.61	22.49	8.37	4.86	9.68	10.21
4	51	8.49	11.87	11.88	14.64	3.81	4.91	5.81	5.02
6	49	8.24	11.50	11.50	15.05	3.84	4.78	5.73	5.33
8	47	8.65	9.30	9.36	14.89	2.96	4.06	5.26	4.84
10	45	8.34	8.46	9.73	14.19	2.98	3.08	4.29	3.63
12	43	8.18	8.18	8.50	13.16	3.07	3.10	4.36	3.70
14	41	8.23	8.85	9.18	13.20	3.04	3.05	4.31	3.79

5.5.2.2 Position-Determination with Rigorous Geometric Models and RFMs

For above SPOT-5 stereo scenes, the tests with rigorous geometric models and rational function models were performed. The test includes direct georeferencing without any GCPs and indirect methods after scene reorientation with different quantity of GCPs. The results are listed in Tables 5.4 and 5.5. The following conclusion can be drawn from Tables 5.4 and 5.5:

1) Employing direct positioning with rigorous geometric processing models or rational function models, the planimetric accuracy is ±10.42 m and ±9.68 m, respectively, about ±1.7 pixels. According to the statistic results of CNES, the direct positioning accuracy of SPOT-5 without considering the relief should be ±10 pixels. The empirical accuracy has outperformed the expected result.

2) When applying the bundle adjustment based on rigorous geometric processing models, the exterior orientation elements are expressed by second-order polynomials, with 36 unknowns to solve for. Using GCPs in four corners of the region, the planimetric and vertical accuracies of indirect positioning reach ±5.83 m and ±5.37 m, about ±1 pixel. The accuracy can be better when more GCPs are used, although the improvement may be very limited. When the number of GCPs exceeds 10, the positional accuracy stabilizes at the level of ±5 m for planimetry and ±4 m for elevation.

3) When applying the bundle adjustment based on rational function models, the planimetric accuracy is increased notably and reached the level of ±1 pixel with four corner GCPs. Although there are 78 RPCs contained in a rational function model, only 6 affine transformation coefficients instead of the initial RPCs were actually used in the adjustment to correct for the systematic errors. This makes the process much more stable and achieves satisfactory results.

4) In this case, the indirect positioning accuracies achieved by two models are at the same level. However, it is worth noting that the test area belongs to a flat region. For the mountainous regions, the accuracy of rational function models may become deteriorated. Rigorous geometric processing models remain the first choice in high-precision geometric location of linear array push-broom satellite remotely sensed imagery.

5.6 Summary

One of the main tasks, also one of the fundamental problems, in photogrammetry and remote sensing is to obtain three-dimensional coordinates of the ground objects (i.e., to extract the non-semantic information) from multi-overlapping aerial or satellite remotely sensed imagery with space intersection method. Beginning from the basic principles of georeferencing with remotely sensed imagery, this chapter gives a comprehensive introduction to the geometric processing models of the frame central projcction and lincar push-broom imagery, including collinearity equations based on perspective projection and rational function models, and accuracy estimation algorithm of photogrammetric point coordinates. These particular methods addressing the problem of photogrammetry with limited ground control points are emphasized. Finally, to analyze the potential georeferencing accuracy achieved by current remote sensing method, several newly acquired aerial or satellite images have been tested. The following conclusions can be drawn by analyzing the empirical results: i) for aerial frame images, the empirical accuracy of traditional bundle block adjustment is the highest. There is no substantial difference between accuracy of GPS-supported bundle block adjustment with one cross-strip on each end of the adjusted block and POS-supported bundle block adjustment with 4 full GCPs in the four corners of the adjusted block. ii) for airborne three-linear imagery, when 4 full GCPs distributed in the four corners are selected and used as the orientation points in the bundle block adjustment, the planimetric accuracy is better than ±1 pixel and the height accuracy is ±0.6 pixels. iii) for SPOT-5 stereo scenes with 6 m ground sampling distance, employing bundle adjustment based on the rigorous geometric processing models or rational function models with 4 full GCPs in the four corners of the adjusted block, the planimetric and vertical accuracies are about ±1.0 pixel, respectively.

Acknowledgements

This work was supported by the National Key Developing Program for Basic Sciences of China (Grant No. 2006CB701302) and National Natural Science Foundation of China (Grant No. 40771176, 40721001). The empirical data acquisition is supported by the Institute of Remote Sensing Applications in Chinese Academy of Sciences, Zhongfei General Aviation Company, Liaoning Jingwei Surveying & Mapping Technology INC, Siwei Aviation Remote Sensing Co. Ltd., and others. These supports are gratefully acknowledged. Dr. Junpeng Yu, Dr. Jianhong Fu and Dr. Yang Ming participated in aspects of the experiments. Prof. Jingxiong Zhang of Wuhan University helped improve the English. The author would like to express his hearty gratitude for their efforts.

References

Amnon K (2000) Accuracy assessment of automatically derived digital elevation models from SPOT images. Photogrammetric Engineering and Remote Sensing 66(8):1017-1024.

Applanix (2008) POSPac^{TM} Air. http://www.applanix.com/products/pospac_airborne_index.php.

Bäumker M, Heimes FJ (2002) New calibration and computing method for direct georeferencing of image and scanner data using the position and angular data of a Hybrid inertial navigation system. Proceedings of OEEPE Workshop on Integrated Sensor Orientation, Hanover, Germany.

Di K, Ma R, Li R (2003) Rational functions and potential for rigorous sensor model recovery. Photogrammetric Engineering and Remote Sensing 69(1):33-41.

Dowman I, Michalis P (2003) Generic rigorous model for along track stereo satellite sensors. ISPRS Workshop on High Resolution Mapping from Space, Hanover, Germany.

Ebner H, Kornus W, Ohlhof T (1992) A stimulation study on point determination for the MOMS-02/D2 space project using an extended functional model. APRS 29(B4):458-464.

Fraser C, Hanley H (2003) Bias compensation in rational functions for IKONOS satellite imagery. Photogrammetric Engineering and Remote Sensing 69(1):53-58.

Fraser C, Hanley H (2004) Bias-compensated RPCs for sensor orientation of high-resolution satellite imagery. ASPRS Annual Conference, Denver, USA.

Friess P (1991) Aerotriangulation with GPS-methods, experience, exception. Proceedings of Photogrammetric Week, Stuttgart, 43-49.

GB 7930-87 (1998) 1:500 1:1,000 1:2,000 Topographical maps specifications for aerophotogrammetric office operation. Standard Press of China, Beijing (in Chinese).

GB/T 13990-92 (1993) 1:5,000 1:10,000 Topographical maps specifications for aerophotogrammetric office operation. Standard Press of China, Beijing (in Chinese).

GB 12340-90 (1991) 1:25,000 1:50,000 1:100,000 Topographical maps specifications for aerophotogrammetric office operation. Standard Press of China, Beijing (in Chinese).

Geo-vision (2009) JX-4C digital photogrammetric system. http://www.jx4.com

Grodecki J, Dial G (2001) IKONOS Geometric Accuracy. Proceeding of ISPRS Working Groups on High Resolution Mapping from Space 2001, Hanover, Germany.

Hu Y, Tao V, Croitoru A (2004) Understanding the rational function model: methods and applications. International archives of photogrammetry remote sensing and spatial information sciences 35(4):663-668.

Huang YQ (1998) The method of compute the exterior orientations elements of SPOT image based on ridge estimation. Journal of the Institute of Surveying and Mapping 15(1):25-27.

Jacobsen K, Passini R (2003) Analysis of. SPOT HRS stereo data. ISPRS Workshop on High resolution mapping from Space, Hanover, Germany.

Kornus W, Ebner M, Schoeder M (1999) Photogrammetric block adjustment using MOMS-2P imagery of the three intersecting stereo-strips. ISPRS Workshop on Integrated Sensor Calibration and Orientation, Portland, Maine, USA.

Kratky V (1989) Rigorous photogrammetric processing of SPOT images at CCM Canada. ISPRS Journal of Photogrammetry and Remote Sensing 44(2):53-71.

Lee C, Theiss H., Bethel J. (2000) Rigorous mathematical modeling of airborne pushbroom imaging systems. Photogrammetric Engineering and Remote Sensing 66(4):385-392.

Li DR, Zheng ZB (1992) Analytical photogrammetry. The Publishing House of Surveying and Mapping, Beijing, China (in Chinese).

Li DR, Yuan XX (2002) Error processing and reliability theory. The Publishing House of Wuhan University, Wuhan, China (in Chinese).

Poli D (2002) General model for airborne and spaceborne linear array sensors. International Archives of Photogrammetry and Remote Sensing 34(B1):177-182.

Shao JL, Wang SG, Li DR (2000) Determination of satellite sensor orientation. Journal of Wuhan Technical University of Surveying and Mapping 25(4):329-333 (in Chinese).

Tao CV, Hu Y (2001) A comprehensive study on the rational function model for photogrammetric processing. Photogrammetry Engineering and Remote Sensing 67(12):1347-1357.

Toutin T (2004) Spatiotriangulation with multi-sensor VIR/SAR images. IEEE Transactions on Geoscience and Remote Sensing 42(10):2096-2103.

Westin T (1990) Precision rectification of SPOT imagery. Photogrammetric Engineering and Remote Sensing (56):247-253.

Wang ZZ (1990) Principles of photogrammetry (with remote sensing). Press of Wuhan Technical University of Surveying and Mapping, Wuhan, China.

Yang X (2000) Accuracy of rational function approximation in photogrammetry. Proceeding of ASPRS Annual Convention, Washington., USA.

Yuan XX (2008) A novel method of systematic error compensation for a position and orientation system. Progress in Natural Science 18(8):953-963.

Zhao SM, Li DR (2006) Experimentation of adjustment math model for ADS 40 sensor. Acta Geodetica et Cartographica Sinica 35(4):342-346 (in Chinese).

Zhao SM (2007) Block combined adjustment for airborne three line scanner sensors. Ph D Dissertation, Wuhan University, Wuhan, China (in Chinese).

Chapter 6

MAPPING HIGH-RESOLUTION LAND SURFACE RADIATIVE FLUXES FROM MODIS: ALGORITHMS AND PRELIMINARY VALIDATION RESULTS

Shunlin Liang, Kaicun Wang, Wenhui Wang, Dongdong Wang, Sheng Gui, Xiaotong Zhang, Jeremy Mirmelstein, Xiufang Zhu, Hye-yun Kim, Juan Du, Steven Running, John Townshend, Si-Chee Tsay, Robert Wolf, Crystal Schaaf, Alan Strahler

6.1 Introduction

Land surface radiative fluxes are needed to address a variety of scientific and application issues related to climate changes, hydrologic and biogeophysical modeling, solar energy applications, and agriculture. The Earth's surface radiation budget (SRB) is the key quantity that determines global climate and climate change from elevated greenhouse gases, air pollution (Wang K. *et al.* 2009), and land cover and land use changes (Wang *et al.* 2007b). The SRB is also important to life and to the use of clean renewable solar energy to improve the quality of the environment.

Altering surface radiation force will lead to a significant adjustment in surface temperature, moisture, and fluxes during the consequent complex land surface thermodynamic and hydrological processes. It affects the surface heat and moisture budget as well as biological productivity. The observed reduction in land surface radiation over the last several decades (1960-1990), the so-called "dimming effect," and the more recent evidence of a reversal in "dimming" over some locations beyond 1990 suggest several consequences on climate, notably on the hydrological cycle (Liepert and Romanou 2005, Wild *et al.* 2005, 2007). Such a reduction in radiation should imply reduced surface temperature and precipitation. Overestimation of the incoming solar radiation over land has major impacts on the climate over land (Betts *et al.* 1996, Dickinson 1995, Garratt *et al.* 1993). Viterbo and Beljaars (1995) found that excessive net radiation at the surface forced excessive surface evaporation, and dried out the soil moisture during data assimilation in the ECMWF (European Centre for Medium-Range Weather Forecasts) global model.

D. Li et al. (eds.), *Geospatial Technology for Earth Observation*,
DOI 10.1007/978-1-4419-0050-0_6,

There are several global radiative flux data sets derived from either satellite observations or GCM (General Circulation Model) reanalysis. The CERES (Clouds and the Earth's Radiant Energy System) Science Team (Wielicki *et al.* 1998) uses cloud and aerosol information from MODIS (Moderate-Resolution Imaging Spectroradiometer) and top-of-atmosphere (TOA) broadband fluxes as a constraint to produce radiative fluxes at the spatial resolution of 140 km. The ISCCP (International Satellite Cloud Climatology Project) has produced a new 18-year (1983-2000) global radiative flux data product called ISCCP FD, every three hours on a 280-km equal-area global grid (Zhang *et al.* 2004). ISCCP FD is calculated using a radiative transfer model from the Goddard Institute for Space Studies (GISS) GCM with the atmosphere and surface properties primarily from the TIROS Operational Vertical Sounding (TOVS) data. The GEWEX SRB Release 2 has a spatial resolution of 1° x 1° and temporal resolution from 3-hour data, daily to monthly from July 1983 to 2007 (Pinker *et al.* 2003,http://www.gewex.org/ssg-21/2009-SSG_GRP-status.pdf). The reanalysis datasets are usually of coarse spatial resolutions (>1°) and fine temporal resolutions, such as those from the NASA DAO (Data Assimilation Office), ECMWF (ERA-40) and NCEP/NCAR (National Centers for Environmental Prediction/ National Center for Atmospheric Research.

Since almost all of these products focus on radiative fluxes at different atmospheric profiles from surface to the TOA, the surface radiative fluxes may not have the accuracy required for land applications. For example, the CERES team uses the predefined albedo and emissivity maps to calculate surface radiative fluxes, which cannot account for their dramatic variations. More importantly, the spatial resolutions of current surface radiative flux products are too coarse to be used for many land applications. For example, the MODIS science team currently has to use the DAO coarse-resolution reanalysis solar radiation product as the forcing data to produce the 1-km PSN/NPP (Photosynthesis/Net Primary Production) product (MOD17) (Zhao *et al.* 2006). The resulting accuracy is not satisfactory and 1-km incident PAR data are a critical need. The hydrological product (Mu *et al.* 2007, Cleugh *et al.* 2007) from MODIS data (MOD16) is also of 1-km resolution, but the meteorological forcing data currently available are quite coarse. As Berg *et al.* (2005) pointed out, bias to many of the reanalysis fields limits their use for hydrological modeling.

One relevant issue is that for coarse spatial resolutions, existence of unresolved sub pixel clouds are crucial. It is actually extremely difficult to quantify the error bounds and uncertainties of the current coarse-resolution SRB products because of the inevitable mismatches with ground "point" measurements in both space and time for validation. The current products have large uncertainties. On a daily basis, the estimate of the solar incoming shortwave radiation from GOES (Geostationary Operational Environmental Satellite) data has an uncertainty of approximately 10%, but at shorter time increments, for example hourly, the uncertainty is much greater (on the order of 20-30%), especially for partly cloudy conditions. The SRB errors for shortwave and longwave over snow and ice surfaces and for longwave in persistently cloudy regions are larger than those in other regions (Pinker *et al.*

2003). Raschke *et al.* (2006) examined ISCCP and GEWEX (Global Energy and Water Cycle Experiment)-SRB data sets and found that there are serious errors in ancillary data that lead to large uncertainty in the final products and they recommended complete reprocessing. Hicke *et al.* (2005) found that mean global NCEP solar radiation exceeded that from GISS by 16%, likely due to lower cloudiness within the NCEP reanalysis compared to satellite observations. Locally, relative differences were up to 40% in the mean and 10% in the trend of solar radiation and NPP, and varied in sign across the globe.

Because of the improved spatial and spectral resolutions of the MODIS solar and IR imager and an available suite of high-level land and atmosphere standard products, we anticipate a significant improvement of radiative flux estimates from the kilometer-scale MODIS observations. They will be extremely useful in driving high-resolution land ecosystem/carbon cycle and hydrological models, and validating the coarse resolution SRB products for improved water and energy fluxes. However, the EOS (Earth Observing System) standard products from MODIS include only certain components of SRB (e.g., broadband albedo, spectral emissivity and skin temperature).

In this chapter, we describe the algorithms for generating all land surface radiative fluxes from MODIS data, including incident insolation and PAR (Photosynthetically-Active Radiation), shortwave net radiation, clear-sky longwave downward and upward radiation, and all-wave all-sky net radiation. The validation results are also presented. The overview of these algorithms is shown in Figure 1, and the individual algorithms are discussed below in details. Measurements collected at about 100 sites globally are being used to validate the algorithms described here. These measurements are collected from different networks, including SURFRAD (Surface Radiation Budget Network), ARM (Atmospheric Radiation Measurement), FLUXNET (Ameriflux, Euroflux and Asiaflux), GAME ANN (GEWEX Asian Monsoon Experiment, in which GEWEX stands for Global Energy and Water cycle Experiment, Asian Automatic Weather Station Network) and AERONET (AErosol RObotic NETwork). The land cover types of these sites include desert, cropland, grassland, and forest. The locations of the sites vary from tropical to temperate and polar areas, with elevations ranging from near sea level to about 5 km high on the Tibetan Plateau.

6.2 All-Sky Incident PAR and Insolation

6.2.1 Background

Incident solar radiation, either PAR in the visible spectrum (400-700 nm) or insolation in the total shortwave (300-4000 nm), is a key variable required by almost all land surface models. Many ecosystem models calculate biomass accumulation linearly proportional to incident PAR. Information on the spatial and temporal distribution of PAR, by control of the photosynthesis and transpiration process, is required for modeling the hydrological cycle and for estimating global oceanic and terrestrial net NPP.

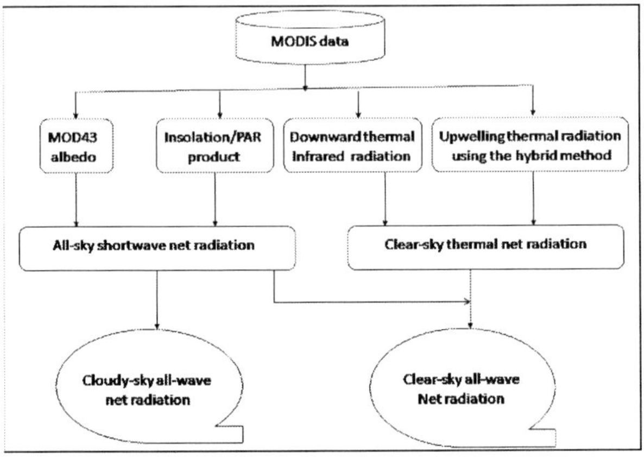

Fig. 6.1 Flowchart of the estimation of net radiation flux from MODIS

The only practical means of obtaining incident PAR/insolation at spatial and temporal resolutions appropriate for most modeling applications is through remote sensing. The existing products normally have high temporal resolutions (around 3 hours) but coarse spatial resolutions (around 1°). They are primarily suitable for climate modeling and analysis.

Because the high-resolution incident PAR/insolation over land surfaces is not a standard EOS product, the MODIS science team has to disaggregate the NASA data assimilation PAR product of three-hourly 1° by 1.5° spatial resolution to produce 1-km NPP and net PSN products using the BIOME-BGC model (Running *et al.* 1999, Running *et al.* 2004). The simplified PAR flux models have been used to produce the incident PAR flux over oceans using MODIS data (Carder *et al.* 1999) and SeaWiFS (Sea-viewing Wide Field-of-view Sensor) data (Frouin *et al.* 2000). No PAR product has been routinely generated from MODIS data over land by NASA. The incident PAR product constitutes a major uncertainty in carbon cycle modeling.

Satellite TOA observations contain information from both the atmosphere and the land surface. The current methods (Carder *et al.* 1999, Frouin *et al.* 2000, Gu *et al.* 2004, Pinker and Laszlo 1992, Pinker *et al.* 2003) assume that either the atmospheric information is available from other sources (e.g., the ISCCP PAR product with atmospheric climatology data as input) or water surface reflectances are known (e.g., SeaWiFS and MODIS PAR products over oceans).

If the atmospheric parameters and surface spectral albedos are known, the existing algorithms for MODIS (e.g., van Laake and Sanchez-Azofeifa 2004, 2005) may be directly applied. However, these models are based on rudimentary two-stream approximations for multiple scattering or even simpler schemes. Computation is speeded up significantly with the simplified models, but they usually are not accurate in calculating multiple scattering that dominates when the atmospheric optical depth is large (hazy or cloudy) and/or the surface is very bright (snow/ice). Simple models are easy to implement, and suitable for cases where no accurate input parameters are available (e.g., processing GOES data alone). Increasing the model complexity increases difficulties of implementation for regional and global applications. Furthermore, the MODIS aerosol optical depth products have much coarser spatial resolution (~10km) and contain many gaps over land surfaces (Remer *et al.* 2005, 2008). Therefore, the method of van Laake and Sanchez-Azofeifa is not suitable for estimating incident PAR from MODIS data at 1-km resolution.

Another issue is the separation of direct and diffuse PAR radiation. The volume of shade within vegetation canopies is reduced by more than an order of magnitude on cloudy and/or very hazy days compared to clear, sunny days because of an increase in the diffuse fraction of the solar radiance. In a recent study, Gu *et al.* (2002) found that: (1) diffuse radiation results in higher light use efficiencies by plant canopies; (2) diffuse radiation has much less tendency to cause canopy photosynthetic saturation; (3) the advantages of diffuse radiation over direct radiation increase with radiation level; and (4) temperature as well as vapor pressure deficit can cause different responses in diffuse and direct canopy photosynthesis, indicating that their impacts on terrestrial ecosystem carbon assimilation may depend on radiation regimes and thus sky conditions. Wang K. *et al.* (2008b) provided observational evidence that diffuse radiation results in not only higher light use efficiencies but also higher evaporative fractions. These findings call for different treatments of diffuse and direct radiation in models of global primary production, and studies of the roles of clouds and aerosols in the global carbon cycle (Gu *et al.* 2003). In fact, many land surface process models, such as SiB2 (Simple Biosphere Model, Sellers *et al.* 1996) and CLM (Community Land Model, Dai *et al.* 2003), separate direct and diffuse solar radiation. However, none of the existing PAR products separate these two components.

A new algorithm for estimating incident PAR and insolation from MODIS data using the look-up table (LUT) approach has been developed (Liang *et al.* 2006). It has been revised to map incident PAR over China (Liu *et al.* 2008). This algorithm has also been extended to GOES data (Zheng *et al.* 2008, Wang D. *et al.* 2009b) and is being altered for other satellite data (Liang *et al.* 2007). Although this algorithm was primarily designed for mapping incident PAR, insolation is one of the outputs. We have further refined the algorithms specifically for PAR and insolation, and the details are given below.

6.2.2 All-sky PAR

6.2.2.1 Instantaneous PAR

Incident PAR depends mainly on atmospheric properties, but also to a lesser extent on surface reflectance. It can be demonstrated by the following formula for calculating downward spectral flux $F(\mu_0)$ at the solar zenith angle θ_0 ($\mu_0 = \cos(\theta_0)$) over a Lambertian surface (Chandrasekhar 1960, Liang, 2004):

$$F(\mu_0) = F_0(\mu_0) + \frac{r_s \overline{\rho}}{1 - r_s \overline{\rho}} \cdot E_0 \cdot \mu_0 \cdot \gamma(\mu_0) \tag{6.1}$$

where $F_0(\mu_0)$ is the downward flux without any contributions from the surface, r_s is surface reflectance, $F(\mu_0)$ spherical albedo of the atmosphere, E_0 is the extraterrestrial solar irradiance, $\gamma(\mu_0)$ is the total transmittance (direct plus diffuse) in the solar illumination direction. It is clear that the surface contribution to the incident PAR mainly depends on $r_s \overline{\rho}$. If the atmosphere is very clear (i.e., $\overline{\rho}$ is small) or the surface reflectance r_s is low, the surface contribution to the incident PAR is relatively small.

All radiometric variables are functions of wavelength. PAR is the integration of the downward spectral fluxes from 400 nm to 700 nm:

$$PAR(\mu_0) = \int_{400}^{700} F_\lambda(\mu_0) d\lambda \tag{6.2}$$

We use the energy unit Wm^{-2} in our LUT. Note that many ecosystem process models typically employ PAR data in quantum units (photosynthetic photon flux density, μ mol m^{-2} s^{-2}) and the conversion between the energy units to the quantum units was discussed by Dye (2004).

Our method first estimates surface reflectance from multi-temporal imagery and then appraises PAR flux for each imagery. The basic procedure is composed of two steps, including (1) determination of the surface reflectance from observations under ‘clearest atmospheric conditions during a temporal window and (2) calculation of incident PAR from the determined surface reflectance and TOA radiance/reflectance using the LUT approach. Determining the length of the temporal window needs to be done carefully. Obviously, it must be short enough so the surface properties do not change dramatically, but long enough to include adequate observations under clear conditions. In all our case studies, it seems a period of one to three months is a reasonable choice.

Most PAR algorithms do not consider surface topography. Evaluating the significance of this omission, Winslow *et al.* (2001) found that the ISCCP-PL PAR product in mountainous regions was underestimated in comparison to long-term radiation climatologies. This is particularly relevant for carbon cycle modeling when the spatial resolution increases since many forests are distributed over mountainous regions. We could add a new dimension to the LUT, but it would make the process much more complicated. We have incorporated topographic correction in our PAR mapping from GOES (Zheng *et al.* 2008), and the new elevation adjustment to the instantaneous PAR from MODIS is presented here.

In the PAR spectral region, surface elevation influences the atmospheric mass that in turn affects Rayleigh scattering transmittance. In creating the LUT to estimate PAR, surface elevation was set at sea level. Ignoring the influence of elevation may introduce substantial errors in PAR retrieval. Since the impacts of aerosol and clouds are considered in the original algorithm, the elevation effect on Rayleigh scattering is considered. To quantify this influence, the normalized transmittance is defined as follow:

$$T_n = \frac{T_{ray,z}}{T_{ray,0}} \tag{6.3}$$

where $T_{ray,z}$ is the Rayleigh scattering transmittance at an elevation of z in km and $T_{ray,0}$ is the Rayleigh scattering transmittance at sea level. We calculate the normalized transmittance using an atmospheric radiative transfer model, 6S (Vermote *et al*. 1997).

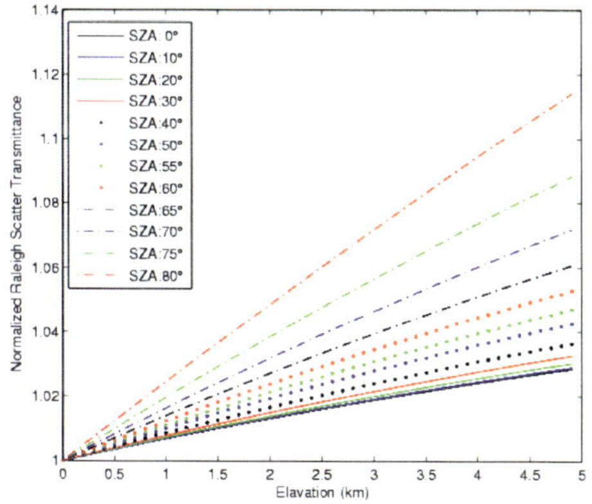

Fig. 6.2 Normalized Rayleigh scattering transmittance (Equation (3)) as a function of surface elevation and solar zenith angle

Figure 6.2 shows that the normalized transmittance is the function of surface elevation and the solar zenith angle. It is shown that PAR can be under-estimated up to 10% for highlands. This effect has not been addressed in the current algorithms to estimate surface solar radiation and PAR. Study has shown that the systematic underestimation of solar radiation products over the Tibetan Plateau (elevation is higher than 4000 m for the most part) perhaps resulted from ignoring the elevation effect (Yang *et al*., 2006). Figure 6.2 illustrates that it is essential to incorporate the effect of surface elevation. However, since including surface elevation into an LUT explicitly as one dimension will substantially increase the computational time in searching for the LUT, we apply an empirical regression to correct for the influence of elevation on Rayleigh scattering:

$$T_n = 1 + \frac{1}{18.56 + 121.55 \cdot \cos(\mu_s)} \cdot z - 0.00035 \cdot z^2 \tag{6.4}$$

where μ_s is the cosine of the solar zenith angle (less than 85°), and z is surface elevation in km. Equation (6.4) predicts the normalized transmittance with a correlation coefficient of 0.9998, a bias of 0.000085 and a standard deviation of 0.00037. Surface elevation data used here are from GTOPO30 at a spatial resolution of 30 arc seconds (http://edc.usgs.gov/products/elevation/gtopo30/gtopo30.html).

Although PAR flux is less dependant on surface reflectance, we need to consider an abrupt significant change of surface reflectance. The extreme case is snowfall and snowmelt. The snow observations have to be labeled and separated from non-snow observations. They must be processed separately but in the same fashion. The MODIS team is routinely producing snow and ice cover maps. These maps can be used to distinguish snow and cloud, which has been an issue in previous studies (e.g., Pinker *et al.*, 2003). Our algorithm separates snow and non-snow observations using the MODIS snow cover maps and applies the same procedures to these two groups of observations separately.

The validation results had been published (Liang *et al.* 2006, 2007). The validation exercises were conducted using ground measurements of incident PAR at six FLUXNET sites (Fort Peck, Montana; Lost Creek, Wisconsin; Oak Ridge, Tennessee; Walker Branch, Tennessee; Santa Rem, Brazil; and Black Hills, South Dakota) from 2002-2004. For each site, a 3km x 3km window (9 pixels) of the MODIS TOA radiance (MOD02) and angular values were extracted from the MODIS datasets ordered from the EOS Gateway. The ground measurements collected every half-hour or one hour were compared with the retrieved values. The measurement values closest to the time of MODIS data acquisition were used to compare with the value of the central pixel in the extracted 9-pixel window without any interpolation.

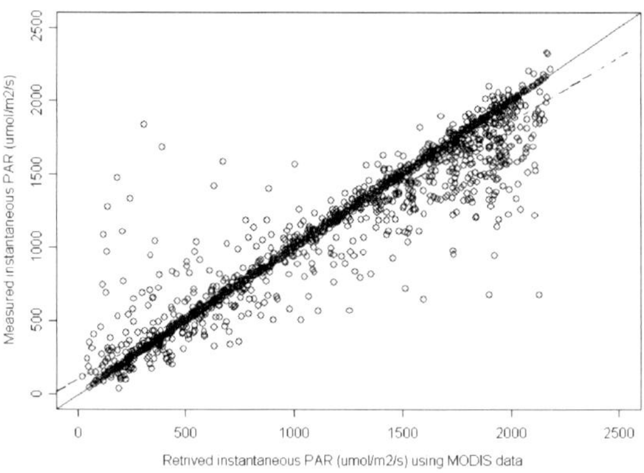

Fig. 6.3 Validation of MODIS-derived instantaneous PAR data at six FLUXNET sites. The solid line is the 1:1 line, the dashed line is the best fit line based on least squares regression.

6.2.2.2 Daily PAR

Daily-integrated PAR is more useful than instantaneous PAR in many cases because many ecosystem models require a daily or coarser temporal resolution. MODIS acquires data from both Terra and Aqua satellites, and the number of overpasses depends on geographical location. Multiple observations are used for calculating daily PAR. A regression method was proposed in our original algorithm (Liang *et al.*, 2006), but we found later that the LUT method performs much better (Wang, D. *et al.*, 2009).

When estimating instantaneous PAR, atmospheric visibility is stored in the LUT as an intermediate quantity. The visibility values at any other times are interpolated based on these recorded visibility values for the instantaneous PAR. If there is only one observation on a given day, the visibility value estimated at the overpass time is used for the entire day. If there are more than one observation per day, the first visibility value is used for the time before the first overpass time and the last visibility value is used for the time after the last overpass time; the visibility value at other times is linearly interpolated:

$$VV_t = VV_{T_n} \cdot (1-p) + VV_{T_{n+1}} \cdot p$$
$$p = \frac{t - T_n}{T_{n+1} - T_n} \tag{6.5}$$

where VV_t is the interpolated visibility value at time t, T_n and T_{n+1} are the two satellite overpass times immediately before and after t, and VV_{T_n} and $VV_{T_{n+1}}$ are the corresponding visibility values estimated from satellite observations.

After obtaining the visibility value at a given time, we can estimate PAR at the moment. To maximize computation efficiency, we calculate the instantaneous PAR values at 30-minute intervals. Daily PAR is obtained by integrating the 30-minute instantaneous PAR values:

$$DailyPAR = \sum_{t=1}^{N} InstPAR(t)\Delta t$$
$$N = (T_{sunset} - T_{sunrise}) / \Delta t + 1 \tag{6.6}$$

where N is the number of instantaneous PAR values used to calculate daily PAR, Δt is the time interval, $InstPAR(t)$ is the instantaneous PAR at time t, calculated with the interpolated visibility value.

The validation results have been presented elsewhere in details (Wang D. *et al.* 2009). Figure 6.4 shows a comparison of field-measured and MODIS-estimated daily-integrated total PAR at ten AmeriFlux stations using the LUT method (Table 6.1), which is better than the sinusoidal method. It demonstrates that MODIS data alone adequately capture the diurnal variations of incident PAR at high latitudes where the number of MODIS overpasses is large. Generally, the estimated values correlate well with field-measured values, and the average R^2 value is 0.927. The average RMSE is 1121.8 kJ/m^2/day.

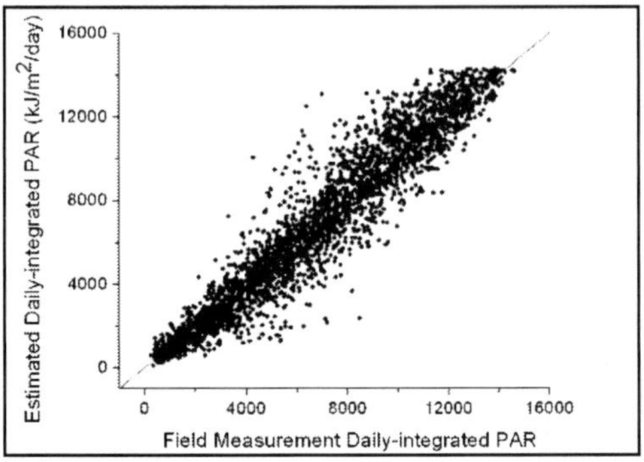

Fig. 6.4 Scatterplots of field-measured and MODIS daily-integrated PAR at 10 AmeriFlux stations

Table 6.1 Comparison of validation results for the adjusted sinusoidal interpolation (SIN) and look-up table (LUT) methods at the ten AmeriFlux stations

Station	R square		RMSE		Bias		Relative Error	
No	SIN	LUT	SIN	LUT	SIN	LUT	SIN	LUT
1	0.887	0.891	1392.8	1090.0	645.8	21.5	14.7%	12.6%
2	0.863	0.873	2125.5	1544.2	1535.3	911.3	26.4%	19.1%
3	0.933	0.937	1138.2	955.8	420.3	-227.6	11.9%	10.8%
4	0.936	0.941	1643.7	1079.9	1125.4	508.9	19.8%	14.2%
5	0.953	0.950	1266.1	863.2	668.6	-45.1	12.2%	10.2%
6	0.952	0.949	1854.5	1234.4	1363.8	719.8	25.3%	18.2%
7	0.927	0.925	1313.6	1252.3	83.2	-560.9	19.8%	19.5%
8	0.942	0.940	1305.0	985.4	541.3	-74.1	19.2%	15.9%
9	0.947	0.943	1137.4	1030.7	279.4	-383.2	15.7%	15.8%
10	0.927	0.925	1319.3	1181.9	291.4	-502.3	19.3%	19.2%
Average	0.927	0.927	1449.6	1121.8	695.4	36.8	18.4%	15.6%

The temporal-scaling approach of the LUT is able to scale both diffuse and direct portions accurately as well. Diffuse PAR is not a standard measurement at AmeriFlux stations, so we cannot perform an extensive validation for diffuse radiation. We chose the Bartlett Experimental Forest, New Hampshire station (44°3'52.7" N, 71°17'17.1" W) to evaluate the performance of our LUT method. An experimental forest of the US Forest Service, the Bartlett Forest is a northern temperate hardwood forest consisting primarily of American beech, red maple, paper birch and hemlock trees. We used the diffuse and total PAR measured in 2005 at Bartlett Forest to validate our algorithm (Figure 6.5). The validation result for total PAR is similar to PAR at the other stations. Diffuse radiation was not as accurately estimated at Bartlett Forest compared to total radiation based on R^2 and bias.

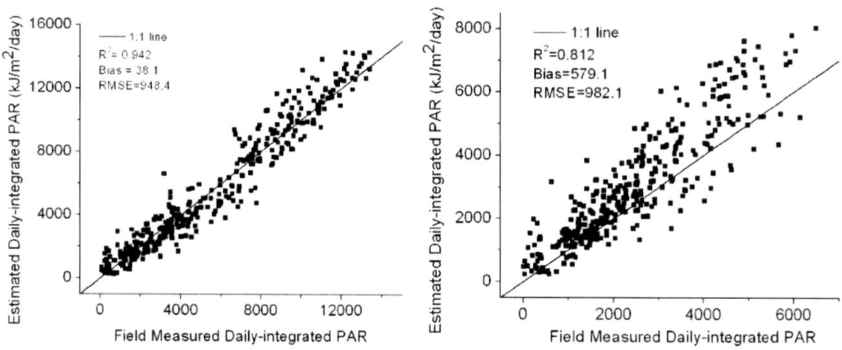

Fig. 6.5 Comparison of field-measured and MODIS-estimated daily-integrated PAR at Bartlett Experimental Forest: total PAR (left), and diffuse PAR (right).

The LUT method has been employed to produce a three-year (2003-2005) daily-integrated PAR product over North America that is being distributed through ORNL DAAC (Oak Ridge National Laboratory Distributed Active Archive Center). Figure 6.6 is the map of daily-integrated PAR compared with GOES PAR for June 3-6, 2003. PAR values are generally high because the solar zenith angle is small and the number of daylight hours is large. Clouds have a strong effect on PAR values. Cloud patterns from MODIS daily-integrated PAR correlate well with GOES patterns. The GOES PAR map is relatively smooth because GOES observes the surface with greater frequency, thus fragmented clouds are usually averaged over time. We calculated the relative difference, D , to better illustrate the difference between MODIS PAR and GOES PAR:

$$D=\frac{2(PAR_{MODIS}-PAR_{GOES})}{(PAR_{MODIS}+PAR_{GOES})} \tag{6.7}$$

Figure 6.6 shows that there is little relative difference between MODIS and GOES for most areas. Although daily-integrated GOES and MODIS PAR correlate well, there are differences between the two datasets. The greatest difference occurs near cloud boundaries because cloud patterns are highly variable near cloud boundaries. A small number of MODIS observations per day cannot accurately record variable daily cloud conditions. If the sky is overcast at the time of a single MODIS overpass, the algorithm classifies the entire day as overcast and daily-integrated PAR is underestimated. Overestimation of daily-integrated PAR occurs when the sky is clear at the time of the MODIS overpass, but cloudy at other times during that day. Figure 6.7 gives an example of the monthly incident PAR over North America in 2003.

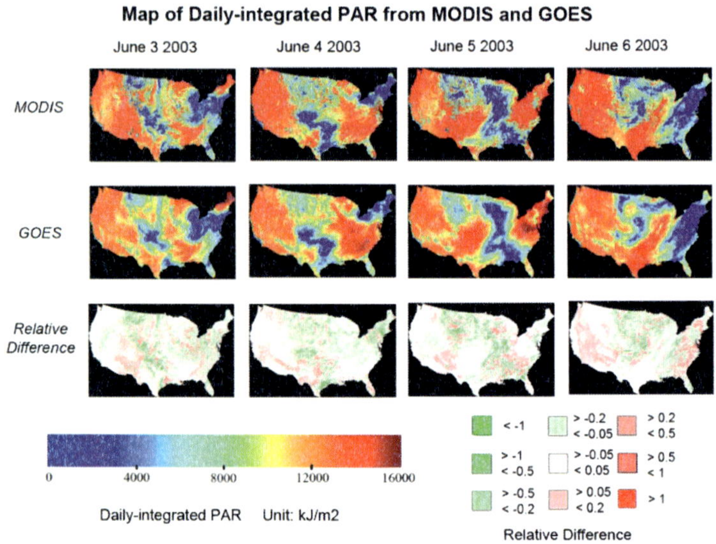

Fig. 6.6 Comparison of MODIS and GOES daily-integrated PAR and relative difference maps of the conterminous United States for June 3-6, 2003.

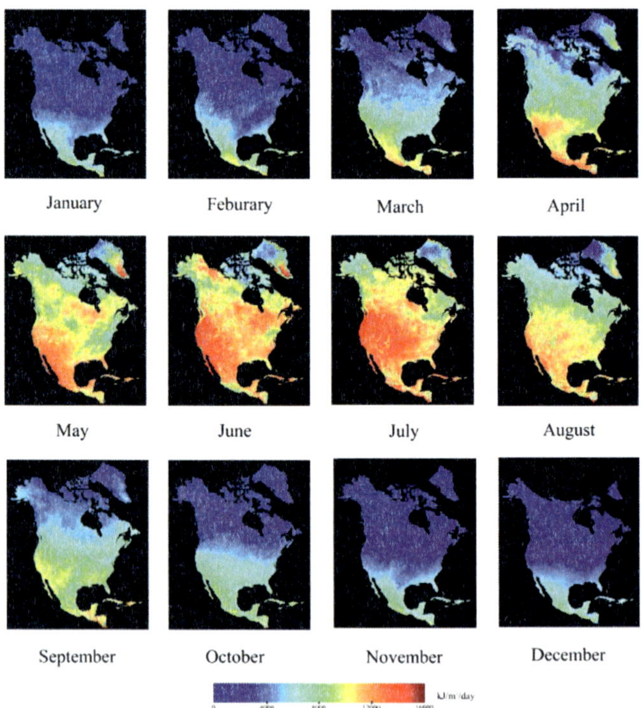

Fig. 6.7 Monthly incident PAR over North America continent in 2003

6.2.3 All-sky Insolation

Insolation is outputted from the LUT in the same way as incident PAR. Here we further consider atmospheric water vapor effects. We could add atmospheric water vapor as another dimension in the LUT, but a simpler parameterization is desirable. The atmospheric water vapor transmittance for insolation has been examined for quite some times, and many parameterizations can be found in the literature, for example,

(1) water vapor transmittance is calculated as (Duchon and Malley 1999):

$$T_W = 1 - 0.077 \cdot (u \cdot m)^{0.3} \tag{6.8}$$

where u is water vapor amount in cm, m is atmospheric mass at surface:

$$m = \frac{35}{\sqrt{1224 \cdot \cos(\theta_S)^2 + 1}} \tag{6.9}$$

where θ_S is solar zenith angle.

(2) water vapor transmittance is calculated as (Annear and Wells 2007):

$$T_W = 1 - \frac{2.4959 \cdot (u \cdot m)}{(1 + 79.034 \cdot (u \cdot m))^{0.6828} + 0.6385 \cdot (u \cdot m)} \tag{6.10}$$

We evaluated the above parameterizations with the atmospheric radiation model 6S (Vermote *et al.* 1997) simulation. Figure 6.8 shows a comparison between the 6S simulated water vapor transmittance and simplified Equation (6.8). It appears equation (6.8) has a systematic overestimation compared with the 6S simulation. Figure 6.9 shows a comparison between the 6S simulated water vapor transmittance and simplified equation (6.10) that appears to have a systematical underestimation.

Because of their significant differences with the 6S simulation results, we have derived a new formula that works much better than the previously mentioned methods:

$$T_W = \sqrt{0.8197 - 0.07066 \cdot \lg(u \cdot m)} \tag{6.11}$$

Figure 6.10 shows a comparison between the 6S simulated water vapor transmittance and simplified equation (6.11). There is no bias, and the RMSE is 0.0036.

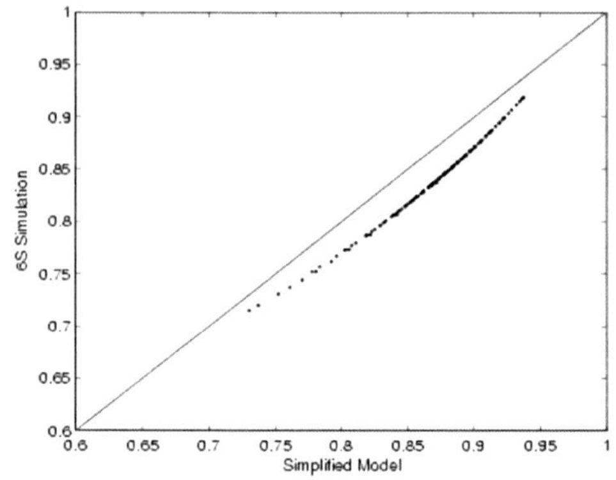

Fig. 6.8 Comparison between the 6S simulated water vapor transmittance and equation (6.7). The bias is 0.0304, and the RMSE is 0.0041.

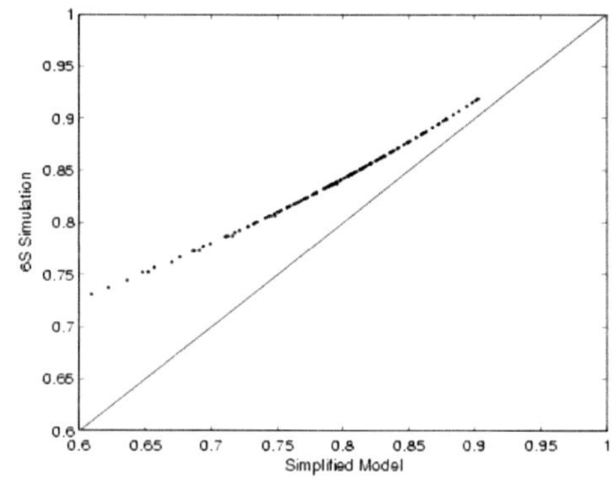

Fig. 6.9 Comparison between the 6S simulated water vapor transmittance and equation (6.9). The bias is -0.0481, and the RMSE is 0.0239.

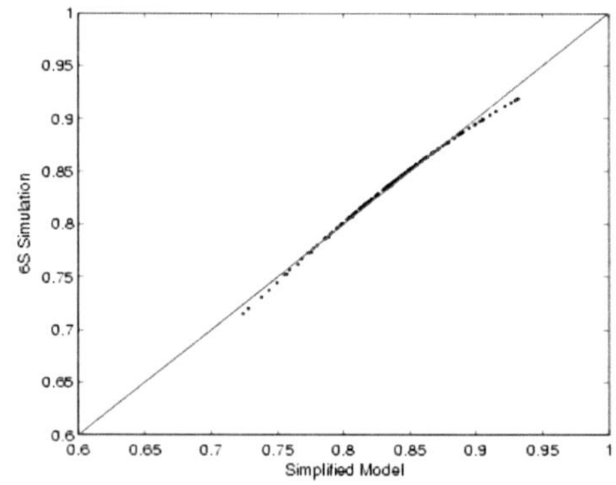

Fig. 6.10 Comparison between the 6S simulated water vapor transmittance and simplified equation (6.11).

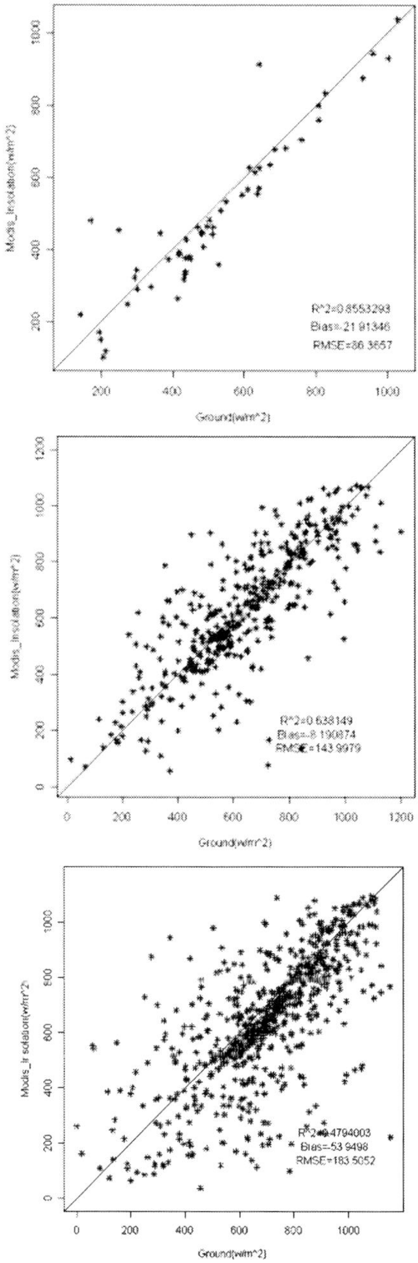

Fig. 6.11 Comparison of MODIS insolation retrieval (y-axis) and ground measurements (x-axis) in 2000 at Dunhuang (top), D66 (middle) and Amdo in 2000-2003 (bottom).

Yang *et al.* (2008) evaluated some state-of-the-art satellite products of incident insolation over the Tibetan Plateau. Their results show that the current satellite products have substantial uncertainty over the Tibetan Plateau area. Here we present our preliminary validation results over the Tibetan Plateau. Ground measurements at three sites were selected: (1) Amdo (32.241N, 91.625E) has a surface elevation of 4700 m and a land surface type of grassland, (2) Dunhuang (40.15N, 94.683E) has a surface elevation of 1140 m and is located in a desert oasis (grassland and cropland), and (3) D66 (35.523N, 93.785E) has a surface elevation of 4600 m and is located in desert. The 30-minute-average insolation data were downloaded from (http://aan.suiri.tsukuba.ac.jp/aan-center.html). We compared the 30-minute measurements with our MODIS retrievals, and the results are shown in Figure 6.11. Table 6.2 summarizes the evaluation results of our MODIS insolation products and other satellite insolation products reported by Yang *et al.* (2008). The GEWEX insolation product clearly exhibits substantial biases, because this product does not account for the elevation effect on the insolation as discussed in Section 6.2.3. Table 6.2 shows that the MODIS insolation product has a much smaller RMSE compared to other satellite insolation products over the Tibetan Plateau.

Table 6.2 Summary of the validation results of current satellite products of hourly incident insolation over the Tibetan Plateau. *: calculated from Table 3 in Yang *et al.* (2008)

Products	Spatial resolution	Temporal resolution	Relative bias	Relative RMSE
GEWEX V2.5*	1°	3 hour	-19.9%	41.4%
GEWEX V2.81*	1°	3 hour	6.5%	33.7%
ISCCP-FD*	2.5°	3 hour	-4.7%	35.4%
University of Maryland -SRB*	0.125°	1 hour	-0.4%	36%
MODIS	1 km	Instantaneous	-4.4%	21.9%

6.2.4 All-sky shortwave net radiation

Surface shortwave net radiation can be calculated from insolation (F_d^s) and upward radiation or the total shortwave broadband albedo (α):

$$S_n = F_d^s - F_u^s = F_d^s(1-\alpha)$$

Assuming we can accurately map insolation, calculation of shortwave net radiation now becomes albedo mapping. Global surface broadband albedo is a standard MODIS product and has been generated routinely from MODIS data every 8 days (MOD43) (Schaaf *et al.*, 2002). We can combine it with the insolation estimate to obtain incident shortwave net radiation.

6.3 Clear-sky Downward Longwave Radiation

Downward longwave radition (F_d^l) is a critical component in energy balance calculations. There have been several comprehensive reviews of methods for estimating the surface thermal radiation (Diak *et al.* 2004, Ellingson 1995, Schmetz 1989). The downward longwave radiation algorithms include three types (Darnell *et al.* 1986, Gupta *et al.* 1992). *The first* is empirical functions using satellite-derived meteorological parameters, for example, the near-surface temperatures and water vapor burden. *The second* is to calculate the radiation quantities with radiative transfer models using satellite-derived soundings. A strong feature of this approach is the validity of the physics. *The third* is to use satellite observed radiances directly to avoid the propagation of retrieval errors of meteorological parameters into the final radiation estimate (Lee and Ellingson 2002). It embeds the physical merits of the radiative transfer within the parameterization of nonlinear functions of observed radiance.

It is straightforward to calculate F_d^l using atmosphere profiles and a radiative transfer model. However, the physical method is sensitive to the errors in the atmosphere profile. F_d^l is dominated by the radiation from a shallow layer that is close to the earth surface. The contribution from the atmosphere above 500 m from the surface only accounts for 16-20% of total F_d^l (Schmetz 1989). The vertical resolution of the MODIS-retrieved atmosphere profiles is too coarse. Only 5 layers are available between 1000 - 800 hPa pressure levels and the detailed structures of the atmosphere cannot be captured (Seemann *et al.* 2003). Our recent study (Wang W. and Liang 2009) showed that F_d^l cannot be estimated with acceptable accuracy using the MODIS atmospheric profile product, especially over high elevation surfaces. For example, we compared F_d^l calculated from the MODIS atmospheric profile product and ground-based measurements at two SURFRAD sites: Sioux Falls (473 m) and Boulder (1689 m). The errors at both sites are larger than 20 W/m^2, and the RMSE at the Boulder site is 37.26 W/m^2.

Consequently, we developed a new hybrid algorithm that combines extensive radiative transfer simulation (physical) and regression analysis (statistical). Validation results indicate that it performs very well (Wang W. and Liang, 2009).

6.3.1 Radiative Transfer Simulation

The MODIS Terra atmosphere product was used to establish the atmosphere profiles database needed in radiative transfer simulation of TIR (Thermal Infrared) TOA radiances and F_d^l to derive statistical relationships. The database consists of more than 8000 representative profiles. Ts, surface pressure, column water vapor, elevation, and Ta (derived from the temperature profile) corresponding to each profile were also included in the database to facilitate method development.

The plant function type for each atmosphere profile was determined using the co-located MODIS land cover product. The Johns Hopkins University (JHU) emissivity spectra in the ASTER (Advanced Spaceborne Thermal Emission and Reflection Radiometer) Spectral Library (ASTER, 1999) were used to approximate spectral emissivity for each plant function type.

MODTRAN4 (MODerate spectral resolution atmospheric TRANSsmittance, Berk *et al.* 1999) was used in the radiative transfer simulation. F_d^l, spectral F_d^l, thermal path radiance, and surface-TOA transmittance were simulated for each atmosphere profile. Thermal path radiance and transmittance were calculated at 0°, 15°, 30°, 45°, and 60° sensor view zenith angles. Surface temperature was assigned using T_s from the MODIS atmosphere product. A Lambertian surface was assumed in the simulation. Spectral surface emissions were calculated using Planck's function and emissivity spectra. MODIS TOA radiances were synthesized using spectral F_d^l, spectral surface emission, emissivity spectra, thermal path radiance, surface-TOA transmittance, and MODIS Terra spectral response function.

6.3.2 Regression Analysis

Previous studies indicate F_d^l can be modeled using a linear combination of thermal-IR TOA radiances (Smith and Wolfe 1983, Morcrette and Deschamps 1986). MODIS channels 27-29 and 31-34 are most valuable for predicting F_d^l. This is consistent with the physics that govern F_d^l: 27 and 29 are water vapor channels; 33 and 34 are air temperature profile channels; 29, 31, and 32 are used for retrieving land surface skin temperature (T_s). Near-surface air temperature (T_a) is one of the dominant factors for clear-sky F_d^l. Ts channels are important for estimating F_d^l because it is closely correlated to T_a. However, the daytime and nighttime relationships between T_a and T_s differ, especially over bare ground and high elevation surfaces. Therefore, separate models were developed to predict F_d^l for daytime and nighttime.

Surface pressure is also an important factor in estimating F_d^l because of the effect of the pressure broadening of the spectral lines (Lee and Ellingson 2002). Surface pressure has not been included in the F_d^l models in previous studies. We used elevation (H) as a surrogate for surface pressure to account for the surface pressure effect. The nonlinear F_d^l model was developed to account for the nonlinear effect (Wang W. and Liang, 2009):

$$F_d^l = L_{Tair}\left(a_0 + a_1 L_{27} + a_2 L_{29} + a_3 L_{33} + a_4 L_{34} + b_1 \frac{L_{32}}{L_{31}} + b_2 \frac{L_{33}}{L_{32}} + b_3 \frac{L_{28}}{L_{31}} + c_1 H \right) \tag{6.12}$$

where L_{Tair} equals L_{31} in the nighttime models and equals L_{32} in the daytime models; a_i, b_i, and c_1 are regression coefficients. The nighttime nonlinear model explains more than 93 % of the variation, with standard errors less than 14.90

(nighttime) and 15.20 (daytime) W/m^2. Table 6.3 shows nonlinear model fitting results and the regression coefficients of the nonlinear models are given in Table 6.4.

Table 6.3 Nonlinear downward longwave radiation model fitting results (Standard error unit W/m^2).

θ	Linear			
	Daytime		Nighttime	
	R^2	Std. Err.	R^2	Std. Err.
0°	0.939	14.44	0.943	13.98
15°	0.938	14.47	0.942	14.01
30°	0.938	14.55	0.943	14.10
45°	0.936	14.74	0.940	14.32
60 °	0.932	15.20	0.936	14.79

Table 6.4 Nonlinear downward longwvae radiation model regression coefficients.

	Daytime					Nighttime				
	0°	15°	30°	45°	60 °	0°	15°	30°	45°	60 °
a0	150.20	153.15	162.14	180.91	214.23	84.14	87.07	95.44	112.65	142.44
a1	4.45	4.34	3.91	3.12	2.13	5.37	5.27	4.90	4.18	3.05
a2	-1.74	-1.80	-1.99	-2.41	-3.28	-1.78	-1.83	-1.99	-2.37	-3.20
a3	-21.03	-20.37	-18.46	-14.02	-3.72	-15.51	-14.87	-13.07	-8.88	0.43
a4	32.22	31.68	30.23	-26.55	16.93	27.08	26.52	25.07	21.51	13.06
b1	-150.87	-154.97	-167.04	-192.69	-239.24	-106.53	-110.08	-119.87	-140.71	-177.34
b2	33.18	34.01	35.64	40.59	53.68	62.67	63.05	63.20	64.90	69.79
b3	-26.81	-25.89	-22.38	-16.07	-6.78	-40.55	-39.73	36.61	-30.99	-21.95
c1	-1.91	-1.91	-1.90	-1.91	-1.99	-1.98	-1.98	-1.97	-1.96	-2.00

6.3.3 Validation Results

The nonlinear F_d^l model was first evaluated using MODIS Terra TOA radiances and co-located SURFRAD ground data. To make the MODIS-derived and ground-measured F_d^l comparable, the model was adapted to predict F_d^l in the spectral range of 4 - 50 μm. The error caused by the spectral range difference between the F_d^l model and ground instruments is less than 0.5%, smaller than ground instrument uncertainty. F_d^l for different view zenith angles was derived using linear interpolation. Clear-sky observations were identified using the MODIS cloud product. We also examined all data points manually to exclude cloud-contaminated pixels with unreasonably low TOA radiance values in MODIS channel 31.

The statistics of validation results using Terra data are summarized in Table 6.5. Figure 6.12 shows the validation plots. RMSEs vary from 14.35 to 20.35 W/m^2 and biases from -6.88 to 9.72 W/m^2. Analysis of the preliminary spatial scaling shows that RMSEs were further reduced by 2 W/m^2 after the nonlinear model-predicted F_d^l was aggregated to 5 km and beyond. Our predicted F_d^l is more accurate than the existing F_d^l products (21-33 W/m^2, monthly) (Wielicki *et al.* 1996, ASDC 2006, Francis and Secora 2004).

Table 6.5 Summary of validation results using MODIS terra and aqua data (unit: W/m2)

Sites	Terra		Aqua	
	Bias	RMSE	Bias	RMSE
Bondville	-9.49	17.75	-7.36	14.98
Sioux Falls	-13.55	17.87	-12.81	17.74
PennState	-7.32	12.72	-5.21	10.92
DesertRock	-21.93	25.03	-20.21	23.94
FortPeck	-8.88	17.46	-7.33	15.11
Boulder	-4.62	19.24	-5.50	18.76
Mean	-10.97	18.35	-9.74	16.91

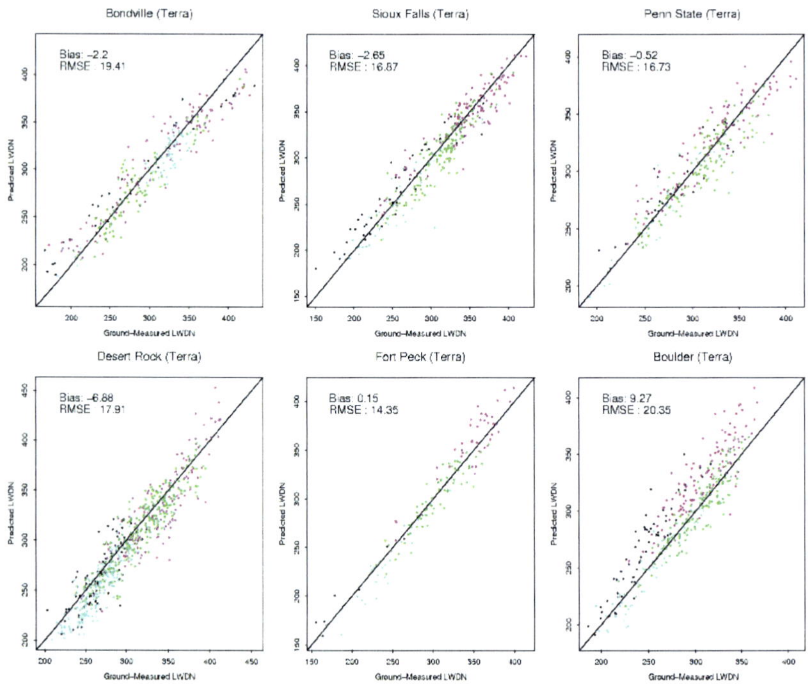

Fig. 6.12 Nonlinear LWDN (downwelling longwave radiation) models validation results using MODIS Terra data (black-fallwinter/day; cyan-fallwinter/night; magenta – springsummer/day; green- spring summer/night).

The spatial mismatch between the MODIS footprint (1 km at nadir) and the ground PIR footprint (~ 200 m) may be a potential cause of the larger scattering during daytime in the model validation results. The Earth's land surface behaves almost as an isothermal and homogeneous surface during nighttime. During daytime, T_s can exceed T_a by more than 20°K (Wan and Dozier 1996). Although MODIS channel 32, rather than channel 31, was used in the daytime nonlinear models, it is still sensitive to T_s.

Cloud contamination may be a significant source of error in this study. The MODIS cloud product cannot mask clouds in all cases, especially for cirrus clouds. Some pixels used in the study may be contaminated by clouds even after manual screening. The TOA radiances of these pixels are mixtures of surface and cloud top contributions and will be lower than true clear-sky values (Wang *et al.* 2007a, Wan 2008). This is especially true at the Desert Rock site. Air traffic out of Los Angeles produces a considerable amount of cirrus over this site.

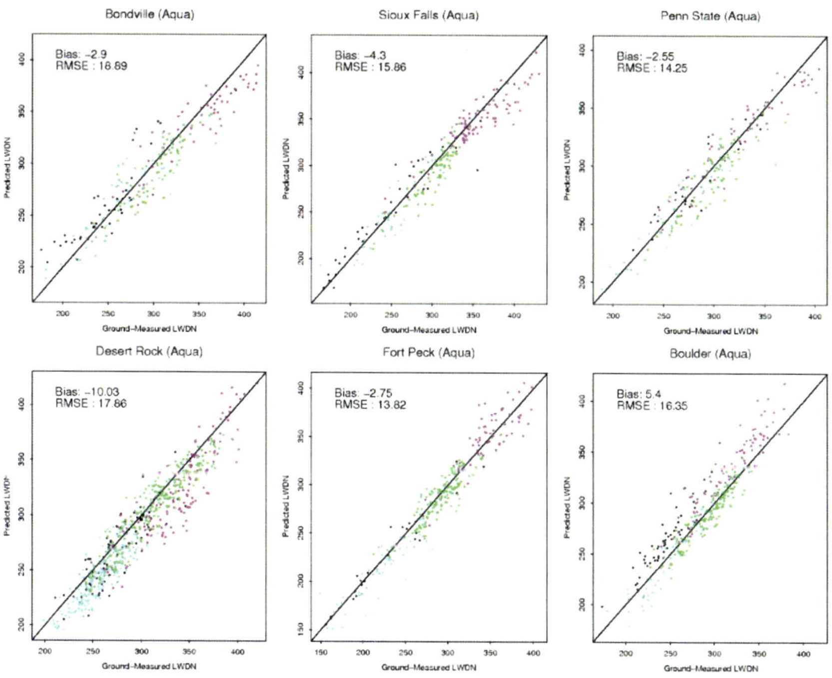

Fig. 6.13 Nonlinear LWDN (downwelling longwave radiation) models validation results using MODIS Aqua data (black: fall-winter/day; cyan: fall-winter/night; magenta: spring-summer/day; green: spring-summer/night).

Although the F_d^l model was developed using MODIS Terra retrieved atmosphere profiles and spectral response functions, it was also applied (without modification) to two years of MODIS Aqua data (2005 and 2006) because the two sensors have similar designs. Table 6.7 also summarizes the validation results using Aqua data. Figure 6.13 shows the validation plots. The nonlinear models' biases range from -10.03 to 5.40 W/m^2 and RMSEs range from 13.82 to 18.89 W/m^2. Terra and Aqua have different satellite overpass times: 10:30 am and 10:30 pm versus 1:30 pm and 1:30 am (local time). The atmosphere and land surface conditions are different when these two sensors overpass. However, the nonlinear model biases and RMSEs are generally smaller for Aqua. Liu *et al.* (2008) show that the Aqua sensor has smaller systematic errors than Terra in thermal channels, which may contribute to the small errors in Aqua-derived F_d^l. The Aqua validation results indicated that the F_d^l model developed in this study is sufficiently general to be applied for both Terra and Aqua observations.

6.4 Clear-Sky Upward Longwave Radiation

The upward longwave radiation emitted in the spectral wavelengths greater than 4 μm is sometimes referred to as "terrestrial radiation" or "infrared radiation". If we know surface skin temperature (T_S) and spectral emissivities (ε_i) that are converted to broadband emissivity (ε), we can easily calculate this quantity:

$$F_u^l = F_d^l(1-\varepsilon) + \sigma\varepsilon T_s^4 \tag{6.13}$$

where F_d^l is downward longwave radiation and σ is the Stefan-Boltzmann's constant. We recently published the formulae for converting the MODIS spectral emissivities to broadband emissivity (Wang *et al.* 2005, Jin and Liang 2006).

MODIS has routinely generated skin temperature products globally. The MODIS land team produces the land surface temperature (MOD11) using both a split-window algorithm and day/night algorithm (Wan *et al.* 2002). The MODIS atmospheric team is also producing surface skin temperature at 5 km resolution (MOD07). We have recently validated the MODIS skin temperature products (MOD7 and MOD11) using FLUXNET sites (Wang W. *et al.* 2008) and found that both products agreed reasonably well with ground measurements over vegetated sites, with biases as large as 3.4° K and root-mean square errors (RMSE) 3.8° K in all eight sites investigated. The uncertainties from both skin temperature and broadband emissivity may result in larger errors in the calculated upward longwave radiation.

Our validation results using ground measurements did not prove that this method is better than the hybrid algorithm discussed in Section 6.3 for downward radiation (Wang W. *et al.* 2008). Therefore, we propose to apply the hybrid algorithm in this study.

In fact, Smith and Woolf (1983) used a hybrid method to estimate both F_u^l and F_d^l from the NOAA geostationary satellite's Visible Infrared Spin Scan Radiometer (VISSR) Atmospheric Sounder (VAS) TOA radiance at the 1000 hPa pressure level. Meerkoetter and Grassl (1984) used the hybrid method to estimate F_u^l and surface net longwave radiation from the Advanced Very High Resolution Radiometer (AVHRR) split-window radiance. However, previous studies have focused on estimating surface longwave radiation over sea surfaces and constant emissivity was usually assumed. We used the hybrid method in this study to derive clear-sky F_u^l models for MODIS data over land surfaces. The emissivity effect is explicitly considered in the radiative transfer simulation process.

The simulated databases were analyzed to develop models for predicting F_u^l from TOA radiance. Multiple regression analysis was first employed to develop a linear F_u^l model. It is found that MODIS bands 29, 31, and 32 predict F_u^l best, consistent with the physics that governs F_u^l. All three bands are sensitive to F_u^l variations. Moreover, these three bands differ in water-vapor absorption. Bands 31 and 32 are used to retrieve column water vapor, an important factor in estimating F_u^l from satellite data. The derived linear model is as follows:

$$F_u^l = a_0 + a_1 L_{29} + a_2 L_{31} + a_3 L_{32} \tag{6.14}$$

where a_0, a_1, a_2, and a_3 are regression coefficients (Table 6), and L_{29}, L_{31}, L_{32} are MODIS channels 29, 31, and 32 TOA radiance. Statistical analysis indicates that the linear model accounts for more than 99% of the variation in the simulated databases, with standard errors of 4.89 W/m^2 (0° sensor view zenith angle) to 6.11 W/m^2 (60° sensor view zenith angle).

We evaluated some non-linear models using both TOA radiance and brightness temperature to reduce the nonlinear effect in the residuals, but the models were not significantly improved.

Table 6.6 Summary of linear model fitting results (θ – sensor viewing zenith angle; unit of standard error: w/m^2)

θ	Linear Regression					
	a_0	a_1	a_2	a_3	R^2	Std. Err
0°	102.7589	10.4963	121.3973	-100.4079	0.993	4.89
15°	104.5829	10.6894	123.4974	-103.0277	0.993	4.94
30°	110.4514	11.4267	129.9471	-111.2339	0.992	5.10
45°	122.3125	13.5455	141.1782	-126.4748	0.991	5.41
60 °	146.0408	20.5749	157.2946	-152.6469	0.990	6.11

The validation results were recently published (Wang *et al.* 2009). Figure 6.14 shows the validation results with MODIS Terra data. The linear model has biases ranging from -4.62 to -21.93 W/m^2 and RMSEs ranging from 12.72 to 25.03

W/m^2. The validation method for the Aqua data is the same as that used for Terra data. It has biases ranging from -5.21 to -20.21 W/m^2 and RMSEs ranging from 10.92 to 23.94 W/m^2.

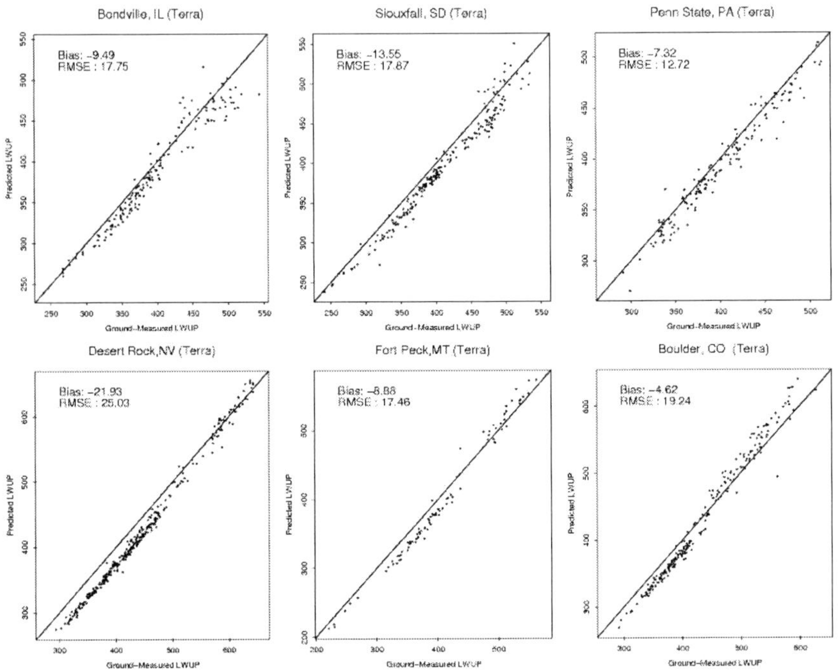

Fig. 6.14 Linear model method validation results using MODIS Terra TOA radiances.

6.5 Clear-Sky Net Longwave Radiation

MODIS-derived surface net longwave radiation validation used the same SURFRAD sites as F_d^l (Wang W. and Liang 2009). Table 6.7 summarizes the validation results. Figure 6.15 illustrates validation results from Terra MODIS data at individual sites. The averaged RMSEs were 17.72 W/m^2 (Terra) and 16.88 W/m^2 (Aqua); the averaged biases are -2.08 W/m^2 (Terra) and 1.99 W/m^2 (Aqua). The RMSEs over all sites are less than 20 W/m^2, the acceptable accuracy for the instantaneous surface longwave radiation budget. Larger scattering exists for daytime observations during spring and summer (high temperature and moisture conditions), similar to the error patterns in F_d^l and surface upward longwave radiation validation results.

Table 6.7 Surface net longwave radaition validation results (unit W/m^2)

Sites	Terra		Aqua	
	Bias	RMSE	Bias	RMSE
Bondville	3.97	19.12	7.67	19.74
Sioux Falls	-2.43	17.08	-1.00	15.60
PennState	1.37	18.92	6.19	18.87
DesertRock	-3.88	18.40	-0.09	16.90
FortPeck	-1.79	13.44	4.70	14.44
Boulder	-9.69	19.38	-5.51	15.75
Mean	-2.08	17.72	1.99	16.88

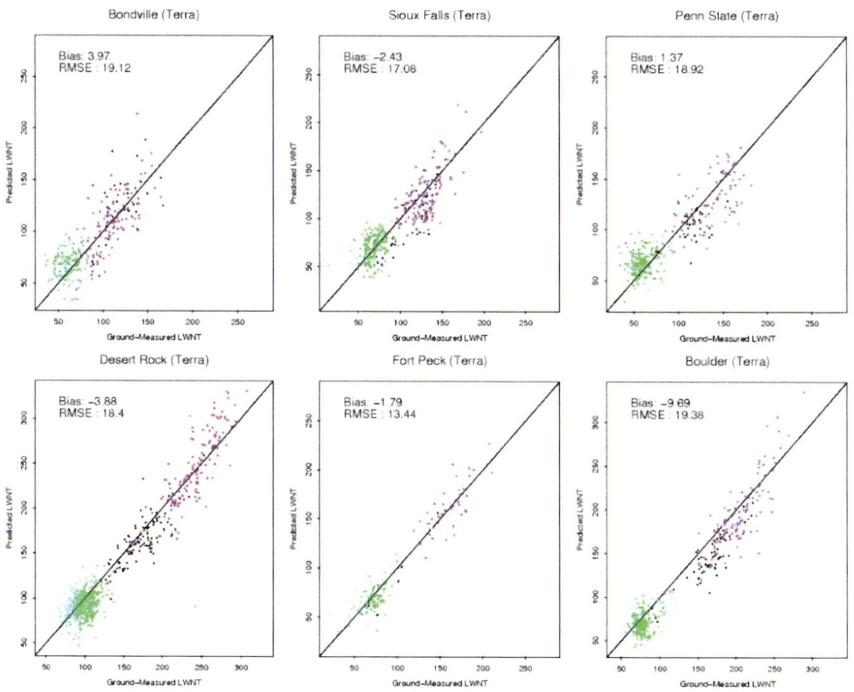

Fig. 6.15 MODIS Terra-derived surface net longwave radiation validation results.

Figure 6.16 compares the MODIS net longwave radiation using the algorithms described above and the corresponding CERES product over the Washington DC and Baltimore Metropolitan region. The MODIS product reveals much more spatial details than the CERES product that has only several values due to its coarse spatial resolution. Note the CERES mean value is also about 23 W/m^2 smaller.

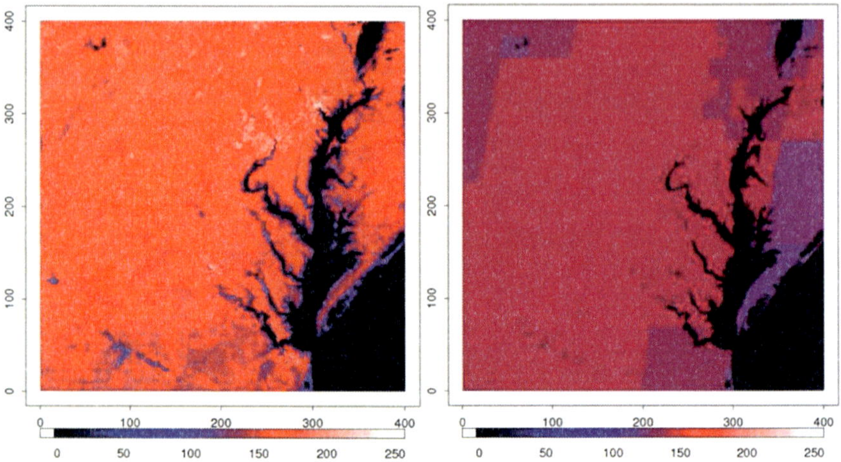

Fig. 6.16 MODIS-derived from Aqua satellite (left) vs. CERES-derived(right) instantaneous clear-sky longwave net radiation (W/m2) images over the Washington D.C. - Baltimore Metropolitan Area (April 10, 2007 18:10 UTC)

6.6 Cloudy-Sky Net Radiation

Sections 6.4 – 6.5 discussed the algorithms for mapping downward and upward longwave radiation under clear-sky conditions. For cloudy-sky conditions, there are many algorithms that have been proposed in the literature (e.g., Diak *et al.* 2004, Zhou and Cess 2001), but we lack confidence that these algorithms are sufficiently mature to generate the products from MODIS during the course of this study. Two empirical formulae have been developed recently.

6.6.1 Instantaneous Value Estimation

An empirical regression method has been developed to estimate instantaneous all-wave net radiation from shortwave net radiation under FLUXNET data under cloudy conditions (Kim, 2008). Cloud-sky measurements at 13 FLUXNET sites were filtered out and their relationship is shown in Figure 6.17. It is highly linear (R_n=0.8347S_n+20.1898), with R^2 = 0.9469 and RMSE =35.1 Wm^{-2}, but the scatters are also significant.

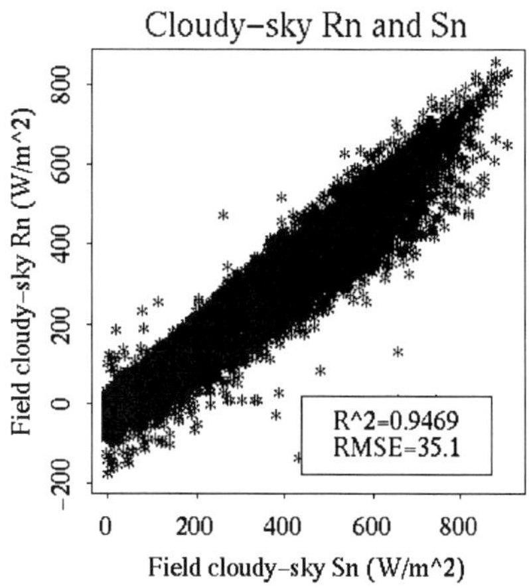

Fig. 6.17 Measured all-wave net radiation and the shortwave net radiation under cloudy-sky conditions

To improve the relationship, we group data in different cover types and add the enhanced vegetation index (EVI) as a new independent variable. The resulting formula is expressed as follows:

$$R_n = a_0 + a_1 EVI + a_2 S_n + a_3 S_n \cdot EVI \tag{6.15}$$

The coefficients for different land cover types are listed in Table 6.8. The estimates are appreciably improved for all cover types except grasslands (see Figure 6.18) because EVI in grass is usually low and the exposed soil and dead grass could increase the outgoing thermal radiation and decrease accuracy.

Table 6.8 Regression coefficients used to estimate cloudy-sky all-wave net radiation

PFT	a_0	a_1	a_2	a_3	R^2	RMSE
Broadleaf crop	*18.57*	*8.05*	*0.76*	*0.19*	*0.9729*	*27.4*
Evergreen needleleaf	*11.84*	*-6.19*	*0.86*	*0.03*	*0.9736*	*28.9*
Grass	*-35.46*	*36.39*	*0.66*	*0.46*	*0.9339*	*43.1*
Deciduous broadleaf	*-14.40*	*-15.32*	*0.74*	*0.24*	*0.9605*	*31.9*

6.6.2 Daily Value Estimation

Kjaersgaard *et al.* (2007) compared six existing models for calculating daytime Rn from solar shortwave radiation using meteorological data at two temperate sites, and concluded that local calibration of the models with at least five years of data is essential to obtain stable calibration coefficients. However, all six methods failed to consider surface characteristics (such as vegetation cover fraction) that have substantial effects on surface energy partitioning into latent and sensible heat fluxes (Wang *et al.* 2007c, Wang and Liang 2008a) which in turn, affect T_s and F_u^l (Wang *et al.* 2006, Wang *et al.* 2007b).

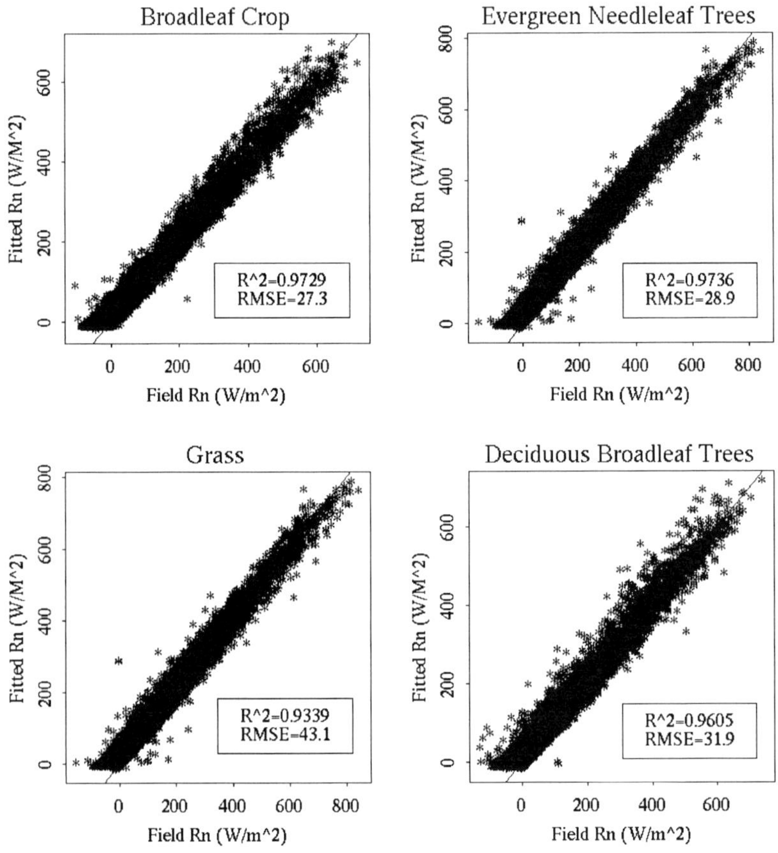

Fig. 6.18 Cloudy-sky all-wave net radiation fitting dependant on plant functional types

Therefore, based on analysis of available long-term measurements, we proposed a robust method to estimate R_n from all-sky shortwave radiation (Wang K. and Liang 2009a). The equation is as follows:

$$\left(\frac{R_n}{S_n}\right) = a_0 + a_1 \cdot T_{\min} + a_2 \cdot DTR + a_3 \cdot VI + a_4 \cdot RH \tag{6.16}$$

where T_{min} is daily minimum T_a (or T_s), DTR is daily T_a (or T_s) range, VI is the MODIS global $NDVI$ (Normalized Difference Vegetation Index) or EVI product with a spatial resolution of 1 km and a 16-day temporal resolution, and RH (Relative Humidity) is relative humidity. To incorporate the elevation contribution, the T_{min} used in equation (6.16) is corrected to sea level by decreasing temperature 6.5 °C for each 1-km increase in elevation.

Since the relationships between R_n/S_n and DTaR (Diurnal air Temperature Range), Tamin, VI, and RH are not exactly linear, none of these parameters can individually account for the variance in R_n/S_n although the four parameters in combination estimate the variance in R_n/S_n with greater accuracy. We also performed the regression analysis with the square of the terms in addition to the linear terms but could not improve the results substantially.

In the equation, near surface air temperature and DTaR are input parameters. These parameters are difficult to obtain at satellite pixel scale with reasonable accuracy. Fortunately, the algorithm is not very sensitive to these parameters and they can be obtained from GMAO (Globally Modeling and Assimilation Office , http://gmao.gsfc.nasa.gov/).

The strength of this method is that it accurately estimates R_n for a wide variety of land cover types and climates, and a range of surface elevations, without local calibration.

The validation results are described in a recent paper (Wang K. and Liang, 2009a). The data collected at twenty-four sites of three networks (ARM, SURFRAD and FLUXNET) have been used for validation. We used the Amdo, Bondville, Desert Rock, EF02, EF07, EF12, EF15, EF18, EF19, Fort Peck, Penn State, and Sioux Falls data to derive the coefficients in equation (6.16), and validated the coefficients with the Boulder, Gaize, EF04, EF08, EF09, EF13, EF20, EF22, EF26, EF27, Goodwin Creek and Wind River data. Table 6.9 lists the derived coefficients.

Table 6.9 The fitted parameters in equation (6.16) and the statistics for all 24 sites. Equation (6.16) is used to predict daytime Rn at all 24 sites using the parameters shown in the following columns. The correlation coefficients and RMSE between measured and predicted daytime Rn are given in the last two lines. DTsR: Diurnal land surface Temperature Range; DTaR: Diurnal air Temperature Range

Combinations of parameters	NDVI, Ts min, DTsR, RH	NDVI, Ta min, DTaR, RH	EVI, Ts min, DTsR, RH	EVI, Ta min, DTaR, RH
a0	0.5749	0.5129	0.5842	0.5195
a1	0.0026	0.0025	0.0026	0.0024
a2	-0.0018	0.0000	-0.0018	0.0001
a3	0.1299	0.1401	0.1813	0.1944
a4	0.2053	0.2604	0.2063	0.2651
Correlation coefficient	0.99	0.99	0.99	0.99
RMSE	16.9(6%)	17.6 (7%)	17.0 (6%)	17.8 (7%)

Equation (6.16) and Table 6.9 coefficients accurately estimate R_n for a range of land cover types, surface elevations, and climates without local calibration. The bias varies from -7.8 W m^{-2} to +9.7 W m^{-2} (± 3% in relative value) for different sites. The RMSE varies from 12.8 W m^{-2} to 21 W m^{-2} (from 5% to 9% in relative value) for different sites and an average of 16.9 W m^{-2} (6% in relative value) for all the sites, and the correlation coefficient is about 0.99 for all the sites.

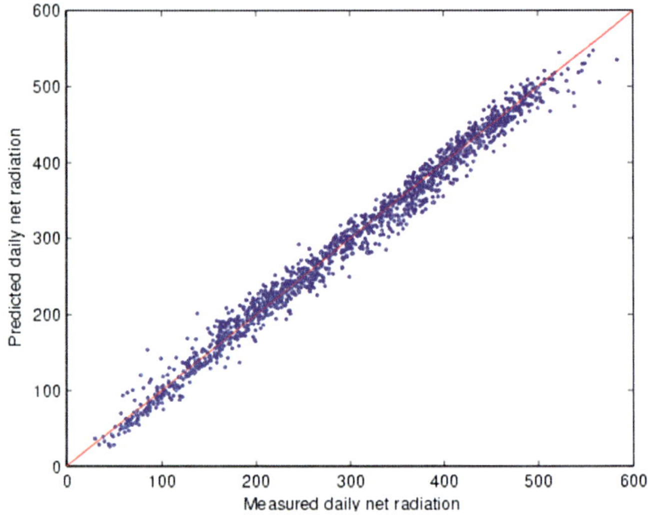

Fig. 6.19 Scatterplot of measured and predicted daytime R_n calculated with Equation (6.16) using daily minimum land surface temperature (Ta min), daily land surface temperature range (DTR), relative humidity (RH), and NDVI using data collected at Pawhuska, Oklahoma (EF12) from 2002 to 2006.

Both EVI and NDVI can be used to estimate R_n. Our previous studies also show that T_s is directly related to F_u^l while the relationship between T_a and F_u^l is indirect (Wang *et al.*, 2005, 2007c, 2008a). Since satellite T_s retrieval is not available under cloudy conditions, we provide an equation using T_a under cloudy conditions.

6.7 All-Wave All-Sky Net Radiation

When we calculate all-wave all-sky net radiation, the errors associated with each individual component may not be a simple addition. To assess the uncertainty of the final product, SURFRAD ground measurements were also used. Figure 6.20 illustrates the preliminary validation results from six SURFRAD sites. It seems the calculated product matches the ground measurements very well. There is a slightly nonlinear pattern, but more extensive validation is still going on.

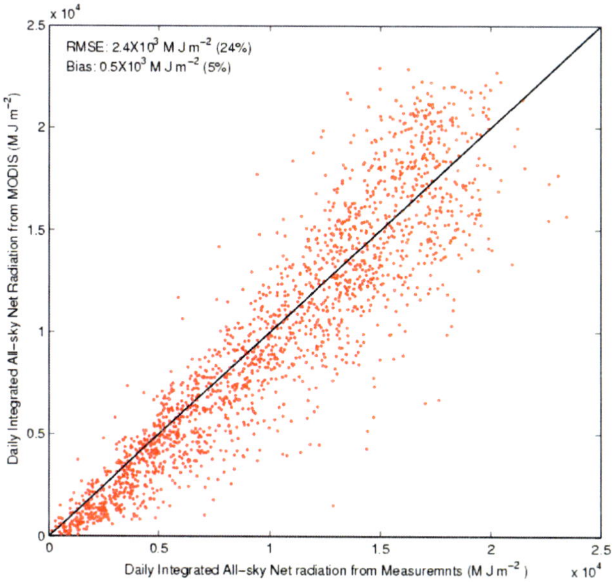

Fig. 6.20 Validation of all-sky all-wave net radiation at six SURFRAD sites in 2005

6.8 Summary

High-resolution land surface radiative fluxes are greatly desired by many land surface models and applications. We have developed a series of algorithms for estimating 1-km daily radiative fluxes from MODIS data, which include incident all-sky PAR insolation and shortwave net radiation, clear-sky upward and downward longwave radiation, cloudy-sky longwave net radiation, and all-wave all-sky net radiation. All of the algorithms have been published. Extensive validation activities have been carried out and the results indicate that these algorithms can be used for generating global high-resolution radiation fluxes accurately. Some of the algorithms have already been used to produce these radiation fluxes on the regional scale. More extensive validation activities will continue.

Those users wishing to characterize land surface biogeophysical and biogeochemical processes at fine spatial resolution should consider using these data sets because they are the only data sets available at such a spatial resolution.

Acknowledgements

This study is supported by NASA under grant NNX08AC53G. Several students have also contributed to data processing and product validation, including Mark Plavnieks, Erica Hafer, and Jennifer Himmelstein.

References

Annear, R. L., and S. A. Wells (2007), A comparison of five models for estimating clear-sky solar radiation, *Water Resource Research*, 43, W10415, doi:10.1029/2006WR005055.

ASDC.(2006), http://eosweb.larc.nasa.gov/HPDOCS/projects/rad_budg.html [visited Dec.22, 2006].

ASTER. (1999), http://speclib.jpl.nasa.gov/ [visited Nov. 30, 2005].

Berg, A.A., J.S. Famiglietti, M. Rodell, R.H. Reichle, U. Jambor, S.L. Holl and P.R. Houser (2005), Development of a hydrometeorological forcing data set for global soil moisture estimation, *International Journal of Climatology*, 25(13): 1697-1714.

Berk, A., Bernstein, L. S., Anderson, G. P., Acharya, P. K., Robertson, D. C., Chetwynd, J. H., & Adler-Golden, S. M. (1998), MODTRAN cloud and multiple scattering upgrades with application to AVIRIS, *Remote Sensing of Environment*, 65, 367– 375.

Betts, A.K., Ball, J.H., Beljaars, A.C.M., Miller, M.J. and Viterbo, P.A., (1996). The land surface-atmosphere interaction: A review based on observational and global modeling perspectives. *Journal of Geophysical Research*, 101(D3): 7209-7225.

Bisht, G., V. Venturini, S. Islama, L. Jiang (2005), Estimation of the net radiation using MODIS (Moderate Resolution Imaging Spectroradiometer) data for clear sky days, *Remote Sensing of Environment*, 97, 52 – 67.

Carder, K., Hawes, S. and Chen, R. (1999), Instantaneous photosynthetically available radiation and absorbed radiation by phytoplankton, Version 5. ATBD-MOD-20.

Chandrasekhar, S. (1960), Radiative Transfer, Dover, Mineola, N. Y.

Charlock, T., (2006), CAVE: Clouds & Radiative Swath (CRS) Footprint Validation, NOAA. Available: http://www-cave.larc.nasa.gov/cave/.

Charlson, R. J., F. P. J.Valero, J. H. Seinfeld (2005), In search of balance. *Science*, 308, 806-807.

Cleugh, H. A., Leuning, R., Mu, Q., & Running, S. W. (2007), Regional evaporation estimates from flux tower and MODIS satellite data. *Remote Sensing of Environment*, 106, 285–304.

Coll, C., E. Valor, R. Niclòs, J. M. Sánchez, R. Rivas, V. Caselles, and J. M. Galve (2005), Ground measurements for the validation of land surface temperatures derived from AATSR and MODIS data, *Remote Sensing of Environment*, 97, 288-300.

Dai, Y. *et al.*, (2003), The Common Land Model (CLM), *Bulletin of American Meteorological Society*, 84(8), 1013.

Darnell, W.L., Gupta, S.K. and Staylor, W.F. (1986), Downward longwave surface radiation from sun-synchronous satellite data: Validation of methodology. *Journal Climate and Applied Meteorology*, 25, 1012-1021.

Diak, G.R. *et al.* (2004), Estimating land surface energy budgets from space - Review and current efforts at the University of Wisconsin-Madison and USDA-ARS. *Bulletin of the American Meteorological Society*, 85(1), 65-75.

Dickinson, R.E. (1995). Land atmosphere interaction. *Reviews of Geophysics*, Suppl.: 917-922.

Duchon, C. E., and O. Malley (1999), Estimating Cloud Type from Pyranometer Observations. *Journal of Applied Meteorology*, 38, 132-141.

Dye, D. (2004), Spectral composition and quanta-to-energy ratio of diffuse photosynthetically active radiation under diverse cloud conditions. *Journal of Geophysical Research*, 109, D10203, doi:10.1029/2003JD004251.

Eck, T. F., and D. G. Dye (1991), Satellite estimation of incident photosynthetically active radiation using ultraviolet reflectance, *Remote Sensing of Environment*, 38, 135– 146.

Ellingson, R.G. (1995), Surface longwave fluxes from satellite observations: A critical review, *Remote Sensing of Environment*, 51, 89-97.

Forsythe, W. C., Rykiel, E. J., Stahl, R. S., Wu, H.-i. and Schoolfield, R. M. (1995), A model comparison for daylength as a function of latitude and day of year. *Ecological Modelling*, 80, 87-95.

Francis, J., and J. Secora. (2004), A 22-year dataset of surface longwave fluxes in the Arctic. *Fourteenth ARM Science Team Meeting Proceedings*, March 22-26, at Albuquerque, NM.

Frouin, R. & Pinker, R. T. (1995), Estimating photosynthetically active radiation (PAR) at the earth's surface from satellite observations. *Remote Sensing of Environment*, 51, 98-107.

Frouin, R., Lingner, D. W., Gautier, C., Baker, K. S., & Smith, R. C. (1989). A simple analytical formula to compute clear sky total and photosynthetically available solar irradiance at the ocean surface. *Journal of Geophysical Research*, 94(C7), 9731– 9742.

Frouin, R., Franz, B. and Wang, M., (2000). Algorithm to estimate PAR from SeaWiFS data, Version 1.0 - Documentation.

Garratt, J.R., Krummel, P. and Kowalczyk, E.A. (1993), The surface energy balance at local and regional scales - A comparison of general circulation model results with observations. *Journal of Climate*, 6, 1090-1109.

Gu, L. *et al.*, (2002). Advantages of diffuse radiation for terrestrial ecosystem productivity. *Journal of Geophysical Research*, 107(D6), 4050, doi:10.1029/2001JD001242.

Gu, L.H. *et al.* (2003), Response of a deciduous forest to the Mount Pinatubo eruption: Enhanced photosynthesis. *Science*, 299: 2035-2038.

Gu, J., Smith, E. A., Cooper, H. J., Grose, A., Liu, G., Merritt, J. D., Waterloo, M. J., de Araújo, A. C., Nobre, A. D., Manzi, A. O., Marengo, J., de Oliveira, P. J., von Randow, C., Norman, J., & Silva Dias, P. (2004), Modeling Carbon Sequestration over the Large-Scale Amazon Basin, Aided by Satellite Observations. Part I: Wet- and Dry-Season Surface Radiation Budget Flux and Precipitation Variability Based on GOES Retrievals. *Journal of Applied Meteorology*, 43(6), 870–886.

Gupta, S.K., Darnell, W.L. and Wilber, A.C. (1992), A parameterization for longwave surface radiation from satellite data: Recent improvements. *Journal of Applied Meteorology*, 31, 1361-1367.

Hicke, J.A. (2005), NCEP and GISS solar radiation data sets available for ecosystem modeling: Description, differences, and impacts on net primary production, *Global Biogeochemical Cycles*, 19, GB2006, doi:10.1029/2004GB002391.

Inamdar, A.K., and V. Ramanathan, Clouds and the Earth's Radiant Energy System (CERES) algorithm theoretical basis document: estimation of longwave surface radiation budget from CERES (subsystem 4.6.2), 1997.

Jin, M., S. Liang (2006), Improved emissivity parametrization for land surface modeling using global remote sensing observations, *Journal of Climate*, 19(12), 2867-2881.

Kaufman, Y. J., D. Tanre´, L. Remer, E. F. Vermote, A. Chu, and B. N. Holben (1997a), Operational remote sensing of tropospheric aerosol over the land from EOS-MODIS., *Journal of Geophysical Research*, 102(D14), 17,051 –17,068.

Kaufman, Y. J., A. Wald, L. A. Lorraine, B. C. Gao, R. R. Li, and L. Flynn (1997b), Remote sensing of aerosol over the continents with the aid of a 2.2 um channel, *IEEE Transactions on Geoscience and Remote Sensing*, 35, 1286– 1298.

Kim, H. Y., (2008), Estimation of Land Surface Shortwave Radiation Budget from MODIS Data, Ph.D dissertation, Department of Geography, University of Maryland

Kjaersgaard, J. H., R. H. Cuenca, F. L. Plauborg, S. Hansen (2007), Long-term comparisons of net radiation calculation schemes. Boundary-Layer Meteorology, 123, 417–431.

Lee, H.T. and Ellingson, R.G. (2000), Development of a nonlinear statistical method for estimating the downward longwave radiation at the surface from satellite observations, *Journal of Atmospheric and Oceanic Technology*, 19(10), 1500- 1515.

Liang, S. (2004), Quantitative Remote Sensing of Land Surfaces, 534 pp., John Wiley, Hoboken, N. J.

Liang, S., T. Zheng, R. Liu, H. Fang, S. C. Tsay, S. Running, (2006), Estimation of incident Photosynthetically Active Radiation from MODIS Data, *Journal of Geophysical Research*, 111, D15208,doi:10.1029/2005JD006730.

Liang, S. T. Zheng, D. Wang, K. Wang, R. Liu, S. C. Tsay, S. Running, J. Townshend, (2007), Mapping High-Resolution Incident Photosynthetically Active Radiation over Land from Polar-Orbiting and Geostationary Satellite Data, *Photogrammetric Engineering and Remote Sensing*, 73(10), 1085-1089.

Liepert, B.G. and Romanou, A. (2005), Global dimming and brightening and the water cycle. *Bulletin of the American Meteorological Society*, 86(5), 622-623.

Liu, R., S. Liang, H. He, J. Liu, and T. Zheng (2008), Mapping Photosynthetically Active Radiation from MODIS Data in China. *Remote Sensing of Environment*, 112, 998-1009.

Meerkoetter, H. and H. Grassl (1984),Longwave net flux at the ground from radiance at the top, presented at *IRS '84 current problems in atmospheric radiation; proceedings of the International Radiation Symposium*, Perugia, Italy, 1984.

Morcrette, J. J., and P. Y. Deschamps. (1986). Downward longwave radiation at the surface in clear sky atmospheres: comparison of measured, satellite-derived and calculated fluxes. *Proc. ISLSCP Conf*, at Rome, ESA SO-248M Darmstadt, Germany.

Mu, Q., F. A. Heinsch, M. Zhao, S. W. Running (2007), Development of a global evapotranspiration algorithm based on MODIS and global meteorology data. *Remote Sensing of Environment*, 111, 519-536.

Pinker, R.T. and Laszlo, I. (1992), Modeling surface solar irradiance for satellite applications on a global scale, *Journal of Applied Meteorology*, 31: 194-211.

Pinker, R. T., *et al.* (2003), Surface radiation budgets in support of the GEWEX Continental-Scale International Project (GCIP) and the GEWEX Americas Prediction Project (GAPP), including the North American Land Data Assimilation System (NLDAS) project, *Journal of Geophysical Research*, 108(D22), 8844, doi:10.1029/2002JD003301.

Prince, S.D. and S.N. Goward (1995), Global primary production: a remote sensing approach, *Journal of Biogeography*, 22(4-5), 815-835.

Raschke, E., Bakan, S. and Kinne, S., 2006. An assessment of radiation budget data provided by the ISCCP and GEWEX-SRB. *Geophysical Research Letters*, 33, L07812, doi:10.1029/2005GL025503.

Remer, L.A. *et al.* (2005) The MODIS aerosol algorithm, products, and validation. *Journal of the Atmospheric Sciences*, 62(4), 947-973.

Remer, L. A., *et al.* (2008), Global aerosol climatology from the MODIS satellite sensors, *Journal of Geophysical Research*, 113, D14S07, doi:10.1029/2007JD009661.

Running, S.W., Nemani, R., Glassy, J.M. and Thornton, P. (1999), MODIS PSN (net photosynthesis) and NPP (net primary productivity) products, Version 3.0. MOD17 PSN/NPP Algorithm Technical Basis Document.

Running, S., Nemani, R.R., Heinsch, F.A., Zhao, M., Reeves, M., & Hashimoto,H. (2004), A continuous satellite-derived measure of Global terrestrial primary production. *Bioscience*, 54, 547–560.

Salomonson, V., W. Barnes, P. Maymon, H. Montgomery, and H. Ostrow (1989), MODIS: advanced facility instrument for studies of the Earth as a system. *IEEE Transactions on Geoscience and Remote Sensing*, 27, 145–153.

Schaaf, C., Gao, F., Strahler, A., Lucht, W., Li, X., Tsung, T., Strugll, N., Zhang, X., Jin, Y., Muller, P., Lewis, P., Barnsley, M., Hobson, P., Disney, M., Roberts, G., Dunderdale, M., Doll, C., d'Entremont, R., Hu, B., Liang, S., Privette, J., & Roy, D. (2002). First operational BRDF, albedo nadir reflectance products from MODIS. *Remote Sensing of Environment, 83*, 135-148.

Schmetz, J., (1989), Towards a surface radiation climatology: retrieval of downward irradiances from satellites, *Atmospheric Research*, 23, 287-321.

Seemann, Suzanne W., Jun Li, W. Paul Menzel, and Liam E. Gumley (2003), Operational retrieval of atmospheric temperature, moisture, and ozone from MODIS infrared radiances, *Journal of Applied Meteorology*, 42(8), 1072-1091.

Sellers, P.J. *et al.* (1996), A revised land surface parameterization (SiB2) for Atmospheric GCMs. Part II: The generation of global fields of terrestrial biophysical parameters from satellite data. *Journal of Climate*, 9, 706-737.

Smith, W. L., and H. M. Wolfe (1983), Geostationary satellite sounder (VAS) observations of longwave radiation flux. *The Satellite Systems to Measure Radiation Budget Parameters and Climate Change Signal*, 29 Aug - 2 Sep, at Igls, Austria.

Van Laake, P. E., & Sanchez-Azofeifa, G. A. (2004). Simplified atmospheric radiative transfer modelling for estimating incident PAR using MODIS atmosphere products. *Remote Sensing Environment*, 91, 98–113.

Van Laake, P. E., & Sanchez-Azofeifa, G. A. (2005). Mapping PAR using MODIS atmosphere products. *Remote Sensing of Environment*, 94(4), 554– 563.

Viterbo, P. and Beljaars, C. (1995), An improved land surface parametrization scheme in the ECMWF model and its validation, *Journal of Climate*, 8, 2716-2748.

Vermote, E. F., Tanré, N. Z., Deuzé, J. L., Herman, M., & Morcette, J. J. (1997), Second simulation of the satellite signal in the solar spectrum: An overview. *IEEE Transactions on Geoscience and Remote Sensing*, 35, 675-686.

Wan, Z. and J. Dozier. (1996), A generalized split-window algorithm for retrieving land-surface temperature from space. *IEEE Transactions on Geoscience and Remote Sensing*, 34(4), 892-905.

Wan, Z. (1999), MODIS land-surface temperature algorithm theoretical basis document (LST ATBD): version 3.3, University of California, Santa Barbara, Santa Barbara, CA, 1999.

Wan, Z., Y.-L. Zhang, Q.-C. Zhang, and Z.-L. Li (2002), Validation of the land surface temperature products retrieved from terra moderate resolution imaging spectroradiometer data, *Remote Sensing of Environment*, 83, 163–180.

Wan, Z., Y. Zhang, Q. Zhang, and Z. L. Li, (2004), Quality assessment and validation of the MODIS global land surface temperature *International Journal Remote Sensing*, 25, 261-274.

Wan, Z. (2008), New refinements and validation of the MODIS land-surface temperature/emissivity products, *Remote Sensing of Environment*, 112, 59-74.

Wang, D., S. Liang and T. Zheng, (2009), Estimation of daily-integrated PAR from sparse satellite observations: comparison of temporal scaling methods, *International Journal of Remote Sensing*, in press.

Wang, K., Z. Wan, P. Wang, M. Sparrow, J. Liu, X. Zhou, and S. Haginoya (2005), Estimation of surface long wave radiation and broadband emissivity using Moderate Resolution Imaging Spectroradiometer (MODIS) land surface temperature/emissivity products, *Journal of Geophysical Research*, 110, D11109, doi:10.1029/2004JD005566.

Wang, K., Z. Li, and M. Cribb (2006), Estimation of evaporative fraction from a combination of day and night land surface temperature and NDVI: A new method to determine the Priestley–Taylor parameter, *Remote Sensing of Environment*, 102, 293-305.

Wang, K., Z. Wan, P. Wang, M. Sparrow, J. Liu, and S. Haginoya (2007a), Evaluation and improvement of the MODIS land surface temperature/emissivity products using ground-based measurements at a semi-desert site on the western Tibetan Plateau. *International Journal Remote Sensing*, 28, 2549 - 2565.

Wang, K., J. Wang, P. Wang, M. Sparrow, J. Yang, and H. Chen, (2007b), Influences of urbanization on surface characteristics as derived from the Moderate-Resolution Imaging Spectroradiometer: A case study for the Beijing metropolitan area. *Journal of Geophysical Research*, 112, D22S06, doi:10.1029/2006JD007997.

Wang, K., P. Wang, Z. Li, M. Cribb, and M. Sparrow (2007c) A simple method to estimate actual evapotranspiration from a combination of net radiation, vegetation index, and temperature, *Journal of Geophysical Research*, 112, D15107, doi:10.1029/2006JD008351.

Wang, K., and S. Liang (2008), An improved method for estimating global evapotranspiration based on satellite determination of surface net radiation, vegetation index, temperature, and soil moisture, *Journal of Hydrometeorology*, 9, 712-727.

Wang, K., R. E. Dickinson, and S. Liang (2008), Observational evidence on the effects of clouds and aerosols on net ecosystem exchange and evapotranspiration, *Geophysical Research Letter*, 35, L10401, doi:10.1029/2008GL034167.

Wang, K., R. E. Dickinson, S. Liang, (2009), Clear Sky Visibility Has Decreased over Land Globally from 1973 to 2007, *Science*, 323, 1468-1470.

Wang, K., S. Liang (2009a), Estimation of daytime net radiation from shortwave radiation measurements and meteorological observations, *Journal of Applied Meteorology and Climatology*, 48:634-643.

Wang, K. and S. Liang, (2009b), Evaluation of ASTER and MODIS land surface temperature and emissivity products using surface longwave radiation observations at SURFRAD sites, *Remote Sensing of Environment*, 113:1156-1165.

Wang, K., S. Liang, T. Zheng and D. Wang (2009). Simultaneous estimation of surface photosynthetically active radiation and albedo from GOES, *Remote Sensing of Environment*, revised.

Wang, W., S. Liang, and T. Meyer, (2008), Validating MODIS land surface temperature products, *Remote Sensing of Environment*, 112, 623-635

Wang, W., S. Liang, J. A. Augustine, (2009), Estimating clear-sky land surface upward longwave radiation from MODIS data, *IEEE Transactions on Geoscience and Remote Sensing*, 47(5):1555-1570, DOI: 10.1109/TGRS.2008.2005206.

Wang, W. & S. Liang, (2009), Estimating High-Spatial Resolution Clear-Sky Land Surface Downwelling Longwave Radiation from MODIS Data, *Remote Sensing of Environment*, 113:745-754.

Wielicki, B.A. *et al.* (1998), Clouds and the Earth's Radiant Energy System (CERES): Algorithm overview. *IEEE Transactions on Geosciences and Remote Sensing*, 36, 1127-1141.

Wild, M. *et al.* (2005), From dimming to brightening: Decadal changes in solar radiation at Earth's surface. *Science*, 308, 847-850.

Wild, M., A. Ohmura, and K. Makowski, (2007), Impact of global dimming and brightening on global warming, *Geophysical Research Letters*, 34, L04702, doi:10.1029/2006GL028031

Winslow, J.C., Hunt, E.R.J. and Piper, S.C. (2001), A globally applicable model of daily solar irradiance estimated from air temperature and precipitation data. *Ecological Modelling*, 143, 227-243.

Yang, K., Koike, T., Stackhouse, P., Mikovitz, C., & Cox, S. J. (2006). An assessment of satellite surface radiation products for highlands with Tibet Instrumental data. *Geophysical Research Letter*, 33, L22403, doi: 10.1029/2006GL027640.

Yang, K., R. T. Pinker, Y. Ma, T. Koike, M. M. Wonsick, S. J. Cox, Y. Zhang, and P. Stackhouse (2008), Evaluation of satellite estimates of downward shortwave radiation over the Tibetan Plateau, *Journal of Geophysical Research*, 113, D17204, doi:10.1029/2007JD009736.

Zhang, Y.C., Rossow, W.B., Lacis, A.A., Oinas, V. and Mishchenko, M.I., (2004), Calculation of radiative fluxes from the surface to top of atmosphere based on ISCCP and other global data sets: Refinements of the radiative transfer model and the input data. *Journal of Geophysical Research*, 109, D19105, doi:10.1029/2003JD004457.

Zhang, Y., W. B. Rossow, and P. W. Stackhouse Jr., (2007), Comparison of different global information sources used in surface radiative flux calculation: Radiative properties of the surface. *Journal of Geophysical Research*, 112, D01102, doi:10.1029/2005JD007008.

Zhao, M., S. Running and R. Nemani, (2006), Sensitivity of Moderate Resolution Imaging Spectroradiometer (MODIS) terrestrial primary production to the accuracy of meteorological reanalyses, *Journal of Geophysical Research*, 111: G01002, doi:10.1029/2004JG000004.

Zheng, T., Liang, S. and Wang, K., (2008), Estimation of incident PAR from GOES imagery. *Journal of Applied Meteorology and Climatology*, 47, 853-868.

Zhong, B., S. Liang, and B. Holben, (2007), Validating a New Algorithm for Estimating Aerosol Optical Depths over Land from MODIS Imagery, *International Journal of Remote Sensing*, 28(18), 4207-4214.

Zhou, Y.P. and Cess, R.D. (2001), Algorithm development strategies for retrieving the downwelling longwave flux at the Earth's surface. *Journal of Geophysical Research*, 106(D12): 12477-12488.

Chapter 7

SPECTRAL INFORMATION CONTENT OF REMOTE SENSING IMAGERY

Rudiger Gens

7.1 Physical Properties of Material and Energy-Matter Interactions

The Earth surface materials show phenomenal diversity in composition, density, grain size, porosity, dielectric constant, surface texture, roughness, orientation, geometry, etc. that in turn control the reflectance, emittance and related physical properties that are particularly significant for remote sensing of Earth's surface.

7.1.1 Physical Properties

Reflectance and related properties: Reflectance and reflectance related properties are in general inconsistently defined and used within the remote sensing community. Directional attributes of reflectance, in particular, are used in loose and unspecified ways in published literature. An excellent summary of reflectance quantities and ambiguities in their usage is presented by Schaepman-Strub *et al.* (2006).

To understand reflectance related measurements, we need to first revisit the concept of energy of a photon or quanta. It is a well accepted fact in the field of electromagnetic radiation studies that energy is not continuous along the wave but concentrated in discrete packets called photon or quanta. *Radiant energy* is the energy of a photon or quanta and is usually measured in joules (J) or kilowatt hour (kWh). *Radiant flux*, also known as power or radiant power, is usually measured in watts (W), which equals joule per second. Radiant energy can be calculated by integrating radiant flux over time.

Radiant flux density is the radiant flux per unit area at a point on a surface, where the surface can be real or imaginary (a mathematical plane). Standard unit of all forms of radiant flux density is watts per square meter (W/m^2). An imaginary surface can, for example, be used to calculate or measure radiant flux density anywhere in space. Depending on the direction there can be two cases:

D. Li et al. (eds.), *Geospatial Technology for Earth Observation*,
DOI 10.1007/978-1-4419-0050-0_7,

total amount of radiant flux incident upon a point on a surface from all directions above the surface is *irradiance*, and the total amount of radiant flux leaving a point on a surface into all directions above the surface is known as *exitance*. In remote sensing, it is common to consider radiation measurements specific to a particular wavelength, i.e. to make the measurements monochromatic instead of spanning over the entire or large parts of the electromagnetic spectrum. Such wavelength dependent measurements are typically denoted with a prefix spectral to the measurement term.

Radiance is by far the most important measurement for quantitative analysis of directional effects of radiation. Radiance is the radiant flux density or the total amount of radiant energy measured at a particular angle and azimuth (particular solid angle). Radiance can be incoming/incident radiance or outgoing radiance, depending on the direction we measure. It is measured in $W/m^2/sr$. A surface where outgoing radiance is uniform in all directions and does not vary with angle is referred to as a Lambertian surface. All real world surfaces are, to varying degrees, non-Lambertian.

Outgoing radiance is assumed to be dependent on the incoming radiance onto the surface. The ratio of the radiant exitance with the irradiance is known as reflectance. Being a ratio, reflectance has no unit and its value ranges from 0 to 1.

The *Bidirectional Reflectance Distribution Function* (BRDF) describes the directional dependence of reflected radiation. It defines the radiance reflected into a specific view direction as a result of the radiant flux incident upon a surface (Roberts 2001). However, the *Bidirectional Reflectance Factor* (BRF) relates the reflectance from a target surface to the reflectance that would be observed from a Lambertian surface located at the target. BRF is unitless and is equal to π BRDF.

Emittance and related properties: The power thermally emitted (given out) by a surface, primarily by virtue of its temperature, is often referred to as the emitting ability or *emittance* of the surface. The amount and the characteristics of electromagnetic radiation emitted by the surface are controlled by two factors: emissivity and temperature.

Emissivity is the emittance of a real material compared to that of an ideal black body (a body that absorbs all energy incident on it, and later emits all energy that it has absorbed) at the same temperature. It depends on the nature of the object's surface. Emissivity denoted by epsilon (ε) is a ratio and varies between 0 and 1. Emissivity has a strong spectral control and directional dependence. If emissivity is independent of wavelength, then the surface is referred to as a gray body.

Temperature (kinetic) is a measure of the amount of heat energy contained in a body. It is measured in different units, such as in Kelvin (K); degrees Centigrade (°C); degrees Fahrenheit (°F).

7.1.2 Electromagnetic Spectrum and Dominant Energy-Matter Interactions

Within the wide range of wavelengths constituting the electromagnetic spectrum, the wavelength ranges from 0.4 μm (i.e. 400 nm) through 14 μm, and from 0.75 cm through 1 m are popularly utilized for terrestrial remote sensing.

Following the concept of total energy, all energy incident upon a surface is either reflected off, transmitted, or absorbed (and later emitted) by the surface. It is indeed this interaction of energy with matter that makes it possible for us to make meaningful measurements for remote sensing. Fig. 7.1 shows the dominant processes operating in different parts of the electromagnetic spectrum.

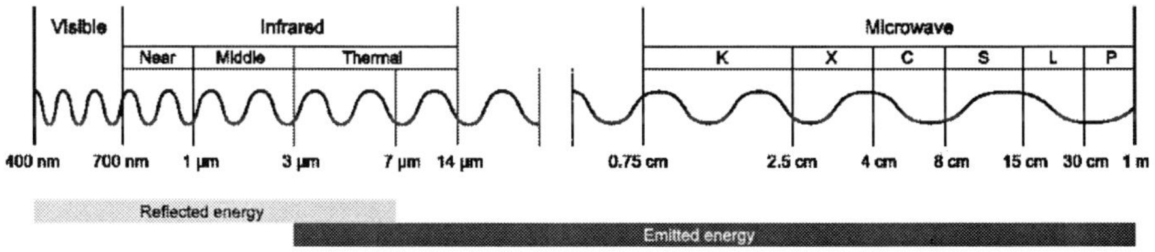

Fig. 7.1 Regions of the electromagnetic spectrum used for remote sensing and dominant processes operating in these regions.

7.1.3 Energy Interaction with Earth Surface Materials

As mentioned earlier the three primary processes when energy interacts with matter are reflection, transmission, and absorption (see Fig. 7.2).

Reflection: In principle, a flat surface can be extremely smooth so that energy incident on the surface is bounced away (reflected) from the surface following the standard laws of reflection. Such reflection is referred to as specular reflection (Fig. 7.3A). On the other hand, a surface can cause the incident energy to bounce off equally in all directions, so that reflected radiation is uniform in all directions. Such a surface that causes reflected energy to have no directional dependence is known as a Lambertian surface (Fig. 7.3B). Natural surfaces are neither completely specular nor Lambertian and cause non-uniform (anisotropic) scattering of incident energy. Such scattering is a result of incident energy being reflected off in different amounts in different directions (Fig. 7.3C).

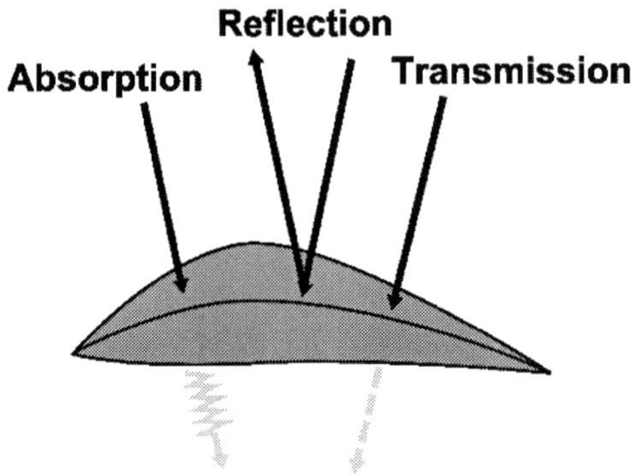

Fig. 7.2 Basic concept of reflection, transmission and absorption

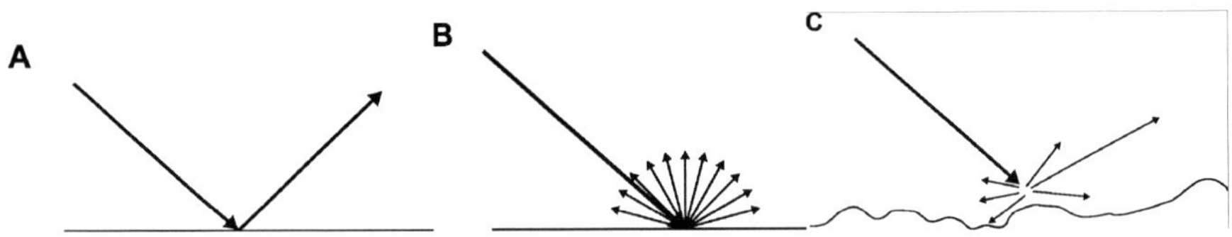

Fig. 7.3 Specular (A), Lambertian (B) and non-uniform (anisotropic) (C) scattering of electromagnetic signals

The process of scattering is also dependent on the relationship between the wavelength of the incoming radiation and the wavelength range of the target material.

In microwave remote sensing, for example, when the wavelength of the incoming signal is the same order of magnitude, or much smaller than the wavelength of the target material, surface scattering is dominant. However, as the wavelength of the incoming signal becomes longer, the signal can penetrate the target/medium, and volume scattering starts coming to play. In volume scattering, the radar signal undergoes multiple scattering events, before it finally emerges out of the medium and is captured by the sensor. This is very common in vegetation canopies (Fig. 7.4). In microwave remote sensing, volume scattering is also a function of moisture content of the medium, as the moisture content changes the electrical properties (specifically the dielectric constant) of a medium, which in turn controls the amount of incoming signal reflected, absorbed and transmitted by the medium. In principle, volume scattering decreases rapidly and surface scattering takes prominence with increasing moisture content.

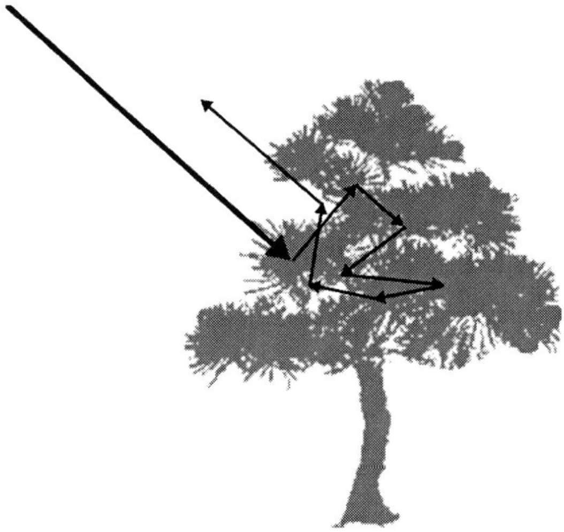

Fig. 7.4 Volume scattering of active microwave signals

One related term, commonly used in microwave remote sensing, is *backscatter*. Backscatter received by the sensor for a particular pixel is the sum of responses of all individual scatterers on the ground within the respective resolution cell.

Transmission: When incident energy interacts with matter, a part of it can transmit (or pass through) material, following the basic laws of refraction. This process, called transmission, is also dependent on the relationship between the wavelength of the incoming radiation and the wavelength range of the target material, and the transmissivity of the target material. Remote sensing is typically carried out in wavelength ranges where the intervening atmosphere between the energy source and target is most transmissive.

Absorption: Absorption is inversely proportional to transmission. All Earth surface material absorbs electromagnetic radiation to some extent. Clear water, for example, absorbs most of the energy incident on it. The absorbed energy raises the temperature of the target material and the warm target later emits the absorbed energy, albeit at a longer wavelength. This process of emission from natural surfaces of the Earth is most dominant in the thermal infrared portion of the electromagnetic spectrum. Emitted energy can also be dominant in shorter wavelength regions when target temperatures are very high, such as for forest fires, volcanic eruptions, and coal mine fires. Advances in sensor technology, now also enable us to record the small amount of emitted energy in the microwave region, if the surface over which we make the integrated measurement is sufficiently large.

7.2 Remote Sensing Techniques

Remote sensing techniques can be classified based the regions of the electromagnetic spectrum, the number of spectral bands, the spectral band widths, based on the source of energy (Table 7.1).

Table 7.1 Remote sensing data sources with their respective characteristics.

Band		Passive				Active	
		Visible	Near- and shortwave infrared	Thermal infrared	Microwave	LIDAR (near-infrared)	SAR
Broad	Panchromatic	•					
	Broadband			•			
Narrow	Multispectral	•	•	•			
	Hyperspectral	•	•	•			
Single frequency					•	•	•

When the remote sensing systems depend on natural sources of energy, such as the incoming radiation from the sun or the emitted radiation from the Earth's surface, remote sensing is classified as *passive remote sensing*. When energy pulses required for remote sensing are generated by the sensor itself, then remote sensing is called active remote sensing.

7.2.1 Passive Remote Sensing

Traditional remote sensing was all passive remote sensing. The remote sensing systems, almost always, relied on the sun as the source of energy. Remote sensing was, therefore, restricted to day time. Earliest images were black-and-white (grey scale) and acquired only in the visible part of the spectrum in one broad panchromatic band. With advancement in camera systems, multi-band color images could be acquired and soon after near-infrared imaging capability was added. This took passive remote sensing from panchromatic to multispectral (many spectral bands/wavelength ranges) mode. Modern day sensors can image in a much wider span of the electromagnetic spectrum, in multi-spectral mode as well as hyperspectral mode. Multispectral sensors operate in spectral bandwidths that are in the order of about 100 nm wavelength range. Hyperspectral sensors typically have many more very narrow spectral bands that are an order of magnitude narrower than the multispectral bands.

Multispectral and hyperspectral remote sensing images can span over visible and infrared regions, including the thermal infrared region. However, remote sensing in the thermal infrared region was earlier restricted to single broad band mode as the earlier sensors were not powerful enough to record a signal from the low amount of energy emitted from the Earth's surface. These days, sensor technology has advanced to the extent that if we integrate the signal over a very large area (radius ranging from about 7 to 25 km) we can capture emitted energy from the Earth's surface even in the longer microwave wavelength regions, making passive microwave remote sensing possible.

7.2.2 Active Remote Sensing

The pulse of energy generated for active remote sensing is typically at a specific frequency (and, therefore, specific wavelength).

Light Detection And Ranging (LiDAR) remote sensing operates in the near-infrared wavelength. Though typically such systems provide high quality point data, more recent systems also provide intensity images in the near infrared range of the electromagnetic spectrum. The airborne LiDAR systems acquire high density altimetry data from which digital elevation models are generated (Wehr and Lohr 1999).

Synthetic Aperture Radar (SAR) is another class of active remote sensing where imaging is carried out in the microwave portion of the electromagnetic spectrum (c.f. Fig. 7.1). In principle, in SAR remote sensing, we measure the time and magnitude of the signal backscattered from the ground to the radar antenna. As SAR images can be acquired at any time of the day, and in all weather-conditions, it gives tremendous advantage to these data, especially for applications in geographic regions that have frequent cloud cover issues and in polar regions where there are extended periods of darkness.

Covering the entire application range of passive and active remote sensing is beyond the scope of this chapter. Here we restrict our discussions to extracting spectral information from active and passive remote sensing data for six selected application areas. These include remote sensing investigations of urban areas; vegetation; biodiversity and coral reefs; geology and soils; water bodies; and cryosphere.

7.3 Extracting Spectral Information Using Remote Sensing

The following six sections provide a comprehensive overview of the current state and practices in information extraction for different application areas. The diversity of topics within each broad section, and constant influx of new satellite data and data processing and analysis techniques, makes these sections serve just as examples to demonstrate the tremendous potential of remote sensing images acquired in different parts of the spectrum.

7.3.1 Remote Sensing in Urban Areas

Monitoring urban areas by means of remote sensing has been carried out for decades at various scales. Small (2006) offered three reasons why the studying the urban climate is of particular interest: the ever rising urban population, the partly predictable dynamics, and the potential of influencing it with specific land use patterns.

The characterization of *impervious surfaces* in the urban infrastructure requires imagery with (sub-)meter resolution. Roads have been tracked by satellite data in the optical (Geman and Jedynak 1996) and microwave range (Tupin *et al.* 2002). Traditionally, the primary image source for impervious surface studies has been aerial photography. The automatic extraction of three-dimensional man-made objects is a field of research in itself within the computer vision community (Katartzis and Sahli 2008). Mayer (1999) reviewed automatic building extraction techniques. With the much improved resolution of the data, especially of optical sensors, the use of satellite imagery has significantly increased and with that the challenges of using the best combination of data, tools and techniques for scale of interest (Dell'Acqua and Gamba 2006). The integration of data sets from different sources, e.g. thermal and optical data (Lu and Weng 2006), hyperspectral and thermal data (Jung *et al.* 2008) or hyperspectral and multispectral data (Weng *et al.* 2008), has gained more attention. The discrimination of urban residential and industrial areas as well as the tracking of *urban sprawl* is performed with data at the 5-m to 20-m resolution (Shaban and Dikshit 2002). Urban multispectral land cover characterization and vegetation analysis is typically carried out at 30-m resolution level.

Fig. 7.5 is an example of the use of Landsat multispectral data to visualize urban areas and urban sprawl around Rio de Janeiro, Brazil. Even at the 30 meter spatial resolution of the data, the densification in urban areas and expansion in the urban fringe between 1988 and 2001 is clearly visible. *Urban heat islands* are studied on various scales, from macroscale mapping on the 1-km level (Kassomenos and Katsoulis 2006), through mesoscale modeling (Hamdi and Schayes 2008) all the way to microscale at discrete surface material level (Martilli 2007). Combining the results of the various scales, e.g. the relationship between microscale urban surface thermal upwelling and the mesoscale regional thermal pattern, is still work in progress (Gluch *et al.* 2006). Especially the geometric processing of data, comprehensively reviewed by Toutin (2004), becomes an important issue, particularly with higher resolution imagery.

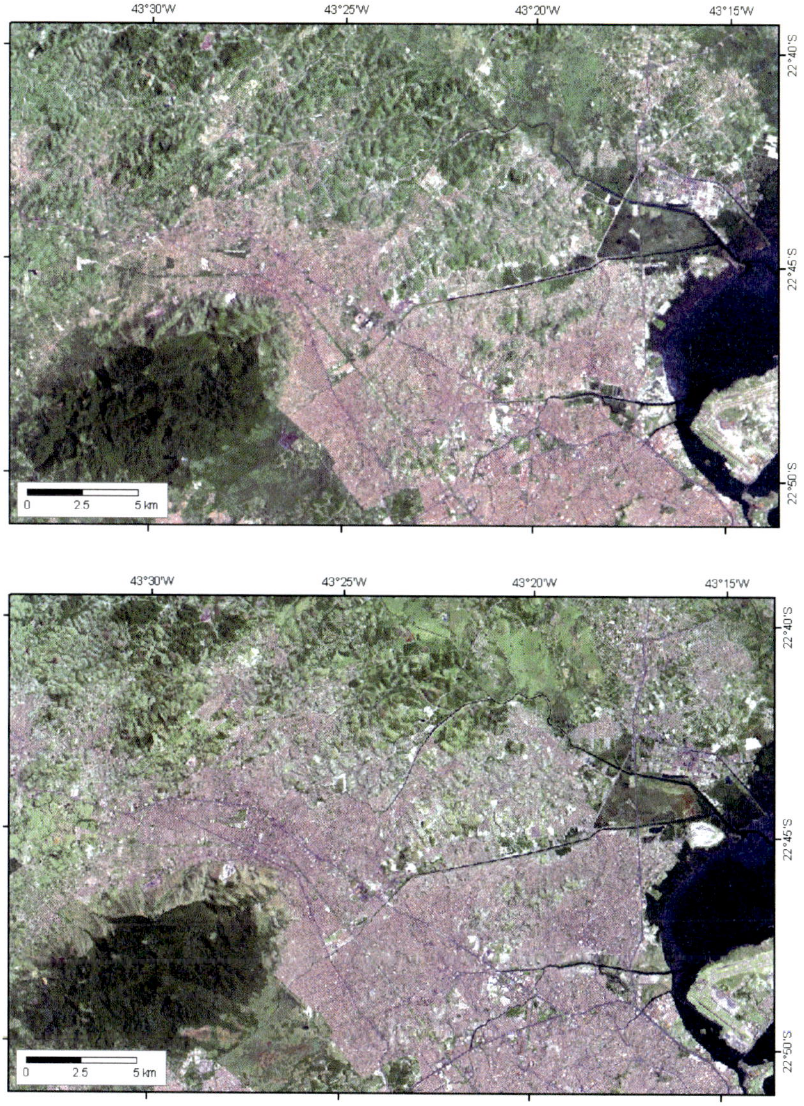

Fig. 7.5 Landsat TM 5 False Color Composites (bands 7, 5 and 3 depicted in red, green and blue, respectively) of parts of Rio de Janeiro, Brazil. (top) Image from 1988 (bottom) Image from 2001. The densification in urban areas and the urban sprawl is evident. Image courtesy of USGS.

Recently the use of SAR data for studies of urban areas has significantly increased, mostly through the development of advanced interferometric permanent scatterer technique that allows the detection and long-term monitoring of small-scale subsidence measurements at sub-centimeter level (Ferretti *et al.* 2000).

Dell'Acqua and Gamba (2006) used multi-scale textural features for urban mapping. Stereoscopic airborne radar data has been used to extract rectangular buildings (Simonetto *et al.* 2005).

7.3.2 Remote Sensing of Vegetation

Mapping the *forest* structure with remote sensing helps with a number of applications such as wildlife habitat analysis (Hyde *et al.* 2006), forest inventories (McRoberts and Tomppo 2007) as well as fire management and fire effects assessment (Chuvieco and Kasischke 2007). However, the forest structure itself has been in the focus of research for many years. Recently, Roberts *et al.* (2007) provided a comprehensive overview of remote sensing techniques for forest structure assessment, reviewing the role of both active and passive sensors at varying spatial scales. Remote sensing data is used to extract attributes such as leaf area index, volume, biomass, basal area, forest age class, tree height, crown diameter and stand density.

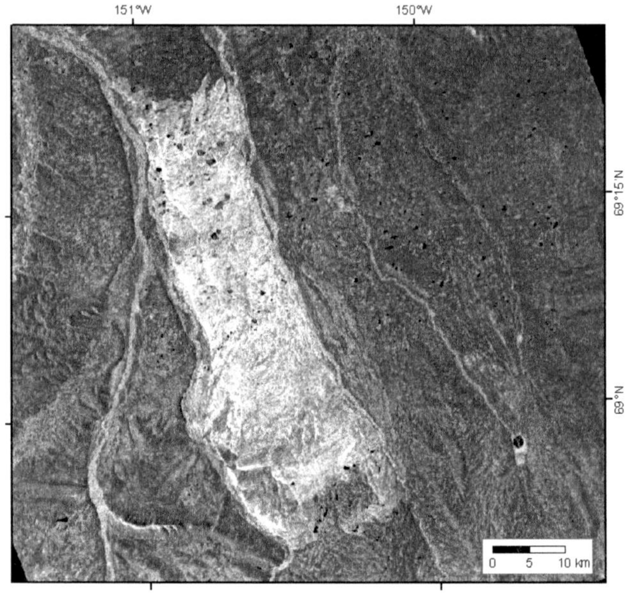

Fig. 7.6 Burn scars of the Anaktuvuk River Fire from July 2007, captured by an ERS-2 SAR image. The re-vegetated burned area shows a significantly higher backscatter than the surrounding areas assisting delineation of fire affected areas. Image © European Space Agency, 2008.

SAR data has been used for forest applications for decades. Fig. 7.6 shows a SAR image of a forested area, where the central part of the image was affected by forest fires. This post fire image is a classic example of the high backscatter signature from re-vegetating burn scars. For forest applications, SAR data with different polarizations has also proved to be particularly useful (Durden *et al.* 1989). Initially, polarimetric data was mainly acquired with multi-frequency airborne systems (Chauhan *et al.* 1991). Later on, shuttle data helped to estimate biophysical characteristics (Dobson *et al.* 1995). The development of polarimetric SAR interferometry (Cloude and Papathanassiou 1998) generated a lot of interest as it allowed to extract forest heights and ground topography based on interferometric coherence using a random volume over ground coherent mixture model (Lee *et al.* 2003).

Arroyo *et al.* (2008) recently reviewed the role of remote sensing for fire models and methods to map fuel types and pointed out that most of the challenges and difficulties in mapping fuel types, originally comprehensively summarized by Keane *et al.* (2001), are still present. However, the advancements in data resolution and analysis techniques have significantly improved the accuracy of the results. Especially, the combination of the various data sources and an object oriented analysis approach has led to promising results (Arroyo *et al.* 2008). Falkowski *et al.* (2005) integrated advanced spaceborne thermal emission and reflection radiometer data with gradient modeling methods for mapping fuel layers. Riaño *et al.* (2007) estimated shrub height for fuel mapping for orthophotos and LiDAR data. Finally, Varga and Asner (2008) combined LiDAR and hyperspectral data to model the fuel load.

Remote sensing can, by measuring spectral reflectance and thermal emittance properties of soils and *crops*, provide valuable information about agronomic and biophysical characteristics. Combined with improved position location technology, information on plant characteristics forms the basis of the various aspects of advanced crop management (water, nutrients and pests) (Pinter *et al.* 2003). The basic research in this field has reached a level of maturity that more crop specific studies become feasible, e.g. Erickson *et al.* 2004, Chakraborty *et al.* 2005, Beeri and Peled 2006, and Abdel-Rahman and Ahmed 2008. Especially, hyperspectral data with its very detailed spectral information content and microwave data with its sensitivity to moisture are the main remote sensing data sources.

7.3.3 Remote Sensing for Biodiversity and Coral Reefs

The use of remote sensing for *biodiversity* studies has been reviewed a number of times in recent years (Gillespie *et al.* 2008, Vierling *et al.* 2008, Duro *et al.* 2007, Gottschalk *et al.* 2005, McDermind *et al.* 2005, Turner *et al.* 2003, and Nagendra 2001). The direct remote sensing techniques map the distribution of species assemblages, whereas indirect approaches use the remote sensing data to model species distribution and their diversity (Gillespie *et al.* 2008). Citing a number of case studies, Vierling *et al.* (2008) showed the use of LiDAR remote sensing for characterizing the three-dimensional habitat structure of terrestrial and aquatic environments in fine detail across broad areas. They concluded that incorporating LiDAR into studies of animal-habitat relationships will help to improve models used for species management and conservation. Duro *et al.* (2007) summarized studies of biodiversity indicators at different scales. They proposed a conceptual approach to map and monitor biodiversity at a country level, with Canada as their example. Especially the issues of inaccessibility of parts of the terrain and scalability can be effectively addressed by incorporating refined remote sensing techniques (Duro *et al.* 2007). McDermind *et al.* (2005) collected examples for habitat mapping at local, regional and national/continental scales. A more specific review of use of remote sensing imagery for habitat analysis and modeling has been given by Gottschalk *et al.* (2005), looking at studies on avian habitat relationships. Nagendra (2001) discussed habitat studies in the context of scale, looking at radiometric, spatial, spectral and temporal resolution of remote sensing data. The advancement of remote sensing with respect to spatial and spectral resolution of sensors significantly increased the feasibility for some types of biodiversity studies (Turner *et al.* 2003). Turner *et al.* (2003) provide a comprehensive list of ecological variables and data sources that they considered useful for quantifying and modeling biodiversity. They pointed out that there is still a slight disconnect between the remote sensing and the application research concerning the spatial scales. However, this situation is steadily improving. Gillespie *et al.* (2008) stressed the importance of field data and pointed out that remote sensing, even at increased resolutions, is supposed to sustain field data rather than to replace them.

The use of remote sensing for studying *coral reefs* goes back to the early days of optical satellite imagery, when Smith *et al.* (1975) worked on the automated mapping and inventory of reef zonation with Landsat data. Since then remote sensing has proven valuable for collecting information on geomorphologic zones and substrate types for coral reef environments (Mumby *et al.* 1997). A common remote sensing approach is the classification of the benthic areas into reef zones, substrate types and cover by coral and algal associations (Kutser *et al.* 2006). Recently, Knudby *et al.* (2007) reviewed the progress in the use of remote sensing for coral reef biodiversity studies. They concluded that data and techniques for single sensor studies have matured and are able to provide complementary information. The usefulness of remote sensing for coral reef studies will need to be measured by the ability to map the environmental variables that are relevant for the decision making for the conservation of threatened coral reefs. Fig. 7.7 shows a standard color composite of the Lighthouse Reef off the coast of Belize captured by the AVNIR sensor onboard the Advanced Land Observation Satellite (ALOS). With its 10 meter spatial resolution in multispectral mode, it can capture the coral reef structure and associated environmental variables in far more detail than the Landsat satellite. Andréfouët *et al.* (2003) analyzed high-resolution optical data for coral reef habitat mapping. The inherent diversity of the selected sites and the acquisition conditions prevented a similar processing approach, which made a quantitative comparison difficult. The study, however, showed the potential of high-resolution optical remote sensing data for this kind of studies.

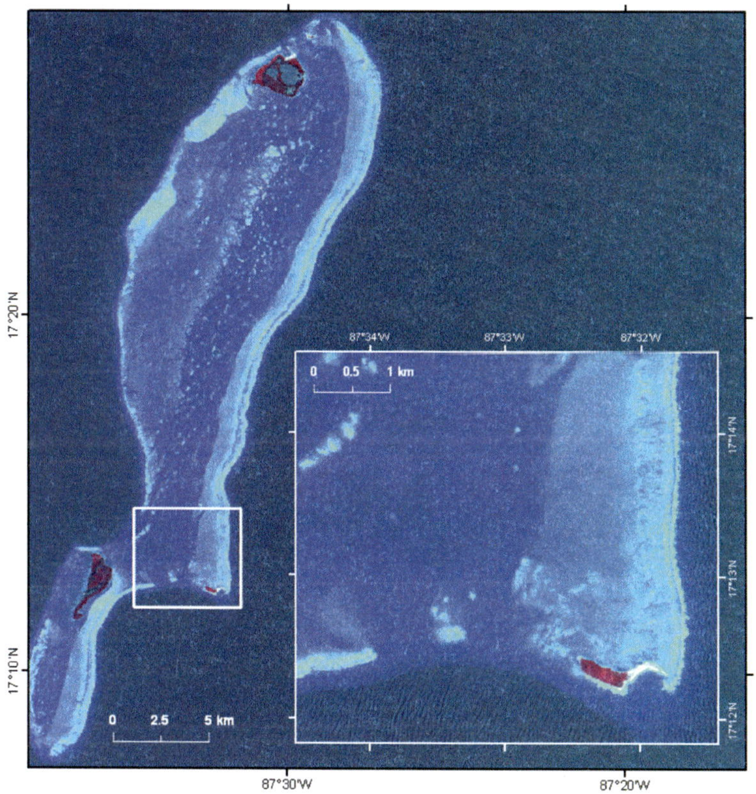

Fig. 7.7 Lighthouse Reef off the coast of Belize, shown as a false color composite of ALOS AVNIR data. The inset shows structural details of the Half Moon Caye. Image © Japan Aerospace Exploration Agency, 2007.

Asner *et al.* (2008) investigated the use of hyperspectral remote sensing data for studying native and invasive species in a tropical forest environment. The hyperspectral spectra express the biochemical and structural properties of the vegetation but not the species composition. It requires an increased level of understanding of the spectral separability, expressed by various leaf and canopy properties. Lawrence and Laub (2003) examined the use of subcanopy spatial resolution hyperspectral imagery for differentiating Douglas-fir trees attacked by the Douglas-fir beetle. Their results showed that the classification and regression tree analysis was superior to the step-wise discriminant analysis and that hyperspectral imagery is suitable to detect and map tree stress.

7.3.4 Remote Sensing of Geology and Soils

Remote sensing serves as an invaluable tool for regional scale mapping of surficial geology, especially in remote, rugged terrains where extensive field based mapping is logistically not feasible. Relatively barren rock and soil exposures, and lithologically controlled vegetation patterns provide clues to underlying geology. Fig. 7.8 shows a standard FCC from Landsat Thematic Mapper data over parts of the Delong Mountains, Alaska, showing classic folded and faulted structures.

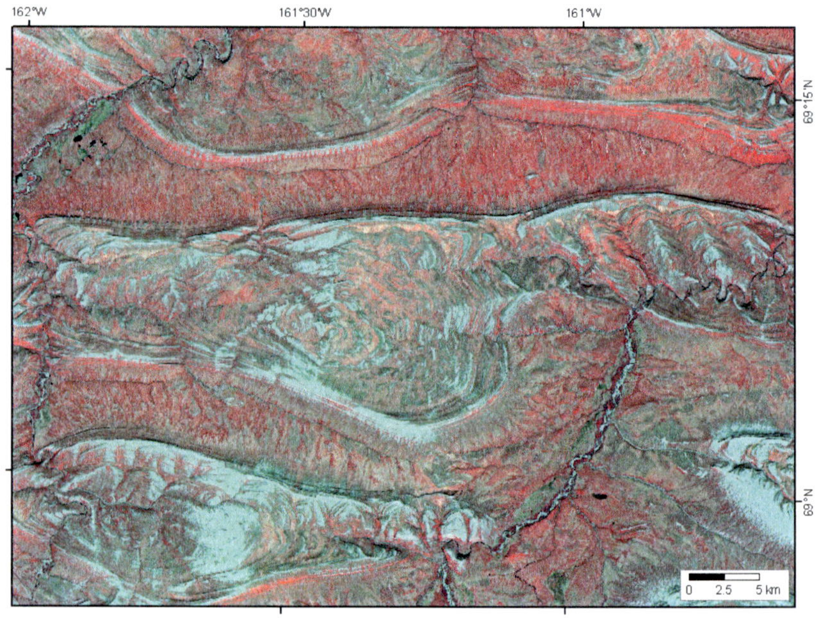

Fig. 7.8 Delong Mountains, Alaska: False color composite (bands 4, 3 and 2) of Landsat ETM, acquired in July 2000. A sequence of folded anticlines and synclines with their major axes trending in the east west direction can be seen. Offsets in rock exposures (cyan blue) are due to several faults that are present in this area. Image courtesy USGS.

Sabins (1999) comprehensively reviewed the use of remote sensing for *mineral exploration*, showcasing case studies on mapping of surficial geology and fracture patterns at regional and local scales, mostly by optical and radar data, as well as the recognition of hydrothermally altered rocks that may be associated with mineral deposits, mostly by hyperspectral data. Baghdadi *et al.* (2005) summarized the contribution of radar satellite imagery to geological exploration, particularly the suitability of the longer wavelengths in the L-band with their ground penetration capability in arid climate for subsurface geological mapping. Rajesh (2004) gave an overview about the application of remote sensing in *mineral resource mapping*. As it is clearly pointed out in these reviews, remote sensing can provide complementary information for geological applications that are not available in any other form. However, it needs to be understood that remote sensing will not replace field mapping. Kruse *et al.* (2003) compared airborne with spaceborne hyperspectral data for mineral mapping purposes and concluded that the spaceborne hyperspectral data demonstrated the ability to remotely map basic surface mineralogy. Due to the superior signal-to-noise of the airborne system, by an order of magnitude, the level of mapping detail is not comparable. However, future satellite missions will need to have higher spatial and spectral resolution to increase the already very promising potential of spaceborne hyperspectral data for mineral mapping.

Synthetic Aperture Radar has shown its strong potential for retrieving *soil moisture* maps at regional scales due to its sensitivity to the dielectric property that is directly related to moisture content in soil. This has been studied for a long time (Dubois *et al.* 1995) with the whole variety of SAR sensors (e.g. Paloscia *et al.* 2008, Makkeasorn *et al.* 2006, Le Hégarat-Mascle *et al.* 2002). Verhoest *et al.* (2008) summarized the issues related to parameterizing the soil roughness parameters used to derive the soil moisture. Soil moisture retrieval techniques on a global scale have been established using passive microwave radiometers (Reichle *et al.* 2007, Kerr *et al.* 2001). Metternicht and Zinck (2003) reviewed the potentials and constraints of remote sensing of *soil salinity*. Even though remote sensing can play an important role in detecting, mapping and monitoring salt-affected surface features, the ground measurements are still the link to determine the salt type and degree of salinity in soil and groundwater. Especially, SAR data (Aly *et al.* 2007) and hyperspectral data (Howari 2003) show some promise in quantifying the salt-affected features.

7.3.5 Remote Sensing of Water Bodies

The *ocean* has been monitored by remote sensing for decades. The primary quantities measured are color, temperature, salinity, roughness and height (Robinson 2004). Dickey *et al.* (2006) reviewed the optical oceanography, primarily concerned with the ocean color but also with derived parameters such as chlorophyll (Nair *et al.* 2008, Kutser 2004), suspended particles (Miller and McKee 2004) and bathymetry (Adler-Golden *et al.* 2005). With sea surface temperature being the primary measure (Barton 2001, Wentz *et al.* 2000), researchers try to derive the mixed layer depth (Zawada *et al.* 2006) and the skin temperature (Donlon *et al.* 2002). The sea surface salinity is measured with microwave image radiometers (Zine *et al.* 2008). The sea surface roughness can be monitored with active (Mouche *et al.* 2006) and passive microwave remote sensing (Long *et al.* 1993). Fig. 7.9 shows an example of wind speed derived from SAR imagery. The ambiguity of the wind direction can be resolved by the use of scatterometer data (Monaldo *et al.* 2001). Finally, the sea surface height has been determined by radar altimetry for decades. From the surface slope other parameters such as geostrophic currents (Cunningham and Pavic 2007), ocean geoid (Craig and Sandwell 1988) and sea floor bathymetry (Ramillien and Cazenave 1997) can be derived.

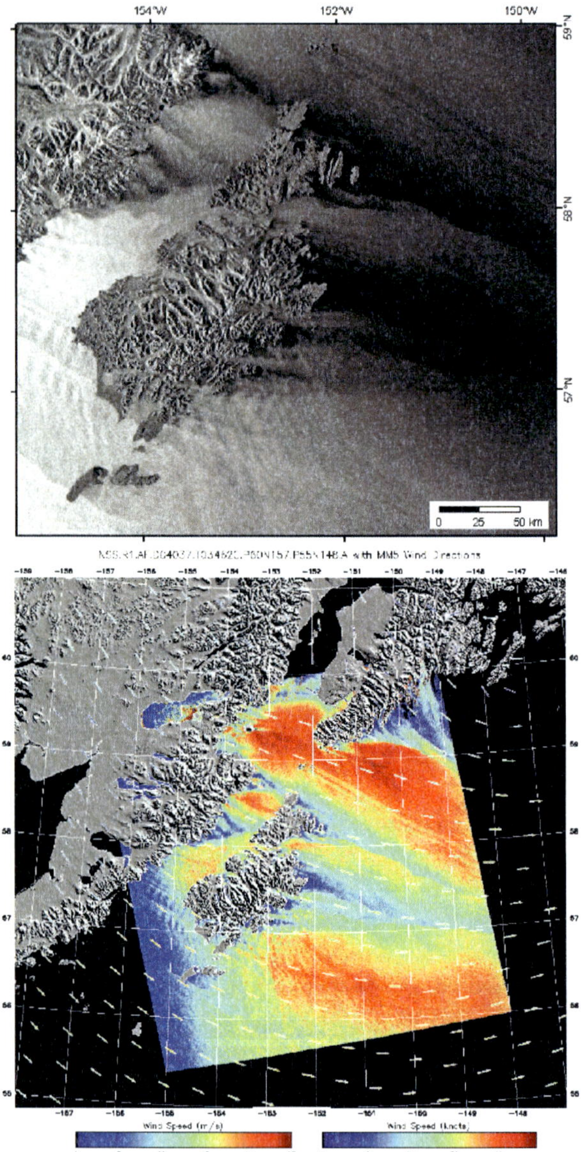

Fig. 7.9 (Top) Radarsat image acquired around Kodiak Island, Alaska, showing the ocean surface manifestation of gap winds. Image © Canadian Space Agency, 2004. (Bottom) Wind speed generated from the Radarsat image overlaid with wind directions from the MM5 forecast model. Image © Johns Hopkins University Applied Physics Laboratory.

For the remote sensing of *lakes and rivers* we consider these environments in this section in their unfrozen state. Frozen lakes and rivers are dealt with in the next section. Studies on lakes in their unfrozen state focus mostly on aspects such as water clarity (Olmanson *et al.* 2008, Koponen *et al.* 2002) and submerged aquatic vegetation (e.g. Yuan and Zhang 2008, Williams *et al.* 2003). Marcus *et al.* (2008) assessed the recent advances in optical remote sensing of rivers and concluded that with the vastly improved spatial resolution of recent optical sensors the mapping of in-stream habitats, depths, algae, wood, stream power and other features should be possible at sub-meter resolution, as long as there is no obstruction in terms of water clarity or of view. Another comprehensive review of the use of remote sensing in riverine landscapes was given by Mertes (2002).

Becker (2006) assessed the potential for satellite remote sensing of *ground water*. It is apparent from this review and other studies (e.g. Brunner *et al.* 2007) that remote sensing can be used most effectively in ground water studies in combination with ground water modeling. Hyperspectral data has been used to study the *water quality* of coastal areas (Brando and Dekker 2003), lakes (Giardino *et al.* 2007) and rivers (Ritchie *et al.* 2003). SAR data is useful in dealing with *flooding events*, as the data is particularly helpful for mapping flood extents. This information is valuable input for flood inundation modeling and runoff predictions (Matgen *et al.* 2007, 2006, Schumann *et al.* 2007).

7.3.6 Remote Sensing of the Cryosphere

The constituents of the cryosphere include the snow cover, ice sheets and glaciers, icebergs, sea ice, freshwater ice, and permafrost (Rees, 2006). As permafrost can not be directly studied with remote sensing, it is not covered in this section.

There are a number of *snow* parameters that can be measured or derived by remote sensing at various scales. The most obvious parameter is the snow covered area. Snow cover maps are operationally generated by the Moderate Resolution Imaging Spectroradiometer (Hall *et al.* 2002). A number of other snow indices have been proposed that can also be used for this purpose (e.g. Shimamura *et al.* 2006). Alternatively, the snow cover can be monitored with SAR data (Luojus *et al.* 2006). A number of algorithms based on passive microwave data have been developed to extract snow water equivalent and snow depth (e.g. Pulliainen 2006). Recently, several groups proposed a combination of optical data and complementary passive microwave data to improve the accuracy of the estimates for snow cover area and snow water equivalent (Liang *et al.* 2008, Durand *et al.* 2008). Other physical properties of the snow pack that are monitored by remote sensing are albedo (König *et al.* 2001), grain size (Painter *et al.* 2003) and temperature (Stroeve *et al.* 1996).

For studying *sea ice* a number of parameters are of relevance. The sea ice extent (Parkinson *et al.* 1999), concentration (Andersen *et al.* 2006) and ice type

(Walker *et al.* 2006) are routinely measured by passive microwave data. Langlois and Barber (2007) reviewed the issues related to seasonally snow-covered sea ice. Some of the ambiguities in the sea ice type determination can be resolved by incorporating microwave radiometer data (Voss *et al.* 2003). The Moderate Resolution Imaging Spectroradiometer offers daily sea ice extent and sea ice temperature products (Hall *et al.* 2004). Passive microwave data and scatterometer data is used to study large scale ice dynamics. SAR imagery with its much higher spatial resolution offers far more detail for the monitoring of ice motion (Lindsay and Stern 2003). Fig. 7.10 shows an example of tracking ice motion using Radarsat imagery on Arctic basin scale. SAR data can help in determining lake depth and ice thickness (Duguay and Lafleur 2003, Jeffries *et al.* 1996).

Fig. 7.10 Ice motion derived by the Radarsat Geophysical Processor System (RGPS). The ice flow is tracked over the season on Arctic basin wide scale, showing the cell deformation of the initially regular grid at 10 km interval.

Visible and near-infrared data as well as SAR data play a major role in the remote sensing of *terrestrial ice masses*, viz. glacier, ice sheets and ice shelves. Kargel *et al.* (2005) summarized the contribution of multispectral imagery to the global land ice measurements from space. Gao and Liu (2001) reviewed the applications of remote sensing in glaciology. SAR data has a long history in monitoring glaciers and has been used for large scale mapping missions (Jezek *et al.* 2003, Jezek 1999). Especially, the application of SAR interferometry (InSAR) showed its potential in measuring glacier velocities with unprecedented detail (Goldstein *et al.* 1993). Rignot (2002) used the InSAR technique to determine the mass balance of a number of glaciers in the Antarctica. Jaenicke *et al.* (2006) used optical and radar data for the mass balance estimate. Finally, Hubbard *et al.* (2000) combined digital elevation models from two different dates with the surface velocity field to map the mass balance.

7.4 Summary

The use of remote sensing techniques for extraction of spectral information content began in the 1960s, when the theoretical background related to the physical properties of material and energy-matter interactions was established. The demonstration of the feasibility of remote sensing and development of the actual image processing techniques picked up with the availability of satellite imagery on a regular basis. Ever since the number and variety of passive as well as active remote sensing instruments, that have been deployed, has steadily increased. With the advancement of technical capabilities, the performance of remote sensing sensors with respect to spectral and spatial resolution has improved. This drove the need for more powerful, accurate and flexible image processing and analysis techniques. While future advancements in sensor technology and processing techniques need to progress in parallel, the increasing archive of remote sensing data will expand application research from simple mapping to time series analysis and integrated observation promoting interdisciplinary research.

References

Abdel-Rahman EM, Ahmed FB (2008) The application of remote sensing techniques to sugarcane (Saccharum spp. hybrid) production: a review of the literature. International Journal Of Remote Sensing 29(13):3753-3767

Adler-Golden SM, Acharya PK, Berk A, Matthew MW, Gorodetzky D (2005) Remote bathymetry of the littoral zone from AVIRIS, LASH, and QuickBird imagery. IEEE Transactions on Geoscience and Remote Sensing 43(2):337-347

Aly Z, Bonn FJ, Magagi R (2007) Analysis of the backscattering coefficient of salt-affected soils using modeling and RADARSAT-1 SAR data. IEEE Transactions On Geoscience And Remote Sensing 45(2):332-341

Andersen S, Tonboe R, Kern S, Schyberg H (2006) Improved retrieval of sea ice total concentration from spaceborne passive microwave observations using numerical weather prediction model fields: An intercomparison of nine algorithms. Remote Sensing Of Environment 104(4):374-392

Andréfouët S, Kramer P, Torres-Pulliza D, Joyce KE, Hochberg EJ, Garza-Perez R, Mumby PJ, Riegl B, Yamano H, White WH, Zubia M, Brock JC, Phinn SR, Naseer A, Hatcher BG, Muller-Karger FE (2003) Multi-site evaluation of IKONOS data for classification of tropical coral reef environments. Remote Sensing Of Environment 88(1-2):128-143

Arroyo LA, Pascual C, Manzanera JA (2008) Fire models and methods to map fuel types: The role of remote sensing. Forest Ecology And Management 256(6):1239-1252

Asner GP, Jones MO, Martin RE, Knapp DE, Hughes RF (2008) Remote sensing of native and invasive species in Hawaiian forests. Remote Sensing Of Environment 112(5):1912-1926

Baghdadi N, Grandjean G, Lahondere D, Paillou P, Lasne Y (2005) The contribution of radar satellite imagery to geological exploration in and areas. C. R. Geosci. 337(8):719-728

Barton IJ (2001) Interpretation of satellite-derived sea surface temperatures. In: Tsuchiya K (ed) Elsevier Science Bv, pp 165-170

Becker MW (2006) Potential for satellite remote sensing of ground water. Ground Water 44(2):306-318

Beeri O, Peled A (2006) Spectral indices for precise agriculture monitoring. International Journal Of Remote Sensing 27(9-10):2039-2047

Brando VE, Dekker AG (2003) Satellite hyperspectral remote sensing for estimating estuarine and coastal water quality. IEEE Transactions On Geoscience And Remote Sensing 41(6):1378-1387

Brunner P, Franssen HJH, Kgotlhang L, Bauer-Gottwein P, Kinzelbach W (2007) How can remote sensing contribute in groundwater modeling? Hydrogeol. J. 15(1):5-18

Chakraborty M, Manjunath KR, Panigrahy S, Kundu N, Parihar JS (2005) Rice crop parameter retrieval using multi-temporal, multi-incidence angle Radarsat SAR data. Isprs Journal Of Photogrammetry And Remote Sensing 59(5):310-322

Chauhan NS, Lang RH, Ranson KJ (1991) Radar Modeling Of A Boreal Forest. IEEE Transactions On Geoscience And Remote Sensing 29(4):627-638

Chuvieco E, Kasischke ES (2007) Remote sensing information for fire management and fire effects assessment. J. Geophys. Res.-Biogeosci. 112(G1)

Cloude SR, Papathanassiou KP (1998) Polarimetric SAR interferometry. IEEE Transactions On Geoscience And Remote Sensing 36(5):1551-1565

Craig CH, Sandwell DT (1988) GLOBAL DISTRIBUTION OF SEAMOUNTS FROM SEASAT PROFILES. Journal of Geophysical Research-Solid Earth and Planets 93(B9):10408-10420

Cunningham S, Pavic M (2007) Surface geostrophic currents across the Antarctic circumpolar current in Drake Passage from 1992 to 2004. Prog. Oceanogr. 73(3-4):296-310

Dell'Acqua F, Gamba P (2006) Discriminating urban environments using multiscale texture and multiple SAR images. International Journal Of Remote Sensing 27(18):3797-3812

Dickey T, Lewis M, Chang G (2006) Optical oceanography: Recent advances and future directions using global remote sensing and in situ observations. Reviews Of Geophysics 44(1)

Dobson MC, Ulaby FT, Pierce LE, Sharik TL, Bergen KM, Kellndorfer J, Kendra JR, Li E, Lin YC, Nashashibi A, Sarabandi K, Siqueira P (1995) Estimation Of Forest Biophysical Characteristics In Northern Michigan With Sir-C/X-Sar. IEEE Transactions On Geoscience And Remote Sensing 33(4):877-895

Donlon CJ, Minnett PJ, Gentemann C, Nightingale TJ, Barton IJ, Ward B, Murray MJ (2002) Toward improved validation of satellite sea surface skin temperature measurements for climate research. J. Clim. 15(4):353-369

Dubois PC, Vanzyl J, Engman T (1995) Measuring Soil-Moisture With Imaging Radars. IEEE Transactions On Geoscience And Remote Sensing 33(4):915-926

Duguay CR, Lafleur PM (2003) Determining depth and ice thickness of shallow sub-Arctic lakes using space-borne optical and SAR data. International Journal Of Remote Sensing 24(3):475-489

Durand M, Molotch NP, Margulis SA (2008) Merging complementary remote sensing datasets in the context of snow water equivalent reconstruction. Remote Sensing Of Environment 112(3):1212-1225

Durden SL, Vanzyl JJ, Zebker HA (1989) Modeling And Observation Of The Radar Polarization Signature Of Forested Areas. IEEE Transactions On Geoscience And Remote Sensing 27(3):290-301

Duro D, Coops NC, Wulder MA, Han T (2007) Development of a large area biodiversity monitoring system driven by remote sensing. Progress In Physical Geography 31(3):235-260

Erickson BJ, Johannsen CJ, Vorst JJ, Biehl LL (2004) Using remote sensing to assess stand loss and defoliation in maize. Photogrammetric Engineering And Remote Sensing 70(6):717-722

Falkowski MJ, Gessler PE, Morgan P, Hudak AT, Smith AMS (2005) Characterizing and mapping forest fire fuels using ASTER imagery and gradient modeling. Forest Ecology And Management 217(2-3):129-146

Ferretti A, Prati C, Rocca F (2000) Nonlinear subsidence rate estimation using permanent scatterers in differential SAR interferometry. IEEE Transactions On Geoscience And Remote Sensing 38(5):2202-2212

Gao J, Liu YS (2001) Applications of remote sensing, GIS and GPS in glaciology: a review. Progress In Physical Geography 25(4):520-540

Geman D, Jedynak B (1996) An active testing model for tracking roads in satellite images. IEEE Trans. Pattern Anal. Mach. Intell. 18(1):1-14

Giardino C, Brando VE, Dekker AG, Strombeck N, Candiani G (2007) Assessment of water quality in Lake Garda (Italy) using Hyperion. Remote Sensing Of Environment 109(2):183-195

Gillespie TW, Foody GM, Rocchini D, Giorgi AP, Saatchi S (2008) Measuring and modelling biodiversity from space. Progress In Physical Geography 32(2):203-221

Gluch R, Quattrochi DA, Luvall JC (2006) A multi-scale approach to urban thermal analysis. Remote Sensing Of Environment 104(2):123-132

Goldstein RM, Engelhardt H, Kamb B, Frolich RM (1993) Satellite Radar Interferometry For Monitoring Ice-Sheet Motion - Application To An Antarctic Ice Stream. Science 262(5139):1525-1530

Gottschalk TK, Huettmann F, Ehlers M (2005) Thirty years of analysing and modelling avian habitat relationships using satellite imagery data: a review. International Journal Of Remote Sensing 26(12):2631-2656

Hall DK, Key JR, Casey KA, Riggs GA, Cavalieri DJ (2004) Sea ice surface temperature product from MODIS. IEEE Transactions On Geoscience And Remote Sensing 42(5):1076-1087

Hall DK, Riggs GA, Salomonson VV, DiGirolamo NE, Bayr KJ (2002) MODIS snow-cover products. Remote Sensing Of Environment 83(1-2):181-194

Hamdi R, Schayes G (2008) Sensitivity study of the urban heat island intensity to urban characteristics. Int. J. Climatol. 28(7):973-982

Howari FM (2003) The use of remote sensing data to extract information from agricultural land with emphasis on soil salinity. Australian Journal Of Soil Research 41(7):1243-1253

Hubbard A, Willis I, Sharp M, Mair D, Nienow P, Hubbard B, Blatter H (2000) Glacier mass-balance determination by remote sensing and high-resolution modelling. Journal Of Glaciology 46(154):491-498

Hyde P, Dubayah R, Walker W, Blair JB, Hofton M, Hunsaker C (2006) Mapping forest structure for wildlife habitat analysis using multi-sensor (LiDAR, SAR/InSAR, ETM plus, Quickbird) synergy. Remote Sensing Of Environment 102(1-2):63-73

Jaenicke J, Mayer C, Scharrer K, Munzer U, Gudmundsson A (2006) The use of remote-sensing data for mass-balance studies at Myrdalsjokull ice cap, Iceland. Journal Of Glaciology 52(179):565-573

Jeffries MO, Morris K, Liston GE (1996) A method to determine lake depth and water availability on the north slope of Alaska with spaceborne imaging radar and numerical ice growth modelling. Arctic 49(4):367-374

Jezek KC (1999) Glaciological properties of the Antarctic ice sheet from RADARSAT-1 synthetic aperture radar imagery. Annals Of Glaciology, Vol 29, 1999 29:286-290

Jezek KC, Farness K, Carande R, Wu X, Labelle-Hamer N (2003) RADARSAT 1 synthetic aperture radar observations of Antarctica: Modified Antarctic Mapping Mission, 2000. Radio Sci. 38(4)

Jung A, Tökei L, Kardevan P (2008) Application of Airborne Hyperspectral and Thermal Images to Analyze Urban Microclimate. Applied Ecology and Environmental Research 5(1):165-175

Kargel JS, Abrams MJ, Bishop MP, Bush A, Hamilton G, Jiskoot H, Kaab A, Kieffer HH, Lee EM, Paul F, Rau F, Raup B, Shroder JF, Soltesz D, Stainforth D, Stearns L, Wessels R (2005) Multispectral imaging contributions to global land ice measurements from space. Remote Sensing Of Environment 99(1-2):187-219

Kassomenos PA, Katsoulis BD (2006) Mesoscale and macroscale aspects of the morning Urban Heat Island around Athens, Greece. Meteorol. Atmos. Phys. 94(1-4):209-218

Katartzis A, Sahli H (2008) A stochastic framework for the identification of building rooftops using a single remote sensing image. IEEE Transactions On Geoscience And Remote Sensing 46(1):259-271

Keane RE, Burgan R, van Wagtendonk J (2001) Mapping wildland fuels for fire management across multiple scales: Integrating remote sensing, GIS, and biophysical modeling. Int. J. Wildland Fire 10(3-4):301-319

Kerr YH, Waldteufel P, Wigneron JP, Martinuzzi JM, Font J, Berger M (2001) Soil moisture retrieval from space: The Soil Moisture and Ocean Salinity (SMOS) mission. IEEE Transactions On Geoscience And Remote Sensing 39(8):1729-1735

Knudby A, LeDrew E, Newman C (2007) Progress in the use of remote sensing for coral reef biodiversity studies. Progress In Physical Geography 31(4):421-434

König M, Winther JG, Isaksson E (2001) Measuring snow and glacier ice properties from satellite. Reviews Of Geophysics 39(1):1-27

Koponen S, Pulliainen J, Kallio K, Hallikainen M (2002) Lake water quality classification with airborne hyperspectral spectrometer and simulated MERIS data. Remote Sensing Of Environment 79(1):51-59

Kruse FA, Boardman JW, Huntington JF (2003) Comparison of airborne hyperspectral data and EO-1 Hyperion for mineral mapping. IEEE Transactions On Geoscience And Remote Sensing 41(6):1388-1400

Kutser T (2004) Quantitative detection of chlorophyll in cyanobacterial blooms by satellite remote sensing. Limnology And Oceanography 49(6):2179-2189

Kutser T, Miller I, Jupp DLB (2006) Mapping coral reef benthic substrates using hyperspectral space-borne images and spectral libraries. Estuarine Coastal And Shelf Science 70(3):449-460

Langlois A, Barber DG (2007) Passive microwave remote sensing of seasonal snow-covered sea ice. Progress In Physical Geography 31(6):539-573

Lawrence R, Labus M (2003) Early detection of Douglas-Fir beetle infestation with subcanopy resolution hyperspectral imagery. West. J. Appl. For. 18(3):202-206

Le Hegarat-Mascle S, Zribi M, Alem F, Weisse A, Loumagne C (2002) Soil moisture estimation from ERS/SAR data: Toward an operational methodology. IEEE Transactions On Geoscience And Remote Sensing 40(12):2647-2658

Lee JS, Cloude SR, Papathanassiou KP, Grunes MR, Woodhouse IH (2003) Speckle filtering and coherence estimation of polarimetric SAR interferometry data for forest applications. IEEE Transactions On Geoscience And Remote Sensing 41(10):2254-2263

Liang TG, Zhang XT, Xie HJ, Wu CX, Feng QS, Huang XD, Chen QG (2008) Toward improved daily snow cover mapping with advanced combination of MODIS and AMSR-E measurements. Remote Sensing Of Environment 112(10):3750-3761

Lindsay RW, Stern HL (2003) The RADARSAT geophysical processor system: Quality of sea ice trajectory and deformation estimates. J. Atmos. Ocean. Technol. 20(9):1333-1347

Long DG, Hardin PJ, Whiting PT (1993) RESOLUTION ENHANCEMENT OF SPACEBORNE SCATTEROMETER DATA. IEEE Transactions on Geoscience and Remote Sensing 31(3):700-715

Lu DS, Weng QH (2006) Use of impervious surface in urban land-use classification. Remote Sensing Of Environment 102(1-2):146-160

Luojus KP, Pulliainen JT, Metsamaki SJ, Hallikainen MT (2006) Accuracy assessment of SAR data-based snow-covered area estimation method. IEEE Transactions On Geoscience And Remote Sensing 44(2):277-287

Makkeasorn A, Chang NB, Beaman M, Wyatt C, Slater C (2006) Soil moisture estimation in a semiarid watershed using RADARSAT-1 satellite imagery and genetic programming. Water Resources Research 42(9)

Marcus WA, Fonstad MA (2008) Optical remote mapping of rivers at sub-meter resolutions and watershed extents. Earth Surface Processes And Landforms 33(1):4-24

Martilli A (2007) Current research and future challenges in urban mesoscale modelling. Int. J. Climatol. 27(14):1909-1918

Matgen P, El Idrissi A, Henry JB, Tholey N, Hoffmann L, de Fraipont P, Pfister L (2006) Patterns of remotely sensed floodplain saturation and its use in runoff predictions. Hydrological Processes 20(8):1805-1825

Matgen P, Schumann G, Henry JB, Hoffmann L, Pfister L (2007) Integration of SAR-derived river inundation areas, high-precision topographic data and a river flow model toward near real-time flood management. International Journal Of Applied Earth Observation And Geoinformation 9(3):247-263

Mayer H (1999) Automatic Object Extraction from Aerial Imagery - A Survey Focusing on Buildings. Computer Vision and Image Understanding 74(2):138-149

McDermid GJ, Franklin SE, LeDrew EF (2005) Remote sensing for large-area habitat mapping. Progress In Physical Geography 29(4):449-474

McRoberts RE, Tomppo EO (2007) Remote sensing support for national forest inventories. Remote Sensing Of Environment 110(4):412-419

Mertes LAK (2002) Remote sensing of riverine landscapes. Freshw. Biol. 47(4):799-816

Metternicht GI, Zinck JA (2003) Remote sensing of soil salinity: potentials and constraints. Remote Sensing Of Environment 85(1):1-20

Miller RL, McKee BA (2004) Using MODIS Terra 250 m imagery to map concentrations of total suspended matter in coastal waters. Remote Sensing of Environment 93(1-2):259-266

Monaldo, F. M., Thompson, D. R., Beal, R. C., Pichel, W. G. and P. Clemente-Colon, 2001, "Comparison of SAR-Derived Wind Speed with Model Predictions and Ocean Buoy Measurements," IEEE Transactions on Geoscience and Remote Sensing, 39(12):2587–2600.

Mouche AA, Hauser D, Kudryavtsev V (2006) Radar scattering of the ocean surface and sea-roughness properties: A combined analysis from dual-polarizations airborne radar observations and models in C band. Journal of Geophysical Research-Oceans 111(C9):18

Mumby PJ, Green EP, Edwards AJ, Clark CD (1997) Coral reef habitat-mapping: how much detail can remote sensing provide? Marine Biology 130(2):193-202

Nagendra H (2001) Using remote sensing to assess biodiversity. International Journal Of Remote Sensing 22(12):2377-2400

Nair A, Sathyendranath S, Platt T, Morales J, Stuart V, Forget MH, Devred E, Bouman H (2008) Remote sensing of phytoplankton functional types. Remote Sensing Of Environment 112(8):3366-3375

Olmanson LG, Bauer ME, Brezonik PL (2008) A 20-year Landsat water clarity census of Minnesota's 10,000 lakes. Remote Sensing of Environment 112(11):4086-4097

Painter TH, Dozier J, Roberts DA, Davis RE, Green RO (2003) Retrieval of subpixel snow-covered area and grain size from imaging spectrometer data. Remote Sensing Of Environment 85(1):64-77

Paloscia S, Pampaloni P, Pettinato S, Santi E (2008) A Comparison of Algorithms for Retrieving Soil Moisture from ENVISAT/ASAR Images. IEEE Transactions On Geoscience And Remote Sensing 46(10):3274-3284

Parkinson CL, Cavalieri DJ, Gloersen P, Zwally HJ, Comiso JC (1999) Arctic sea ice extents, areas, and trends, 1978-1996. Journal Of Geophysical Research-Oceans 104(C9):20837-20856

Pinter PJ, Hatfield JL, Schepers JS, Barnes EM, Moran MS, Daughtry CST, Upchurch DR (2003) Remote sensing for crop management. Photogrammetric Engineering And Remote Sensing 69(6):647-664

Pulliainen J (2006) Mapping of snow water equivalent and snow depth in boreal and sub-arctic zones by assimilating space-borne microwave radiometer data and ground-based observations. Remote Sensing Of Environment 101(2):257-269

Rajesh HM (2004) Application of remote sensing and GIS in mineral resource mapping - An overview. J. Mineral. Petrol. Sci. 99(3):83-103

Ramillien G, Cazenave A (1997) Global bathymetry derived from altimeter data of the ERS-1 Geodetic Mission. J. Geodyn. 23(2):129-149

Rees WG (2006) Remote sensing of snow and ice. CRC Press, Boca Raton, 285 p.

Reichle RH, Koster RD, Liu P, Mahanama SPP, Njoku EG, Owe M (2007) Comparison and assimilation of global soil moisture retrievals from the Advanced Microwave Scanning Radiometer for the Earth Observing System (AMSR-E) and the Scanning Multichannel Microwave Radiometer (SMMR). Journal Of Geophysical Research-Atmospheres 112(D9)

Riaño D, Chuvieco E, Ustin SL, Salas J, Rodriguez-Perez JR, Ribeiro LM, Viegas DX, Moreno JM, Fernandez H (2007) Estimation of shrub height for fuel-type mapping combining airborne LiDAR and simultaneous color infrared ortho imaging. Int. J. Wildland Fire 16(3):341-348

Rignot E (2002) Mass balance of East Antarctic glaciers and ice shelves from satellite data. Annals Of Glaciology, Vol 34, 2002 34:217-227

Ritchie JC, Zimba PV, Everitt JH (2003) Remote sensing techniques to assess water quality. Photogrammetric Engineering And Remote Sensing 69(6):695-704

Roberts G (2001) A review of the application of BRDF models to infer land cover parameters at regional and global scales. Progress In Physical Geography 25(4):483-511

Roberts JW, Tesfamichael S, Gebreslasie M, van Aardt J, Ahmed FB (2007) Forest structural assessment using remote sensing technologies: An overview of the current state of the art. South. Hemisph. For. J. 69(3):183-203

Robinson IS (2004) Measuring the oceans from space. Springer, Berlin, 669 p.

Sabins FF (1999) Remote sensing for mineral exploration. Ore Geol. Rev. 14(3-4):157-183

Schaepman-Strub G, Schaepman ME, Painter TH, Dangel S, Martonchik JV (2006) Reflectance quantities in optical remote sensing-definitions and case studies. Remote Sensing Of Environment 103(1):27-42

Schumann G, Matgen P, Hoffmann L, Hostache R, Pappenberger F, Pfister L (2007) Deriving distributed roughness values from satellite radar data for flood inundation modelling. J. Hydrol. 344(1-2):96-111

Shaban MA, Dikshit O (2002) Evaluation of the merging of SPOT multispectral and panchromatic data for classification of an urban environment. International Journal Of Remote Sensing 23(2):249-262

Shimamura Y, Izumi T, Matsuyama H (2006) Evaluation of a useful method to identify snow-covered areas under vegetation - comparisons among a newly proposed snow index, normalized difference snow index, and visible reflectance. International Journal Of Remote Sensing 27(21):4867-4884

Simonetto E, Oriot H, Garello R (2005) Rectangular building extraction from stereoscopic airborne radar images. IEEE Transactions On Geoscience And Remote Sensing 43(10):2386-2395

Small C (2006) Comparative analysis of urban reflectance and surface temperature. Remote Sensing Of Environment 104(2):168-189

Smith V, Rogers R, Reed L (1975) Automated mapping and inventory of Great Barrier Reef zonation with LANDSAT data. Oceans 7:775-780

Stroeve J, Haefliger M, Steffen K (1996) Surface temperature from ERS-1 ATSR infrared thermal satellite data in polar regions. Journal Of Applied Meteorology 35(8):1231-1239

Toutin T (2004) Geometric processing of remote sensing images: models, algorithms and methods. International Journal Of Remote Sensing 25(10):1893-1924

Tupin F, Houshmand B, Datcu M (2002) Road detection in dense urban areas using SAR imagery and the usefulness of multiple views. IEEE Transactions On Geoscience And Remote Sensing 40(11):2405-2414

Turner W, Spector S, Gardiner N, Fladeland M, Sterling E, Steininger M (2003) Remote sensing for biodiversity science and conservation. Trends In Ecology & Evolution 18(6):306-314

Varga TA, Asner GP (2008) Hyperspectral and lidar remote sensing of fire fuels in Hawaii Volcanoes National Park. Ecol. Appl. 18(3):613-623

Verhoest NEC, Lievens H, Wagner W, Alvarez-Mozos J, Moran MS, Mattia F (2008) On the soil roughness parameterization problem in soil moisture retrieval of bare surfaces from synthetic aperture radar. Sensors 8(7):4213-4248

Vierling KT, Vierling LA, Gould WA, Martinuzzi S, Clawges RM (2008) Lidar: shedding new light on habitat characterization and modeling. Frontiers In Ecology And The Environment 6(2):90-98

Voss S, Heygster G, Ezraty R Improving sea ice type discrimination by the simultaneous use of SSM/I and scatterometer data. Polar Research 22(1):35–42

Walker NP, Partington KC, Van Woert ML, Street TLT (2006) Arctic sea ice type and concentration mapping using passive and active microwave sensors. IEEE Transactions On Geoscience And Remote Sensing 44(12):3574-3584

Wehr A, Lohr U (1999) Airborne laser scanning - an introduction and overview. Isprs Journal Of Photogrammetry And Remote Sensing 54(2-3):68-82

Weng QH, Hu XF, Lu DS (2008) Extracting impervious surfaces from medium spatial resolution multispectral and hyperspectral imagery: a comparison. International Journal Of Remote Sensing 29(11):3209-3232

Wentz FJ, Gentemann C, Smith D, Chelton D (2000) Satellite measurements of sea surface temperature through clouds. Science 288(5467):847-850

Williams DJ, Rybicki NB, Lombana AV, O'Brien TM, Gomez RB (2003) Preliminary investigation of submerged aquatic vegetation mapping using hyperspectral remote sensing. Environ. Monit. Assess. 81(1-3):383-392

Yuan L, Zhang LQ (2008) Mapping large-scale distribution of submerged aquatic vegetation coverage using remote sensing. Ecol. Inform. 3(3):245-251

Zawada DG, Zaneveld JRV, Boss E, Gardner WD, Richardson MJ, Mishonov AV (2005) A comparison of hydrographically and optically derived mixed layer depths. Journal of Geophysical Research-Oceans 110(C11):13

Zine S, Boutin J, Font J, Reul N, Waldteufel P, Gabarro C, Tenerelli J, Petitcolin F, Vergely JL, Talone M, Delwart S (2008) Overview of the SMOS sea surface salinity prototype processor. IEEE Transactions On Geoscience And Remote Sensing 46(3):621-645

Chapter 8

ALGORITHMS AND APPLICATIONS FOR LAND COVER CLASSIFICATION – A REVIEW

Björn Waske, Mingmin Chi, Jón Atli Benediktsson, Sebastian van der Linden, Benjamin Koetz

8.1 Introduction

8.1.1 Land Cover Classification in Remote Sensing

During the last decades the manner how the Earth is being observed was revolutionized. Earth Observation (EO) systems became a valuable and powerful tool to monitor the Earth and had significant impact on the acquisition and analysis of environmental data (Rosenquist *et al.* 2003). Currently, EO data play a major role in supporting decision-making and surveying compliance of several multilateral environmental treaties, such as the Kyoto Protocol, the Convention on Biological Diversity, or the European initiative Global Monitoring for Environment and Security, GMES (Peter 2004, Rosenquist *et al.* 2003, Backhaus and Beule 2005). However, the need for such long-term monitoring of the Earth's surface requires the standardized and coordinated use of global EO data sets, which has led, e.g., to the international Global Earth Observation System of Systems (GEOSS) initiative as well as to the Global Climate Observation System (GCOS) implementation plan (GCOS 2004, GEO 2005). The evolving EO technologies together with the requirements and standards arising from their exploitation demand increasingly improving algorithms, especially in the field of land cover classification.

In the context of international policies and global change issues, land cover classifications of remote sensing data are the most commonly used EO product and the development of adequate classification methods is an ongoing research field. In this chapter, state-of-the-art image classification concepts are reviewed. These concepts are discussed in the context of EO and relate both to well-known approaches and implications of the evolving observation techniques in this field. Following this selected applications are presented in order to demonstrate qualities and characteristics of these classification concepts.

D. Li et al. (eds.), *Geospatial Technology for Earth Observation*,
DOI 10.1007/978-1-4419-0050-0_8,

8.1.2 Image Classification Concepts

Currently, the development of adequate image classification concepts is mainly driven by algorithms based on the field of machine learning and pattern recognition. However, in the early years of satellite remote sensing, methods that originated from the field of signal processing were modified for image data processing, e.g., the Gaussian maximum likelihood classifier (Cardillo and Landgrebe 1966, Fu *et al.* 1969). Thus, many early methods were based on relatively simple data models and concepts, which might be a limitation for current data, due to the rapid development in EO technologies (Richards 2005). Nevertheless, such methods still experience frequent application in operational EO data processing.

Enhanced EO systems, including numerous bands, high spatial resolution and increased repetition rates as well as the availability of more diverse remote sensing data sources improve the potential of EO applications. However, such improvements demand more sophisticated classification concepts. Moreover, continuously increasing performance requirements, such as near-real time applications with high accuracies, also call for development of adequate techniques (Richards 2005, Jain *et al.*, 2000). Undoubtedly, the rapidly increased computer power made available during the last decades was an important factor for faster and more sophisticated processing of huge and diverse data sets. Today, the available literature on image classification is huge and users can choose among a multitude of different classification approaches. The decision on a classification approach should ideally be driven by the input imagery, the available information, the final application and the available processing power. Unfortunately, commercial EO data processing software does not offer the full range of state-of-the-art concepts. Therefore, such approaches are still not fully established.

Generally, classification methods are categorized, depending on the information available for the classifier design. The two main dichotomies are *unsupervised* and *supervised* approaches. During unsupervised classification, the data are aggregated into natural clusters or classes that have similar properties, without having any additional knowledge, for instance from ground truth data. Widely used clustering methods are for example the ISODATA and the k-means algorithms (Richards and Jia 2006, Duda *et al.* 2001). A general problem of unsupervised algorithms is that data may comprise clusters of different shapes and sizes. In this context, the definition of clusters and the selection of an adequate similarity measure are difficult (Jain *et al.*, 2000). Another drawback for these algorithms is that the number of classes is usually unknown, because the clustering process considers feature classes instead of final information classes, i.e., land cover types. On the other hand, supervised classification approaches rely on *a priori* knowledge of the desired information classes. This *a priori* knowledge is used to design the classifier and hence often referred to as training data. Some techniques employ training data as well as unlabeled data, Such *semi-supervised* techniques are discussed in Section 8.3.

Supervised classification concepts can be split into parametric and non-parametric approaches. The statistical maximum likelihood classifier is the most widely used supervised classification strategy and the most prominent example of a parametric classifier. The approach is based on the assumption that the form of a multivariate probability density function is known (e.g., a Gaussian distribution) and the function can be defined for each class by computing the mean vector and covariance matrix of the classes in a training data set. If this information is not available, the density must be approximated. In such cases, a Parzen window can, for example, be used to estimate a density function for a classical statistical Bayes classifier. If such estimates are not possible or when the assumption of a specific distribution is known to be wrong, alternative approaches need to be taken. Such alternative approaches that are not constrained by assumptions on the input data distribution include artificial neural networks (ANN) (Benediktsson *et al.* 1990), k-nearest neighbor (kNN), self-learning decision trees (DT) and support vector machines (SVM). These *non-parametric* methods are not constrained by prior assumptions on the distribution of input data as is the case for *parametric* approaches and can be applied, even if class conditional densities, class means or deviations cannot be estimated reliably.

Over the past years, a categorization into pixel- and object-based approaches has entered the field EO data processing. The latter relates to approaches where data preprocessing is extended by a spatial generalization of adjacent pixels, i.e. image segmentation, which results in the generation of additional features such as textural or contextual information. Despite possible disadvantages for spectral classification accuracy (van der Linden *et al.* 2007), these additional features often help overcoming spectral deficiencies (Shackleford and Davis 2003, Bruzzone and Carlin 2006). Another efficient approach to conquer such problems is the use of morphological profiles (Benediktsson et al., 2003, 2005; Fauvel et al., 2008). Against the background of this chapter, however, object-based approaches only constitute a modification of the available input data without direct consequences on the classification algorithms. Therefore, they are not discussed further here.

8.2 Supervised Classification Algorithms

At the beginning of the (image) classification process, patterns (or pixels) are represented by d-dimensional random variables or vectors $x = (x_1, x_2,\dots x_d)$ where d denotes the number of available features (e.g., the image bands). These vector representations in the feature space $\Re^d$ are used by classification algorithms to assign pixels to one of c land cover classes $\Omega = \{\omega_i\}$, for i=1,..., c. This constitutes the ultimate goal of image classification. In the case of supervised classification these assignments are based either on the similarity to a certain class or on the relative position to decision boundaries. The bases for comparison of similarity or the positions of decision boundaries are determined through training patterns (or pixels).

Training pixels are represented by vectors in the same feature space $\Re^d$ and the class they belong to is known a priori.

8.2.1 Maximum Likelihood Classifier

Although the maximum likelihood classification (MLC) is one of the most common supervised classification techniques in the field of remote sensing, a brief introduction is given below for a better understanding of differences to the non-parametric concepts that will be described subsequently. The MLC is derived from the Bayes rule which says that classes have equal priorities (i.e, equal prior probabilities). Following the *Bayes decision rule* a sample $\boldsymbol{x}$ is classified to ω_i if

$$P(\omega_i \mid x) > P(\omega_j \mid x) \tag{8.1}$$

with $p(\omega_i|\boldsymbol{x})$ being the probability that the true class of $\boldsymbol{x}$ is ω_i. However, density functions $p(\omega_i|\boldsymbol{x})$ are generally unknown and need to be derived from the so-called training data. In the case of EO applications, such training data may exist in the form of maps or information from field survey and are often referred to as *ground truth* or *reference data*. The underlying assumption in this approach is that features are represented by specific probability densities, which are conditioned on the observed land cover classes. Consequently, a sample x belonging to class ω_i is assumed to be an observation drawn randomly from the *class-conditional probability density* function $p(\boldsymbol{x}|\omega_i)$. Thus, differences between two classes ω_1 and ω_2 would be described by differences in the corresponding density functions $p(\boldsymbol{x}|\omega_1)$ and $p(\boldsymbol{x}|\omega_2)$ (Duda *et al.* 2001). Moreover, the approach assumes some prior knowledge regarding the occurrence of each class, described by the *a priori probabilities* $p(\omega_i)$. If, for instance it is known that 30% of a considered image is covered by class ω_i then the prior would be $p(\omega_i)$=0.3 and there would be 30% probability of finding a pixel x belonging to class ω_i. Following the *Bayes formula* the desired *posteriori probabilities* $p(\omega_i|\boldsymbol{x})$ can be derived by:

$$p(\omega_i \mid x) = \frac{p(x \mid \omega_i) \cdot p(\omega_i)}{p(x)} \tag{8.2}$$

where $p(\boldsymbol{x})$ denotes the probability of the sample $\boldsymbol{x}$. This term is also called *evidence*, and is just a scaling factor that guarantees that the sum of all posteriori probabilities is one (Duda *et al.* 2001). In general $p(\boldsymbol{x})$ is a common factor and can be removed. Thus, $\boldsymbol{x}$ is classified as ω_i if the following inequlity is fulfilled

$$\mathit{if}\ \ p(x \mid \omega_i) p(\omega_i) > p(x \mid \omega_j) p(\omega_j) \tag{8.3}$$

According to this general decision rule, the selected class is the one which has the greatest posteriori probability. At the same time, the probability of making an error is minimized. Other approaches aim at the reduction of the risk and loss, respectively, instead of aiming at the error rate which is important, if the costs, of misclassification vary with different types of classes. In that context, a risk function is introduced, which depends on the true class of the sample and the class as which it is classified. In doing so all the possible decisions will be weighted. However, in general a 1/0 risk function is used that weights all misclassifications equally and consequently the decision rule is equal to Eq. 8.3, also called the maximum a posteriori rule (MAP). Thus, a classifier with a minimum error rate can be constructed, if the class-conditional densities and the prior probabilities are known. Usually it is assumed that the priors are known or that they can be estimated by the user. However, generally the likelihoods are known. Often it is assumed that the distributions follow the form of a *multivariate normal (Gaussian) model*. Actually, many processes can be described by a normal model and this type of model simplifies the approach, because the model is only described by the mean and the covariance matrix. Following the maximum likelihood estimation (Duda *et al.* 2001), the mean and the covariance matrix of the distribution can be estimated based on (training) samples, which were taken randomly out of the distributions. However the maximum likelihood estimation for the covariance matrix is generally biased, and consequently (Duda *et al.* 2001) an unbiased version is used for the estimation. The so-called sample covariance matrix is described by (Richards and Jia 2006):

$$\hat{\Sigma} = \frac{1}{n-1}\sum_{i=1}^{n}(x_i - \mu)(x_i - \mu)^T \tag{8.4}$$

with x_i as the i-th training sample out of n, and μ as the mean vector. Finally, after estimating the distributions, the Bayes decision strategy is followed and unknown pixels are classified to the class with the highest likelihood. Although the Bayesian approach can provide very accurate results, its use might be critical in some cases: (1) An adequate number of representative training samples is needed (depending on the dimensionality of our data) for an accurate estimation, and (2) an adequate multivariate statistical model needs to be assumed for the data at hand. However in context of high-dimensional data sets the number of samples could be too low, whereas the assumption of an adequate model might be difficult for multitemporal and multisource data sets. Thus, the performance of the classifier might decrease. However, Richards and Jia (2008) have introduced a concept to handle high dimensional data sets, with a small number of samples. The approach is based on the extension of the training set by neighboring pixels that are likely belonging to the same class. The expand training set enables the derivation class statistics for high dimensional imagery. Another limitation could occur due to differences between spectral and information classes. Whereas the final classification aims at the assignment of each pixel to an information class (i.e., the specified land-cover classes), the classification algorithm can only differentiate classes within the spectral feature space. In this context, it could be possible that a land cover class is represented by different spectral classes, which have to be considered for an adequate training process.

8.2.2 Decision Trees

Decision trees (DT) are a non-parametric classification method, which can handle diverse data sets, including categorical variables. An overview on DT concept is given by Safavian and Landgrebe (1991). The basic idea of this technique differs from other concepts in pattern recognition: Whereas most classifiers use the entire features space at once and make a single membership decision per class, a DT is based on a multistage or hierarchical concept that does not necessarily include the entire feature space. During the training phase the DT successively partition the training data into an increasing number of smaller, more homogenous groups. That is done by producing efficient test rules, estimated from the training data. Due to its nature, this process might lead to a splitting of thematic classes into more than one final node.

The general structure of a DT comprises a root node, which includes all samples, internal or split nodes that constitute the successive tests, and the final leave nodes, representing the different classes. After a tree is generated based on the training samples, an unknown sample is handed-on from one internal node to the other following the rules outputs. It is classified according to the class that corresponds to the leaf node in which it finally ends up.

Most DT are of binary nature, using only one feature at each node (the one with highest discrimination power). In contrast to this, multivariate DT use more than one feature and rules base on combinations of features. However, in the same time the computational complexity is increased compared to the little training times of binary DT. In the following discussion, we focus on more common binary trees.

In general, the handling of DT is rather simple and their training time is relatively low compared to computationally complex approaches such as ANN (Friedl and Brodley 1997, Pal and Mather 2003). Another advantage is the transparent classification scheme, which enables a direct interpretation of the decision with regard to the impact/consequence of individual features, such as spectral regions in multispectral data or acquisition dates in multitemporal datasets.

As mentioned above, the discrimination of the training samples and thus the tree induction, is performed by the decision rules, which are derived from the training data. The fundamental idea underlying the determination of rule T at node N is that of purity: The algorithm aims at generating the descendent nodes as pure as possible. However, a contrary formalization of the strategy is more convenient, using the impurity rather than the purity (Duda *et al.* 2001).

Several measurements exist (Zambon *et al.* 2006) and they generally follow similar concepts. If all samples within a specific node belong to the same class, the node is pure and the measurement value is 0; the value is large, if all samples classes are equally distributed over the classes. A common criterion in this context is the *Gini index* (Breiman *et al.* 1984), which is dominantly peaked in cases with equal class distributions (Duda *et al.* 2001). It is described as:

$$g(N) = 0.5 \cdot \left[1 - \sum_{j=1}^{nc} p^2\left(\omega_j\right) \right] \tag{8.5}$$

The maximum value is defined by the number of classes and is given by (1-n/n_c). By using an impurity measurement as the *Gini index*, it is possible to select an appropriate test, i.e., the test that decreases the impurity as much as possible, resulting in the purest possible descendent nodes. This can be defined as:

$$\Delta g(T) = g(N) - P(N_1) \cdot g(N_1) - P(N_2) \cdot g(N_2) \tag{8.6}$$

where N_1 and N_2 denote the two descendent nodes with their corresponding impurity measures g(N_1) and g(N_2). $P(N_1)$ and $P(N_2)$, with $P(N_1)+P(N_2)=1$, indicate the fraction of samples from the initial node N that will assign to the two new, descendent nodes, when test T is used. The best split rules is maximizing Eq. 8.6, which consequently minimizes the impurity of the descendent node. Nevertheless, one should bear in mind that such an approach leads to a local optimization, and of course it is not guaranteed that the decision also results in a global optimum.

Like other classifiers (e.g., neural networks) decision trees can be easily overfitted and different approaches exist, which can apply during or after the tree generation to avoid complex tree structures (Esposito *et al.*, 1997). In general, the tree induction is stopped, when 1) all training samples in a node belong to the same class, 2) have the same attribute values, respectively, 3) the number of samples within a node is under a certain threshold, and 4) the improvement in the error rate is too low, respectively. In contrast to this (pre-)pruning methods, (post-)pruning methods are applied after the tree was fully grown. They eliminate inefficient and weak branches of a tree to produce a more compact one, which is consequently less overfitted. This is done by evaluating the tree performance with the training samples and an additional pruning test set respectively. Some of these techniques start at the final leaves and proceed towards the node (bottom-up approach), whereas top-down approaches begin at the root node (Esposito *et al.* 1997). In Pal and Mather (2003), four different pruning methods were compared in context of remote sensing. Although the authors preferred Quinlan's error-based pruning method, the overall accuracies achieved by the differently pruned trees show only small varieties.

8.2.3 Support Vector Machines

Support vector machines (SVM) are well known in the field of machine learning and pattern recognition (Vapnik 1998, Schölkopf and Smola 2002). The approach is based on an optimal linear separating hyperplane (OSH), which is fitted to the training samples of two classes within a multi-dimensional feature space. The optimization problem that must be solved is based on structural risk minimization. It aims at maximizing the margins between the hyperplane and the closest training samples, the so-called support vectors (Vapnik 1998). Hence, the classifier only requires samples close to the class boundary and works well with small training sets, even when high dimensional data sets are classified (Melgani and Bruzzone 2004, Pal and Mather 2006). Other valuable properties of SVM are their ability to handle noisy patterns and multimodal feature spaces. A detailed introduction to the general concept of SVM is given by Burges (1998) and Schölkopf and Smola (2002).

For a binary classification problem in a d-dimensional feature space $\Re^d$, $x_i \in \Re^{d,}$ i=1,..., n is a training set of n samples with their corresponding class labels $y_i \in \{-1,+1\}$. The optimal separating hyperplane $f(x)$ is described by a normal vector $w \in \Re^d$ and the bias b, where $|b|/||w||$ is the distance between the hyerplane and the origin, with $||w||$ as the Euclidean norm from w:

$$f(x) = w \cdot x + b \tag{8.7}$$

The support vectors lie on two canonical hyperplanes $w \cdot x + b = \pm 1$ that are parallel to the optimal separating hyperplane. The margin maximization leads to the following optimization problem:

$$\min \left[\frac{w^2}{2} + C \sum_{i=1}^{L} \xi_i \right] \tag{8.8}$$

where the slack variables ξ_i and the regularization parameter C are introduced to deal with misclassified samples in a non separable case. The constant C is used as a penalty for cases that lie on the wrong side of the hyperplane. Effectively it controls the shape of the solution of the decision boundary. Thus, it affects the generalization capability of the SVM, e.g. a large value of C might cause an over-fitting to the training data.

Kernel methods enable the above linear SVM approach to be extended for non-linear separable cases. Based on a non linear mapping of the data into a higher dimensional feature space, an OSH, which appears non linear in the original feature space, more complex classes that are not linearly separable in the original feature space can be fit to such complex classes. The input sample x can be described by $\Phi(x)$ in the new high-dimensional space. The computationally extensive transform into a higher dimensional space is reduced by using a positive definite kernel k, which meets Mercers conditions (Schölkopf and Smola 2002):

$$\left(\Phi(x_i)\cdot\Phi(x_j)\right)=k(x_i,x_j) \tag{8.9}$$

Thus, the final hyperplane can be defined as:

$$f(x)=\left(\sum_{i=1}^{L}\alpha_i y_i k(x_i,x_j)+b\right) \tag{8.10}$$

with α_i Lagrange multipliers. A detailed step by step derivation of the equations is given, for example, in Burges (1998).

The kernel-trick enables to implicitly work within the newly transformed feature space. Explicit knowledge of Φ is not required and only the kernel function is needed. Consequently, the training process involves the estimation of the parameter(s) of the kernel function in addition to the regularization parameter C. In the literature, different concepts for an automatic model selection have been introduced, which are usually based on a cross-validation procedure (Chapelle *et al.* 2002, Chung *et al.* 2003). Widely used kernel functions are the polynomial of various orders and the Gaussian radial basis function (RBF) kernel (Schölkopf and Smola 2002). The latter is perhaps the widest used in remote sensing. An RBF kernel can handle more complex, non-linear class distributions, when compared to a simple linear kernel, which is only a special case of the Gaussian RBF kernel (Keehrti and Lin 2003). On the other hand, a polynomial kernel is based on more parameters than a Gaussian kernel and the computational complexity of the model is increased. The RBF kernel is defined as:

$$k(x_i,x_j)=\exp\left[-\gamma\left\|x_i-x_j\right\|^2\right] \tag{8.11}$$

where the parameter γ controls the width of the kernel.

In contrast to other algorithms, which directly provide a class label as classifier output (e.g., DT) and probabilities of class memberships (e.g., MLC) respectively, an SVM provides distances of each pixel to the hyperplane. These values are used, depending on the multiclass strategy, to determine the final classification result.

SVM are designed for binary classification problems, which normally do not exist in the context of remote sensing applications. In the literature, several multiclass strategies have been introduced, whereas two main strategies exist: The *one-against-one* strategy and the *one-against-rest* strategy, which are based on the separation of the multiclass problem in several binary classifications (Foody and Mathur 2004).

Let $\Omega = \{\omega_i\}$ with i=1,...,c be a set of c possible class labels (i.e., land cover classes). The one-against-one strategy trains $c(c$-1)/2 individual binary SVM, one for each possible pair of classes ω_i and ω_j ($\omega_i \neq \omega_j$). The sign of the distance to the hyperplane is used for the one-against-one voting scheme. For the final decision, the score function S_i is computed for each class ω_i which sums all positive (i.e., sgn=+1) and negative (i.e., sgn=-1) votes for the specific class. The final class for sample x is predicted by a simple majority vote. Using the $c(c$-1)/2 SVM outputs, a simple majority vote is applied to compute the final class membership:

$$S_i(x) = \sum_{\substack{j=1 \\ j \neq i}}^{n} \mathrm{sgn}\left(f_{ij}(x)\right) \tag{8.12}$$

In the case of the one-against-rest approach, a set of c binary classifiers is trained to separate each class from the remaining ones and the maximum decision value of the individual SVM classifier is used to define the class membership. In contrast to these multiclass strategies one-shot SVM are an alternative approach (Seebald and Bucklew 2001, Hsu and Lin 2002, Mathur and Foody 2008).

More recently, approaches have been introduced to convert the rather abstract decision measures of SVM, i.e., distances of data points to the hyperplane, into more intuitive class probabilities (Wu *et al.* 2004). Based on the idea of logistic regression a sigmoidal function is fit to the distance values of each binary SVM and transferred into probability values for individual classes (Lin *et al* 2003). This way, class memberships can be derived from probability values (similar to the concept of maximum likelihood classification) and possible ambiguities after majority vote are avoided.

8.3 Semi-Supervised Approaches

One of the critical issues of supervised techniques is the classification of high-dimensional data, with a relatively small number of training samples. As a matter of fact, an increasing number of features could decrease the classification accuracy. This effect is known as the Hughes phenomenon (Hughes 1968). The second critical issue consists in problems where spatial correlation exists, because training and test samples are acquired in the same region. On the other hand, the spectral signature of classes in the entire spatial domain of the scene is usually non-stationary if training and test samples are collected from different areas. To deal with such problems, *semi-supervised learning* (SSL) can be used, which is particularly interesting in the context of hyperspectral imagery (Shahshahani and Landgrebe 1994, Bandos *et al.* 2006, Chi and Bruzzone 2007). These methods, such as self-labeling approaches (Shahshahani and Landgrebe 1994, Joachims

1999), low density separation SSL approaches (Chapelle and Zien 2005, Chi and Bruzzone 2007) and graph-based approaches (Bandos *et al.* 2006, Zhu *et al.* 2003, Zhou *et al.* 2004), consider both labeled and unlabeled samples during the training process. A detailed overview is given in Zhu *et al.* (2003) and Chi (2006).

The first explicit use of unlabeled samples for semi-supervised learning in the context of remote sensing was proposed by Shahshahani and Landgrebe (1994). This is an iterative process, referred to the self-labeling approach in the SSL setting, where both labeled and unlabeled samples are used to estimate covariance matrices with the aid of the Expectation-Maximization (EM) algorithm (Dempster 1977). The work was improved by a robust parameter estimation method in an iterative framework (Tadjudin and Landgrebe 2000, Jackson and Landgrebe 2002). The aforementioned approaches are based on generative models (i.e., a full probability model of all variables) using Gaussian distribution. The self-labeling approach is further extended to the discriminative framework with kernel-based methods, such as SVM (Joachims 1999, Chi and Bruzzone 2005, Bruzzone *et al.* 2006).

The generalization of SVM with limited labeled samples is poor and semi-supervised algorithms based on SVM (S^3VMs for short) were proposed in (Vapnik 1998, Chapelle and Zien 2005, Chapelle *et al.* 2006, Chi and Bruzzone 2007). An alternative strategy for SVM with limited labeled samples was given by Foody and Mathur (2006), which is based on the generation of small, but synthetic training sets, containing mixtures of samples. In contrast to a conventional SVM classifier, the S^3VM approach aims at finding a hyperplane that separates both labeled and unlabeled data with maximum margins. However, with the additional penalization term of unlabeled samples integrated to the objective function of SVM, the objective function of S^3VM is non-convex such that the solution may be stuck in many local minima. This makes the optimization problems difficult. Nonetheless, S^3VM is implemented on the cluster assumption, i.e., the samples in the same cluster belong to a single class and so the obtained decision boundary is set in low density regions. This is referred to as low density separation approach.

An alternative popular SSL implementation can be carried out on manifold assumption, which is graph-based semi-supervised method. Graph-based semi-supervised methods define a graph where the vertices are labeled and unlabeled samples and edges represent the similarity/dissimilarity among samples in the datasets. Many graph-based semi-supervised methods can be viewed as estimating a function over the graph (Zhou *et al.* 2004, Zhu 2005). Usually, the function has the following two properties: 1) It should be close to given labels on the labeled data; and 2) should be smooth on the whole graph. This can be expressed in a regularization framework, where the former is related to the loss function and the latter is a regularizer. The smoothness is a key issue for the successful semi-supervised graph-based methods and is usually measured by the graph Laplacian. Usually, there exists a closed-form solution, especially in the normalized Laplacian. Different graph-based semi-supervised methods differ due to the choice of the loss function and a regularizer. This approach was used for the classification of hyperspectral data (Bandos *et al.* 2006).

A more detailed introduction to semi-supervised SVM is given in (Chapelle *et al.* 2006). A brief introduction is given below. In contrast to a common supervised SVM, the semi-supervised approach considers both labeled $(x_i)_{i=1}^{n}$, i.e., the class membership is known, and unlabed $(x_i)_{i=n+1}^{n+m}$ training samples within the cost function:

$$\min_{\mathbf{w},\xi}\left\{\frac{1}{2}\|\mathbf{w}\|^2 + C\sum_{i=1}^{n}\xi_i + C^*\sum_{i=n+1}^{n+m}\xi_i\right\} \tag{8.13}$$

$$s.t.\begin{cases}\forall_{i=1}^{n}: y_i(\mathbf{w}^{\mathrm{T}}\mathbf{x}_i + b) \geq 1-\xi_i, \xi_i \geq 0\\ \forall_{i=n+1}^{n+m}: \left|\mathbf{w}^{\mathrm{T}}\mathbf{x}_i + b\right| \geq 1-\xi_i, \xi_i \geq 0.\end{cases}$$

Similarly to the supervised SVM, to be able to handle non-separable training and unlabeled data, the slack variables $(\xi_i)_{i=1}^{n}$ and $(\xi_i)_{i=n+1}^{n+m}$, and the associated penalty values C and C^* are introduced.

Joachims (1999) proposed a heuristic iterative algorithm for the quadratic implementation using the Lagrange theory; an implementation algorithm was intrdouced as S^3VM^{Light}[1]. This is also referred to the self-labeling approach based the discriminative approach and further developed to be applied in remote sensing data (Bruzzone et al. 2006). Due to the non-convexity of the cost function in Eq. 8.13, we can use different optimization techniques on the primal formulation directly. Firstly, all the constraints for labeled and unlabeled samples are put in the objective by rewriting Eq. 8.13 as follows:

$$\frac{1}{2}\mathbf{w}^{\mathrm{T}}\mathbf{w} + C\sum_{i=1}^{n}H\left(y_i\mathbf{w}^{\mathrm{T}}\mathbf{x}_i\right) + C^*\sum_{i=n+1}^{n+m}H(\mathbf{w}^{\mathrm{T}}\mathbf{x}_i) \tag{8.14}$$

We can replace the loss of labeled samples as quadratic one and the loss of unlabeled samples (i.e., max(0,1-|t|)) in Eq. 8.14 by the Gaussian approximate symmetric hinge loss $\exp(-st^2)$ with s=3 and then a local optimization technique, e.g., conjugate gradient descent is applied for the solution of Eq. 8.14, which is denoted as ∇S^3VM in this chapter. Although the objective function of S^3VM is non-convex (Joachims 1999, Chapelle and Zien 2005), the objective function in

[1] soft code is available at: http://svmlight.joachims.org.

Eq. 8.14 of S^3VM is well-designed since it combines the powerful regularization with S^3VM carried out on the cluster assumption (Chapelle and Zien 2005). Accordingly, for the suboptimal problem different implementation techniques can yield significantly different results. For this reason, it is important to better analyze the optimization problem of S^3VM (Chapelle and Zien 2005).

In the following, a global optimization technique (e.g., the continuation approach (Wu 1996)) is introduced in more detail to carry out the S^3VM in the primal, denoted as cS^3VM (Chapelle and Zien 2005). The basic idea is as follows: the original objective function is smoothed with a large enough smoothing degree γ_0 such that there only exists a global minimizer, where the optimization is easy and the global minimum can be found. Then, the smoothing degree γ_i, i=1,…,N decreases to obtain a less smoothed version of objective function and the new minimum is computed on such function. Note that the local optimization search starts from the minimizer provided by the previous local search. The algorithm iterates until there is no smoothing[2]. Formally, we let the regularized functional of S^3VM be convoluted with a multivariate Gaussian[3]

$$g(\mathbf{w}) = \frac{1}{(2\pi\gamma)^{d/2}} \exp(-\frac{\mathbf{w}^T\mathbf{w}}{2\gamma}) \tag{8.15}$$

The convolution is done only on w. Since it is easy to convolve two Gaussians together, we replace the loss of unlabeled samples by the Gaussian approximate symmetric hinge loss as ∇S^3VM and keep using the hinge loss for labeled samples, i.e., we let the following functional

$$\mathsf{L}(\mathbf{w}) = \frac{1}{2}\mathbf{w}^T\mathbf{w} + C\sum_{i=1}^{n} H\left(y_i\mathbf{w}^T\mathbf{x}_i\right) + C^* \sum_{i=n+1}^{n+m} \tilde{H}(\mathbf{w}^T\mathbf{x}_i) \tag{8.16}$$

be the original objective function. Therefore, we have the convolved loss in the form

$$\begin{aligned} \mathsf{L}^\gamma(\mathbf{w}) &:= \int \mathsf{L}(\mathbf{w} - \Delta\mathbf{w}) g(\Delta\mathbf{w}) d\Delta\mathbf{w} \\ &= \frac{1}{2}\mathbf{w}^T\mathbf{w} + \frac{1}{2}\gamma d + C\sum_{i=1}^{n} \frac{\gamma\|\mathbf{x}_i\|}{\sqrt{2}} \left[\frac{\exp(-b_i^2)}{\sqrt{\pi}} - b_i \operatorname{erfc}(b_i)\right] + \\ &+ C^* \sum_{i=n+1}^{n+m} \frac{1}{\sqrt{a_i}} \exp\left(\frac{-s(\mathbf{w}^T\mathbf{x}_i)^2}{a_i}\right) \end{aligned} \tag{8.17}$$

---

[2] when $\gamma_i \rightarrow 0$ the original function is recovered.

[3] the multivariate Gaussian has a zero mean and identical variance $\sqrt{\gamma}$ i.e., N $(0, \sqrt{\gamma})$

$$\text{with} \begin{cases} a_i = 1 + 2\gamma s \|\mathbf{x}_i\|^2 \\ b_i = \dfrac{y_i(\mathbf{w}^\mathrm{T}\mathbf{x}_i) - 1}{\sqrt{2\gamma}\|\mathbf{x}_i\|} \end{cases}$$

where erfc() denotes the complementary error function $\mathrm{erfc}(y) = \frac{2}{\pi}\int_y^{\infty} e^{-t^2} dt$.

8.4 Multiple Classifier Systems

Instead of using only one classifier, one may use sets of classifiers to generate the final results. Land cover classifications based on such multiple classifier systems have been performed successfully during recent years (Benediktsson *et al.* 2007). They are particularly interesting for multisource and high-dimensional data sets (Benediktsson and Swain 2002, Benediktsson and Kanellopoulos 1999, Briem *et al.* 2002, Ham *et al.* 2005. Waske and Benediktsson 2007, Waske and van der Linden 2008) It has been shown both experimentally and theoretically that the classification accuracy can be increased by combining different independent classifiers in a multiple classifier system (Schapire 1990, Turmer and Gosh 1996). In addition, the variance and the bias of the classifier might be reduced in doing so. The general idea of this concept is based on the assumption that independent classifiers produce individual errors, which are not produced by the majority of the other classifiers. Consequently, the diversity between the classifiers is an important fact in order to improve the classification accuracy (Polikar 2006).

Sets of independent classifiers can be achieved in various ways. On the one hand, the input data may be resampled in terms of feature or sample selection. On the other, the same set of training data may be used to train different classification algorithms. In either case, such procedures lead to different outputs, which are combined to create the final results. In order to produce the final results a simple majority vote based on the individual outputs is often sufficient, nevertheless more sophisticated voting strategies such as weighted voting or meta classifiers are possible.

Whereas many multiple classifier systems are based on the same classifier, other approaches combine different algorithms (Benediktsson and Kanellopoulos 1999, Steel 2000, Waske and van der Linden 2008). In doing so, the different advantages of the algorithms can be joint and perhaps one classifier is more adequate to describe one part of the feature space whereas another method is performs more accurate in another part of the feature space.

However, many classifier systems use variants of the same classifier method, the so called *base classifier*. Although these ensemble techniques that base on resampled input data are not restricted to decision trees (Hansen and Salamon 1990, Kim *et al.* 2003), DT classifiers are particularly interesting due to their simple handling and fast training time. Various strategies to resample training data have been introduced. *Bagging* or *boosting* relate to modifications of the set of training samples, whereas the term *random feature selection* or *attribute bagging* is used to describe approaches that alter the sets of features used with the training data. In the following an overview of the main strategies is given: *Boosting* was originally introduced by Schapire (1990) to improve the performance of a weak classifier, where *weak* means slightly better than random guessing. The concept underwent various developments and the AdaBoost.M1 approach is nowadays a widely used boosting concept in pattern recognition (Freund and Schapire 1996). Boosting is an iterative process that successively concentrates on the more difficult training samples, i.e. misclassified samples. In the beginning a base classifier is trained on the samples x, which have the equal weight $1/n$, where n denotes the number of training samples. During the boosting process, the weights of the training samples are modified after each iteration, using the error rate of the classifier. A new weight for correct classified training sample x in iteration I is defined as:

$$\omega_{Ix} = \varepsilon/(1-\varepsilon) \tag{8.18}$$

where the error rate is defined by the sum of the weights of all misclassified samples in this iteration:

$$\varepsilon_I = \frac{1}{n}\sum_j \omega_j \,. \tag{8.19}$$

After each iteration the weights are normalized so that the sum is one. The next classifier is trained on these newly weighted samples. The final classification is generated by a weighted voting over the individual outputs, where a classifier C_I is weighted by $\log(1/\omega_I)$. However, a minimum accuracy is required for the base classifier and the boosting process would stop if the classification error is greater than 50%. The technique tends to exhibit virtually no overfitting when the data is noiseless. Moreover boosting can reduce both the variance and the bias of the classification. On the other hand, the sensitivity to noisy training data is a disadvantage of boosting.

Bootstrap aggregating (*bagging*) by Breiman (1996) describes the random generation of training sample subsets also known as bootstrapped aggregates or bags. The approach is based on the random and uniform selection – with replacement - of n samples from a training set of same size n. This way some training samples can be selected several times in the same training set and perhaps other samples are not considered in this particular bag. Afterwards an individual classifier, e.g., a decision tree, is trained on each of these training sample sets, resulting in various classifier predictions. In Briem *et al.* (2002) bagging was used to classify multi-source remote sensing data. Although the overall accuracy of single classifiers was significantly increased, boosting usually outperforms the bagging technique in terms of accuracy.

In other studies subsets of features are generated, instead of using training sample subsets – a technique referred to as *random feature selection* and *attribute bagging* (Ho 1998, Bryll *et al.* 2003). In contrast to the two aforementioned data partitioning strategies, the set of training samples remains unchanged by this concept. For the base classifier, a user defined number of feature subsets is created. In contrast to the conventional bagging, the method usually selects a user defined number of features without replacement. In Waske *et al.* (2006) this concept was used for classifying a time series of SAR data. In doing so noise in the classification results was significantly decreased and a more accurate map could be generated.

In *Random Forests* (RF), which is a classifier based on decision trees, both strategies, i.e., the bagging of samples and the random feature selection, are combined (Breiman 2001). Each tree within the ensemble is trained on a subset of the original training samples; in addition the split rule at each tree node is determined, using only a randomly selected feature subset of the input data. In doing so, a diversity of trees with different split rules is generated. Contrary to normal decision trees, the trees in a random forest are usually fully grown and not pruned. Again, different (preliminary) classification maps are obtained by majority vote. The number of selected features within the subset is user-defined, and the parameter is usually set to the square root of the number of input features. The computational complexity of the individual DT classifier is simplified, by reducing the number of features at each split. This enables random forests to handle high-dimensional data sets. In addition, the correlation between the classifiers is decreased, which generally improves the performance of a classifier system. The approach has been successfully used in diverse remote sensing applications, and has often performed more accurately or at least equally well compared to other classifiers, including sophisticated strategies such as support vector machines and other ensemble strategies (Pal 2005, Gislason *et al.* 2006, Waske and van der Linden, 2008).

8.5 Applications and Experiments

Evolving technologies and data availability lead to increased information dimensionality provided by current and future Earth Observation systems monitoring the complex and dynamic land cover of the Earth. Advanced and reliable algorithms that fully exploit this enhanced Earth Observation information are needed to deliver consistent data sets of the Earth land cover describing its spatial distribution and change over time.

In the following subsection the power of advanced classifiers is presented for three case studies exploiting the enhanced information dimensionality of multi-temporal, hyperspectral and multi-source EO data sets.

8.5.2 Classification of Multitemporal SAR Data

8.5.2.1 Data Set and Preprocessing

The study site for the first experiment is located near the city of Bonn in Germany. The almost flat area is dominated by agriculture and thus characterized by the typical spatial patterns that result from differences in the phenology of planted crops. The size of the agricultural parcels is approximately 5 ha, with cereals and sugar beets being the main crops. The experiments were conducted on a set of multitemporal SAR images, consisting of 9 Envisat ASAR alternating polarization (HH/HV) and ERS-2 precision images. Hence, the multitemporal SAR data set comprised information from varying phenological stages and different data types. While it seemed difficult to distinguish between specific classes on a single date, the use of the entire multitemporal data set appeared more useful. The SAR data was calibrated to backscatter intensity following the procedures suggested by Lauer *et al.* (2002). In addition, an enhanced Frost filter was applied to reduce the speckle noise. Finally, the SAR images were orthorectified with a spatial accuracy of approximately one pixel, using a digital elevation model, orbit parameters and a multispectral reference image.

8.5.2.2 Methods

In this experiment the following eight classes were investigated: *Arable crops*, *Cereals*, *Forest*, *Grassland*, *Orchards*, *Rapeseed*, *Root crops*, and *Urban*. Independent training and validation sample sets of respective 300 and 500 samples per class were generated from the field survey areas using equalized random sampling. Two multiple classifier systems were applied to the data set: (1) a boosted decision tree and (2) Breimans' Random Forests. The results were compared to the results achieved by a traditional maximum likelihood classifier and an SVM classifier. Random Forests application is based on a FORTRAN code by Breiman and Cutler[4]. For the boosted DT a C4.5 implementation was used as a base classifier. The number of features in the Random Forest was set to 3 and up to 500 iterations were performed for both classifier systems.

The SVM was trained, following the OAO approach. The training of the SVM classifier with a Gaussian kernel was performed using *imageSVM*[5] (Janz *et al.*

---

[4] Software available at: (http://www.stat.berkeley.edu/~breiman/RandomForests)

[5] Software is available at http://www.hu-geomatics/imageSVM

2007), a freely available IDL/ENVI implementation, that uses the *LIBSVM*[6] approach by Chang and Lin (2001) for the training of the SVM. The kernel parameters of C and γ were determined via grid-search, using a cross validation. Possible combinations of C and γ were tested in user defined ranges and the optimal combination for γ and C was selected to train the final SVM classifier.

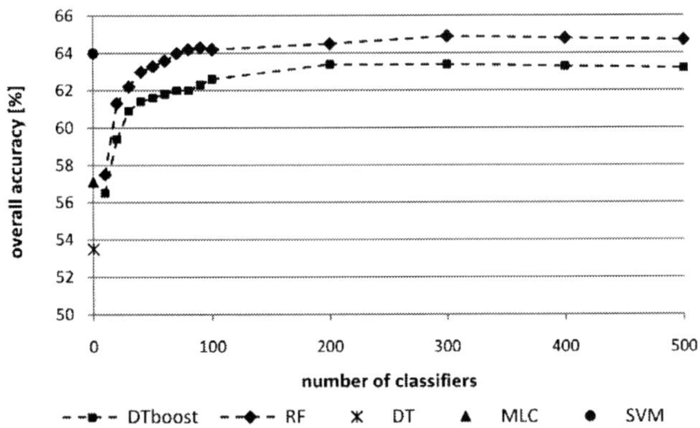

Fig. 8.1 Overall accuracy [%], using different classification methods and number of classifiers within the ensembles (i.e., boosting and RF)

8.5.2.3 Results and Discussion

The accuracy assessment shows that the ensembles of DT classifiers significantly improve the results of a single DT (Fig. 8.1). They perform clearly better than a maximum likelihood classifier (MLC) and similar to SVM. A simple DT achieves 53.5%, whereas the overall accuracy of a simple maximum likelihood classifier is 57.1%. The main reason for the weak performance of the MLC might be the assumption of a Gaussian distribution, which is not fulfilled in context of multitemporal SAR imagery. A boosted decision tree achieves 63% overall accuracy. However, the inherent noise in the SAR data might be a reason for the weaker performance of the boosted DT, compared to the RF and SVM. Boosting is performed in series and is iteratively focusing on difficult samples. This fact makes the concept relatively sensitive to noise and boosting tends to overtrain a classifier. In contrast to this RF and SVM are relatively unaffected by this data inherent noise and yield highest accuracies (64.7% and 64.0%). The increase of overall accuracy with a growing number of classifiers also demonstrates the positive effect of including several classifiers within the ensemble: Whereas a relatively small RF system (10 classifier) achieves an accuracy below 58% the accuracy is significantly increased by including additional classifiers and already 30 RF iterations result in an accuracy above 62%. Overall, the accuracy increases with the number of classifiers within the ensemble and asymptotically reaches

[6] Software is available at http://www.csie.ntu.edu.tw/~cjlin/libsvm

values up to approximately 65%. The boosted DT shows a similar behavior, when more iterations are performed.

A visible assessment demonstrates the better performance of the RF and SVM as well as the positive effect of increasing the number of classifiers within an ensemble. The map produced by a maximum likelihood classifier and a small classifier ensemble respectively shows the general structures of the study site (Fig. 8.2). However the entire map appears noisy, sometimes to a degree that the true land cover class of a homogeneous parcel cannot be assigned to the entire area. Boundaries – particularly along individual agricultural fields – appear blurred; consequently they are hard to be identified. This confusion and noise is best reduced by sophisticated algorithms and in general features appear with clear edges and can easier be assigned to a specific land cover class.

Irrespectively of the classifier, the class accuracies show relatively high variances and generally the classification accuracy for the land cover classes arable crops and orchards is lower compared to other classes (not presented in detail). A reason for the lower accuracies could be the diversity within these two land cover classes. Both classes are characterized by spatial varieties, e.g. the surface beneath the orchards is generally covered by *grassland*. Moreover *orchards* are often characterized by a relatively open canopy. Thus this class can appear as a mixture between grassland and forest. Classes like *cereals* and *root crops*, on the other hand, are less variable and consist mainly of winter wheat and sugar beets. Consequently, *cereals* and *forests* are classified with relatively high accuracies, followed by *grassland*.

Overall Random Forests demonstrate the full advantage of multiple classifier ensembles and are probably the most powerful assembly of DTs. It has previously been seen that small ensembles improve the classification accuracy significantly. The simple handling and fast training time – compared to boosting and SVM - are some positive side effects, which makes RF particularly interesting in the context of operational monitoring systems. Nevertheless, a further improvement of the classification results would be appropriate. In this context, the integration of multi-frequency data and multispectral imagery could be advantageous, as well as the use of spatial contextual information. The concept of RF classifier ensembles will remain applicable and beneficial for such large multi-source data stacks (Waske and van der Linden 2008). The need for efficient classifiers on SAR imagery becomes even more important in regard to the current and upcoming missions with various polarizations and increasing repetition rates.

8.5.3 Semi-Supervised Classification of Hyperspectral Data

Within the semi-supervised land cover study we are focusing on the mapping of hyperspectral data. For the comparison, we conducted experiments with the supervised SVM and dual S^3VM, i.e., S^3VM^{Light} and primal S^3VMs including ∇S^3VM and cS^3VM.

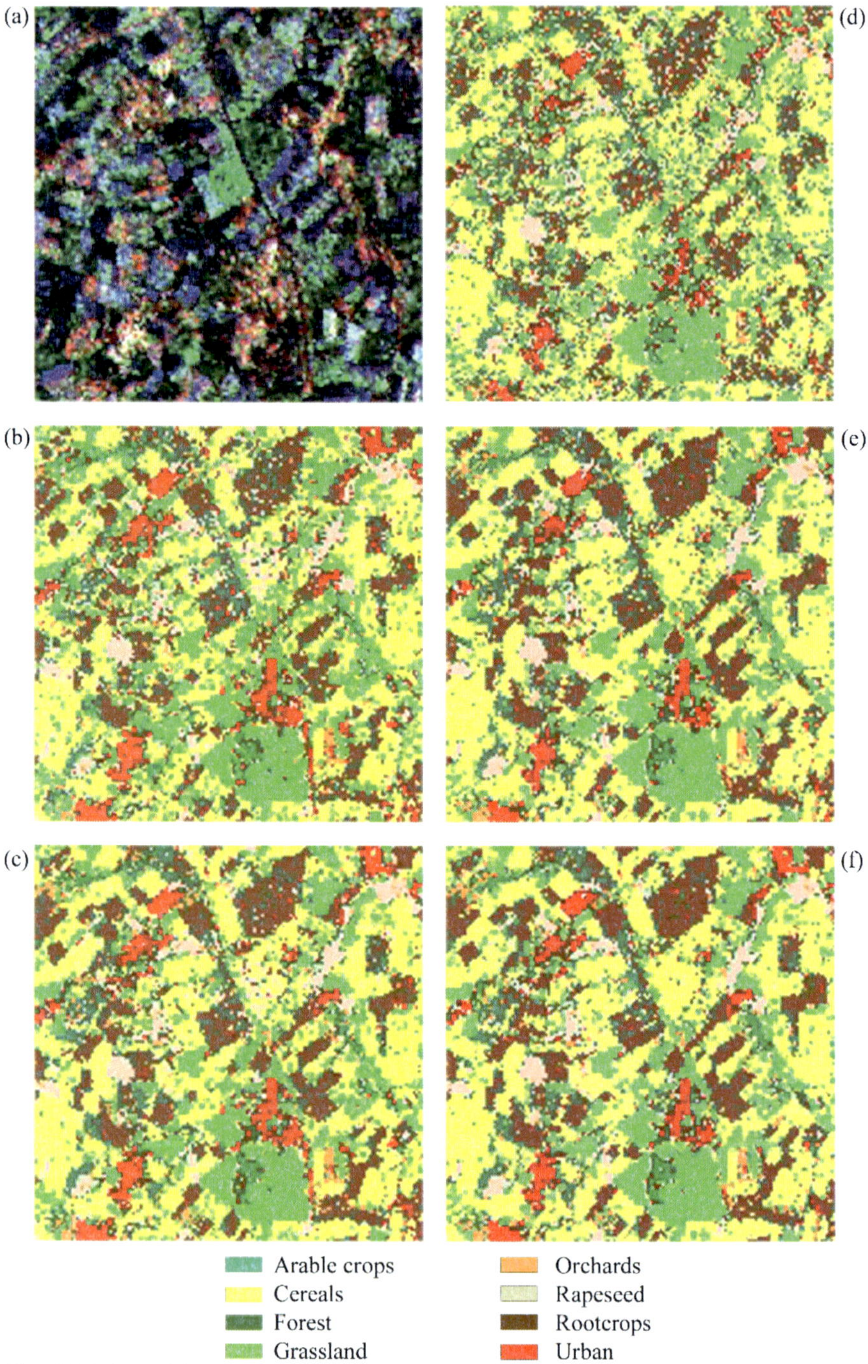

Fig. 8.2 Subset of the original SAR time series (a) and classification maps using MLC (b), SVM (b) and a RF with 5 (d), 50 (e) and 500 iterations (f).

8.5.3.1 Data Set and Preprocessing

The data analyzed in this study were acquired on May 31 2001 by the NASA EO-1 satellite over the Okavango Delta, Botswana. The Hyperion sensor on EO-1 acquired the data at 30m spatial resolution over a 7.7 km strip in 242 bands ranging from 400-2500 nm of the electromagnetic spectrum. The study area is highly affected by seasonal floods (Chi and Bruzzone 2007). The experiment is focusing on 14 land cover classes (see Table 8.1). For the experiments the reference data was split into a Spatially Disjoint (SD) and Spatially Adjoin (SA) data sets. In the beginning the image data was preprocessed, normalizing the data in a range [-1,1], for optimizing the search strategy of the kernel parameters (Chi 2006).

Table 8.1 The original training (reference) dataset, Spatially Disjoint (SD) test dataset and Spatially Adjoin (SA) test dataset

Land cover class	Reference Set	SD Test Set [7]	SA Test Set [8]
Water	270	126	68
Hippo grass	101	162	26
Floodplain grasses 1	251	158	63
Floodplain grasses 2	215	165	54
Reeds	269	168	68
Riparian	269	211	68
Firescar	259	176	65
Island interior	203	154	51
Acacia woodlands	314	151	79
Acacia shrublands	248	190	62
Acacia grasslands	305	358	77
Short mopane	181	153	46
Mixed mopane	268	233	67
Exposed soils	95	89	24

8.5.3.2 Methods

To study the quality and quantity problems, two datasets are investigated, based on the selection of the test dataset (Chi 2006, Chi and Bruzzone, 2007). One is the Spatially Adjoin (SA) dataset where the training and test data are collected in the same area. The second one is the Spatially Disjoint (SD) dataset where test and training data are acquired in the different areas (see Table 8.3).

In case of the SA dataset with 3248 labeled samples, ten randomly sampled partitions of the reference data set were drawn and subdivided into sets for training and testing of 75% and 25% (818 samples) respectively. In order to investigate

[7] This spatially disjoint test set was not included in the training set.

[8] This test set is generated from the labeled set. It covers 25% samples of labeled set.

the impact of the quantity of labeled data on classifier performance and to underline the effectiveness of semi-supervised algorithms, experiments were conducted on small training sample sets. Thus the training data were then sub-sampled to obtain ten splits comprised of 50%, 30%, 15%, and 5% of the original labeled data. For our problems, only small-size training samples made up of 5% (156 samples) originally labeled data are taken into account for the learning. For this dataset, these 10 sets of 5% labeled samples and 25% test/unlabeled samples are used for the semi-supervised learning. All classifiers were evaluated using the ten sets of test data containing 25% of the original labeled samples. For the comparison, the classification accuracies using supervised classifiers in different ratio were also reported in the following.

The SD data sets were generated in a similar way, except that the test set (containing 2494 samples) was acquired from a geographically separate location within the test site. Therefore, in the SSL setting, the SD training set contain the same labeled training data (5% original training set) as the SA and the test set containing 2494 samples (Chi and Bruzzone 2007). In the semi-supervised setting, the training dataset contains labeled samples, unlabeled samples and test samples (if test and unlabeled samples are different). In the following experiments, we assume that the test data as the unlabeled data for both SA and SD datasets.

Table 8.2 Average overall test accuracies over 10 splits for the SA and the SD test sets using the supervised SVM and the semi-supervised SVMs.

	supervised	semi-supervised		
	SVM	S^3VM^{Light}	∇S^3VM	cS^3VM
SA	90.0	89.8	91.2	90.3
SD	70.7	71.9	74.0	69.8

In the following experiments, 5-fold cross validation is utilized to evaluate the experimental results in semi-supervised learning and supervised learning. Gaussian RBF kernels are chosen for all the experiments since they are good general purpose kernels. In supervised SVMs, there are two hyperparameter σ ($2^{\{0,1,\ldots,5\}}$) and C (=10 and 100). In S^3VM^{Light}, ∇S^3VM and cS^3VM, an additional penalization parameter C^*should be selected for semi-supervised learning. To deemphasize the influence of unlabeled samples, C^* should be smaller than that of labeled samples. Consequently, the proportion of C, defined as C_p, is considered instead of C^* as one of hyperparameters such that $C^* = C\times C_p$ (=0.1 and 1). We used the one-vs-rest combination strategy for all the algorithms. Regarding the SA and SD datasets, average test accuracies over 10 splits for all the experiments are listed in Table 8.2. The model selection was conducted over 10 splits.

8.5.3.3 Results and Discussion

The experimental results underline the good performance of the semi-supervised approach. Using the SA data set the semi-supervised approaches perform at least similar or slightly outperform the supervised SVM. Furthermore, it is demonstrated that the primal approaches performs better than the S^3VM in the dual (S^3VM^{Light}) and ∇S^3VM performs most accurate.

Regarding the SD data set, the classification performance is significantly decreased, due to the different characteristics of training and test datasets. Again, compared to the supervised SVM the S^3VMs perform similar or better compared to the SVM in terms of accuracy. As in the first case, ∇S^3VM obtained the best classification results.

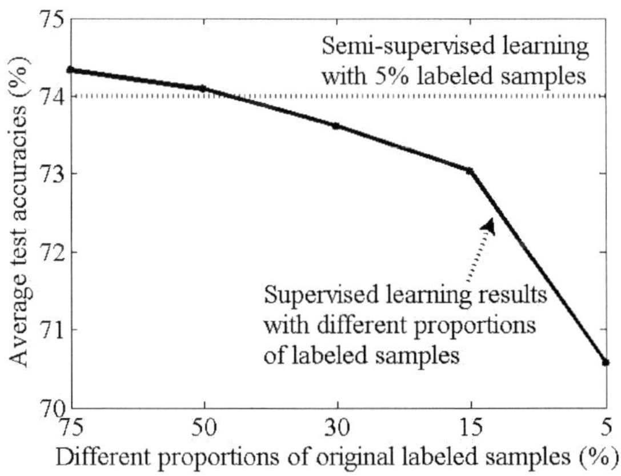

Fig. 8.3 Average test accuracies over 10 splits provided by supervised SVMs for different proportion of original labeled set (e.g., 5%, 15%, 30%, 50% and 75%) with comparison to the test accuracy obtained by primal S^3VM, i.e., ∇S^3VM.

In order to further observe the effectiveness of the semi-supervised algorithms, i.e., using the SD dataset, experiments were also conducted by supervised SVMs with different proportion of labeled samples, which contains 5%, 15%, 30%, 50% and 75% original 3248 labeled samples. A conventional supervised SVM is trained using each of these sample sets. The results are compeared to the classifcation accuracy achived by a semsi-supervised SVM, which is trained with 5% of the orignal labeled set. The average test accuracy achieved by the SVM classifiers, using different sample set sizes, is shown in Fig. 8.3. The accuracy obtained by ∇S^3VM is reported as a baseline in Fig. 8.3. The results clearly demonstrate the advantagous of semi-supervised SVM. One can see from the figure, in the semi-supervised setting, the average overall test accuracy by 5% labeled samples is almost equal to that with 50% labeled samples in the supervised setting.

Although SVM classifiers have shown good results in terms of accuracies in several applications, the performance can be limited in context of large data sets, particularly when only small reference data sets are available. The study presented here, have addressed this problem. The experimental results demonstrate that semi-supervised approaches can handle this problem and S^3VM are particularly interesting in context of small sample sets. A semi-supervised approach with uses 125 labeled samples performs similar than a supervised SVM that was trained on 1246 training samples. This fact makes the approach particularly interesting for operational applications with limited ground truth data.

8.5.4 Classification of Multisource Data Sets

8.5.4.1 Data Set and Preprocessing

Within this multisource land cover classification study we focus on mapping forest fire fuel types over a landscape characteristic for wildland-urban interfaces and peri-urban woodlands. We have chosen this complex environment in order to highlight the synergistic capabilities of the combined data sets of an imaging spectrometer and a LiDAR system for land cover classification applications.

An airborne survey was conducted over a Mediterranean site south of Aix-en-Provence, France, in early October 2006 (Koetz *et al.* 2008). The covered site comprised typical Mediterranean vegetation intermixed with urban structures, thus forming the typical fire prone wildland urban interface in the Mediterranean Basin. The employed LiDAR (ALTM3100) and the imaging spectrometer (AISA/Eagle) were mounted together with a very high-resolution photogrammetric camera (10 cm spatial resolution) on a helicopter operated by the company HELIOGS (Sgonico, Italy). The common platform for the two sensor systems provided the means to acquire simultaneous observations and cost efficient data acquisition, which were both essential for the proposed multisource land cover classification. After pre-processing, LiDAR derivatives and surface reflectance of the imaging spectrometer were co-registered and jointly considered for the classification as a layer stack.

8.5.4.2 Methods

The multi-source land cover classification was performed by SVM classification. SVM were chosen for their insensitivity to small sample sizes and high dimensional data sets. Furthermore, they have demonstrated good performance in terms of accuracies when applied to multi-source data sets, as for instance SAR data and multispectral imagery (Waske and Benediktsson 2007, Waske and van der Linden 2008).

SVM classification was applied to three different data sets to study the classification performance relative to different input sources and information dimensionality. Separate SVM were trained for three different remote sensing inputs: (a) single source imaging spectrometry data consisting of 97 spectral bands between 0.4 and 0.94 μm (from now on referred to as *IS*), (b) 7 layers of single source Li-

DAR data (*LiDAR*), and (c) a 104 band multi-source stack of IS and LiDAR data (*IS & LiDAR*). The 7 input layers retrieved from the LiDAR represented simple derivatives describing the vertical and horizontal geometric properties of the surface. For this purpose the relative vertical density distribution in six equidistant layers, as well as the canopy height model, were derived from the LiDAR returns (Naesset 2002, 2005).

The training of the SVM classifier with a Gaussian kernel was done using imageSVM (Janz *et al.* 2007) for the training of the SVM. The kernel parameters were determined via grid-search and cross validation. The one-against-all (OAA) strategy was applied to derive a multiclass SVM classifier. For the training of the SVM, a clustered sampling strategy was performed (100 samples per class). The selected samples were labeled based on very high-resolution aerial photographs acquired along with the remote sensing data. For discrimination of vertical properties for certain classes the LiDAR canopy height model (CHM) was consulted. An independent validation set was collected by unstratified randomized sampling (900 samples), which was complemented to ensure a minimum of 100 samples per class.

8.5.4.3 Results and Discussions

The method presented in this study is capable of a joint one-step SVM classification for the fusion of multi-source remote sensing data provided by an imaging spectrometer and a LiDAR. The specific advantage of the employed SVM classifier was to exploit efficiently the increased information content in the (hyper)spectral and the three-dimensional spatial domain. It was shown that SVM generalize well, even when only small training sets were available for the classification of the high dimensional data provided by the multiple data sources. The advantages for land cover mapping of each sensor system and the improvement of the multisource fusion has been assessed on three separate SVM classifications based on the different remote sensing inputs. The SVM classification of the imaging spectrometer data provided acceptable results in terms of overall accuracy and kappa coefficient (Table 8.3). Class specific assessment reveals significant confusion between spectrally similar classes such as the three vegetation classes, which leads to moderate user accuracies for these classes.

The overall classification performance of the pure LiDAR data was poor, but nevertheless provided significant user accuracies for classes with properties in the vertical dimension, e.g. for roof tiles and tree canopy (Table 8.3). The different vertical properties of these two classes caused very distinct signatures in the LiDAR derivatives, which significantly supported their separation.

Table 8.3 Class and overall accuracies and kappa coefficient for the three classifications

Class	*IS*	*LiDAR*	*IS & LiDAR*
Bare soil	0	84.9	86.7
Bare rock	19.3	67.7	75
Swimming pool	13.0	100	100
Ground fuel	0	50.4	60.6
Shrub	40.8	61.4	62.6
Tree Canopy	66.4	65	80.6
Road asphalt	20	64.2	76.7
Road gravel	12	86.8	78.7
Roof tile	95.1	91	96.8
Overall accuracy [%]	**69.1**	**31.7**	**75.4**
Kappa coefficient	**0.645**	**0.226**	**0.716**

The joint classification of the multi-source imaging spectrometer and LiDAR data set leads to a significant improvement in terms of overall accuracy and kappa (Table 8.3). Most of the achieved improvement in accuracy for the multiple-source classification can be explained by the decreased confusion between the vegetation classes. The vertical information content of the LiDAR observations was especially helpful to separate the classes ground fuel and tree canopy improving their user accuracy by 10.2% and 15.65%, respectively. LiDAR provides no additional information on the class shrub due to issues related to vertical separability of laser returns in low and dense vegetation. In addition, spectrally similar classes such as bare soil and roof tiles made of similar material could be better separated by the vertical information provided by the LiDAR. This effect was not revealed by the confusion matrix but is visible in the land cover maps (Fig. 8.4). A visual inspection of differences in resulting land cover maps reveals in addition a far more continuous and thus realistic discrimination of closed tree canopies in the multi-source classification.

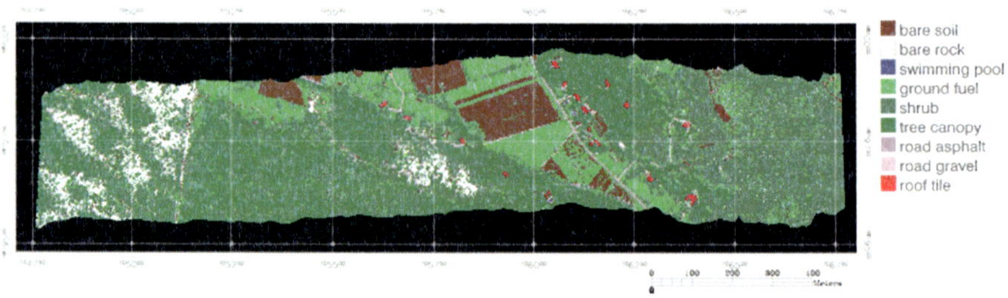

Fig. 8.4 Land cover map based on the multisource input based on imaging spectrometry and LiDAR (after Koetz *et al*. 2008)

The commonly used approach of land cover classification based on multispectral data is limited by the spectral similarity of certain surface types. Moreover, three dimensional features of important surface types, such as built-up urban construction or vertical vegetation structure, are not directly inferable from the spectral information content provided by passive optical sensors. Due to this underdetermined and partly indirect relationship, the interpretation of remote sensing data for land cover monitoring should rely on as many independent observations as possible. This conclusion leads to the combined exploitation of multiple information sources as provided for example by complementary sensor systems. The increased dimension and complexity of such information also requires new classification methods to adequately interpret the data of multiple information sources as presented here.

8.6 Conclusions

In this chapter on algorithms and applications of land cover classification, diverse state-of-the-art image classification concepts were introduced. Examples of how such approaches can be applied to currently typical remote sensing data sets were shown. Recently the potential of remote sensing land cover applications increased due to the rapid development of EO systems, resulting in the availability of such multitemporal, multi-source or high-dimensional data sets.

The different experiments presented dealt with the mapping of multitemporal SAR data, semi-supervised classification of hyperspectral imagery and the fusion of a LIDAR and a hyperspectral data sets. The experimental results underline that recent approaches handle such complex data sets efficiently and robust. Problems that results from limitations of traditional methods are avoided and results are thus more accurate.

A further increase in the number of EO instruments and in the variety of data types can be expected for the future. TerraSAR-X, CosmoSky-Med, RapidEye or EnMAP are good examples for this trend and the growing demand of EO products will lead to a higher degree of operational use of multi-source data. Therefore, methodological development is assumed to continue in the direction outlined in this chapter.

Acknowledgements

This work was partially supported by the Research Fund of the University of Iceland and The Icelandic Research Fund.

References

Backhaus R, Beule B (2005): Efficiency evaluation of satellite data products in environmental policy. Space Policy 21:173-183.

Bandos T, Zhou D, Camps-Valls G (2006). Semi-supervised Hyperspectral Image Classification with Graphs. IEEE International Geoscience and Remote Sensing Symposium, IGARSS'06.

Benediktsson JA, Swain PH, Ersoy OK (1990), Neural Networks Versus Statistical Methods in Classification of Multisource Remote Sensing Data. IEEE Trans. Geosci. and Remote Sens. 28: 540-552.

Benediktsson JA, Swain PH (1992). Consensus Theoretic Classification Methods. IEEE Trans. Syst., Man and Cyb. 22: 688-704.

Benediktsson JA, Kanellopoulos I (1999) Classification of Multisource and Hyperspectral Data Based on Decision Fusion. IEEE Trans. Geosci. and Remote Sens. 37: 1367-1377.

Benediktsson JA, Pesaresi M, Arnason K (2003). Classification and feature extraction for remote sensing images from urban areas based on morphological transformations. IEEE Trans. Geosci. and Remote Sens. 41: 1940-1949.

Benediktsson JA, Palmason JA, Sveinsson J (2005). Classification of hyperspectral data from urban areas based on extended morphological profiles. IEEE Trans. Geosci. Remote Sens. 43: 480–491.

Benediktsson JA, Chanussot J, Fauvel M (2007) Multiple Classifier Systems in Remote Sensing: From Basics to Recent Developments. In: Haindl M, Kittler J, Roli F (Ed.) Multiple Classifier Systems. 7th International Workshop, MCS 2007, Prague, Czech Republic, May 23-25, 2007, Proceedings. Lecture Notes in Computer Science 4472, Springer.

Breiman L, Friedman J, Olshen RA, Stone CJ (1984): Classification and Regression Trees CRC Press.

Breiman L (1996) Bagging predictors. Machine Learning 24:123-140.

Breiman L (2001) Random forests. Machine Learning, 45: 5-32.

Briem GJ, Benediktsson JA, Sveinsson JR (2002) Multiple classifiers applied to multisource remote sensing data. IEEE Transactions on Geoscience and Remote Sensing 40: 2291-2299.

Bruzzone L, Carlin L (2006) A multilevel context-based system for classification of very high spatial resolution images. IEEE Transactions on Geoscience and Remote Sensing 44: 2587-2600.

Bruzzone L, Chi M, Marconcini M (2006). A Novel Transductive SVMs for semisupervised classification in ill-posed problems. IEEE Trans. Geosci. and Remote Sensing, 44(11):3363-3373.

Bryll R, Gutierrez-Osuna R, Quek F (2003) Attribute Bagging: improving accuracy of classifier ensembles by using random feature subsets. Pattern Recognition 36: 1291-1302.

Burges CJC (1998) A tutorial on support vector machines for pattern recognition. Data Mining and Knowledge Discovery 2:121-167.

Cardillo G, Landgrebe D (1966) On pattern recognition. Purdue University, West Lafayette, IN, LARS Tech. Note 101 866.

Chang C-C ,Lin C-J (2001): LIBSVM: a library for support vector machines, 2001. Software available at: http://www.csie.ntu.edu.tw/~cjlin/libsvm.

Chapelle O, Vapnik V, Bousquet O, Mukherjee S (2002): Choosing multiple parameters for support vector machines. Machine Learning 46:131-159.

Chapelle O, Zien A (2005): Semi-Supervised Classification by Low Density Separation. Tenth International Workshop on Artificial Intelligence and Statistics.

Chapelle O, Chi M, Zien A (2006). A Continuation Method for Semi-Supervised SVMs. International Conference on Machine Learning.

Chi M (2006): Advanced Semi-Supervised Techniques for the Classification of Remote Sensing Data. PhD thesis, University of Trento, Trento, Italy.

Chi M, Bruzzone L (2005) A semilabeled-sample-driven bagging technique for ill-posed classification problems. IEEE Geosci. and Remote Sensing Lett., 2(1):69-73.

Chi M, Bruzzone L (2007) Semi-Supervised Classification of Hyperspectral Images by SVMs Optimized in the Primal. IEEE Trans. Geosci. and Remote Sensing, 45(6):1870-1880.

Chung KM, Kao WC, Sun CL, Wang LL, Lin CJ (2003): Radius margin bounds for support vector machines with the RBF kernel. Neural Computation 15: 2643-2681.

Cortes C, Vapnik V (1995) Support Vector Networks. Machine Learning, 20(3):273- 297.

Dempster A, Laird N, Rubin D (1977) Maximum likelihood from incomplete data via the EM algorithm. The Royal Statistical Society, Series B.

Duda RO, Hart PE, and Stork DG (2001) Pattern Classification. 2nd edn. Wiley-Interscience.

Esposito F, Malerba D, Semeraro G, Kay J (1997) A comparative analysis of methods for pruning decision trees. IEEE Transactions on Pattern Analysis and Machine Intelligence 19:476-491.

Fauvel M, Benediktsson JA, Chanussot J, Sveinsson JR (2008). Spectral and Spatial Classification of Hyperspectral Data Using SVMs and Morphological Profiles. IEEE Trans. Geosci. and Remote Sensing, 46:3804-3814.

Foody G.M. and Mathur A. (2006) The use of small training sets containing mixed pixels for accurate hard image classification: Training on mixed spectral responses for classification by a SVM. Remote Sensing of Environment 103: 179-189.

Freund Y, Schapire RE (1996) Experiments with a new boosting algorithm. Proc. 13th Int. Conference of Machine Learning, Bari, Italy.

Friedl MA, Brodley CE (1997) Decision tree classification of land cover from remotely sensed data. Remote Sensing of Environment 61: 399-409.

Fu KS, Landgrebe DA, Phillips TL (1969) Information processing of remotely sensed agricultural data. IEEE Proceedings 57:639–653.

GCOS (2004) Implementation Plan for the Global Observing System for Climate in Support of the UNFCCC, GCOS-92.

GEO (2005): Global Earth Observation System of Systems (GEOSS) 10-Year Implementation Plan. Reference Document, available at: http://www.earthobservations.org.

Gislason PO, Benediktsson JA, and Sveinsson JR (2006). Random Forests for land cover classification. Pattern Recognition Letters, 27: 294-300.

Ham J, Chen YC, Crawford MM, and Ghosh J (2005): Investigation of the random forest framework for classification of hyperspectral data. IEEE Transactions on Geoscience and Remote Sensing, 43: 492-501.

Hansen LK, Salamon P. (1990) Neural network ensembles. IEEE Transactions on Pattern Analysis and Machine 12:993-1001.

Ho TK (1998) The random subspace method for constructing decision forests. IEEE Transaction on Pattern Analysis. and Machine Intelligence 20:882-844.

Hsu CW, Lin CJ (2002) A comparison of methods for multi-class support vector machines. IEEE Trans. Neural Networks, 13: 415-425.

Hughes GF (1968) On the mean accuracy of statistical pattern recognition. IEEE Trans. Inform. Theory, IT-14:55-63.

Jackson Q, Landgrebe DA (2002) An adaptive method for combined covariance estimation and classification. IEEE Trans. Geosci Remote Sensing, 40(5):1082-1087.

Jain AK, Duin RPW, Mao JC (2000) Statistical pattern recognition: A review. IEEE Transactions on Pattern Analysis and Machine Intelligence, 22:4-37.

Janz A, Schiefer S, Waske B, Hostert P (2007) A user-oriented tool for advanced classification of hyperspectral data using support vector machines. Proc. of the 5th workshop of the EARSeL Special Interest Group on Imaging Spectroscopy, 23-25 Apr., Bruges, Belgium.

Joachims T (1999). Transductive inference for text classification using support vector machines. International Conference on Machine Learning, ICML 1999.

Keehrti SS and Lin CJ (2003): Asymptotic behavior of support vector machines with Gaussian kernel. Neural Computation 15:1667-1689.

Koetz B, Morsdorf F, van der Linden S, Curt T, Allgöwer B (2008) Multi-source land cover classification for forest fire management based on imaging spectrometry and LiDAR data. Forest Ecology and Management 256(3): 263-271.

Laur H, Bally P, Meadows P, Sanchez J, Schaettler B, Lopinto E, Esteban D (2002): Derivation of the backscattering coefficient σ0 in ESA ERS SAR PRI products. ESA Document ES-TN-RE-PM-HL09, Issue 2, Rev. 5d, 2002.

Lin H.-T., Lin C.-J., Weng R. C. (2003): A note on Platt's probabilistic outputs for support vector machines. Technical report, Department of Computer Science, National Taiwan University, URL http://www.csie.ntu.edu.tw/~cjlin/papers/plattprob.ps.

Hyun-Chul Kim, Shaoning Pang, Hong-Mo Je, Daijin Kim, Sung Yang Bang (2003) Constructing support vector machine ensemble. Pattern recognition 36: 2757-2767.

Mathur A, Foody GM (2008) Multiclass and Binary SVM Classification: Implications for Training and Classification Users. IEEE Geoscience and Remote Sensing Letters 5:241245

Melgani F, Bruzzone L (2004): Classification of hyperspectral remote sensing images with support vector machines. IEEE Transactions on Geoscience and Remote Sensing 42:1778-1790.

Naesset E (2002). Predicting forest stand characteristics with airborne scanning laser using a practical two-stage procedure and field data. Remote Sensing of Environment 80: 88-99.

Naesset E (2005). Assessing sensor effects and effects of leaf-off and leaf-on canopy conditions on biophysical stand properties derived from small-footprint airborne laser data. Remote Sensing of Environment 98: 356-370.

Pal M, Mather PM (2003) An assessment of the effectiveness of decision tree methods for land cover classification. Remote Sensing of Environment 86:554-565.

Pal M, Mather PM (2005) Support vector machines for classification in remote sensing. International Journal of Remote Sensing 26:1007-1011.

Pal M and Mather P (2006) Some issues in the classification of DAIS hyperspectral data International Journal of Remote Sensing 27: 2895-2916.

Pal M (2005) Random forest classifier for remote sensing classification. International Journal of Remote Sensing. 26:217-222.

Peter N (2004): The use of remote sensing to support the application of multilateral environmental agreements. Space Policy 20:189-195.

Polikar R. (2006) Ensemble Based Systems in Decision Making. IEEE Circuits and Systems Magazine, 6:21-45.

Richards JA (2005): Analysis of remotely sensed data: The formative decades and the future. IEEE Transactions on Geoscience and Remote Sensing 43:422-432.

Richards JA, Jia X (2006) Sensing Digital Image Analysis: An Introduction. 4th edn. Springer, New York,

Richards JA, Jia X (2008) Using Suitable Neighbors to Augment the Training Set in Hyperspectral Maximum Likelihood Classification. IEEE Geosci. and Remote Sensing Letters 5: 774-777.

Rosenqvist A, Milne A, Lucas R, Imhoff M, Dobson C (2003) A review of remote sensing technology in support of the Kyoto Protocol. Environmental Science & Policy 6:441-455.

Safavian SR, Landgrebe DA (1991) A Survey of Decision Tree Classifier Methodology. IEEE Transactions on Systems, Man, and Cybernetics 21:660-674.

Schapire RE (1990) The strength of weak learnability. Machine Learning 5:197-227.

Schölkopf B, Smola A (2002) Learning with Kernels. MIT Press, Cambridge, MA.

Sebald DJ, Bucklew JA (2001) Support vector machines and the multiple hypothesis test problem," IEEE Trans. Signal Processing 49: 2001.

Shackelford AK, Davis CH (2003) A combined fuzzy pixel-based and object-based approach for classification of high-resolution multispectral data over urban areas. IEEE Transactions on Geoscience and Remote Sensing 41:2354-2363.

Shahshahani BM and Landgrebe DA (1994) The effect of unlabeled samples in reducing the small sample size problem and mitigating the Hughes phenomenon. *IEEE Trans. Geosci Remote Sensing*, 32(5):1087-1095

Steele BM (2000) Combining multiple classifiers: An application using spatial and remotely sensed information for land cover type mapping. Remote Sensing of Environment 74: 545-556.

Tadjudin S and Landgrebe DA (2000) Robust parameter estimation for mixture model. IEEE Trans. Geosci Remote Sensing, 38(1):439-445.

Turmer K and Gosh J (1996) Anaylsis of Decision Boundaries in Linearly combined neural classifiers. Pattern Recognition Letters 29:341-348.

Vapnik VN (1998) Statistical Learning Theory. Wiley & Sons, New York.

van der Linden S, Janz A, Waske B, Eiden M., und Hostert P (2007) Classifying segmented hyperspectral data from a heterogeneous urban environment using support vector machines. Journal of Applied Remote Sensing 1.

Waske B, Schiefer S, Braun M (2006) Random feature selection for a decision tree classification of multi-temporal SAR data. IEEE Geoscience and Remote Sensing Symp. IGARSS 2006, Denver, Colorado, USA.

Waske B, Benediktsson JA (2007) Fusion of Support Vector Machines for Classification of Multisensor Data. IEEE Trans. Geosci. and Remote Sens. 45: 3858-3866.

Waske B, van der Linden S (2008): Classifying multilevel imagery from SAR and optical sensors by decision fusion. IEEE Trans. on Geoscience and Remote Sensing 46: 1457-1466.

Wu Z (1996) The effective energy transformation scheme as a special continuation approach to global optimization with application to molecular conformation. SIAM Journal on Optimization, 6(3):748-768.

Wu T.-F., Lin, C.-J., Weng R. C. (2004) Probability estimates for multi-class classification by pairwise coupling. Journal of Machine Learning Research, 5:975-1005.

Zambon M, Lawrence R, Bunn A, Powell S (2006) Effect of alternative splitting rules on image processing using classification tree analysis. Photogrammetric Engineering and Remote Sensing 72:25-30.

Zhou D, Bousquet O, Lal TN, Weston J, Schölkopf B (2004). Learning with Local and Global Consistency. In Thrun S, Saul L, Schölkopf B (eds) 321-328, Cambridge, MA, USA, 2004. MIT Press.

Zhu X, Ghahramani Z, Lafferty J (2003) Semi-Supervised Learning Using Gaussian Fields and Harmonic Functions. The Twentieth International Conference on Machine Learning, ICML 2003.

Zhu X (2005) Semi-Supervised Learning Literature Survey. Technical report, 1530, Computer Sciences, University of Wisonsin-Madison.

Chapter 9

ANALYSIS OF HYPERSPECTRAL REMOTE SENSING IMAGES

Liangpei Zhang, Yanfei Zhong

9.1 Introduction

Hyperspectral remote sensing, or known as imaging spectroscopy, is a recently developed technique since the last two decades of the 20th century (Chang 2003). Imaging spectroscopy is a relatively fully-fledged experimental tool that has been successfully used in the laboratory by physicists and chemists for over 100 years for identification of materials and their composition. Absorption features accord to the special chemical bound of a material, which can be calculated by imaging spectroscopy. With the demand of earth observation, imaging spectroscopy technique has extended to the detection and mapping of materials by satellite imagery. Since the 1980s, geologists have used sensors on the man-made satellite to obtain the spectrum of every position in a large scale in the ground which combines a datacube which combines the imaging and spectroscopy in a single system. That's to say that hyperspectral remote sensing not only contain spatial features but also spectral features of the ground objects. But it doesn't refer to the remote sensing imagery with only several bands, such as Landsat TM or Modis imagery. In fact, the most significant difference between hyperspectral remote sensing and these multispectral remote sensing is that it has much more bands with much higher spectral resolution. Hyperspectral remote sensing usually has over one hundred bands with a spectral resolution of under 10 nm. Fig. 9.1 shows the concept about hyperspectral remote sensing imagery which usually comprises of datacube with a series of images. In this case, it provides a better discrimination among similar targets. On the other hand, subtle spectral differences would be hidden in spectra acquired with multispectral remote sensing with broad spectral band sensors. Hyperspectral remote sensing has been widely used in many civil and military applications such as geology, agriculture, and global change, defense, intelligence, and law enforcement. The aim of the chapter is to discuss the basic data processing and analysis techniques for hyperspectral remote sensing such as feature selection, classification, mixed pixel unmixing etc.

D. Li et al. (eds.), *Geospatial Technology for Earth Observation*,
DOI 10.1007/978-1-4419-0050-0_9,

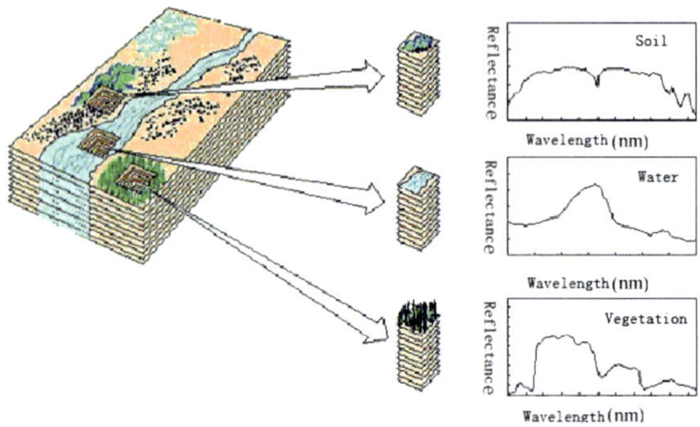

Fig. 9.1 Datacube of hyperspectral imagery

9.2 Feature Selection for Hyperspectral Imagery

As hyperspectral sensors acquire images in very narrow spectral channels, the resulting high-dimensional feature sets may contain redundant information. As a result of this, the number of features given as input to a classifier can be reduced by dimensionality reduction without a significant loss of information (Fukunaga 1990). In addition, dimensionality reduction in high-dimensional data space can decrease the computational cost and may also improve the accuracy during the classification process (Chang and Wang 2006). For instance, when a supervised classifier is applied to classification problems in high-dimensional feature spaces, the Hughes phenomenon (Hughes 1968) can be observed; that is, when the number of input features exceeds a given limit for a fixed training-sample size, the classification accuracy will decrease.

Dimensionality reduction can be achieved in essentially two ways: feature extraction and feature selection (Webb 2002). Feature extraction finds the transformation from a higher dimension to a lower dimensional feature space with most of the desired information content preserved (Lee and Landgrebe 1993). This transformation may be a linear or nonlinear combination of the original variables and may be supervised or unsupervised. In contrast to the feature extraction techniques, feature selection identifies the variables that do not contribute to the classification process (Sebastiano and Lorenzo 2001). In a discrimination problem, those variables that do not contribute to class separability would be neglected.

The problem of feature selection is shown in Fig. 9.2 and the details are as follows. A hyperspectral remote-sensing dataset $X = \{x^1, x^2, \ldots, x^{N_b}\}^T$ through N_b bands is observed. The objective of feature selection is to reduce the number of features utilized to characterize patterns by selecting, through optimization in terms of a criterion function F (e.g. maximization of a separability index or minimization of an error bound), a good subset S of N_m features, with $N_m < N_b$, without significantly degrading the performance of the resulting classifier:

$$S = \{s^1, s^2, \cdots, s^{N_m}\}^T \qquad S \subset X \tag{9.1}$$

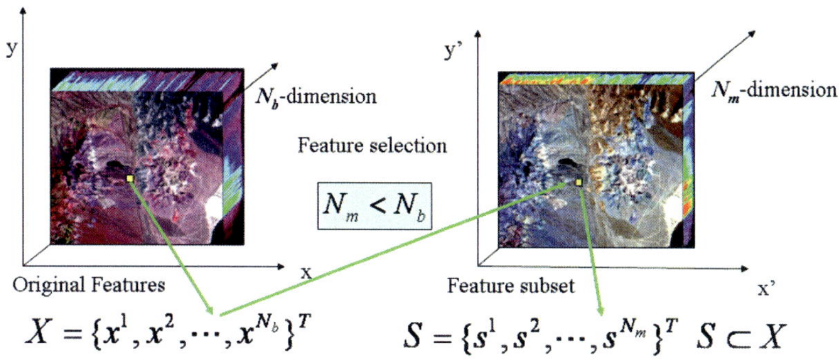

Fig. 9.2 Feature Selection for hyperspectral remote sensing imagery

9.2.1 Feature Selection Algorithms

Given a set of feature selection criterion, F, different feature selection techniques including optimal and suboptimal search algorithms have been proposed. Optimal search algorithms, such as an exhaustive search and branch and bound method (Narendra and Fukunaga 1977), are not suitable for use in hyperspectral space because of their heavy computational costs. In such a case, suboptimal algorithms will be considered for searching for an appropriate feature subset of hyperspectral remote sensing imagery.

9.2.1.1 Best Individual N_m

The simplest method, and perhaps the one giving the poorest performance, for choosing the best N_m features is to assign a discrimination power estimate to each of the features in the original set, X, individually (Webb 2002). Thus, the features are ordered, so that

$$F(x^1) \geq F(x^2) \geq \cdots \geq F(x^{N_b}) \tag{9.2}$$

and we select as our best set of N_m features, these features with the best individual scores

$$\{x^i \mid i \leq N\} \tag{9.3}$$

In some cases this method can produce reasonable feature sets, especially if the features in the original set are uncorrelated, since the method ignores multivariate relationships. However, if the features of the original set are highly correlated, i.e. hyperspectral remote sensing image, the chosen feature set will be suboptimal as some of the features will be adding little discriminatory power. There are cases when the N_m best features are not the best N_m features even when the variables are independent (Hand 1981).

9.2.1.2 Sequential Forward Selection (SFS)

Sequential forward selection (SFS) is a bottom-up search procedure that adds new features to a feature set one at a time until the final feature set is reached. Suppose we have a set of N_d features, X_d. For each of the feature x^j not yet selected (i.e. in $X - X_d$) the criterion function $F_j = F(X_d + x^j)$ is evaluated. The feature that yields the maximum value of F_j is chosen as the one that is added to the set X_d. Thus, at each stage, the variable is chosen, that, when added to the current set, maximizes the selection criterion. The feature set is initialized to the null set. When the best improvement makes the feature set worse, or when the maximum allowable number of feature is reached, the algorithm terminates (Jain and Zongker 1997). The main disadvantage of the method is that it does not include a mechanism for deleting features from the feature set once they have been added should further additions render them unnecessary (Webb 2002).

9.2.1.3 Sequential Backward Selection (SBS)

Sequential backward selection (SBS) is the top-down analogy to SFS. Variables are deleted one at a time until d measurements remain. Starting with the complete set, the variable x^j is chosen for which $F(X - x^j)$ is the largest (i.e. x^j decreases F the least). The new set is $\{X - x^j\}$. This process is repeated until a set of the required cardinality remains. The procedure has the disadvantage over SFS that it is computationally more demanding since the criterion function F is evaluated over larger sets of variables (Webb 2002).

9.2.1.4 Floating Search Methods

Floating search methods, sequential forward floating selection (SFFS) and sequential backward floating selection (SBFS), may be regarded as a development of the SFS algorithm that they may change at different stages of the selection procedure (Pudil *et al.* 1994).

Suppose that at stage k we have a set of subsets $X_1, \cdots, X_k$ of sizes 1 to k respectively. Let the corresponding values of the feature selection criteria be F_1 to F_k, where $F_i = F(X_i)$, for the feature selection criterion, $F(\cdot)$. Let the total set of features be X. At the k-th stage of the SFFS procedure, do the following (Webb 2002).

Step 1. Select the feature x^j from $X - X_k$ that increases the value of F the greatest and add it to the current set, $X_{k+1} = X_k + x^j$.

Step 2. Find the feature, x^r, in the current set, X_{k+1}, that reduces the value of F the least; if this feature is the same as x^j then set $F_{k+1} = F(X_{k+1})$; increment k; go to step 1; otherwise remove it from the set form $X'_k = X_{k+1} - x^r$.

Step 3. Continue removing features from the set X'_k to form reduced sets X'_{k-1} while $F(X'_{k-1}) > F_{k-1}$; $k = k-1$; or $k = 2$; then continue with Step 1.

The algorithm is initialized by setting $k = 0$ and $X_0 = \emptyset$ (the empty set) and using the SFS method until a set of size 2 is obtained.

9.2.2 Clonal Selection Feature Selection Algorithm (CSFS)

Besides these methods above, a new stochastic search strategy for hyperspectral feature selection has been proposed, which is based on the clonal selection algorithm (CSA) (De Castro and Von Zuben 2002) in Artificial Immune Systems (AIS). AIS, which are inspired by the immune systems, use the immunological properties to support a wide range of applications (Dasgupta 1999, De Castro and Timmis 2002, Carter 2000, Zhong *et al.* 2006). CSA, derived from the clonal selection theory (Burnet 1959, 1978), is an effective method of AIS and has been successfully applied to pattern recognition, multi-modal optimization and classification (Dasgupta 1999, De Castro and Timmis 2002, De Castro and Von Zuben 2002, Zhong *et al.* 2007)). In this research, a new feature selection algorithm, namely clonal selection feature selection algorithm (CSFS), has been developed based on the clonal selection theory for feature selection in hyperspectral space (Zhang *et al.* 2007a). The algorithm is able to find the feature subset in feature space using immune operators, such as clone, selection, mutation, and replacement.

To apply the clonal selection algorithm to hyperspectral feature selection, the entire set of features is represented by a discrete binary space. In this search space, each point represents an individual band. Value "0" in the *i*-th position indicates that the *i*-th feature is not included in the corresponding feature subset; value "1" in the *j*-th position indicates that the *j*-th feature is included in the corresponding feature subset. For example, in a simple case with $N_m = 6$ and $N_b = 10$ features, the binary vector b=(1,1,0,1,0,1,0,1,0,1) indicates the feature subset has the first, second, fourth, sixth, eighth, and tenth features. The criterion function *F* can be viewed as a scalar function defined in the discrete binary space. Without loss of generality, there is a case in which the criterion function has to be maximized. Thus, the dimensionality reduction problem of selecting N_m out of N_b features can be suitably formulated as an optimization problem to find the global maximum of the criterion function. The CSFS algorithm consists of the following steps (Zhang *et al.* 2007a):

9.2.2.1 Initialization

A first antibody population AB including N antibodies is generated with the value of each bit in each antibody $ab_i \in AB$ assigned 1 or 0 according to the number of selected subset features, N_m. The value 0 in the i-th position indicates that the *i*-th feature is not included in the corresponding feature set; the value 1 in the j-th position indicates that the j-th feature is included in the corresponding feature set.

$$ab_i^t = \begin{cases} 1 & if\ t = k \\ 0 & otherwise \end{cases} \quad i = 1,2,\cdots,N;\ t = 1,2,\cdots,N_b \tag{9.4}$$

$$k = Irandom(1, N_b) \tag{9.5}$$

$$\sum_{t=1}^{N_b} ab_i^t = N_m \tag{9.6}$$

where N is the number of the initial antibody population, N_b is the number of features or bands, function Irandom(1, N_b) returns a random integer value within the range [1, N_b] using a uniform distribution. Eq. (9.6) indicates that the sum of selected features is equal to N_m, a user-defined feature subset size.

9.2.2.2 Cycle of the Generations

After initialization, the simulation of the clonal selection process begins. One generation after another is created and each must prove its affinity to the criterion function. In each iteration, several possible solutions are generated by means of applying the immune operators such as clone, mutation, selection in a stochastic process guided by an affinity measure. The algorithm seeks to evolve an optimal solution to the problem.

(1) Calculation of Affinity

According to the initial antibody population, the affinity of all N abs in the antibody population AB are calculated using the criterion function $F = F(ab_i)$.

As a criterion function, the proposed algorithm uses the average Jeffries-Matusita (*JM*) distance (Bruzzone *et al.* 1995, Swain and Davis 1978), which is a common class separability index utilized by the remote sensing community for feature selection in multiclass problems and is a saturating transform of the Bhattacharyya distance (*BD*) (Richards 1986). Assuming that there exist *c* classes with Gaussian distributions, in order to simplify the computation of the Bhattacharyya distance, the average *JM* distance of a feature subset is calculated using the following equation:

$$JM = \sum_{m=1}^{c} \sum_{n=1}^{c} P_m P_n JM_{mn} \tag{9.7}$$

$$JM_{mn} = \sqrt{2(1 - e^{-BD_{mn}})} \tag{9.8}$$

$$BD_{mn} = \frac{1}{8}(M_m - M_n)^T \left(\frac{\Sigma_m + \Sigma_n}{2}\right)^{-1} (M_m - M_n) + \frac{1}{2}\ln\left(\frac{\left|\frac{\Sigma_m + \Sigma_n}{2}\right|}{\sqrt{|\Sigma_m||\Sigma_n|}}\right) \tag{9.9}$$

where c is the number of classes, m and n are the two classes being compared, Pm is a priori probability of the m-th class, Σm and Mm are the covariance matrix and mean vector of the m-th class, respectively, T is the transposition function, and |Σm| is the determinant of Σm.The JMmn distance between classes m and n is an affinity/distance measure of separability. The smaller the JMmn distance, the more difficult it is to separate the classes and vice versa. Assuming $JM_{mm} = 0$, the average JM distance can be written as Eq. (9.10) and the proposed algorithm uses this equation as the affinity function:

$$F(ab_i) = JM = \frac{2}{c(c-1)} \sum_{m=1}^{c-1} \sum_{n=i+1}^{c} P_m P_n JM_{mn} \tag{9.10}$$

(2) Selection

From AB, ‘n’ highest affinity antibodies are selected to compose a new set $AB_{\{n\}}$ of high-affinity antibodies and the highest affinity memory cell is found (memory cell, mc).

(3) Clone

After receiving antibody individuals closer to the solution, the next generation should mainly be derived from the better-fitting individuals. Thus, the n selected abs are cloned based on their antigenic affinities, generating the clone set C. The total number of clones generated, Nc, is defined as follows:

$$N_c = \sum_{i=1}^{n} round(\beta \cdot N) \tag{9.11}$$

where β is a multiplication factor
N is the total number of antibodies
round(·) is the operator that rounds its argument.

This step draws the evolutionary process closer to the goal. It raises the average affinity value and gives the following steps a good chance to further move towards the solution.

(4) Mutation

Provide each *ab* in the clone set *C* with the opportunity to produce mutated offspring C^*. The higher the affinity, the smaller the mutation rate. To adaptively determine the mutation rate according to the affinity of each *ab*, the process is as follows:

Firstly, for each antibody $ab_i \in AB$, normalize its affinity $F(ab_i)$ into the range [0, 1]:

$$F'(ab_i) = \frac{F(ab_i) - \min(F(ab_i))}{\max(F(ab_i)) - \min(F(ab_i))} \quad i = 1,2,\cdots,N_c \tag{9.12}$$

Then, let each ab_i have the chance to mutate; the mutation rate is adaptively calculated as:

$$p_m = \exp(-2 \cdot F'(ab_i)) \tag{9.13}$$

where p_m is the mutation rate of each *ab*, 2 is the empirical value to control the decay, and $F'(ab_i)$ is the affinity according to Eq. (9.12). In Eq. (9.13), the range of the mutation rate is $[0,1]$. The mutation process utilizes the non-uniform mutation operator (Zhang *et al.* 2007a).

Finally, the cloned antibodies are mutated with probability p_m. This step is crucial in the proposed algorithm. It generates random changes of single features of the individual solutions. The value of these changes can be found at the criterion function calculation within the next generation cycle. This helps avoid local maxima and produces new properties of mutated antibodies that can remain if they are successful.

To avoid chaotic development and maintain the best abs for each clone during evolution, one original ab for each clone without mutation during the maturation process is kept; else it would destroy the positive development of the previous step and disable any major development towards the solution.

(5)Re-Calculation of Affinity
Calculate the affinity $F^*(ab_i)$ of the matured clones C^*.

(6)Reselection
From the mature clone set C^*, reselect the *n abs* with the highest affinity to replace the *n abs* with the lowest affinity in *AB*. Select the highest affinity *ab* in C^* to be a candidate memory cell, $mc_{candidate}$. If the affinity of $mc_{candidate}$ is higher than the memory cell, *mc*, then $mc_{candidate}$ will replace *mc* and become a new memory cell.

(7)Displace
In order to replace the *d* lowest affinity *abs* from *AB, d* new antibodies are produced by a random process. This step may increase the diversity of the antibody population.

9.2.3 Stopping Condition

When the number of iterations reaches the user-defined number or the change of memory cell between two consecutive iterations is less than a change threshold, terminate the execution of the algorithm. Otherwise, return to 9.2.2 until the stop criteria are satisfied.

Finally, the proposed algorithm outputs the value of the memory cell and obtains the subset space through transformation from the binary space.

9.2.4 Experiments

Experiments have been conducted to test the performance of the above feature selection algorithms using the hyperspectral remote sensing imagery. Consistent comparisons between CSFS and the traditional feature selection algorithms: SFS (sequential forward selection) and SFFS (sequential forward floating selection), were performed.

The primary running parameters that should be provided by users for CSFS were the number of iterations, antibody population size N, the number of highest affinity ab, n (see also step (3)), clone multiplication factor β (see also Eq. (9.11)), the number of displace antibody, d (see also step (7)) To facilitate the comparison of the proposed algorithm with other traditional algorithms, n is set to N and d to zero. The affinity function is determined by the JM distance in Eq. (9.10). The values of parameters in the experiment were set as: Population size = 50, Maximum iterations = 100, n = 50, β = 0.02, d = 0.

9.2.4.1 Dataset

The dataset used in this experiment was acquired from the Xiaqiao test site, a mixed agricultural area in China, using the Pushbroom Hyperspectral Imager (PHI) produced in China. Eighty bands of the PHI image (340×390 pixels) were tested, and their spectral ranges were from 0.417 to 0.854 μm. Fig. 9.3 shows the experimental PHI image cube. The ground truth spectral data were collected by field spectrometer SE590. Nine representative classes, i.e., corn1 (713 samples), corn2 (217 samples), corn3 (322 samples), vegetable-sweet potato (464 samples), vegetable-cabbage (253 samples), soil (1368 samples), float grass (220 samples), road (662 samples), and water (659 samples) were considered. Fig. 9.4 displays the reflectance curves of the above nine land cover classes. The field map is shown in Fig. 9.5 based on the ground truth data (Zhang *et al.* 2007a). In order to test three algorithms, approximately half of the available samples were used as the training dataset, with the other half of the available samples serving as the test dataset. The number of features in the feature subset is selected from 2 to 70 to test these algorithms.

Fig. 9.3 Xiaqiao PHI image

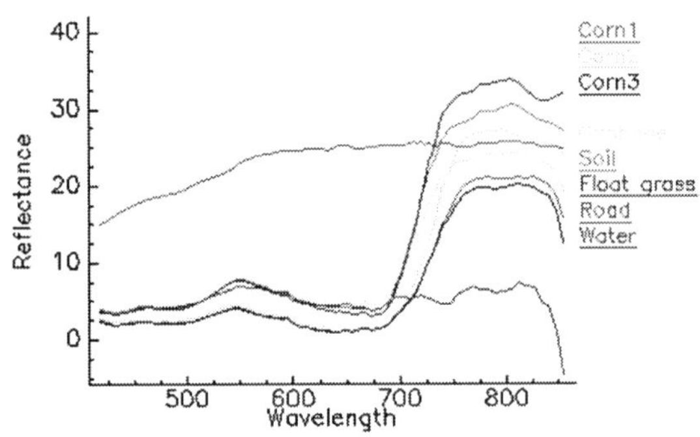

Fig. 9.4 The reflectance of nine land cover classes

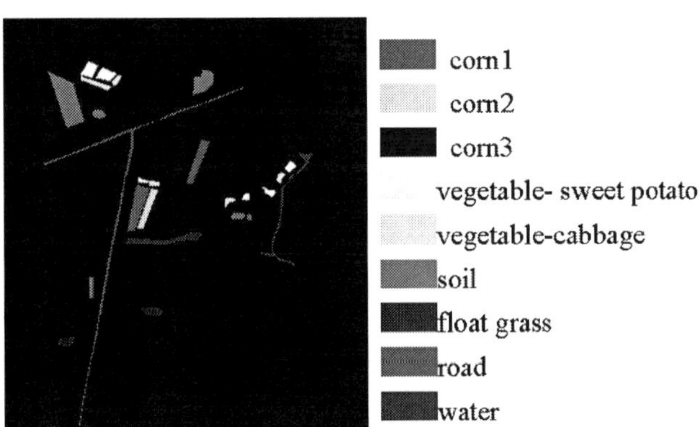

Fig. 9.5 The field map of the Xiaqiao site

9.2.4.2 Results

An important characteristic of the algorithms is their execution time. In our case, it is in the form of processor ticks (1 millisecond (msec)) spent in the user space. Fig. 9.6 shows the execution times with different feature subset sizes provided by the three algorithms. For every number of selected features from 2 to 70, SFS is the fastest and CSFS is slower. When the number of selected features is over 67, SFFS is the slowest.

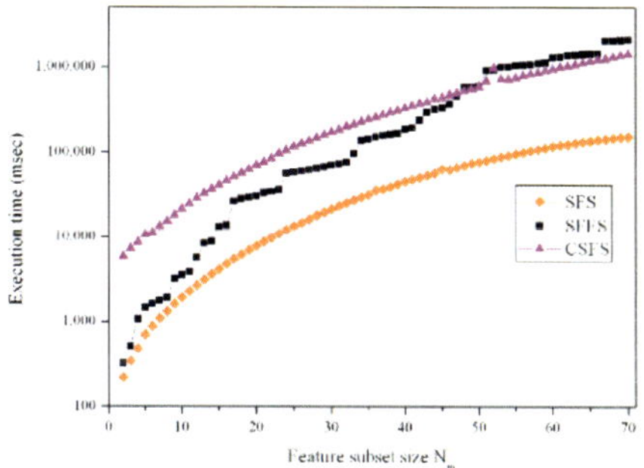

Fig. 9.6 Execution times required by different algorithms

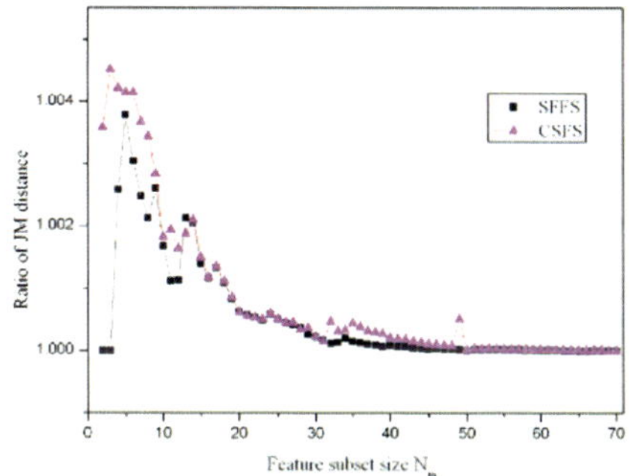

Fig. 9.7 Values of the criterion function for different feature subset sizes provided by different algorithms

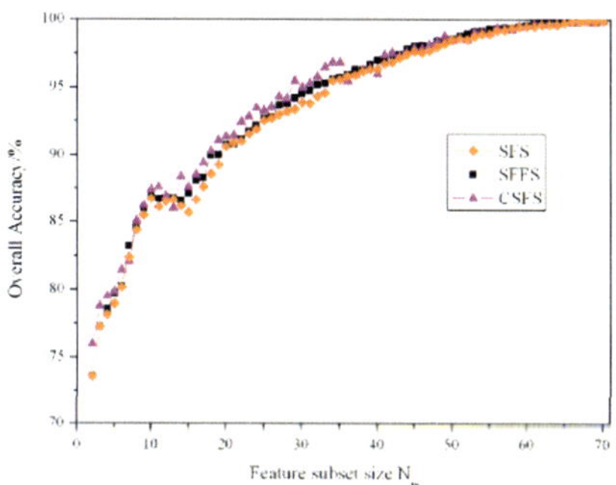

Fig. 9.8 The classification accuracy achieved by the six algorithm using the training data

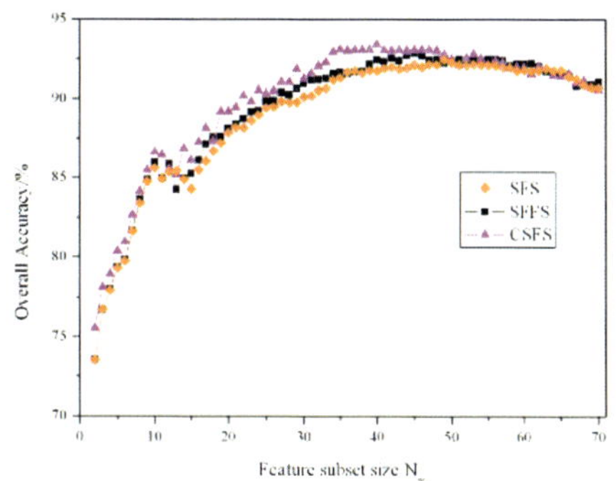

Fig. 9.9 The classification accuracy achieved by the six algorithms: using the test data

Fig. 9.7 depicts the values of the criterion function computed on the subsets in the training process provided by CSFS, SFS, and SFFS to different numbers of selected features from 2 to 70. All algorithms were described by the relationship curves between the predefined number of features in the subset and the corresponding JM distance. To better present the experiment results, the JM distance of SFS was used as a reference; that is, the values of the criterion functions provided by SFFS, and CSFS are divided by the corresponding values obtained by SFS. For example, if the CSFS and SFS provided the same JM distance values, the value on Fig. 9.7 equals to 1.

For a more detailed comparison of different algorithms: SFS, SFFS and CSFS, all the labeled training and test samples were classified using the dimensionality reduction algorithms with the increase of selected feature numbers. The Maximum Likelihood (ML) classifier was employed in this case. Fig. 9.8 and Fig. 9.9 illustrate the overall accuracy for training data and test data, respectively. For convenience of description, SFS-ML denotes the maximum likelihood classifier with the feature subsets provided by the SFS algorithm. Other notations can be inferred by analogy.

As can be found from Fig. 9.7, Fig. 9.8 and Fig. 9.9, the SFFS and CSFS algorithm perform better than SFS for all the selected subset features. Comparing the results of CSFS with those of SFFS, CSFS make some improvements over SFFS. In particular, when the number of selected features is below 50, the improvement is greater. The highest overall accuracies achieved by CSFS-ML using the test data is 93.07%. A comparison between CSFS and SFFS shows that CSFS usually provides better results than those provided by SFS; however, the differences can be neglected when the number of selected features is larger than 50. Specifically, the Hughes phenomenon can be observed: when the number of input features exceeds around 50, the classification accuracy decreases.

9.3 Unsupervised Classification

Various algorithms such as maximum likelihood, parallelepiped, and minimum distance from mean have been employed in the past for classifying hyper-spectral data in a pixel-wise manner (Landgrebe 2002). These algorithms are based on the fact that each class of materials, in accordance to its molecular composition, has its own spectral signature. A vast majority of these are supervised algorithms, which require that the number of classes and the class distribution model be known in advance. Furthermore, these algorithms entail training samples from each class to build models for different classes. Unsupervised classification algorithms are built to solve the site labeling problem without the need for training samples.

9.3.1 Unsupervised Classification Algorithms

Today several different unsupervised classification algorithms are commonly used in hyperspectral remote sensing. The two most frequently used algorithms are the K-mean and the ISODATA (Iterative Self-Organizing Data Analysis Techniques) clustering algorithm.

9.3.1.1 K-means Algorithm

K-means algorithm is one of the simplest unsupervised classification algorithms and classify a given remote sensing imagery through a certain number of clusters (often assume k clusters) fixed a priori by the iterative procedures (Jensen

2005). The objective of K-means algorithm is to minimize the within cluster variability, which the objective function is the sums of square distance (errors) between each pixel and its assigned cluster center (Duda *et al.* 2001). In general, it assigns first an arbitrary initial cluster vector. The second step is to take each pixel belonging to a given data set and associate it to the nearest centroid. In the third step the new cluster mean vectors are calculated based on all the pixels in one cluster. The second and the third steps are repeated until the "change" between the iterations is small. The "change" can be defined in several different ways, either by measuring the distances the mean cluster vector have changed from one iteration to another or by the percentage of pixels that have changed between iterations.

9.3.1.2 ISODATA Algorithm

ISODATA stands for Iterative Self-Organizing Data Analysis Techniques (Hall and Ball 1965). This is a more sophisticated algorithm which allows the number of clusters to be automatically adjusted during the iteration by merging similar clusters and splitting clusters with large standard deviations. Clusters are merged if either the number of pixel in a cluster is less than a certain threshold or if the centers of two clusters are closer than a certain threshold. Clusters are split into two different clusters if the cluster standard deviation exceeds a predefined value and the number of pixels is twice the threshold for the minimum number of members (Jensen 2005).

The ISODATA algorithm is similar to the K-means algorithm with the distinct difference that the ISODATA algorithm allows for different number of clusters while the K-means assumes that the number of cluster is known a priori. So, the ISODATA algorithm is more flexible than the K-means method. But the user has to choose empirically many more parameters.

9.3.1.3 Fuzzy K-Means Algorithm

Fuzzy K-means is an extension of K-means, which is the popular simple clustering technique (Campbell 2002). While K-means discovers hard clusters (a pixel belong to only one cluster), Fuzzy K-means is a more statistically formalized method and discovers soft clusters where a particular pixel can belong to more than one cluster with certain probability. It has the advantage that it more naturally handles situations in which subclasses are formed by mixing or interpolating between extreme examples and the mixing phenomenon is often happening in remote sensing because of the spatial resolution. So it is fitter for remote sensing image clustering. The fuzzy K-means algorithm attempts to find a solution for parameters $y_{ji}(i = 1,\cdots,n;\ j = 1,\cdots,g)$ for which (Webb 2002)

$$J_r = \sum_{i=1}^{n}\sum_{j=1}^{g} y_{ji}^{r} \,|\, x_i - m_j \,|^2 \tag{9.14}$$

is minimized subject to the constraints

$$\sum_{j=1}^{g} y_{ji} = 1 \quad 1 \le i \le n \tag{9.15}$$

$$y_{ji} \ge 0 \quad i = 1,\cdots,n; \; j = 1,\cdots,g \tag{9.16}$$

The parameter y_{ji} represents the degree of association or membership function of the i-th pattern or object with j-th group. r (≥ 1) is a scalar termed the weighting exponent which controls the 'fuzziness' of the resulting clusters. As $r \to 1$, this algorithm tends to the basic k-means algorithm. m_j is the 'centroid' of the j-th group

$$m_j = \frac{\sum_{i=1}^{n} y_{ji}^r x_i}{\sum_{i=1}^{n} y_{ji}^r} \tag{9.17}$$

The basic algorithm is iterative and can be stated as follows (Bezdek 1981).

Step 1. Select r ($1 < r < \infty$); initialize the membership function values y_{ji}, $i = 1,\cdots,n; \; j = 1,\cdots,g$.

Step 2. Compute the cluster centers m_j, $j = 1,\cdots,g$, according to (9.17)

Step 3. Compute the distance d_{ij}, where $d_{ij} = | x_i - m_j |$.

Step 4. Compute the membership function: if $d_{il} = 0$ for some l, $y_{li} = 1$, and $y_{ji} = 0$, for all $j \ne l$; otherwise

$$y_{ji} = \frac{1}{\sum_{k=1}^{g} \left(\frac{d_{ij}}{d_{ik}} \right)^{\frac{2}{r-1}}} \tag{9.18}$$

Step 5. If not converged, go to step 2.

FCM is much more sensitive to the initialization, and easily falls into a local optimum (Jensen 2005).

9.3.2 An Unsupervised Artificial Immune Classifier

In addition to the above clustering algorithms, Bayesian classifiers (Storvik *et al.* 2005) and Markov Random Field (Yamazaki and Gingras 1999) have also been employed to obtain better unsupervised classification results. Recently, there has been considerable interest in applying unsupervised neural networks (Bagan *et al.* 2005) and Genetic Algorithm (GA) (Bandyopadhyay *et al.* 2007), to multi/hyperspectral remote sensing image classification.

In contrast to the above classifiers, a novel Unsupervised Artificial Immune Classifier (UAIC) has proposed to perform hyperspectral remote sensing image classification. Artificial Immune Systems (AIS), which are inspired by the immune systems, use the immunological properties in order to develop adaptive systems to accomplish a wide range of tasks in various areas of research including pattern recognition, intrusion detection, clustering, optimization, and intelligence control (Dasgupta 1999, De Castro and Timmis 2002). In contrast to the conventional statistical classifiers, UAIC is a self-learning algorithm utilizing immunological properties, such as memory property and clonal selection.

The clonal selection followed by the B-cells of the biological immune system is the fundamental mechanism on which UAIC is modeled. To better describe UAIC, the relationship between artificial immune systems and hyperspectral image classificaiton were given as follows. Antigens in UAIC are simulated as feature vectors that are presented to the system during training and testing. In particular, UAIC has its specific representation in remote sensing image classification. The antibodies as candidate clustering centers experience a form of clonal expansion after being presented with input image data (analogous to antigens).

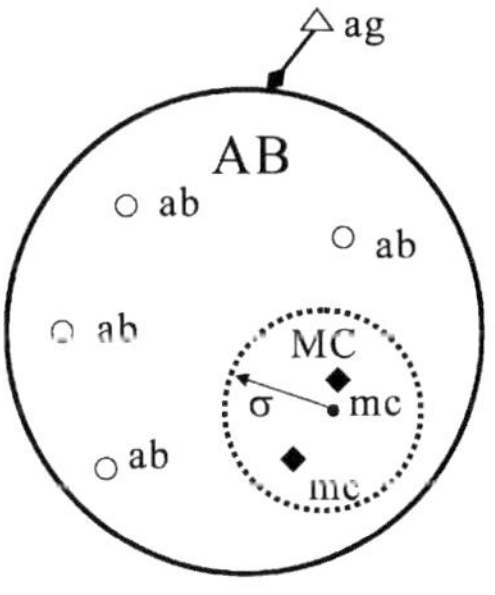

Fig. 9.10 The antibody population (denoted by AB) model of one class (σ represents the AB's scale/radius of influence). (© [2006] IEEE)

With the above notations, the AB model can then be built. Fig. 9.10 shows a diagrammatic representation of the notion of Antibody (AB) set model of one class: there is a certain volume AB in the immune system that contains many antibodies of the class (represented by the circles and denoted by ab) and memory cells (represented by the diamond and denoted by mc). In AB, there is a small surrounding region called memory cell set contained all memory cells of the class, denoted by MC. In remote sensing image classification, the memory set decides the recognizing ability of the whole AB. In Fig. 9.10, σ represents the MC's scale/radius of influence. Within the range of σ, the AB can recognize all antigens. That is, the AB can represent a number of antigens. As can be seen from Fig. 9.10, upon encountering an antigen, antibodies (ab in Fig. 9.10) are stimulated undergoing cloning and mutation. The antigens are then attacked by antibodies and removed from the immune systems. The immune systems maintain and evolve the memory set (MC in Fig. 9.10) so that if ever exposed to the same antigen a quicker response can be elicited against the infection. The proposed algorithm is as follows (Zhong *et al.* 2006).

9.3.2.1 Initialization

UAIC applies the Kaufman approach (KA) (Kaufman and Rousseeuw 1990) to initial memory cell population MC. In this case, the initial memory cell population is obtained by the successive selection of representative instances until C memory cells have been found.

In UAIC, the function $dis(x,y)$ represents the distance between vector x and y. Since UAIC is applied to multi/hyperspectral remote sensing image classification, the distance between x and y, $dis(x,y)$, is calculated using the Spectral Angle Mapping algorithm (SAM) (Yuhas *et al.* 1992). Let vector $x=\{x^1,x^2,\cdots,x^{N_b}\}$ and $y=\{y^1,y^2,\cdots,y^{N_b}\}$, N_b is the band number of the remote sensing image. Then the distance between x and y is given by Eq. (9.19):

$$dis(x,y)=\alpha=\cos^{-1}\left[\frac{\sum_{i=1}^{N_b} x^i y^i}{\left[\sum_{i=1}^{N_b}(x^i)^2\right]^{\frac{1}{2}}\left[\sum_{i=1}^{N_b}(y^i)^2\right]^{\frac{1}{2}}}\right] \tag{9.19}$$

Affinity is inversely proportional to distance in the feature space. In UAIC, affinity is defined as in Eq. (9.20) below according to the antibody population model (Fig. 9.10) so that the affinity between antigens and antibodies or between two antibodies is in the range [0,1] and each AB has its radius of influence.

$$affinity(x,y)=\exp\left(-\frac{dis(x,y)}{2\sigma_i^2}\right) \tag{9.20}$$

σ_i is AB's scale/radius of influence

9.3.2.2 Classification Using UAIC

Once initialization is over, the next step is the iteration of the algorithm. For each iteration, the algorithm performs the following steps to train each antigen *ag* in the remote sensing image.

Step 1: Assign *ag* to *k*-th class

For each *ag* in the image, assign that antigen to one of *nc* classes, where the class is assumed to be the *k*-th class. Given a specific training antigen, *ag*, find the memory cell, *mc,* that has the maximal affinity as follows:

$$mc = \arg\max_{mc \in MC} affinity(ag, mc) \tag{9.21}$$

Then assign that ag to the class of mc, $ag.c \equiv mc.c \equiv k$ $(k \in C = \{1, 2, \cdots, nc\})$.

Step 2: Evolving the antibody population AB^k

After assigning the *ag* to the *k*-th class, evolving the antibody population AB^k and the memory cell pool MC^k are accomplished as follows:

1) Determine the vector f_k that contains the affinity of ag to all the N_{AB} Abs in AB^k, where N_{AB} is the number of the antibody set AB^k.
2) Select the n highest affinity Abs from AB^k to compose a new set $AB^k_{\{n\}}$ of high affinity Abs in relation to ag, where n is the number of the cloned antibodies in AB^k.
3) The n selected Abs independently and proportional to their antigenic affinities, generating a clone set C^k: the higher the antigenic affinity, the higher the number of clones generated for each of the n selected Abs. The number of clones generated for all these n selected antibodies is given by

$$NumClones = \sum_{i=1}^{n} round(Clonal\_rate) \times affinity(ag, ab_i) \tag{9.22}$$

where *NumClones* is the total number of clones generated for ag. The clonal rate, denoted by $Clonal\_rate$, is used to determine how many clones are produced by Abs and memory cells, a typical value is 10, and round() is the operator that rounds its argument toward the closest integer.

4) Submit the clones set C^k to an affinity maturation process inversely proportionally to its antigenic affinity, generating a population MU^k of matured clones: the higher the affinity, the smaller the mutation rate. The mutation rate is determined by Eq.

(9.23) as follows:

$$mutate\_rate = 1 - affinity(ag, ab_i) \qquad (9.23)$$

The mutation of the clones set C^k is performed according to the following equation:

$$MU_i^k = C_i^k + mutation\_rate \times N(0,1) \; i \in [1, NumClones] \qquad (9.24)$$

where N(0,1) is a Gaussian random variable of zero mean and standard deviation of one. As MU_i^k represents a candidate solu tion, it must be within the range of the functions specified do main. If MU_i^k exceeds that, then it is rejected and removed from the population.

5) Redetermine the affinity f_k^* of the matured clones MU^k in relation to antigen ag .
6) Select the highest affinity *ab* in relation to ag to be a candidate memory cell, $mc_{candidate}$, to enter the set of memory antibodies MC^k .
7) Replace the β lowest affinity *ab* from AB^k with d highest affinity from MU^k in order to evolve the antibody population. β is the displace rate.

Step 3: Updating memory cell pool MC^k

The final stage in the training process is the potential introduction of the just-developed candidate memory cell, $mc_{candidate}$, into the set of existing memory cells *MC*.

1) Find the memory cell in MC^k , mc_{match} , that has the following property.

$$mc_{match} = \arg\max_{mc \in MC_k} affinity(ag, mc) \qquad (9.25)$$

2) Calculate the distance threshold(DT).

$$DT = \sum_{i=1}^{N_b} (MAX_i - MIN_i) \qquad (9.26)$$

where MAX_i , MIN_i represent the maximum and minimum

values of the i-th band of the remote sensing image, respectively. N_b is the number of bands of the image.

3) Promote candidate memory cell to memory cell pool MC^k.
The candidate memory cell is added to the set of memory cells only if it has higher affinity in relation to the training antigen, ag, than mc_{match}, where affinity is defined as in Eq. (9.20). If this test is cleared, then if the distance between $mc_{candidate}$ and mc_{match} is less than the product of the affinity threshold and the user-defined distance threshold scalar (DTS), then $mc_{candidate}$ replaces mc_{match} in the set of memory cells.

Once the candidate memory cell has been evaluated for addition into the set of established memory cells, training on this antigen is complete. The next antigen in the multi/hyperspectral image is then selected and the training process proceeds from step 1 to step 3. This process continues until the system has been presented with all antigens in the image.

Step 4: Consolidating and controlling the memory cell pool MC
Subsequent to each iteration, memory cells with identical session data in formation should be merged to limit the memory cell population growth according to their affinity.

9.3.2.3 Stopping Condition

A fixed threshold, i.e., the pixel change threshold, for the proportion of pixels in each class that change class is set as the stop condition. Finally, UAIC outputs the classification result for the remote sensing image.

9.3.3 Experiments

The aforementioned UAIC algorithm was coded in Visual C++6.0 and tested on different images. The experiment using MODIS Image was conducted to test its performance. (For other experiments, refer to the reference Zhong *et al.* 2006.) Consistent comparisons between UAIC and traditional unsupervised algorithms, K-means and ISODATA, were completed. The estimation of classification accuracy for the several classifiers is provided.

In this experiment, the data is an airborne imaging spectrometer (PHI) data shown in Fig. 9.4, 80 bands taken from Xiaqiao test site which is a mixed agricultural area in P.R.China. 80 bands of PHI image (340×390 pixels) were used in this experiment, and their spectral ranges were from 0.417 to 0.854μm.

The chief running parameters that should be provided by users in the classification calculation were as follows: the number of iterations (Ite), the number of classes (C), the number of highest affinity ab (n), clonal rate Clonal_rate, displace rate, and distance threshold scalar (DTS). The values of parameters were set as follows: Ite=20, C=7, n=10, Clonal_rate=15, displace rate = 0.1, DTS=0.85. For a convenient comparison between UAIC and traditional unsupervised algorithms, the pixel change threshold as stop condition is kept at the same value, namely 3%. Fig. 9.11(a) illustrates the classification result using UAIC. Fig. 9.11(b)–(d) illustrate the classification results using K-means, ISODATA and Fuzzy K-means algorithms.

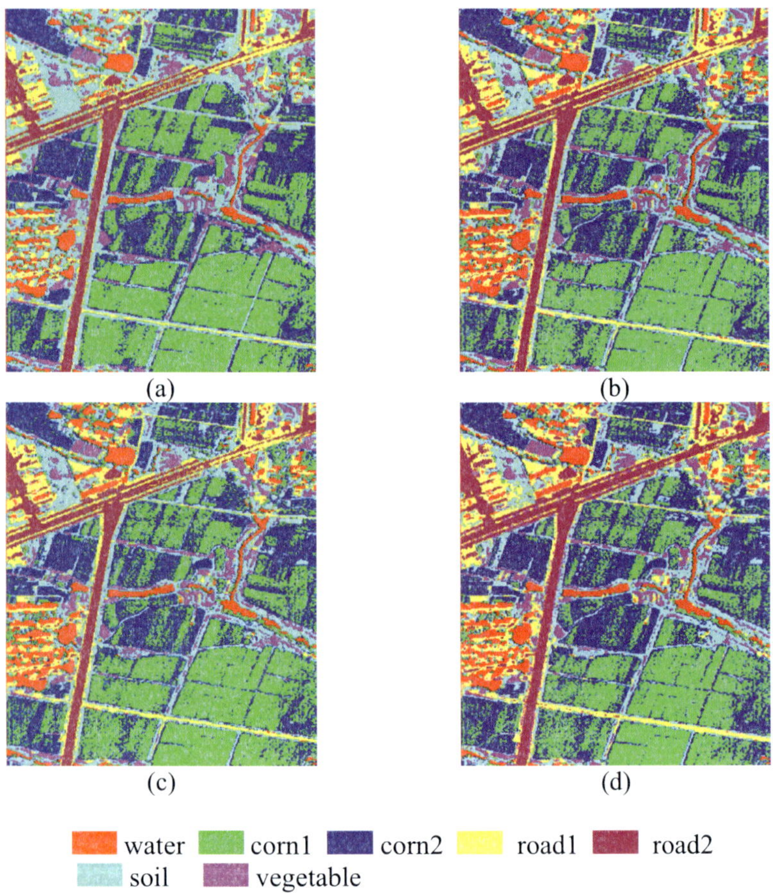

Fig. 9.11 Unsupervised Classification images for Xiaqiao PHI image. (a) UAIC (b) K-means (c) ISODATA (d) Fuzzy K-means (© [2006] IEEE)

As shown in Fig. 9.11, the PHI image was expected to fall into seven classes by ground truth data: water (473 samples), corn1 (460 samples), corn2 (478 samples), road1 (442 samples), road2 (434 samples), soil (434 samples), and vegetable (449 samples).

The visual comparisons of the five cluster classifications in Fig. 9.11 suggest varying degrees of accuracy of pixel assignment. It can be found from the classification images (Fig. 9.11) that four classifiers have similar classification results in the water and road class. K-means and ISODATA create similar classification maps and it is hard to differentiate between corn1 and corn2. While being able to distinguish between corn1 and corn2, Fuzzy K-means fares the worst in soil classification because some soil pixels are misclassified to the road1 class. By contrast, UAIC achieves the best visual accuracy in the corn1 and corn2 class than other classifiers, and also performs satisfactorily to the soil classes. As a result, those using UAIC have better results for four classes.

For a more detailed verification of the results, we compared ground truth data with the classified images and assess the accuracy of each classifier quantitatively using both the overall accuracy measure and the Kappa coefficient. Table 9.1 list the results of comparisons between the ground truth data and classified images obtained by four classifier: UAIC, K-means, ISODATA and Fuzzy K-means.

Table 9.1 Comparison of four classifier performances in the experiment

Accuracy	K-means	ISODATA	Fuzzy K-means	UAIC
Overall accuracy	70.21%	71.33%	75.67%	81.56%
Kappa coefficient	0.6153	0.6756	0.7199	0.7535

It is seen from Table 9.1 that the UAIC classifier produces better classification results than traditional classifiers. The details are as follows: UAIC improves overall classification accuracy from 70.21% to 81.56%, an improvement of 11.35% and Kappa coefficient from 0.6153 to 0.7535, an improvement of 0.1382. This is due to the conventional unsupervised multivariate classifiers requiring ideal conditions. However, because of the complexity of ground substances and the diversity of disturbance, the ideal conditions are not often met in real classification calculations. As a result, these conventional classification methods may have a low precision. On the other hand, UAIC is a data driven self-adaptive method, which can adjust itself to the data without any explicit specification of functional or distributional form for the underlying model. UAIC can approximate any function with arbitrary accuracy by a universal functional approximator. In addition, UAIC is a nonlinear model, which makes it flexible in modeling real, complex relationships. Therefore, the UAIC classifier has the capacities of self-learning and is robust. Based on the above, we can conclude that UAIC is a better classifier for multi-spectral remote sensing image classification.

9.3 Sub-Pixel Data Processing

Hyperspectral imaging spectrometers collect image data reflected from surface materials in many contiguous spectral bands. Owing to the complexity of the land surface and the limited spatial resolution, a pixel in the remotely sensed hyperspectral imagery is typically a mixture of multiple electromagnetic radiances from assorted ground cover materials. Spectral unmixing is a quantitative analysis procedure used to recognize constituent ground cover materials (or endmembers) and obtain their mixing proportions (or abundances) from a mixed pixel.

9.3.1 Linear Mixture Model

A general approach for spectral unmixing is to first build a mathematical model of the spectral mixture and then apply this model to implement spectral unmixing. By modeling pixel signature in different ways, the unmixing methods can be generally grouped into two categories: linear mixture models (LMM) and non-linear mixture models (NLMM). LMM are based on the assumption that each ground cover material only produces a single radiance and the mixed spectrum is a linear combination of ground cover radiance spectra.

In one band of the imagery, each pixel's signal is the sum of the response of different endmembers. So the reflectance of i-th band is expressed as:

$$\rho_i = \sum_{j=1}^{n} (a_{ij} x_j) + e_i \tag{9.27}$$

where ρ_i is the average reflectance of *i*th band of a pixel which contains one or several endmembers;

a_{ij} is the reflectance of the *i*th band of *j*th endmember;

x_j is the abundance of *j*th endmember in the pixel;

m ($i = 1, 2,$, m) is the number of the bands in the imagery.

And the pixel is assumed to contain n endmembers $(j = 1, 2, ..., n)$. Two constrained conditions are defined to (9.27):

$$\sum_{j=1}^{n} x_j = 1 \tag{9.28}$$

$$x_j \geq 0 \tag{9.29}$$

(9.27) can be rewritten as

$$\rho = Mx + e \tag{9.30}$$

where M is a $m \times n$ matrix with its columns composed of endmembers' spectral vectors. By the use of least square method, the abundances can be figured out as:

$$x = (M^T M)^{-1} M^T \rho \tag{9.31}$$

where x is a $n \times 1$ vector and its component is the percentage of each endmember in the mixed pixel ρ.

Hitherto, LMM have remained the dominant method for spectral unmixing analysis owing to their simplicity and generality (Settle and Campbell 1998, Hu *et al.* 1999, Settle 2002). LMM have been successfully applied to the abundances estimation problems in many areas, such as geology, forest and vegetation (Cross *et al.* 1991, Quarmby *et al.* 1992, Gong *et al.*1994, Adams *et al.* 1996). However, LMM may not be appropriate for the cases where multiple scattering results in a nonlinear mixture. Areas where multiple rock types are all visible on the region's surface when viewing striated soils are typical examples of such cases. In these cases, the resultant mixture reflectance spectrum may best be described by assuming that the source radiation is multiply scattered by the randomly distributed endmembers before being collected by the imaging spectrometer.

9.3.2 Nonlinear Mixture Model

Typically, it has been found that NLMM simulates the physical phenomena more accurately (Mustard *et al.* 1998, Zhang and Li 1998); however, the models are usually complicated and application dependent (Borel and Gerstl 1986, Mustard and Pieters 1998). For instance, Mustard *et al.* (1998) showed that the NLMM produced a more accurate result after analyzing the mixture of materials on the lunar surface using both the LMM and the NLMM. Mustard and Pieters (1998) also performed a quantitative analysis of mineral mixture spectra using a NLMM based on the bidirectional reflectance spectroscopy theory (Hapke 1981). In a nutshell, it could be stated that LMM has been widely employed for spectral unmixing analysis as it allows the application of mature mathematical methods, such as least squares estimation (LSE). NLMM is popular for its higher accuracy although there does not exist a simple and generic NLMM that can be utilized in various spectral unmixing applications. A natural problem then turns out to be whether we can utilize the nonlinear characteristics of the spectral mixture to obtain the higher unmixing accuracy whilst simultaneously maintaining the simplicity of LMM.

In this section, we present a nonlinear kernel based least squares regressive algorithm to unmix remote sensing images (Zhang *et al.* 2007b). A kernel-based least squares regressive algorithm performs spectral linear mixing model regression in the feature space induced by a Mercer kernel (Vapnik 1998) and can be used to recursively construct the minimum mean squared-error regressor. The algorithm is formulated in terms of a dot product in order to reduce computation complexity. It is implemented in two steps. The first step is to select the feature vectors by defining a global criterion to characterize the image data structure in the feature space and the second step is the projection of pixels on to the feature vectors and the application of the classical linear regressive algorithm.

9.3.3 Kernel Least Squares Regression

Let r be a $L \times 1$ column image pixel vector in a multispectral or hyperspectral image where L is the number of spectral bands. Assume that M is a $L \times p$ signature matrix, denoted by $M = [m_1, m_2, ... m_p]$, where m_i is a column vector represented by the $i-th$ image endmember signature resident in the pixel vector r, and p is the number of signatures of interest. Let $a = (a_1, a_2, ... a_p)^T$ be an $p \times 1$ abundance column vector associated with the fraction of the $i-th$ signature in the pixel vector r. A LMM assumes that the spectral signature of a pixel vector is linearly superimposed by spectral signatures of image endmembers $m_1, m_2, ... m_p$ present in the pixel vector r and can be described by:

$$r = \sum_{i=1}^{p} a_i m_i + n \tag{9.32}$$

where n is a $L \times 1$ column additive noise vector representing a measurement or model error. Eq. (9.32) is a general LMM and can be solved by interpreting the noise as the error resulting from the goodness of fit in the least squares sense and then minimizing its least squares error. The model's goodness-of-fit is assessed by the length of e, using the sum-of-squared errors (SSE):

$$SSE(\hat{a}) = \|e\|^2 = r^T r - \hat{r}^T \hat{r} = r^T P_M^{\perp} r \tag{9.33}$$

where $P_M^{\perp} = I - M(M^T M)^{-1} M^T$.

LMM has been widely used for spectral unmixing analysis and can generate satisfactory results, i.e., meets the requirements in most applications (Gong *et al.* 1994, Adams *et al.* 1996); however, it is difficult for the method to capture a nonlinear relationship with a linear procedure and hence to reduce the estimated proportional accuracy in most cases. In order to overcome such a limitation, we extended the LMM and LSE to a nonlinear version to keep the simplicity of LMM and also support LSE algorithms. One way for a nonlinear extension is to lift the input space to a higher dimensional feature space by a nonlinear feature mapping and then to find a linear dimension reduction in the feature space. The basic principle behind kernel machines is that a Mercer kernel function, applied to pairs of input vectors, can be interpreted as an inner product in a high-dimensional Hilbert space (the feature space) rather than the data points themselves, thus allowing inner products in feature space to be computed without making dircct reference to feature vectors. The great success of the support vector machine is an example of the effective use of the kernel functions to capture nonlinear data structure. The kernel principal component analysis (PCA) and the kernel linear discriminate analysis (LDA) (Schölkopf *et al.* 1999, Smola and Schölkopf 2000) have also been introduced as nonlinear generalizations of the PCA and the LDA by kernel functions, respectively, and some of their interesting experimental results have been presented.

In order to establish a linear problem in the feature space that corresponds to the nonlinear problem in the input space, how to map input vectors $r \in R^L$ into the feature space and how to handle its possible high dimensionality need to be specified. If an L-dimensional input pixel vector $r \in R^L$ is mapped into a high dimensional space $\Re$ by a nonlinear mapping function ϕ, then

$$\phi : R^L \rightarrow \Re, r \rightarrow \phi(r) \tag{9.34}$$

This high dimensional space $\Re$ is often called feature space. Thus, LMM was rewritten in the following form in the feature space:

$$\phi(r_i) = \sum_{i=1}^{p} \alpha_{\phi(i)} \phi(m_i) + \phi(n_i) \tag{9.35}$$

where $\phi(m_1), \phi(m_2), \cdots, \phi(m_p)$ vectors are endmember spectral in the features, $\alpha_{\phi(i)}$ is a coefficient vector that accounts for the abundances of each endmember spectrum in the feature space. Since the feature space $\Re$ is usually very high, to employ the high dimension data, an inner product in a feature space is calculated by a positive definite kernel function k, which satisfies:

$$\langle \phi(x), \phi(y) \rangle = k(x, y) \tag{9.36}$$

where $\langle .,. \rangle$ denotes the inner product.

Consider $M_p = [m_1, m_2, \cdots, m_p]$ and define the kernel matrix $\boldsymbol{K}$ of dot products as:

$$K = (k_{ij}), 1 \le i \le p, 1 \le j \le p \tag{9.37}$$

where $k_{ij} = \phi^T(m_i)\phi(m_j)$. If the inner product $\langle .,. \rangle$ form of SSE in Eq. (9.33) was replaced by $k(x,y)$, it becomes:

$$SSE(\hat{\alpha}) = \|e\|^2 = r^T r - \hat{r}^T \hat{r}^T = k(r,r) - k^T(r,M_p)K^{-1}k(r,M_p) \quad (9.38)$$

where $k(r,M_p) = [k(r,m_1),k(r,m_2),...,k(r,m_p)]$.

As the dimension of $\Re$ is higher than that of the original space, the selective $\phi(m_1),\phi(m_2),...,\phi(m_p)$ vectors cannot be used to form the basis for characterizing the image data structure in the feature space. Hence, using the kernel trick to obtain the value of $\alpha_{\phi(i)}$ in Eq. (9.35) does not mean 'the achievement of' abundances. In order to obtain the meaningful proportions from each pixel by Eq. (9.35), a feature vector selection and linear regression algorithm was proposed. This algorithm includes two steps. The first step is to select the feature vectors by defining a global criterion to characterize the image data structure in the feature space, and the second step is the projection on to the feature vectors and then applying the classical linear regressive algorithm.

Above all, kernel methods belong to a pattern recognition theory, which focus on approaching the problem by mapping the original data into a higher dimensional feature space, where each co-ordinate corresponds to one feature of the data items. In this way, the data was transformed into a set of points in a Euclidean space. Support vector regression (SVR) refers to a regression application of support vector machine, and build a cost function by a subset of training data. Fig. 9.12 shows the architecture of the proposed algorithm using the feature vectors selection and linear regression.

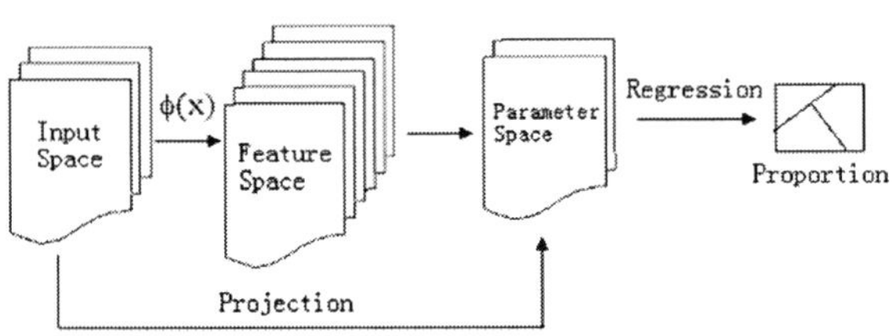

Fig. 9.12 Architecture of the proposed algorithm

An unsupervised method has been developed for a iterative selection of feature vectors in the $\Re$ space so as to capture the image data structure. Initially, a feature vector denoted by m_1 is selected, for which there may be *a priori* knowledge; otherwise, we choose a pixel vector with maximum length as the first feature vector. One reason for this selection is that the brightest pixel may correspond to a pixel containing a material with the largest radiance spectrum in the image scene on the condition of LSE measuring the Euclidean distance, and it is also the largest spectrum in kernel space in that the mapping function is monotone. Then the initial target signature is applied with an orthogonal subspace projector specified by Eq. (9.38) with $M_s = m_1$ to all image pixel vectors. A spectral feature signature is then found, which is denoted by m_2 with the maximum projection in the orthogonal complement space, denoted by $\langle m_1 \rangle^{\perp}$ that is orthogonal to the space, linearly spanned by $\langle m_1 \rangle$. A third spectral feature signature m_3 can be found by applying an orthogonal subspace projector $P^{\perp}_{[m_1,m_2]}$ with $M_s = [m_1, m_2]$ to the original image and a spectral signature that has the maximum projection is selected as m_3. The above procedure is repeated until all the spectral feature signatures are found or a stopping rule is met. A similar algorithm is proposed by Ren and Chang (2000), which performs on the pixel vector rather than on on kernel feature vector as in the procedure above.

Let ε be the prescribed error threshold; the corresponding algorithm is described as follows (Zhang *et al.* 2007b):

1) Initial condition: select the pixel vector with maximum length denoted by m_1. Set $i = 1$ and $M_1 = \{m_1\}$.
2) Apply $P^{\perp}_{m_1}$ via (9.38) to all image pixel vectors in the image.
3) Find another signature, denoted by m_2, which has the maximum orthogonal projection $m_2 = \arg\{\max(\|e\|^2)\}$, and set $i = 2$ and $M_2 = \{m_1, m_2\}$.
4) If $m_1^T P^{\perp}_{[m1,m2]} m_1 < \varepsilon$, stop iterations; otherwise $i = i+1$.
5) Find the $i-th$ target generated at the ith stage by $m_i = \arg\{\max(\|e\|^2)\}$, where $M_{i-1} = [m_1, m_2, ... m_{i-1}]$ is the target signature set generated at the $i-th$ stage.
6) The above procedure repeats until the error threshold is met or a predefined number of feature vectors are obtained.

Once the feature vectors M_s are selected, they define a subspace M_s in $\Re$, $(s \geq p)$. For any pixel vector r_i in the scene, a suitable coefficient vector $w = (w_1, w_2, ..., w_s)^T$ can be found to satisfy the following approximate linear dependence condition:

$$\left\| \sum_{i=1}^{s} w_i \phi(m_i) - \phi(r_i) \right\| \leq \varepsilon \tag{9.39}$$

Note that the smaller ε was, the more feature vectors were selected, which means that ε is an important parameter determining the level of sparsity. Sparsity is related to generalization ability and is considered a desirable property in learning algorithms (Schölkopf and Smola 2002, Vincent and Bengio 2002), as well as in signal processing (Duda *et al.* 2001). The ability of a kernel machine to correctly generalize from its learned experience to new data can be shown to improve as the number of its free variables decreases, which means that sparsification may be used as a regularization instrument. The feature vector selection can also be viewed as the definition of the hidden layer of a multi-layer neural network (Johnson *et al.* 1992). The number of hidden neurons where the data are projected corresponds to the number of feature vectors selected in $\Re$. However, the feature vectors selection (FVS) has the advantage of providing a number of hidden neurons for a given kernel. The kernel trick makes it possible to proceed to select data in the feature space $\Re$ .For a neural network, the feature space remains hidden and is never explored.

Moreover, M_S being basis implies that the matrix $M_S^T M_S$ is full rank and its inverse matrix exists. Hence, an orthogonal basis can be defined:

$$\Omega \equiv M_S (M_S^T M_S)^{-\frac{1}{2}} \tag{9.40}$$

We then apply the dot product projection to obtain the parameter space of original pixels:

$$\pi_\phi(r) = (M_S^T M_S)^{-\frac{1}{2}} M_S^T \phi(r) = K^{-\frac{1}{2}} K(M_S r) \tag{9.41}$$

Since $rank(\pi_\phi) = S$ and the parameter space is linear, this suggests that a linear blind source separation on $\pi_\phi(r)$ based on simultaneous digitalization techniques can be performed to obtain S linear directions of separated nonlinear components in the input space (Harmeling *et al.* 2003). However, the solutions with the highest eigen values may not correspond to the source, but to some function of these sources (Dominique and Alistair 2003). Owing to the complexity of remote sensed data, our experiment did not generate acceptable results by performing linear blind source separation on parameter space. Hence we resorted to regression on w_i to obtain the proportional value:

$$\hat{a}_i = w_i A + b \tag{9.42}$$

Given a set of training data $(w_i, \hat{a}_i)$ where is projection (9.36) onto π_ϕ of the sample r_i, the goal of the algorithm is to estimate the vector $\hat{a}_i$, the output of the function. The mean square error is minimized using the classic LSE technique. It has been shown that Eq. (9.37) leads to the posterior probability estimation and gives the best approximation of the Bayesian decision function (Duda at al. 2001).

9.3.4 Experiments

To evaluate the performance of the proposed algorithm, a dataset was used. It is a simulated dataset based on Hapke's bidirectional approximating reflectance model (Hapke 1981), which is used to test the nonlinear approximating ability of the proposed algorithm. LSE was also implemented for comparison purposes. It has been demonstrated above that LSE is the widely used algorithm for sub-pixel proportion estimation, so it is an important comparison. SVR has been popular for regression problems in the past decade (Smola and Schölkopf 2004). It differs from conventional regressions in that it maps input data into a high-dimensional reproducing kernel Hilbert space and uses an insensitive loss function. As a result, SVR also has a sparse representation of solutions, which lends itself for comparisons. RBF was introduced into the neural network literature in the late 1980s. The RBF has been studied in multivariate approximation theory, particularly in the field of function interpolation, and recently received considerable interest in hyperspectral classification (Du and Chang 1999, Kerri *et al.* 2001) and RBF has been reported to be a useful tool to perform proportion estimation of hyperspectral imagery. Therefore, it was also selected as a comparison algorithm. Root mean square errors (RMSE) is the quadratic sum of the subtraction between estimated value and the real value. It was used as performance measure. (Li *et al.* 2002, Liu and Wu 2005), which is usually used for evaluating total accuracies.

Hyperspectral reflectance spectra were selected from the original push-broom hyperspectral imager image with 80 bands, which was acquired on September 9, 1999 over Xiaqiao Town, Jiangsu Province, China. The wavelength of this image ranges from 0.42 μm to 0.85 μm. The main ground materials in the area are roads, water and vegetation. Their spectra are distinguishable from each other and hence selected as endmembers for simulating data. For each of the three groundcover materials, 15 spectra were chosen to keep the spectral variety. For each class, these spectra were separated into two groups, from which five spectra were selected randomly for system training and the other ten spectra were utilized for system testing. As such, the testing data do not include any information from the training data, which ensures a fair and reliable testing. Fig. 9.13 shows the 45 hyperspectral signals that are utilized for the simulated dataset.

Given three endmember spectra, r_1, r_2 and r_3, and their abundances, a_1, a_2 and a_3, the nonlinear mixed pixel spectra are synthesized based on Hapke's bidirectional approximating reflectance model.

$$R(i,e) = \frac{wH(\mu)H(\mu_0)}{4(\mu + \mu_0)} \tag{9.43}$$

where $R(i,e)$ is bidirectional reflectance, w is single scattering albedo, $\mu = \cos i$, $\mu_0 = \sin i$, i is angle of emergence, $H(\mu)$ is multi-scattering function.

$$H(\mu) = \frac{1+2\mu}{1+2\mu\sqrt{1-w}} \tag{9.44}$$

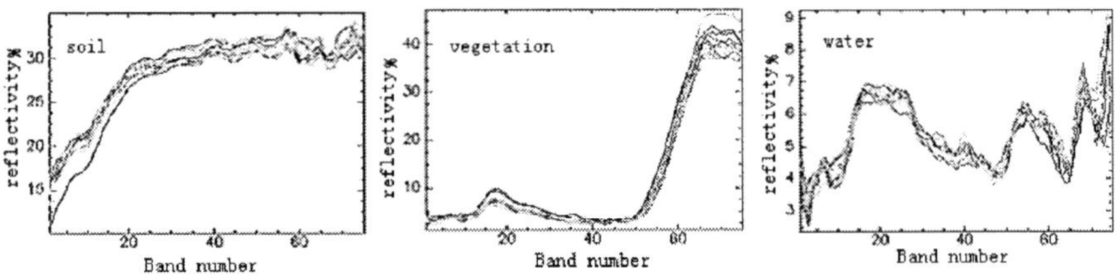

Fig. 9.13 The pushbroom hyperspectral imager reflectance signals of road, water and vegetation utilized for the simulated nonlinear mixture spectra

Table 9.2 The relationship between threshold ε and the number of selected feature vectors.

ε	*0.05*	*0.06*	*0.08*	*0.10*	*0.15*	*0.20*
FVS	64	57	49	38	22	12

FVS= feature vector selection

In all of the following experiments, the approximation to $H(\mu)$ should not greatly affect the abundance estimation. According to Johnson *et al.* (1992), the albedo of a mixture is a linear combination of the single-scattering albedos of its endmember constituents. Thus, if we can convert all of the pixels in an image to albedo, the popular linear unmixing methods for estimating abundance should provide more accurate results (Kerri *et al.* 2001). A mixed pixel spectrum $\widetilde{R}$ can be synthesized as follows:

1. Transfer r_1, r_2 and r_3 to the single scattering albedo w_i using Eq. (9.38).

2. Generate three stochastic values within [0, 1] as a_1, a_2 and a_3 satisfying $a_1 + a_2 + a_3 = 1$; hence, the mixture albedo turns out to be $\rho = \sum_{i=1}^{k} a_i w_i$.

3. Inverse the mixture albedo ρ to the mixture reflectance R, and add additional Gaussian noise N(0, 1).

4. A mixed pixel spectrum, $\widetilde{R}$, can be synthesized as $\widetilde{R} = R(1+\frac{N(0,1)}{SNR})$, where SNR is the signal to noise ratio.

With the simulated data, the kernel function was chosen as the Polynomial kernel with parameter $d = 2$. It can be inferred from Table 9.2 that with the threshold ε decreasing, the number of selected feature vectors increases. Thus, ε is an important parameter to determine the level of sparsity.

A total of 2,500 simulated data points are generated with $SNR = 20:1$. Among them, 500 samples were selected randomly for training and the rest for testing. Since SNR equals to $20:1$, the threshold e is defined as 0.05, to exclude the effect of noise. It can be found from Fig. 9.14(a) that the RMSE of the three materials is 0.016, 0.062, and 0.074, respectively. Points along a 1:1 line on the BDF graph indicate that a prediction matches exactly the real proportion. The smaller the difference between the predicted and the real proportion, the closer the point to the diagonal 1:1 line. Fig. 9.14(a) illustrates that almost all points within the 10% error bound reach the accuracies of 98.8%, 93.6% and 89.2% respectively. The same data were used for traditional linear regression. It can be found from Fig. 9.14(b) that the linear unmixing results are worse than the former, with the RMSE of the three materials being 0.1051, 0.2940, and 0.2589, respectively. Points along a 1:1 line on the BDF graph indicate that a prediction is tremendously partial to the real proportion. The main reason for the partiality is that the simulated data have strong nonlinear properties. This experiment shows that classical linear regression is not suitable for nonlinear data, but our algorithm can perform rather well.

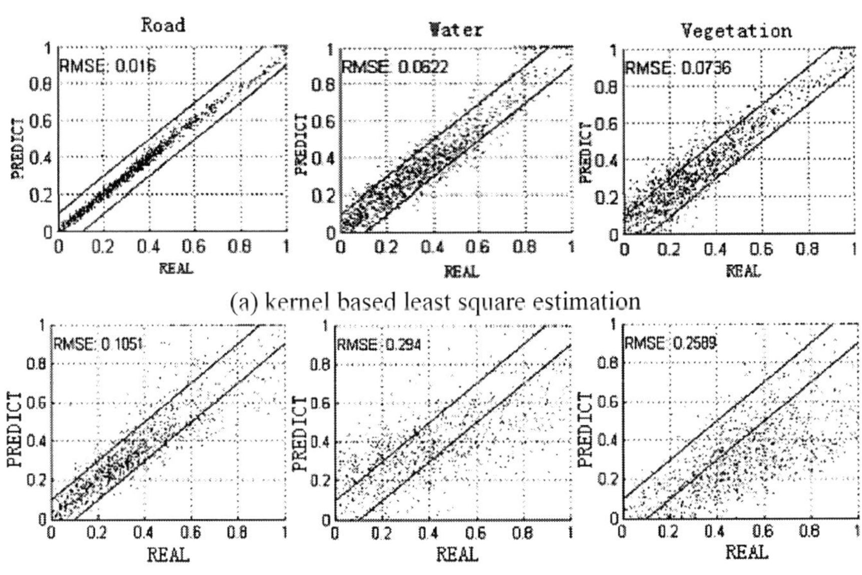

(a) kernel based least square estimation

(b) classic least square estimation

Fig. 9.14 BDF of the simulated test data for road, water and vegetation with the RMSE of each material listed on each chart

9.5 Conclusions

Hughes phenomenon is a challenging problem in hyperspectral remote sensing image processing. Feature selection is an effective method to solve the problem by selecting a subset from a larger number of features or variables used in classification while maintaining an acceptable classification accuracy. Section 2 presents three feature selection algorithms (SFS, SFFS and CSFS) for hyperspectral remote sensing imagery. Specially, a feature subset search algorithm, Clonal Selection Feature Selection Algorithm (CSFS) inspired by the clonal selection theory in Artificial Immune System (AIS) is described. In this new strategy, feature selection is formulated as an optimization problem that searches for an optimum with a fewer number of features in a feature space. In the CSFS, each solution is evolved in binary space and the value of each bit is 0 or 1 indicating that the corresponding feature is selected or removed. Three algorithms are used to feature selection using hyperspectral remote sensing imagery acquired by the Pushbroom Hyperspectral Imager (PHI). Experimental results demonstrate that the accuracy of CSFS outperforms other two algorithms. However, compared with SFS and SFFS in terms of computational costs, CSFS is lowest, and SFS is fastest. So, if time permits, CSFS may provide effective new options for feature selection of hyperspectral remote sensing imagery.

In Section 3, three traditional unsupervised classification algorithms (K-means, ISODATA, Fuzzy K-means and UAIC) for hyperspectral remote sensing imagery have been presented. In addition, a novel algorithm based on the paradigm of nature's immune systems, UAIC, was designed and implemented. The UAIC was successfully applied for classifications of multi/hyperspectral remote sensing images. UAIC was capable of performing data clustering by generating a representative set of memory cells for classification. When compared with three other unsupervised classifiers, K-means, ISODATA, and Fuzzy K-means, the performance of UAIC is better. This evinces that UAIC is applicable for processing of hyperspectral remote sensing images and has high classification precision.

Section 4 has presented sub-pixel processing of hyperspectral imagery by example of a kernel based least squares mixture model. It uses the nonlinear characteristics of spectral mixture to obtain the higher unmixing accuracies. With the simulated data, the proposed method accomplishes perfect results for all the three materials with RMSEs of less than 0.08, while the traditional linear regression is not suitable for estimation of this nonlinear dataset with the maximal RMSE being more than 0.25.

Acknowledgments

The authors would like to thank IEEE and Talyor & Francis publishers for the permissions to reuse materials in our published papers. This work was supported by Major State Basic Research Development Program (973 Program) of China under Grant No. 2009CB723905, the National High Technology Research and

Development Program (863 Program) of China under Grant Nos.2009AA12Z114, 2007AA12Z148,2007AA12Z181, the National Science Foundation of China under Grant Nos. 40771139,40523005, 40721001, Research Fund for the Doctoral Program of Higher Education of China under Grant No. 200804861058 and the foundation of National Laboratory of pattern recognition.

References

Adams JB, Smith MO, Johnson PE (1996) Spectral mixture modeling: A new analysis of rock and soil types at the Viking Lander 1 site. Journal of Geophysical Reserach 91: 2089–8112

Bagan H., Wang Q, Watanabe M, Yang Y, Ma J (2005) Land cover classification from MODIS EVI times-series data using SOM neural network. International Journal of Remote Sensing 26: 4999–5012

Bandyopadhyay S, Maulik U, Mukhopadhyay A (2007) Multiobjective genetic clustering for pixel classification in remote sensing imagery. IEEE Transactions on Geoscience and Remote Sensing 45: 1506–1511

Bezdek JC (1981) Pattern Recognition with Fuzzy Objective Function Algorithms. Plenum Press, New York.

Borel CC, Gerstl SA (1986) Nonlinear spectral mixing models for vegetative and soil surface. Remote Sensing of Environment l: 8098–8112

Bruzzone L, Roli F, Serpico SB (1995) An extension of the Jeffreys-Matusita distance to multi-class cases for feature selection. IEEE Trans. Geosci. Remote Sensing 33: 1318–1321

Burnet FM (1959) The Clonal Selection Theory of Acquired Immunity, Cambridge University Press Cambridge, U.K.

Burnet FM (1978) Clonal selection and after. In: Bell G I, Perelson A S, Pimbley G H eds. Theoretical Immunology : 63–85 Marcel Dekker Inc, New York

Campbell JB (2002) Introduction to Remote Sensing. Taylor & Francis, London

Carter JH (2000) The immune system as a model for pattern recognition and classification. Journal of the American Medical Informatics Association 7: 28–41

Chang CI (2003) Hyperspectral Imaging: Techniques for Spectral Detection and Classification. Orlando. FL: Kluwer, Academic

Chang CI, Wang S (2006) Constrained band selection for hyperspectral imgery. IEEE Trans. Geosci. Remote Sensing 44: 1575–1585

Cross AM, Settle JJ, Drake NA, Paivinen RT (1991) Subpixel measurement of tropical forest cover using AVHRR data. Int. J. Remote Sensing 12: 1119–1129

Dasgupta D (1999) Artificial Immune Systems and Their Applications. Springer, Germany

De Castro LN, Timmis J (2002) Artificial Immune systems: A New Computational Intelligence Approach. Springer-Verlag, London, U.K.

De Castro LN, Von Zuben FJ (2002) Learning and optimization using the clonal selection principle. IEEE Transactions on Evolutionary Computation 6: 239–250

Dominique M, Alistair B (2003) Nonlinear blind source separation using kernels. IEEE Trans. On Neutral Networks 14: 228–234

Du Q, Chang CI (1999) An interference rejection-based radial basis function neural network for hyperspectral image classification. International Joint Conference on Neural Networks 4: 2698–2703

Duda RO, Hart PE, Stork DG (2001) Pattern classification. 2nd edn. John Wiley & Sons

Fukunaga K (1990) Introduction to Statistical Pattern Recognition. 2nd eds. New York: Academic

Gong P, Miller JR, Spanner M (1994) Forest canopy closure from classification and spectral unmixing of scene components - multisensor evaluation of an open canopy. IEEE Trans. Geosci.

Remote Sensing 32: 1067–1080
Hall D, Ball G (1965) Isodata: a novel method of data analysis and pattern classification. Technical report, Stanford Research Institute
Hand, D J (1981) Discrimination and Classification. New York: Wiley.
Hapke B (1981) Bidirectional reflectance spectroscopy 1. Theory. J. Geophys. Res 86: 3039–3054
Harmeling S, Ziehe A, Kawanabe M, Muller KR (2003) Kernel-based nonlinear blind source separation Neural Compute 15: 1089–1124
Hu YH., Lee HB, Scarpace FL (1999) Optimal Linear Spectral Unmixing. IEEE Transactions on Geoscience and Remote Sensing 37: 639–644
Hughes GF (1968) On the mean accuracy of statistical pattern recognizers. IEEE Trans. Inf. Theory IT–14: 55–63
Jain A, Zongker D (1997) Feature selection: evaluation, application, and small sample performance. IEEE Trans. Pattern Anal. Machine Intell. 19: 153–158
Jensen JR (2005) Introductory digital image processing: A remote sensing perspective, 3rd edn. NJ Prentice Hall, Upper Saddle River
Johnson PE, Smith MO, Adams JB (1992) Simple algorithms for remote determination of mineral abundances and particle sizes from reflectance spectra. Journal of Geophysical Research 97(E2): 2649–2657
Kaufman L, Rousseeuw P J (1990). Finding Groups in Data: An Introduction to Cluster Analysis. John Wiley & Sons, New York
Kerri JG, Mark LA, Chang CI (2001) A Quantitative and Comparative Analysis of Linear and Nonlinear Spectral Mixture Models Using Radial Basis Function Neural Networks. IEEE Trans. Geosci. Remote Sensing 39: 2314–2318
Landgrebe DA (2002) Hyperspectral image data analysis. IEEE Signal Processing Magazine 19: 17–28
Lee C, Landgrebe DA (1993) Feature extraction based on decision boundaries. IEEE Transactions on Pattern Analysis and Machine Intelligence. 15: 388–400
Li J, Bruce LM, Mathur A (2002) Wavelet Transform for Dimensionality Reduction in Hyperspectral Linear Unmixing. IEEE Geoscience Remote Sensing Symposium 6: 3513–3515
Liu W, Wu EY (2005) Comparison of non-linear mixture models: sub-pixel classification. Remote Sensing of Environment 94: 145–154
Mustard JF, Li L, He G (1998) Nonlinear spectral mixture modeling of lunar multispec- tral data: Implications for lateral transport. Journal of Geophysical Research 103: 19419–19425
Mustard JF, Pieters CM (1998) Photometric phase functions of common geologic minerals and applications to quantitative analysis of mineral mixture reflectance spectra. Journal of Geophysical Research 94: 13619–13634
Narendra PM, Fukunaga K (1977). A branch and bound algorithm for feature subset selection. IEEE Trans. Comput. C-31: 917–922
Pudil P, Novovicova J, Kittler J (1994) Floating search methods in feature selection. Pattern Recognit. Lett. 15: 1119–1125
Quarmby NA, Townshend JR, Settle JJ, White KH, Milnes M, Hindle TL, Silleos N (1992) Linear mixture modelling applied to AVHRR data for crop area estimation. Int. J. Remote Sensing 13: 415–425
Ren H, Chang CI (2000) A generalized orthogonal subspace projection approach to unsupervised multispectral image classification. IEEE Trans. Geosci. Remote Sensing 39: 2515–2528
Richards JA (1986) An Introduction to Remote Sensing digital image analysis. Springer Verlag
Schölkopf B, Smola AJ (2002) Learning with Kernels. MIT Press, Cambridge, MA.
Sebastiano SB, Lorenzo L (2001) A new search algorithm for feature selection in hyperspectral remote sensing images. IEEE Trans. Geosci. Remote Sensing 39: 1360–1367
Settle J (2002) On constrained energy minimization and the partial unmixing of multispectral images. IEEE Trans. Geosci. Remote Sensing 40: 718–721
Settle J, Campbell N (1998) On the errors of two estimators of sub-pixel fractional cover when

mixing Is linear. IEEE Trans. Geosci. Remote Sensing 36: 163–170

Smola AJ, Schölkopf B (2004) A tutorial on support vector regression. Statistics and Computing 14: 199–222

Smola AJ, Schölkopf B (2000) Sparse greedy matrix approximation for machine learning. International Conference on Machine Learning ICML Stanford, CA.

Storvik G, Fjortoft R, Solberg AHS (2005) A Bayesian approach to classification of multiresolution remote sensing data. IEEE Trans.on Geosci. and Remote Sensing 43: 539–547

Swain PH, Davis SM (1978) Remote sensing: the quantitative approach. McGraw-Hill, New York

Vapnik VN (1998) Statistical Learning Theory. Wiley, New York

Vincent P, Bengio Y (2002) Kernel matching pursuit. Machine Learning Journal 48: 165–187

Webb A R (2002) Statistical Pattern Recognition. 2nd edn. John Wiley & Sons, Inc.

Yamazaki T, Gingras D (1999) Unsupervised multispectral image classification using MRF models and VQ method. IEEE Transactions on Geoscience and Remote Sensing 37: 1173–1176

Yuhas RH, Goetz AFH, Boardman JW (1992) Discrimination Among Semi-Arid Landscape Endmembers Using the Spectral Angle Mapper (SAM) Algorithm. Summaries of the 4th JPL Airborne Earth Science Workshop: 147–149 JPL Publication

Zhang L, Li D (1998) Study of the spectral mixture model of soil and vegetation in PoYang lake area, China. Int. J. Remote Sensing 19: 2077–2084

Zhang L, Zhong Y, Huang B, Gong J, Li P (2007a) Dimensionality Reduction Based on Clonal Selection for Hyperspectral Imagery, IEEE Trans. In Geoscience and Remote Sensing 45: 4172–4185

Zhang L, Wu B, Huang B, Li P (2007b) Nonlinear Estimation of Subpixel Proportion Via Kernel Least Square Regression. Int.J.Remote Sensing 28: 4157–4172

Zhong Y, Zhang L., Huang B, Li P (2006) An Unsupervised Artificial Immune Classifier for Multi/hyper-spectral Remote Sensing Imagery. IEEE Trans. on Geosci. and Remote Sensing 44: 420–431

Zhong Y, Zhang L, Gong J, Li P (2007) A Supervised Artificial Immune Classifier for Remote Sensing Imagery. IEEE Trans. on Geosci. and Remote Sensing 45: 3957–3966

Chapter 10

EFFECTS OF AGGREGATION METHODS ON IMAGE CLASSIFICATION

Peng Han, Zhilin Li, Jianya Gong

10.1 Introduction

A major concern in scale- and resolution-related issues is to develop methods for determining the most appropriate scale and resolution of study and assessing their effects (Cao and Lam 1997). The choice of an appropriate scale, or spatial resolution, for a particular application depends on several factors. These include the desired information about the ground scene, the analysis methods to be used to extract the information, and the spatial structure of the scene itself (Woodcock and Strahler 1987). When an appropriate scale or resolution is determined, the next step is to get the corresponding images. Unfortunately, the resolutions of existing remote sensing satellite images are discrete and one may not be able to obtain an image with desired resolution (e.g. 7m). In this case, resampling techniques are often used to interpolate an image into desired resolution and aggregation is a particular resampling technique widely practiced for "up-scaling" image data from high resolution to low resolution (Bian and Butler 1999). It can be visualized that different aggregation methods may introduce different kinds of noise, create different kinds of mixed pixels and thus lead to different results. Therefore, inferring spatial data across scales is an important challenge faced by scientists (Wang *et al.* 2004).

Scientists have long noted the effect of aggregation or scaling on image classification (Marceau *et al.* 1994, Niemann *et al.* 1997, He *et al.* 2002, Narayanan *et al.* 2002). Marceau *et al.* (1994) evaluated the impact of the measurement scale and the aggregation level on the information content and the classification accuracy of airborne Multi-detector Electro-optical Imaging Scanner II (MEIS- II) images acquired over a mid-latitude temperate forested environment. The original MEIS- II image data were progressively averaged to coarser spatial resolutions, namely 5m, 10m, 20m and 30m. The Bayesian multi-spectral classifier was employed to classify these up-scaled images. The general effects of aggregation on classification has been found but the results indicate that there is no unique spatial resolution appropriate for the detection and discrimination of all geographical enti-

D. Li et al. (eds.), *Geospatial Technology for Earth Observation*,
DOI 10.1007/978-1-4419-0050-0_10,

ties composing a complex natural scene such as a forested environment. He *et al.* (2002) examined random rule-based aggregation and its effect on cover type abundance and landscape patterns, in comparison with the majority rule-based aggregation. They aggregated a classified TM imagery (about 1.5 million ha.) from 30m (4231×3717 pixels) incrementally to 990m resolution (132×116 pixels). Cover type proportion, mean patch size ratio, aggregation index, and fractal dimension were used to assess the effects of aggregation. Their study indicated that these two spatial aggregation techniques led to different results in cover type proportions, making spatial pattern altered in opposite ways. However, their study was restricted only to the effect of two aggregation methods on the cover type proportion and spatial pattern of classified imagery.

In summary, it can be noted here that there is no systematic investigation into the effect of various aggregation methods on image classification. Indeed, the objective of this paper is to do such a study through experimental testing.

Following this introduction is a theoretical analysis of aggregation on classification accuracy (Section 10.2). In this section, five aggregation methods are described and their effects analyzed. Section 10.3 outlines the design of experimental testing on the effect of aggregation methods on classification. Two experimental tests are then reported in Sections 10.4 and 10.5. Finally some conclusions are made in Section 10.6.

10.2 Effect of Aggregation Methods on Classification Accuracy: A Theoretical Analysis

In this section, the aggregation methods used in this study are briefly described and a theoretical analysis of their effect on image classification is made.

10.2.1 Aggregation Methods

Several types of methods are available for image aggregation, such as simple arithmetic, geo-statistical, and transform-based aggregation methods. The simple arithmetic aggregation methods include averaging, central-pixel, median, nearest neighbor, bilinear, bicubic, etc. All these methods extract a value from a $n \times n$ window in the original image as the pixel value in the new image. Geo-statistical methods consider the spatial properties in the operation of aggregation, including variance-weighted, geo-statistical variance estimation, spatial variability-weighted and simulation methods (Hay *et al.* 1997, Collins and Woodcock 1999, Wang *et al.* 2004). Transform-based methods decompose the original image into components with different frequencies and the low-frequency components together compose a smoothed image. Wavelet is a transform widely used for this purpose.

In this study, five aggregation methods:,Arithmetic Average Variability-Weighted (AAVW), Averaging (AV), Bilinear (BL), Bicubic (BC), and Nearest neighbor (NN),,were selected for comparative analysis. The averaging method uses the average value over a $n \times n$ window. The nearest neighbor method simply selects the value of the nearest pixel but does not at all consider the values of other neighboring pixels. In bilinear method, a value is assigned to an output pixel by linear interpolation of four pixel values in two orthogonal directions within the input image. Bicubic method is similar to bilinear method. In bicubic method, cubic interpolation (instead of linear interpolation) is used in two orthogonal directions. Images resampled with bicubic interpolation are smooth and have fewer interpolation artifacts. Thus, bicubic method is often preferred to the bilinear or nearest neighbor method in image resampling when processing speed is not an issue. The AAVW method (Wang *et al.* 2004) is one of the spatial variability-weighted methods. The idea behind the spatial variability-weighted methods is that the local spatial autocorrelation can be measured by square differences between the pixel values and a measure of the dominant spatial feature within a window. The larger the square difference, the further the pixel value is deviated from the dominant feature, and less weight should be given for this pixel in aggregation from a finer to a coarser resolution. The weight can be defined as the reciprocal of the square difference.

10.2.2 Theoretical Analysis of Aggregation Effect

In early 1980s, Markham and Townshend (1981) pointed out that the effect of spatial resolution on the classification accuracy of remotely sensed images was related to two factors. One is the change in the number of mixed pixels which are located near the boundaries among classes and the other is the change of the spectral variations within classes. While the spatial resolution of remotely sensed data becomes finer, the number of mixed pixels will decrease, which is positive for classification accuracy. However, the spectral variation within classes will increase, which is negative for classification accuracy. The net effect of these two factors is a function of the environment of the image scene.

In the aggregation process, some mixed pixels are created. Due to the differences in selection of a value for each pixel of the new image, there must be differences between the effects of the aggregation methods on the increase in mixed pixels. It is noticeable that mixed pixels result from both the boundary and inner region of image objects. For the classes with stable spectral variation, the mixed pixels mainly increase on the boundary of objects. In the case of the classes with relatively unstable spectral variation, the mixed pixels induced by aggregation emerge not only on the boundary, but also in the inner region. Homogeneity is usually used to reflect the spectral variation of classes. The larger the spectral variation, the lower the homogeneity in a class.

That is, for the classes with high homogeneity, the mixed pixels mainly fall on the boundary of objects in the aggregation operation. On the up-scaled images of the NN method, pixel values are directly derived from the original image, thus resulting in a minimal number of mixed pixels. Fig. 10.1 demonstrates the NN interpolation on the classes with high homogeneity. In the case of BC, BL, AV and AAVW, boundary pixels in the aggregated image have a mix of different land surface elements, increasing the confusion in the classification process. For this reason, the NN method is suitable for the land cover types with high homogeneity, like "water" and "concrete".

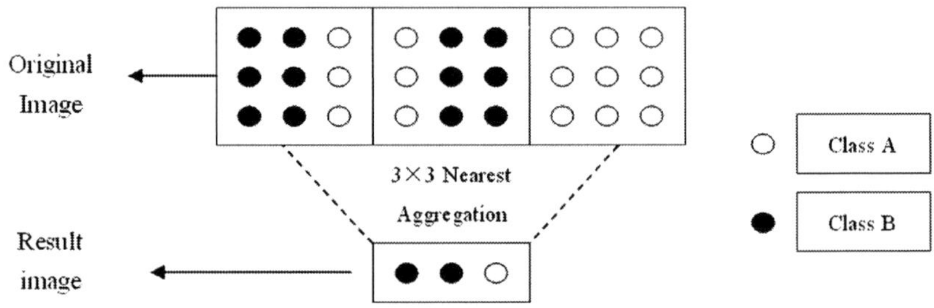

Fig. 10.1 The Nearest Neighbor (NN) aggregation process on class with high homogeneity

For the classes with low homogeneity, due to the unstable spectral variation, the NN method is very likely to result in misleading resample, leading to a decrease in the classification accuracy. Fig.10.2 demonstrates the NN aggregation on the classes with low homogeneity. On the other hand, the BC and BL methods partly overcome these problems because they capture the dominant land cover type within a window. The key to accurately inferring spatial information across scales lies in the capture of dominant spatial features, patterns, and processes of a variable (Wang *et al.* 2004). An inappropriate resample will result in false pixel value which may be far away from the dominant land cover type. For this reason, the BC, BL, AV and AAVW methods should perform better than NN for the land cover types with unstable spectral variation or low homogeneity.

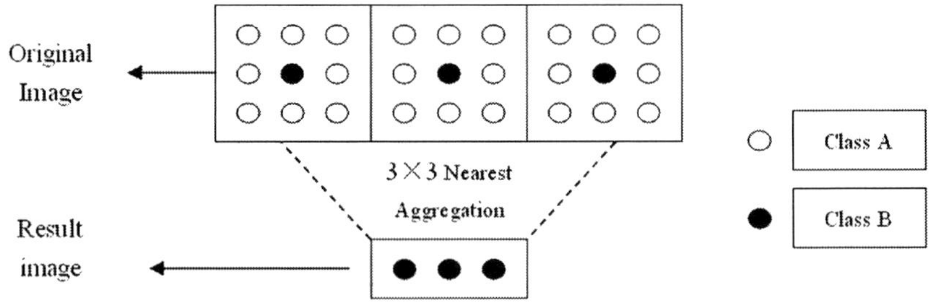

Fig. 10.2 The nearest aggregation process on class with low homogeneity

The overall accuracy of image classification is determined by the accuracies of all land cover types together, especially that of the dominant classes. Therefore, before evaluating the effect of aggregation methods on overall classification accuracy, it is necessary to analyze the spatial distribution pattern of all land cover types in the image so as to identify the dominant classes. It seems reasonable to make the following hypothesis:

If the dominant classes are with high homogeneity, it would be better to employ the NN method for aggregation; otherwise, other methods such as AV, BL, BC and AAVW could be considered.

In this study, two experimental tests on this hypothesis have been conducted and will be reported in the following sections.

10.3 Effect of Aggregation Methods on Classification Accuracy: Experimental Design

To conduct experiments effectively, it is essential to select a set of appropriate image data and make proper design of the experiments.

10.3.1 Selection of Study Area and Test Images

As explained previously, the homogeneity of the dominant land cover types should be considered in the selection of an aggregation method. However, the homogeneity of a type may depend on the resolution of images. For example, in an area, the homogeneity of the "forest" type is high on medium resolution images but may appear to be low on high resolution images. This makes the authors select two types of images, i.e. TM with medium resolution and SPOT with relatively high resolution. The reason for selection of a SPOT image instead of high-resolution images such as the IKONOS image is due to the fact that conventional classification methods may not work well for high-resolution images and thus different classification methods may need to be employed for classification. In such a case, the results might not be directly compared. It is also believed that the images should cover these common types of land covers such as water, forest, farmland, built-up area, and grassland. With these in mind, two images with both urban and rural areas are selected.

The TM image covers Guangzhou and its surrounding areas (Fig.10.3a) and was acquired on 20 November 2001 with spatial resolution of 30m. The image size is 3200×2400 pixels. This TM image was colleted with six bands. Six main classes of land covers are present, i.e., water, forest, farmland, built-up area, grassland and bare soil. Among these, water, forest and farmland are the dominant classes, occupying 68% of the area. An analysis shows that these three classes are all of high homogeneity.

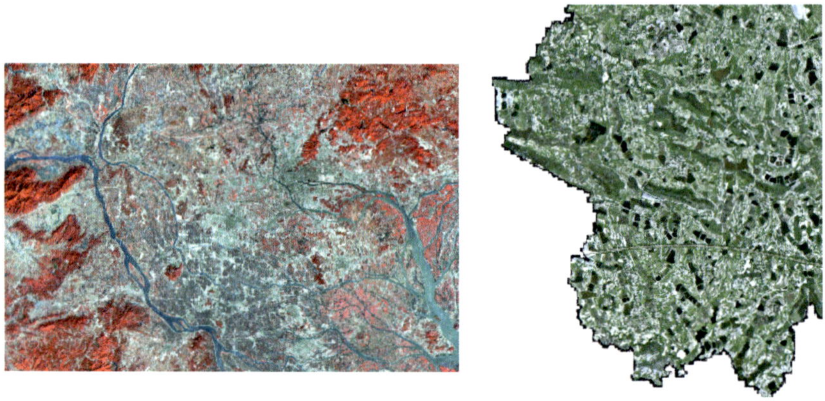

(a) TM image (b) SPOT5 image

Fig. 10.3 The test images

The SPOT5 image, acquired in October 2002, covers an area in Wuhan, China (Fig.10.3b). The image consists of three multi-spectral bands with 2.5m resolution. The image size is 1671×1935 pixels. This SPOT image was the result of fusing the SPOT5 multi-spectral (10m resolution) and SPOT5 panchromatic (PAN, 2.5m resolution) images. In this area, forest and farmland are the dominant classes and these two classes occupy a proportion of about 70%. Investigation indicates that these two classes are of low homogeneity, possibly due to the relatively high resolution.

10.3.2 Strategies for the Experiments

A number of strategies need to be considered such as the proper selection of classification methods, aggregation of images, and the use of ground truth.

All the five aggregation methods are used for aggregation of the images from their original resolution to new images with a range of resolutions. For the TM image, a total of 9 sets of new images are generated with aggregation with different window sizes, i.e. 2×2, 3×3, 4×4, 5×5, 6×6, 7×7, 8×8, 9×9, 10×10. That is, the resolutions of new images are 60m, 90m, …, 300m. In total, including the original TM image, 10×5 images are prepared for classification. Due to the limited space, only the five new images with 300m resolution are shown in Fig.10.4. For the SPOT5 image, only five different window sizes, i.e. 2×2, 4×4, 6×6, 8×8, 10×10, are used for aggregation so as to obtain a total of 5×5 new images. As a result, the resolution ranges from 5 m to 25 m. Also only the five new SPOT images with 25m resolution are shown in Fig.10.5.

It was decided to use the maximum likelihood (ML) classifier for image classification because this is the most commonly used classifier. Six classes are designed for classification, i.e. water, forest, farmland, built-up area, grassland and bare soil.

For accuracy assessment, 728 ground truth points were used as benchmark in the TM experiment. In the case of SPOT5 image, a land cover map in the same region was used as the reference data, which was produced by Land and Resources Bureau of Wuhan. Both overall accuracy and accuracies of individual classes are obtained for analysis. The results of these two experimental tests are reported in the following two sections, respectively.

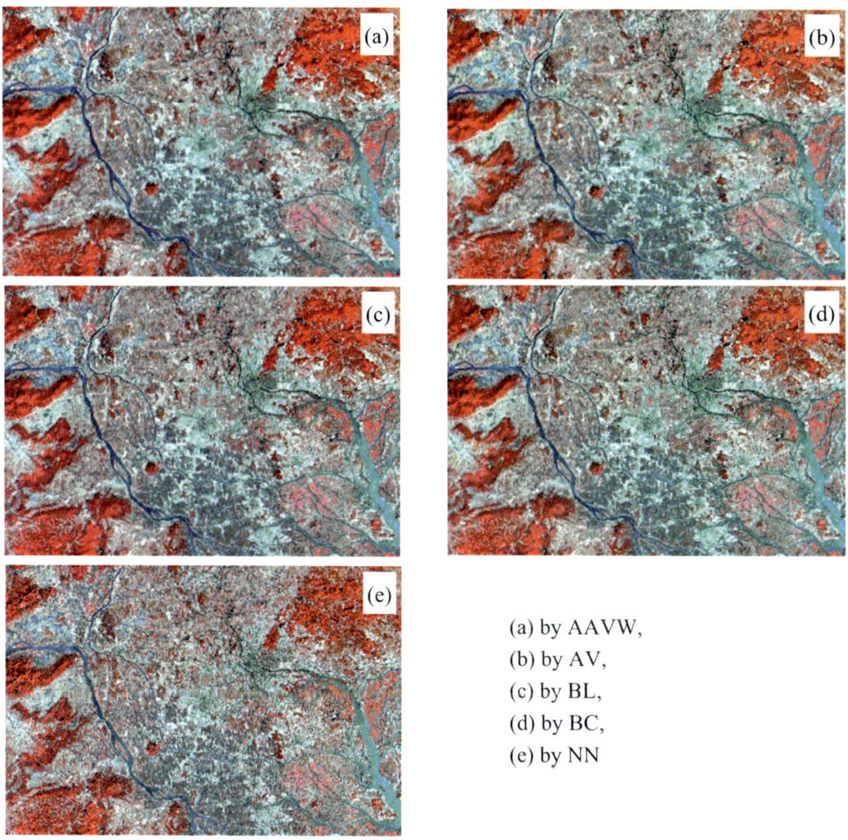

(a) by AAVW,
(b) by AV,
(c) by BL,
(d) by BC,
(e) by NN

Fig. 10.4 The up-scaled TM images at 300m resolution by different aggregation methods

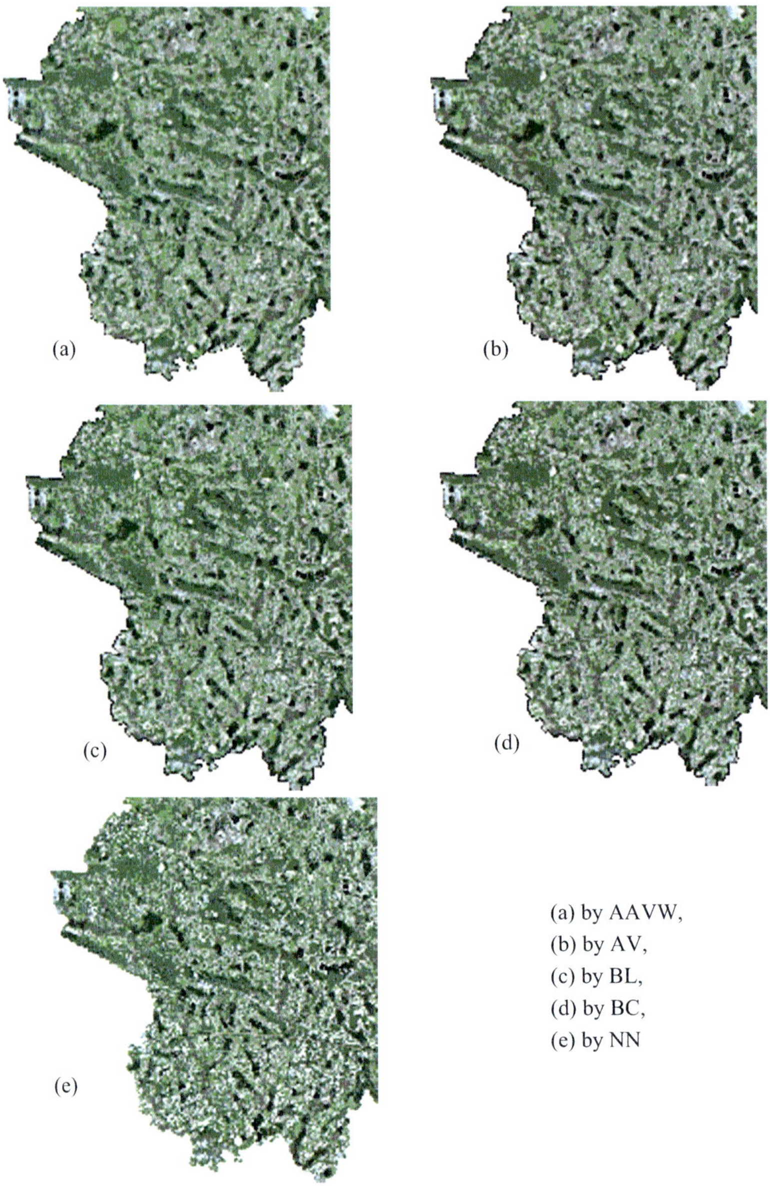

Fig. 10.5 The up-scaled SPOT-5 images at 25m resolution by different aggregation methods

10.4 Effect of Aggregation Methods on Classification Accuracy: Experiment with TM Image

In this section, the test on the effect of aggregation methods on classification accuracy with a TM image will be reported.

10.4.1 Results of Experiment with TM Image

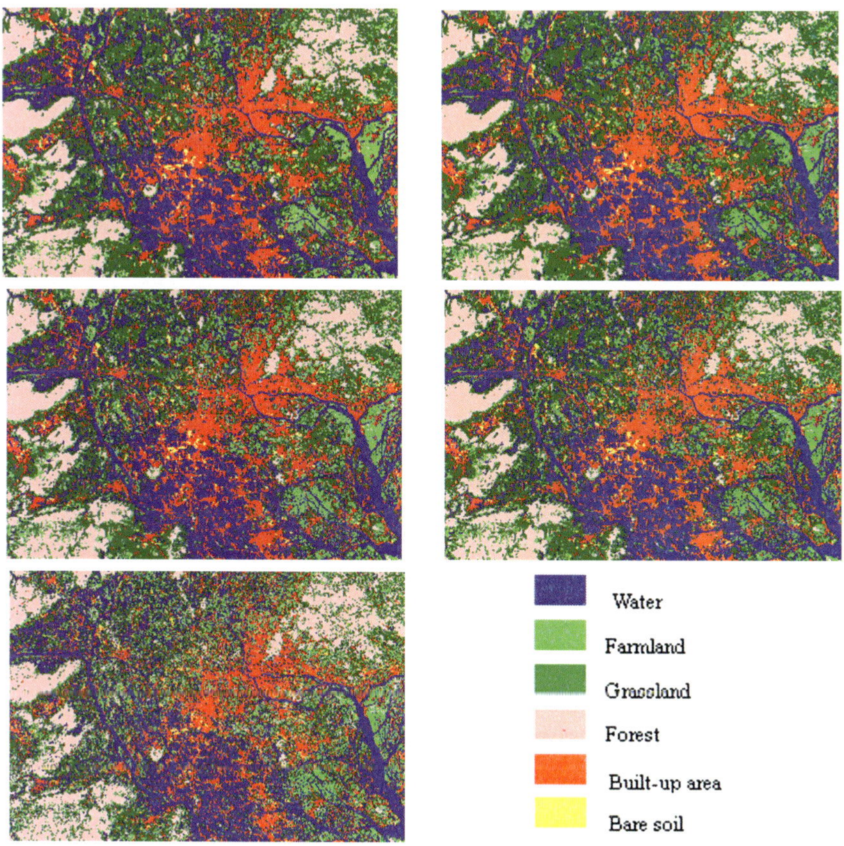

Fig. 10.6 The classified results of TM images at 300m aggregated by 5 methods. (a) by AAVW, (b) by AV, (c) by BL, (d) by BC, (e) by NN

All the 9×5 new images aggregated by 5 methods are classified by using the maximum likelihood (ML) classifier. The original TM image was also classified with a ML classifier. Due to the limitation of space, only the 5 classified results at 300 m resolution are shown in Fig.10.6.

The overall accuracy is shown in Fig. 10.7. From this figure, not much information could be obtained, except a general trend, i.e. a decrease in accuracy after 90m. To make more meaningful analysis possible, the classification accuracies of different types of land cover are also computed and shown in Figs. 10.8-10.13.

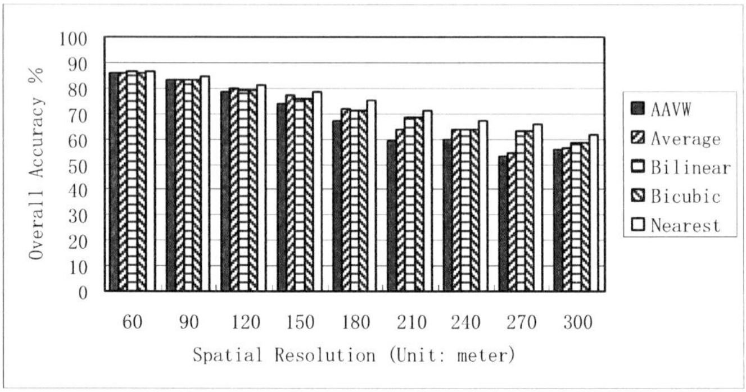

Fig. 10.7 The overall classification accuracy of aggregated TM images at different resolution

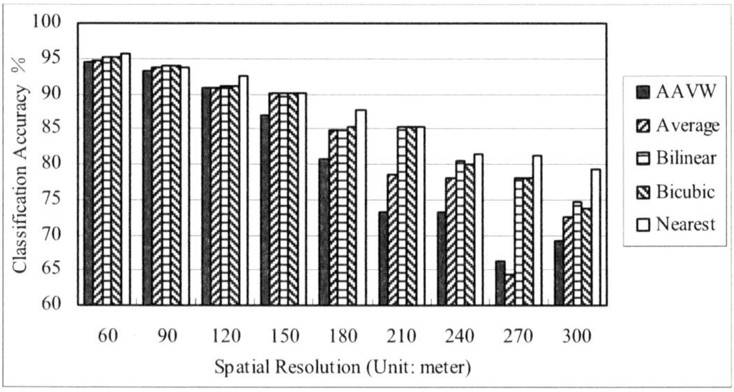

Fig. 10.8 Classification accuracy of water on aggregated TM images at different resolution

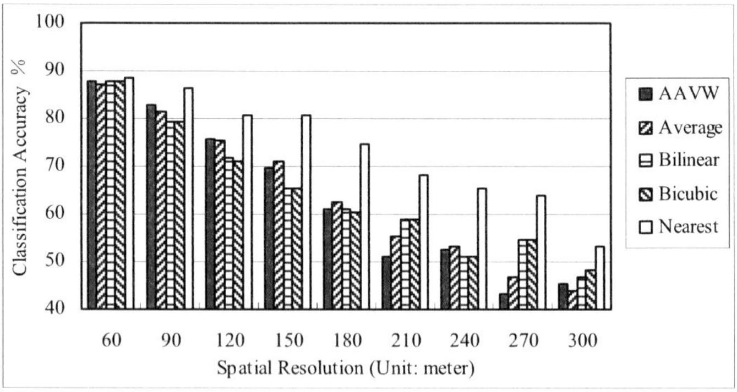

Fig. 10.9 Classification accuracy of forest on aggregated TM images at different resolution

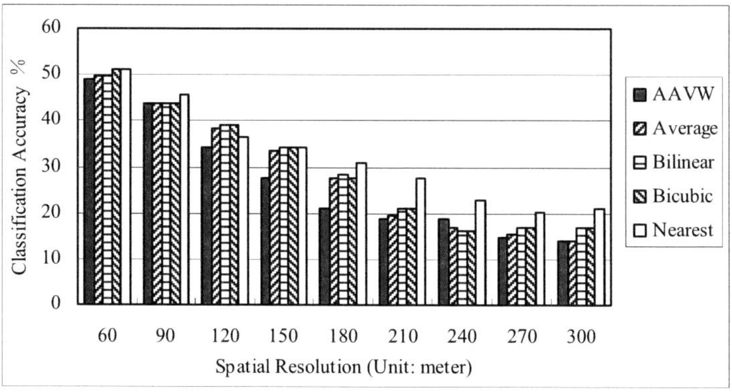

Fig. 10.10 Classification accuracy of farmland on aggregated TM images at different resolutions

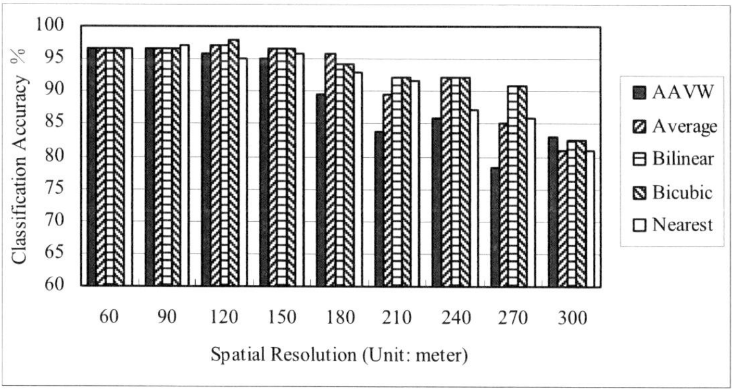

Fig. 10.11 Classification accuracy of built-up area on aggregated TM images at different resolution

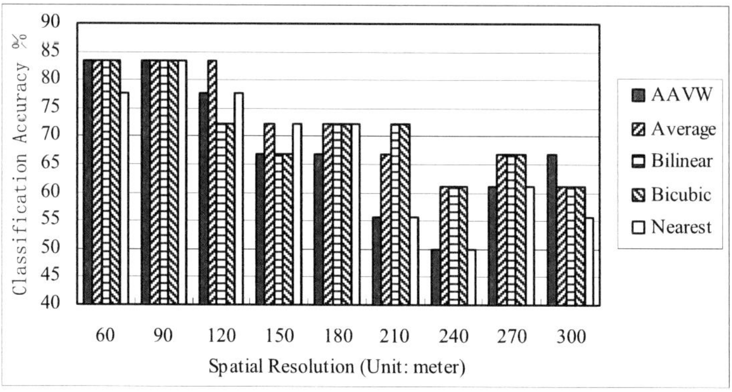

Fig. 10.12 Classification accuracy of grassland on aggregated TM images at different resolution

10.4.2 Effect of Aggregation Methods on the Overall Accuracy

Fig.10.7 shows the impact of the five methods on the overall classification accuracy. Results reveal that in general there is a decrease in accuracy when the resolution of aggregated images is over 90m. It is also observed that the NN method achieves a higher overall accuracy at all resolutions than the other methods, although the increase is not much. The AAVW produces the worst results (i.e. lowest accuracy) at almost all resolutions. The BC and BL methods perform better than AV method at most resolutions.

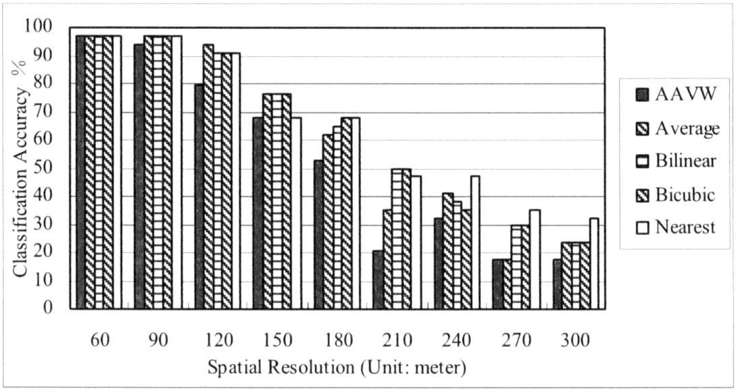

Fig. 10.13 Classification accuracy of bare soil on aggregated TM images at different resolution

The phenomenon that NN produces most impressive results might be explained by the fact that the dominant classes of this area, "water," "forest" and "farmland", are of high homogeneity and these classes occupies a proportion of 68%. The homogeneity of the TM band 1 image is 0.8352 (supported by ENVI 4.4).

10.4.3 Effect of Aggregation Methods on Classification Accuracy

It would be more appropriate to have an analysis of the individual class results. On the whole, there is a decrease in accuracy when resolution is coarser than a certain value. But this value varies with the type of land cover and the trend of decrease is also different. In particular, the rate of decrease for the type "built-up area" is rather slow.

For the class "water", the accuracy of classified images aggregated by NN method is almost higher than the others at each resolution (Fig.10.8). This trend becomes clearer when the resolution becomes coarser. On the other hand, the accuracy of classified images aggregated by AAVW method is the lowest at each resolution. This trend also becomes clearer when the resolution becomes coarser. Similar trends can also be observed from "forest" and "farmland" classes (Fig.10.9 and Fig.10.10).

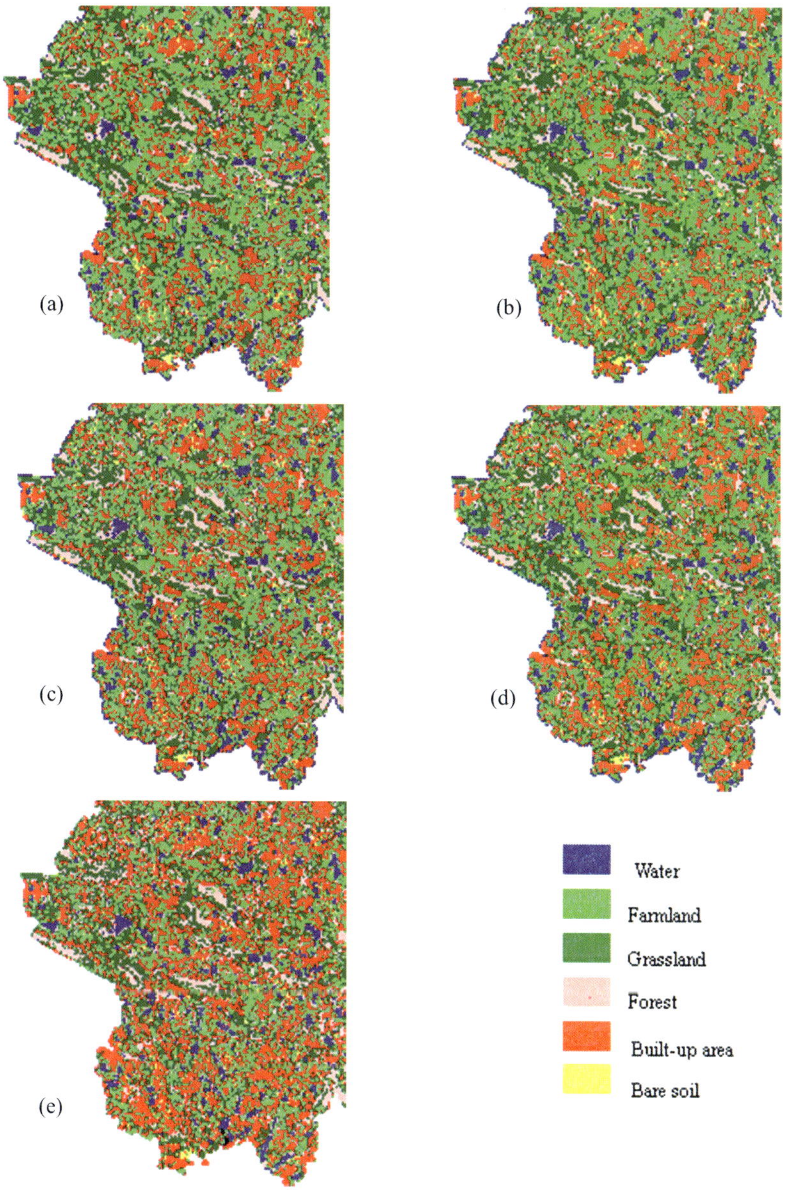

Fig. 10.14 The classified results of SPOT-5 images at 25m, aggregated by different methods. (a) by AAVW, (b) by AV, (c) by BL, (d) by BC, (e) by NN

The results for "built-up area" and "grassland" show another trend (Fig.10.11 and Fig.10.12). That is, the classification accuracies of images aggregated by BC and BL are generally higher than the others, especially when the resolution is very coarse. But occasionally, AV and AAVW prevail.

It is worth noting that the accuracy trend in class "bare soil" (Fig.10.13). It can be found that BC and BL perform better than others at the higher resolutions (60m to 210m), however, the NN outperforms at the coarser resolutions (240m to 300m).

10.5 Effect of Aggregation Methods on Classification Accuracy: Experiment with SPOT Image

In this section, the test on the effect of aggregation methods on classification accuracy with a SPOT image will be reported.

10.5.1 Results of Experiment with SPOT Image

All the 5×5 new images aggregated by 5 methods and the original SPOT5 image are classified by using the maximum likelihood (ML) classifier. Due to the limitation in space, only the 5 classified results at 25m resolution are shown in Fig.10.14. The overall classification accuracy is shown in Fig.10.15, and the classification accuracies of the six classes are shown from Fig.10.16 to Fig.10.21.

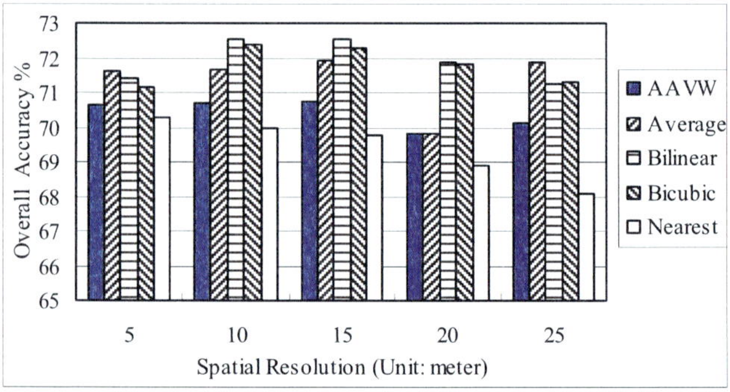

Fig. 10.15 Overall classification accuracy of aggregated SPOT-5 images at different resolutions

10.5.2 Effect of Aggregation Methods on the Overall Accuracy

Fig.10.15 demonstrates the impact of the five aggregation methods on the overall classification accuracy. In general, the classification accuracy is low. This might be because the commonly used ML classifier is not very suitable for the image of such resolution. Indeed, it is found that the homogeneity of the dominant classes in this area ("forest" and "farmland"), which occupies 70% of the area, is low (i.e. 0.6168 compared with 0.8325 for the TM image). Therefore, the NN method produces the worst results at all resolutions and the BC and BL achieve higher accuracies at most resolutions than the other methods. The AV method also performs very well.

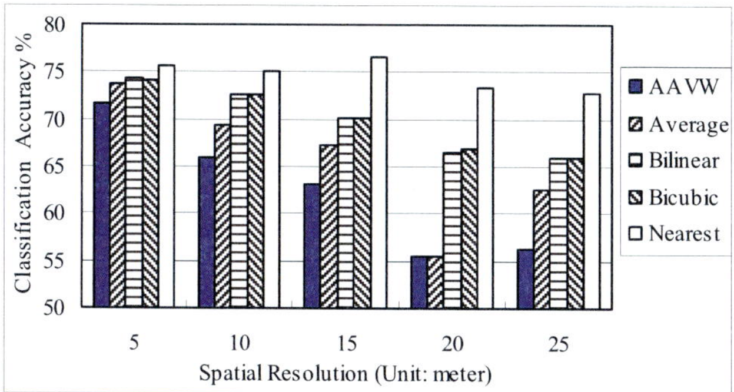

Fig. 10.16 Classification accuracy of water on aggregated SPOT-5 images at different resolutions

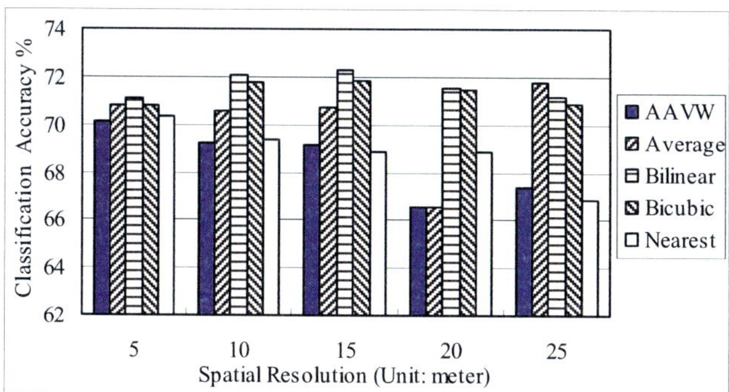

Fig. 10.17 Classification accuracy of forest on aggregated SPOT-5 images at different resolutions

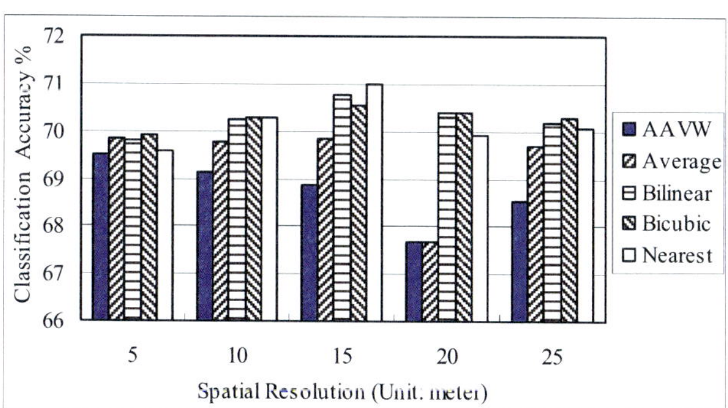

Fig.10.18 Classification accuracy of farmland on aggregated SPOT-5 images at different resolutions

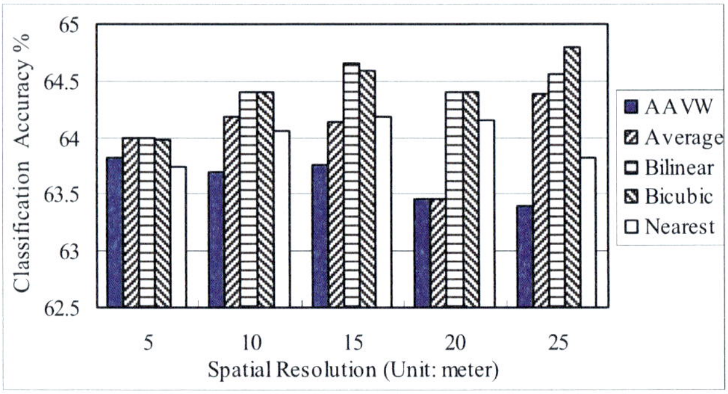

Fig. 10.19 Classification accuracy of built-up area on aggregated SPOT-5 images at different resolutions

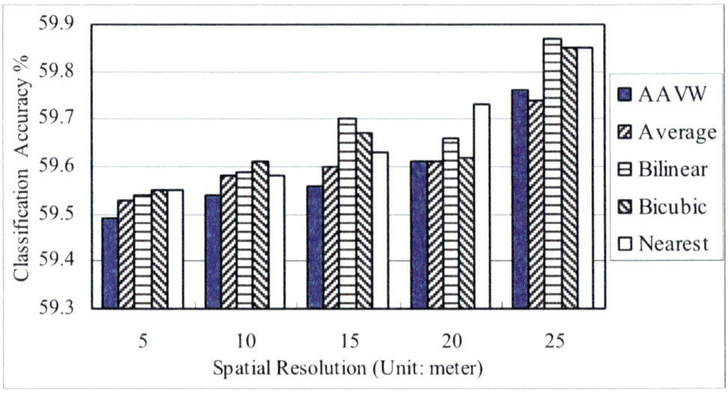

Fig. 10.20 Classification accuracy of grassland on aggregated SPOT-5 images at different resolutions

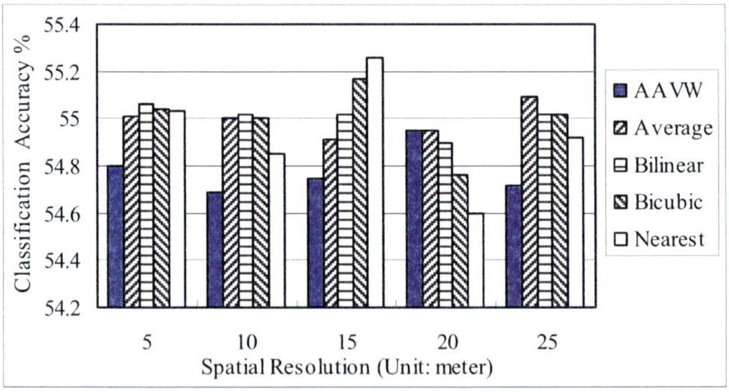

Fig. 10.21 Classification accuracy of bare soil on aggregated SPOT-5 images at different resolutions.

10.5.3 Effect of Aggregation Methods on Classification Accuracy

For individual classes, it is found that the classification accuracy might be improved with aggregation, especially the case for grassland. This also supports the argument that the ML classifier might be not suitable for classifying high-resolution images. Although the results are not very good, we still find out some interesting results from the data.

For the class "water," the NN has the best performance, especially when the resolution becomes coarser. The AAVW method produces worst results (Fig.10.16). The classes "forest" and "farmland" show another trend (Fig.10.17 and Fig.10.18). In these cases, BC and BL produce steady results over the resolution range. For "forest," BC and BL perform better than others except at 25m at which AV outperforms these two methods. In the case of "farmland," BC and BL show steady performance at all resolutions except 15m. Similar trends are shown in "built-up area" and "grassland" classes (Fig.10.19 and Fig.10.20).

The experimental results indicate that the NN method is compatible with the class "water." For "forest," "farmland," "built-up area" and "grassland," the BC and BL have excellent performance. This is because, comparing with "water," the spectral variations in the classes "forest," "farmland," "built-up area" and "grassland" are relatively unstable in the SPOT5 image.

It is also worth noting that the trend in accuracy for class "bare soil" in this experiment (Fig.10.21) is not clear. That is, no single aggregation has consistent performance over the whole range of resolution.

10.6 Conclusions

In this study, the effect of five aggregation methods (AAVW, AV, BL, BC and NN) on the image classification are investigated through experimental testing. Two images with different resolutions (i.e. a TM image and a SPOT-5 image) are used for evaluation. The effect of different aggregation methods on overall classification accuracy and classification accuracies of different land cover types are analyzed and compared. The results indicate that it is sound to avoid the generation of mixed pixels during the selection of aggregation method. Some conclusions can be made from the results:

1) It is not possible to recommend a best aggregation method for common use, as an image may comprise of different types of land cover and each land cover may appear quite differently at images with different resolutions.

2) To achieve relatively higher overall classification accuracy, the spatial distribution patterns of the land cover types on the image should be analyzed when selecting the aggregation method. The homogeneity of the dominant classes determines the aggregation method to be used. If the dominant classes are of high

homogeneity, then the Nearest Neighbour method is the best choice. Otherwise, Bicubic or Bilinear methods could be considered.

These conclusions in this study can provide a foundation for selecting the up-scaling method in the study on the choice of a optimal scale in image classification. However, more investigations are desirable to make these conclusions generalized.

Acknowledgments

This research was supported by a State 973 Project (2006CB701304) of China.

References

Atkinson, P. M. and Aplin, P. (2004). Spatial variation in land cover and choice of spatial resolution for remote sensing. International Journal of Remote Sensing, 25(18): 3687–3702.

Bian, L. and Butler, R. (1999). Comparing effects of aggregation methods on statistical and spatial properties of simulated spatial data. Photogrammetric Engineering and Remote Sensing, 65(1): 73–84.

Cao, C. and Lam, N.S.N. (1997). Understanding the scale and resolution effects in remote sensing and GIS, In: Quattrochi, D.A., and GoodChild, M.F., (ed.) Scale in Remote sensing and GIS. CRC Press, Boca Raton, 57–72.

Chen, D.M. (2001). *Multi-resolution Image Analysis and Classification for Improving Urban Land Use/Cover Mapping Using High Resolution Imagery*. PhD Thesis, University of California.

Collins, J. B. and Woodcock, C.E. (1999). Geostatistical estimation of resolution-dependent variance in remotely sensed image. Photogrammetric Engineering and Remote Sensing, 65: 41-51.

Hay, G. J., Niemann, K. O., and Goodenough, D. G. (1997), Spatial thresholds, image objects, and up-scaling: a multi-scale evaluation. Remote Sensing of Environment, 62: 1–19.

He, H. S., Ventura, S. J., and Mladenoff, D. J. (2002). Effects of spatial aggregation approaches on classified satellite imagery. International Journal of Geographical Information Science, 16(1): 93–109.

Marceau, D. J., Howarth, P.J., and Gratton, D.J. (1994). Remote sensing and the measurement of geographical entities in a forested environment. 1. The scale and the spatial aggregation problem. Remote Sensing of Environment, 49: 93–104.

Markham, B.L. and Townshend, J.R.G. (1981). Land cover classification accuracy as a function of sensor spatial resolution. Proceedings of the 15th International symposium of Remote Sensing of Environment, 1075-1090.

Narayanan, R. M., Desetty, M. K. and Reichenbach, S. E. (2002). Effect of spatial resolution on information content characterization in remote sensing imagery based on classification accuracy. International Journal of Remote Sensing, 23(3): 537-553.

Niemann, K. O., Goodenough, D. J. and Hay, G. J. (1997). Effect of scale on the information content in remote sensing imagery. IGARSS'97, Singarpore, 664-666.

Wang, G., Gertner, G., and A. B. Anderson. (2004). Up-scaling methods based on variability-weighting and simulation for inferring spatial information across scales. International Journal of Remote Sensing, 25 (22): 4961-4979.

Woodcock, C. E. and Strahler, A. H. (1987). The factor of scale in remote sensing. Remote Sensing of Environment, 21: 311–332.

Chapter 11

TOWARDS AUTOMATION OF INFORMATION EXTRACTION FROM AERIAL AND SATELLITE IMAGES

John Trinder, Arcot Sowmya

11.1 Background to Automation in Digital Photogrammetry

Since photogrammetry was developed more than 100 years ago as a technology for map production and the measurement of objects on images, photogrammetrists have attempted to improve the efficiency and accuracy of the process. These attempts commenced with the development of analogue approaches to solving the major computations in converting image coordinates on photographs to coordinates in a ground or object system. Subsequent developments included computer driven instruments, the 'analytical stereoplotter' and in the early 1990s, purely digital systems based on digital image acquisition and processing. The precision of all components of the photogrammetric process was continually improved so that smaller image scales could be taken to achieve the required accuracy on the object, thus improving efficiency and reducing costs of the mapping process. Digital image processing in photogrammetry currently enables the determination of elevations and the production of digital orthophotos more rapidly and with greater efficiency than could be achieved with analogue instruments. However, while a certain level of automation has been achieved in the presentation of roads, buildings and other cultural features for the production of digital map data, the automatic extraction of these features from images has not been achieved. The availability of high resolution digital satellite and multi-spectral aerial images, coupled with the community's increasing need for more detailed, timely and lower cost spatial information for the production of digital maps and GIS (Geographic Information System) databases, has driven research on feature extraction over recent decades. While this research continues to develop new approaches to the extraction of features from images, no system has been so far demonstrated that enables extraction of features reliably under a range of image conditions and scales.

Before progressing, some terms will be defined. In the context of this chapter, 'information' is a general term that refers to knowledge about the content of

D. Li et al. (eds.), *Geospatial Technology for Earth Observation*,
DOI 10.1007/978-1-4419-0050-0_11,

images. Since Earth observation images are recorded over the terrain surface, the details that require interpretation and that represent the 'information' contained in the image are objects such as roads, buildings, terrain cover types, and landscape characteristics. The term 'feature' refers to the individual objects that occur in the images, which are described by various 'attributes', such as color, size and so on. This is at variance with typical terminology used by many researchers in computer science where 'features' are equivalent to 'attributes' used in this chapter.

Levels of automation may vary from relatively simple semi-automatic to fully automatic methods. Semi-automatic methods typically take advantage, in varying degrees, of an observer's ability to interpret images, and the ability of a computer to process information rapidly. A number of semi-automatic methods has been demonstrated for extracting particular features in aerial and satellite images, such as roads and buildings, but few of them have been introduced into production mapping software. Examples of semi-automatic approaches using snakes and simulated annealing are given in Trinder and Li (1995), Gruen and Li (1996), Gruen and Li (1997) and Trinder et al. (2000).

Fully automatic methods for extraction of features from aerial and satellite images in principle should not require input from an observer and would extract features reliably and accurately, but they may not be achieved for some time. For complex images over urban areas, fully automated feature extraction may never be achieved. However, research continues to improve the level of understanding of the processes involved in the automation of information extraction from images. This chapter will review some recent approaches that have been taken in the development of automatic procedures for the extraction of features for digital mapping and compilation of GIS databases.

11.2 Issues in Information Extraction from Aerial and Satellite Images

Aerial and satellite images contain a complex array of features, which are characteristic of specific locations. Identification of these features by humans on aerial or satellite images often requires significant levels of training and/or experience. Replacing this training of humans by automatic computer based methods requires a new approach to understanding the content of images and how features may be extracted automatically from them. There is still no single approach or algorithm that outperforms all other competing methods in all locations and for all image characteristics; each has its own strengths and weaknesses for different tasks in the extraction process and tends to be application specific. A major challenge then, lies in selecting the best data sources and algorithms for a given task, and by extension, selecting the right combination of data and algorithms at the different stages of feature extraction.

Early research on extraction of features for digital mapping and GIS was based on the geometric and radiometric content of images using panchromatic images, due to the limited availability of color or multi-spectral images. However, recent research has incorporated color and multi-spectral images as well as multi-source data, and this trend will continue as the availability and range of image data increases. Given the complexity of aerial and satellite images, it is not surprising that feature extraction has proved to be a difficult process requiring more intelligent methods, many of which are available from the field of computer vision. This has led to an increasing collaboration between photogrammetrists and computer vision scientists to develop more effective methods of feature extraction, including those based on machine learning.

Computer vision techniques apply automation and integration of processes for image interpretation and understanding for object and scene recognition. They employ attribute and relation extraction in an image for interpretation of features by matching *models of objects* with *representations of features* derived from images, usually incorporating artificial intelligence methods. Control strategies may include elements of both bottom-up or data driven, as well as top-down or model driven approaches. Low level processes, such as those for extraction of edges of objects generally require little knowledge about the image, while high level processes, such as those for object recognition, will be based on prior knowledge of the scene and the application. A combination of low level and high level processes will usually be required in a feature recognition system.

In computer vision, shape and appearance of objects have been traditional attributes considered for recognition of objects from images. However, for the extraction of features from remotely sensed images, different cues are usually necessary. Features vary in appearance according to the scale of the image, levels of illumination, shadows, terrain slopes, geometric distortions due to perspective projection, camera tilts and elevation differences, changes in direction of observation of the imaging system, atmospheric conditions, occlusions as well as ground cover types.

The ‘context’ of the images is also an important issue for feature identification (Strat 1995). It is a function of the location of the features imaged and parameters of the image acquisition process, and hence should be described in terms of sufficient descriptors to support the feature identification and extraction. Baumgartner et al. (1999) and Gerke et al. (2003) in extracting roads have defined global context as a general description of the types of regions being studied, urban, forest or rural, while local context describes relations between features such as roads, sidewalks, driveways and even neighboring features being occluded by trees. Gerke et al. (2003) also include information on global context in the databases used to assist in information extraction. There are no definitive descriptions of context, but global context may also include the overall characteristics of the scale and type of imaging since these factors will influence the approach to extraction of features. The method of including context in a feature extraction process will depend on the strategy used. For example, rule based systems will need to specifically define context, while systems based on machine learning may derive context as part of the learning process, by the provision of sufficient training samples which incorporate the context.

Future approaches for extracting features from digital images are likely to include:

- All available image information, including multi-spectral, hyperspectral, multi-resolution and multi-sensor data, as well as other relevant information, such as digital map and GIS databases
- Adequate definition of context of the images and features within the images
- Appropriate methods of modeling objects and matching them with extracted features in images based on artificial intelligence approaches
- 3D coordinate descriptions of features if they can be derived from overlapping images

11.3 Review of Road Extraction Techniques

Recently, advances in remote sensing technology and efforts to fuse data have led to the inclusion of prior knowledge from existing GIS databases and elevations, as well as multi-scale and multi-resolution, multi-temporal, multi-spectral and hyperspectral analyses (Mena 2003). Tests on several methods of road extraction have been described in Mayer et al. (2006), in which completeness ('percentage of the reference data which is explained by the extracted data') and correctness ('percentage of correctly extracted road data') of extraction of roads from aerial and satellite images vary from high values of about 80% to less than 30%, depending on the method and the data used. Current road extraction techniques can be categorized and summarized in the following sections.

11.3.1 Image Processing Approaches

Morphological methods to detect features utilize geometric shape information, based on set operations such as union, intersection and complementation, as well as dilation, erosion, thinning and other derived operations. They have been applied to noise removal (Zhang et al. 1999) and grey-scale image simplification (Amini and Sarahjian 2000). Model-based methods for the automatic extraction of linear features from aerial images (Chanussot and Lambert 1998, Katartzis et al. 2001) and road extraction methods based on edge detection and filters (Chiang et al. 2001) have also been proposed.

11.3.2 Tracking and Growing Approach

Tracking and growing methods utilize automatically detected or manually selected road seeds. Road tracking methods trace a road network from predetermined seeds (Dal Poz et al. 2000, Nevatia and Babu 1980, Sakoda et al. 1993) or profiles perpendicular to the road axis (Quam 1978, Vosselman and Knecht, 1995). Several techniques may also be combined (McKeown and Denlinger 1988). Recent work appearing in (Baumgartner et al. 2002, Zhao et al. 2002, Bonnefon et al. 2002) incorporates automatic searching for seed points in (Zlotnick and Carnine 1993, Wufeng and Qiming 1998) and fully automatic road extraction in (Tesser and Pavlidis 2000, Yoon et al. 2002, Dial et al. 2001). Region growing methods start from preselected seeds and agglomerate points around the seeds that satisfy certain homogeneity criteria. Deformable contour models or snakes (Kass et al. 1988) have been employed as a boundary detector and object extractor (Cohen 1991, Trinder and Li 1995, Gruen and Li 1997, Mayer et al. 1998, Ferraro et al. 1999, Jeon et al. 2000, Chiang et al. 2001) and also in multispectral datasets (Zafiropoulos and Schenk 1998, Laptev et al. 2000). Recently a new region growing approach called the fast marching method (Keaton and Brokish 2003) that has the ability for smart handling of propagating contours, has been used.

11.3.3 Grouping and Clustering Approach

This approach mainly employs supervised machine learning and unsupervised clustering algorithms in order to extract road networks, on the assumption that road regions form a separate cluster in feature space. The main problem with clustering is determining the number of clusters, also known as cluster validity. Chen et al. (2002) utilized unsupervised segmentation based on edge grouping and model selection for automatic road extraction in grey scale images. Maximum likelihood decision techniques on infrared images (Benjamin and Gaydos 1990) and pattern classification methods on airborne images (Roggero 2002) have been applied. These techniques have also been introduced to multispectral images (Faber and Förstner 1999, Agouris et al. 2001).

Ziems et al. (2007) have used knowledge of road classifications from prior information together with a statistical analysis of color space of roads to train the extraction of additional roads. Training is also based on the homogeneity of the road surface and its contrast with respect to its background, using a histogram analysis to build up clusters in feature space. Ravanbakhsh et al. (2008) have used level sets to extract road islands in road junctions with significant success. Level sets are described later in this chapter.

11.3.4 Data Fusion and Learning Approach

This approach attempts to learn to optimize road extraction and to take advantage of complementary classifiers and characteristics of objects recorded by a variety of sensors. Neural networks have been used to optimize road extraction (Doucette et al. 1999) and to improve the performance of a junction extraction system (Barsi et al. 2002). Genetic algorithms have been used by (Shanahan et al. 2000, Bandyopadhyay et al. 2000, Jeon et al. 2000) to generate probabilistic models for high level road classification. A data fusion approach uses two road detectors, one of which is edge based and the other pixel correlation based, and their results are then combined by a mechanism called associative symmetrical summation (Tupin and Maitre 1998).

More recently, a multi-detector method for road extraction in SAR data has been proposed (Dell'Acqua et al. 2003), where the results are drawn from the fusion of three fuzzy algorithms. Multi-view aerial image data sources have been fused (Hinz and Baumgartner 2003) and complementary algorithms for road centerline extraction have been presented (Haverkamp and Poulsen 2002). Information streaming from three different sources has been fused for GIS updating using texture analysis and data fusion in a Dempster-Shafer evidence theory framework (Mena and Malpica 2004). A demonstration of the application of the Dempster-Shafer algorithm is given later in this chapter.

An integrated road extraction system based on multi-detector fusion has been proposed (Jin and Davis 2004). Data fusion, a decade-old nascent field, studies methods to handle data from multiple distinct data sources in order to achieve refined and improved decision-making (Kessler et al. 1992). It deals with fusion at three levels, namely the sensor, feature and decision levels. Recent advances in the machine learning field have given rise to powerful techniques for combining different learning algorithms and classifiers to obtain optimal performance, including ensemble learning (Dzeroski and Zenko 2002) and model selection (Caruana et al. 2004). The idea of combining the results of several algorithms is also known within pattern recognition (Kittler et al. 1998). Applying data fusion on the traditional methods to combine the information from different sources and developing machine learning techniques to adapt the fusion models will lead to more accurate and robust extraction of road networks from remotely sensed images, and has been the goal of research at The University of New South Wales.

11.4 RAIL: Road Extraction by Inductive Learning

We now present RAIL, a road extraction system that combines cluster analysis with inductive learning in a novel fashion (Singh and Sowmya 1998, Chen et al. 2002, Lai et al. 2005, Cai et al. 2005).

11.4.1 RAIL Framework

RAIL is an adaptive and trainable multi-level edge-based road extraction system. Instead of using prior rules, RAIL uses multi-level learning to derive rules that can be applied during segmentation. This approach makes few assumptions and therefore is applicable to images of different scale, content, complexity and quality. Starting with low-level objects (edges), RAIL incrementally builds higher-level objects (road network) at multiple levels defined as follows:

Level 1: Road Edge Pairs - pairs of edges that enclose a segment of road.

Level 2: Linked Road Edge Pairs - adjacent road edge pairs that form continuous roads.

Level 3: Intersections - road edge pairs that meet to form junctions.

Level 4: Road Network- linked roads and intersections.

11.4.2 RAIL Levels 1 and 2

RAIL utilizes a set of classifiers that can be applied to recognize the road structures in Levels 1 and 2. After preprocessing, edge extraction and heuristic filtering, features of potential road edge pairs are extracted and feature subset selection performed. The edge pairs are then clustered in feature space using different clustering techniques. The goal is to identify clusters of road edge pairs at Level 1, and adjacent road edge pairs in Level 2. This goal is achieved using a specially designed inductive technique called Inductive Clustering (IC) that can automatically select parameters for clustering algorithms by learning from clustering examples, as well as select a good algorithm that best fits the data. Reference images are extensively used for training the IC algorithm. These steps are now described.

11.4.2.1 Image Preprocessing

The low-level processing in RAIL aims to extract raw edges from an image with minimal intervention. Since remotely sensed images are inherently noisy, the Canny operator was chosen for its known performance on noisy images. RAIL uses Vista's[1] implementation of the Canny operator as well as other functions. The process is split into four steps:

1. Canny edge detector: edge pixels are detected between two regions with relatively distinct intensity values.
2. Edge linking: edge object is formed by tracking edge pixels using eight-connectivity.

---

[1] Software available at http://www.cs.ubc.ca/nest/lci/vista/vista.html

3. Edge straightening: straight edge objects are formed by dividing linked edge objects at the point where it deviates most from a straight line joining its endpoints.
4. Edge splitting: edges longer than a threshold are split recursively into shorter straight edges.

The parameters used in the Canny edge detector were manually tuned to produce outputs with less noise. This was accomplished by adding noise to the original image prior to Gaussian smoothing with a large standard deviation. Adding artificial noise to the images before blurring removes noise that is inherently present in high resolution images. Specifically, edge orientation as well as edge magnitude maps were produced (orientation = `true'), standard deviation of the Gaussian filter was set to 2.4, and the fraction of edge pixels considered to be edge noise was set to 0.75. This resulted in a dramatic decrease in the number of extracted edges. Removing noisy edges is advantageous because machine learning algorithms may otherwise be distracted by them, and it makes the data size more manageable.

11.4.2.2 Reference Image

In the IC learning pipeline, the results of clustering at each stage must be evaluated by comparison to a reference image that provides ground truth. For road networks in remotely sensed images, such ground truth must come from a geographic information system, or manual generation. RAIL has an operator assisted centerline reference image acquisition function based on (Wiedemann et al. 1998), which can assess the learned outputs more correctly by checking if the extracted pairs have edges that do in fact lie opposite each other near the reference model. The operator is provided with interactive drawing tools to help create the reference images.

11.4.2.3 Feature Extraction for Levels 1 and 2

The features that are extracted at Levels 1 and 2 of RAIL are described in detail below. *In Level 1*, edge pairs are formed by joining two edges. The properties of a road edge pair are listed in Table 11.1, together with the corresponding attributes that measure it. The highlighted attributes are those selected for segmentation using feature subset selection techniques, which are described later.

Pair width, enclosed intensity (mean), bearing and projection form an intuitive feature subset that describes road segments, i.e. roads have similar width and intensity and their opposite sides are almost parallel. Pair length is a good feature because during the preprocessing stage a maximum length for edges was set to 50 pixels. Generally, road sides are long and continuous and are split into smaller segments after preprocessing. When road pairs are formed, their lengths do not vary too much. This is because non-road edges are usually of shorter length.

Enclosed intensity variance did not prove to be a good feature since the area enclosed by an edge pair is small and the intensity is fairly uniform. Length difference between edges was also discarded. Road pairs are expected to have similar edge length but non-road pairs may also have similar edge lengths, thus it does not convey much information. Intensity gradient differences between the two edges do not show consistencies between road pairs and non-road pairs. The assumption that the intensity levels are the same on both the external sides of the roads also proved invalid.

Table 11.1 RAIL Level 1 features (bolded attributes were used for the segmentation.)

Attribute	Description	Road Property Addressed
Width (mean)	Average distance between edges	Roads generally have constant widths
Enclosed Intensity (mean)	**Average intensity between edge pair**	**Roads generally have intensities that differ from the intensity of their surrounding**
Enclosed Intensity (var.)	Intensity variance between edge pair	Roads generally have constant intensity values
Pair Length (centerline)	**Length of edge pair measured by an imaginary centreline**	**Roads are generally long objects**
Length Difference	Length difference between edges	Road segments are usually of constant length
Bearing Difference	**Direction difference between the two edges**	**Road pairs are generally parallel**
Intensity Gradient Difference	Image gradient of the edges should be parallel	Road boundaries have opposite gradient directions
Projection	**Binary value that defines if edges of an edge pair are opposite to each other**	**Road segments generally project onto each other since they are opposite to each other**

In *Level 2*, the short road segments extracted in Level 1 are joined together into longer road segments. Therefore, the attributes at Level 2 are designed to generate all possible twin-linked road pairs, which are chained together after segmentation. The properties of road edge pairs and the corresponding attributes are listed in Table 11.2. The highlighted entries are the attributes used for segmentation.

Linked road pairs should have similar enclosed intensity with little difference. Ideally linked pairs should be minimally separated and have no gap, thus gap intensity and gap separation are excellent features to distinguish between linked road pairs and other linked edge pairs. Roads generally have smooth curves except at intersections; therefore, the bearing difference between linked road pairs should not be very large.

Width features are not good attributes for Level 2 because Level 1 outputs all have similar widths. The same argument applies to length attributes. Enclosed intensity variance and gap intensity variance are not very good features for the same reasons discussed earlier, i.e. intensity levels do not change much in an enclosed edge pair or in a road gap. Again, intensity levels across edges cannot be assumed to be the same on both sides of the linked edge pairs.

Table 11.2 Level 2 features (bolded attributes were used for the segmentation.)

Attribute	Description	Road Property Addressed
Width (mean)	Average distance between the edge pairs	Roads generally have constant widths
Width (var.)	Distance variance of the edge pairs	Roads generally have constant widths
Width Difference	Width difference of the edge pairs, the average values are used	Linked pairs should have similar widths
Enclosed Intensity (mean)	**Average intensity of the edge pairs**	**Roads generally have intensities which differ from intensity of their surroundings**
Enclosed Intensity (var.)	Intensity variance of the edge pairs	Roads generally have constant intensity values
Enclosed Intensity Difference	**Intensity difference of the edge pairs**	**Linked pairs should have similar enclosed intensity values**
Gap Intensity (mean)	**Average intensity of the gap bridging the two road pairs**	**Road gaps should have similar intensity values to edge pair intensities**
Gap Intensity (var.)	**Intensity variance of the gap bridging the two road pairs**	**Road gaps usually have constant intensity values**
Length Combined	Length as measured by a centerline expanding from one edge pair to the other	Roads are generally long objects
Length Difference	Length difference of the edge pairs	Linked pairs should be similar in length
Minimum Gap Separation	**Minimum separation of the edge pairs**	**Linked pairs should be close in proximity**
Maximum Gap Separation	**Maximum separation of the edge pairs**	**Linked pairs should be close in proximity**
Gap Separation (mean)	**Average separation of the edge pairs**	**Linked pairs should be close in proximity**
Bearing Difference	**Direction difference of the edge pairs measured from their centerlines**	**Roads are generally straight or have smooth curves**
Intensity Gradient Difference (left)	Image gradient difference of the left edges	Linked road boundaries should have the same gradient direction
Intensity Gradient Difference (right)	Image gradient difference of the right edges	Linked road boundaries should have the same gradient direction

11.4.2.4 Feature versus Heuristic Preprocessing

Although RAIL attempts to produce less noisy attributes, fairly large datasets for Level 1 and Level 2 are still created. Therefore, RAIL uses additional heuristic preprocessing to reduce the data to a more manageable size. The heuristic rules throw away cases where an expert would agree that a positive classification is impossible. For example, in Level 1 if edges in an edge pair do not project onto each other, they cannot be classified as an edge pair, as they are not opposite each other. This feature has a binary value, therefore using this attribute as a heuristic filter, effectively removes the projection attribute from the feature space. Another heuristic discards any edge pairs that are wider than twice the maximum road width in the images, as road pairs should not be too distant from each other. In Level 2, the heuristic rules involving mean gap separation and bearing difference are applied. Road pairs that are separated by more than twice the maximum road width are unlikely to be linked. A bearing difference of greater than 45° between two road pairs will also mean that the road pairs are unlikely to be next to each other. When a bearing difference heuristic rule is applied, possible junctions will be broken up; junctions are dealt with at Level 3 of RAIL.

11.4.2.5 Feature Subset Selection

In many machine learning problems, using all available features, relevant or not, causes detrimental effects on accuracy. This is because some features may be dependent on other features or features may be noisy. Removing such attributes that may distract a machine learning algorithm will improve accuracy and give a better description of the learned concepts. It will also enable learning to take to place with fewer examples.

The general approach used for feature subset selection (FSS) is analogous to supervised learning - sample data from the classes of interest are used either to derive the best subset or to rank the attributes. Several hundred positive and negative examples from road and non-road classes were selected from different images so that the results were not biased towards the most frequent road type.

The WEKA[2] data mining suite (version 3.4) was used to conduct the FSS experiments. WEKA contains a suite of machine learning algorithms for data mining tasks, and may be run directly or called from Java code. The inputs to WEKA are a collection of instances (dataset) and algorithm specific parameters, which may be set by default or tuned. Each instance must consist of a number of attributes, which may be of different types. The FSS algorithms in WEKA may be grouped into two types: feature subset-based (Type I) and feature-based (Type II).

Type I algorithms were run using three different search methods, namely best first, rank and genetic search based on Goldberg (1989). Default parameter values in WEKA were used, and no attempt at any parameter tuning was made at this stage. For the genetic search, the population size is set to 20, number of generations to 5, probability of crossover to 0.5 and frequency of reporting equal to the number of generations. A total of 21 algorithm-search combinations, arising from the 7 methods of Type I and the 3 search methods, were used. Type I

[2] Software available at http://www.cs.waikato.ac.nz/ml/weka

algorithms select the best subset of features, based on the performance on the labeled examples, and the frequency of each was recorded and averaged to produce a frequency score. Type II algorithms rank the individual attributes by assigning them a relative weight. These were normalized and averaged. The desired subset of features should have a high frequency score in Type I and a high relative weight in Type II. The Type I and Type II results were ranked and the smallest subset in which the features are identical for each type was picked. This has produced good results.

11.4.3 Inductive Clustering Framework

The Inductive Clustering (IC) framework has been designed to perform the following functions:

Algorithm Learning: learn a clustering algorithm most suited to an image type.

Parameter Learning: learn the parameter values that give the best results for that algorithm and image type.

Cluster Learning: identify a cluster of interest.

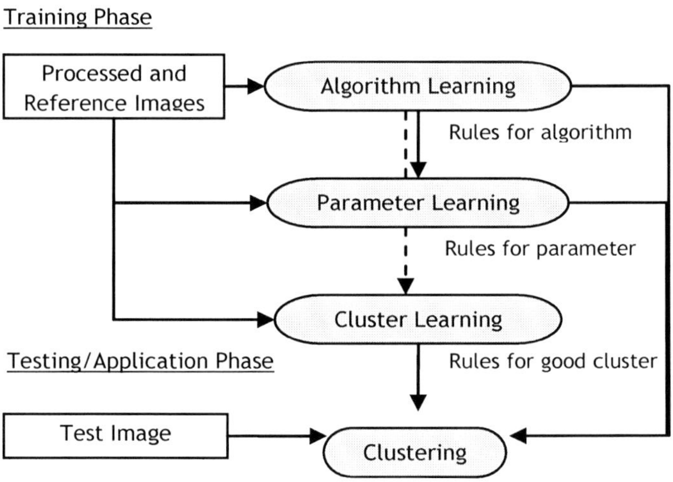

Fig. 11.1 IC framework

Of the above, algorithm and parameter learning address cluster validity and model selection problems, while cluster learning is specific to the application, and may be modified to identify more than one cluster, if appropriate. The goal of IC is to generate a set of rules, based on comparison of competing clustering algorithms on images of different characteristics, which may then drive the automatic choice of an appropriate algorithm and its parameters for a new image, and also identification of the cluster of interest to the application.

Meta information on which algorithm or parameter values to use, and which cluster to select, are learned; this process is termed as 'meta-learning'. At each learning stage of IC, an inductive learning algorithm called C4.5 (Quinlan 1993) that generalizes from positive and negative examples is utilized as the meta-learning algorithm, as it has been found to be good for meta learning (Brazdil and Henery 1994). A multi-level learning strategy is employed, as shown in Fig. 11.1.

IC consists of training and testing phases. In the training phase, a collection of processed images, together with their reference images are input into the learning pipeline. At each stage of the pipeline, rules to select the best algorithm/parameter values/cluster are learned and stored. Parameter learning also utilizes the algorithm learning rules in addition to the image and reference data. In the testing phase, the learned rules may be applied to any other test image, without reference information, to perform road segmentation.

The general learning strategy employed at each stage of the IC learning pipeline is illustrated in Fig. 11.2. In the training phase, each training image is processed and the extracted attributes for all the images are clustered. The resulting clusters are evaluated against the corresponding reference images which provide `ground truth' and the evaluations, together with image attributes and any other relevant data, sent to the meta-learning algorithm. The rules learned may then be applied to test images at the testing phase, to obtain the best algorithm, parameter values or cluster of interest.

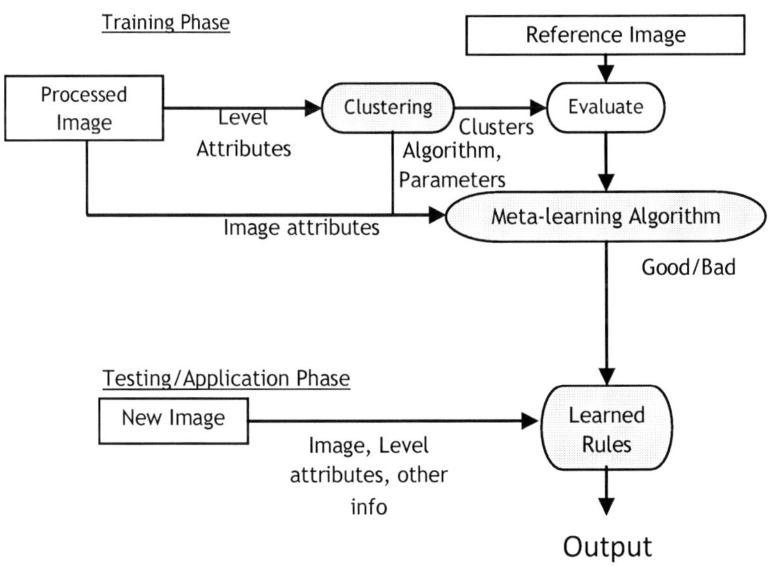

Fig. 11.2 Learning strategy at each stage of IC pipeline

11.4.3.1 IC Setup

Four different clustering algorithms were used in the experiments. They are K-Means (KM), Expectation Maximization (EM), Farthest First (FF) and Self Organizing Map (SOM). C4.5 (release 8)[3] is used as a meta-learner to automate the process of selecting clustering algorithms, tuning their parameters and selecting a cluster of interest.

The measures used for empirical evaluation of the results are based on (Wiedemann et al. 1998). They address two questions: firstly, how complete is the extracted road network, and secondly, how correct is the classification. They are calculated as:

$$completeness = \frac{length\ of\ matched\ reference}{reference\ length} \times 100\% \tag{11.1}$$

$$correctness = \frac{length\ of\ matched\ extraction}{length\ of\ extraction} \times 100\% \tag{11.2}$$

Completeness measures the percentage of the road reference image that has been segmented correctly in the training/test image, known otherwise as ‘recall’. Correctness measures the percentage of the segments that are actual road objects, known otherwise as ‘precision’. A high completeness means that a segmenter has extracted most of the road network, whereas high correctness implies that not too many incorrect road objects have been segmented.

The two measures above are combined into a more general measure of quality, called cxc, which is expressed as:

$$cxc = completeness^2 \times correctness \tag{11.3}$$

Clearly, this measure is biased towards correctness, and may be varied for specific applications. It is used to compare road clusters extracted by different algorithm-parameter combinations for the first two stages of the learning pipeline in IC, with higher values being more desirable.

For the third stage of cluster learning, a set of thresholds was determined by empirical observation of what a ‘good’ road cluster should evaluate to, and is expressed as:

$$completeness \geq 0.85 \quad and \quad correctness \geq 0.25 \tag{11.4}$$

[3] Software available at http://www.rulequest.com/Personal/

Although one goal of IC is to automate the process of parameter tuning, these parameters are still used for cluster definition; however, the number of parameters set manually has been reduced to only 2 overall, compared to a few for each base algorithm.

The three stages of learning in the IC pipeline are now described in detail.

11.4.3.2 Algorithm Learning

The purpose of algorithm learning is to learn which algorithm works best for a given image. The learning attributes are the image characteristics and the algorithms used for clustering (see Table 11.3). The current image characteristic used is the number of edges extracted during low-level processing of the images.

Table 11.3 Algorithm learning attributes

Attribute	Description	Value
Size	Number of edges	Continuous
Algorithm	Clustering algorithm used	KM, EM, FF, SOM
Class	Whether the algorithm produces the best road cluster	Good, Bad

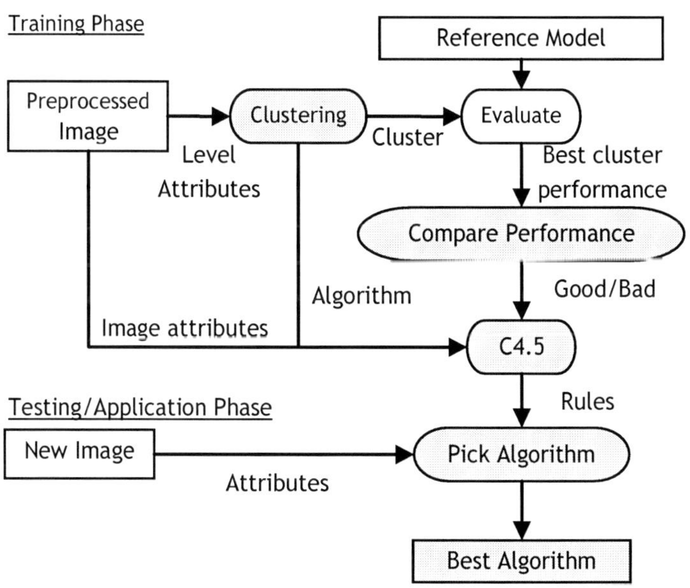

Fig. 11.3 Algorithm learning

The two phases of algorithm learning are shown in Fig. 11.3. In the training phase, RAIL attributes are calculated for each level and clustering performed for each algorithm-parameter combination. The clusters generated are evaluated using the reference image for each image, from which the completeness, correctness and cxc values are obtained. For each algorithm, the highest cxc cluster is selected for comparison. The algorithm with the highest cxc overall is classified as 'Good'. The other algorithms are labeled 'Bad'. Then, using the attributes in Table 11.3, C4.5 is employed to generate a decision tree and the rules for algorithm selection. In the testing phase the learned rules are applied to a new image after extracting the required image attributes.

11.4.3.3 Parameter Learning

The goal of parameter learning is to deduce rules for the required number of clusters, given an image and an algorithm. The learning attributes are the image characteristics, the algorithm used and the number of clusters produced (see Table 11.4). The current image characteristic used is the number of edges extracted during low-level processing of the images.

Table 11.4 Parameter learning attributes

Attribute	Description	Value
Size	Number of edges	Continuous
Algorithm	Clustering algorithm used	KM, EM, FF, SOM
ncluster	Number of clusters	[2,29]
Class	Whether the parameter produces a good road cluster	Good, Bad

The two phases of parameter learning shown in Fig. 11.4 are similar to algorithm learning. In the training phase, RAIL attributes are calculated for each level and clustered with each algorithm-parameter combination. The clusters generated are evaluated using the reference image. Each parameter value that defines a cluster satisfying Eq. 11.3 is labeled a 'Good' parameter. Otherwise, the parameter is labeled 'Bad'. An alternative is to label the highest cxc valued cluster as `Good' and all others as `Bad'. C4.5 is used to deduce rules using attributes in Table 11.4, which are then applied on a new image. The image attribute extracted and the learned algorithm (from the algorithm learning stage) for that image, are used to determine the 'Good' parameters to cluster with. As in the previous steps, in the testing phase the learned rules are applied to a new image after extracting the required image attributes.

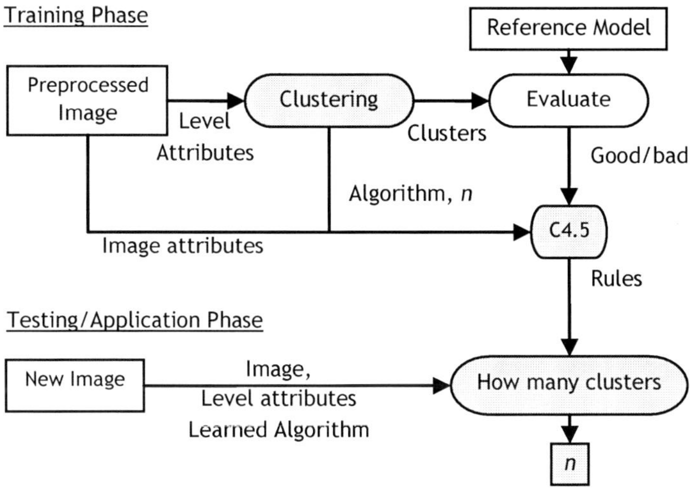

Fig. 11.4 Parameter learning

11.4.3.4 Cluster Learning

The objective of cluster learning is to distinguish the road cluster from the non-road clusters, when given an image, the algorithm and the number of clusters required. The learning attributes are the image characteristics, the algorithm used, the number of clusters and the cluster centroids (see Table 11.5). The current image characteristic used is the number of edges extracted during low-level processing of the images. The cluster centroids are based on the level attributes used for clustering the data.

Table 11.5 Cluster learning attributes

Attribute	Description	Value
Size	Number of edges	Continuous
Algorithm	Clustering algorithm used	KM, EM, FF, SOM
Centroid	Cluster centre (RAIL Level Attributes)	Continuous
Class	Whether the clusters satisfies (3)	Good, Bad

The two phases of cluster learning are shown in Fig. 11.5. In the training phase, RAIL attributes are calculated for each level and clustered with each algorithm-parameter combination. The clusters generated are evaluated using the reference model for that image. Each cluster that satisfies Eq. 11.4 is labeled a ‘Good’ cluster, and other clusters are labeled as ‘Bad’. C4.5 is used to deduce rules using attributes in Table 11.5, which are then applied on a new image. The image attribute extracted along with the learned algorithm and learned parameter (from the previous learning stages) for that image are used to determine the 'Good' road cluster and used in the testing phase on other images.

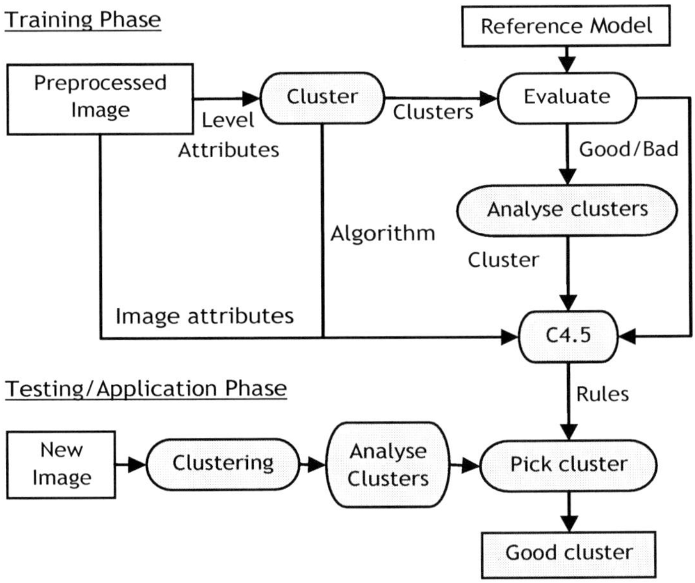

Fig. 11.5 Cluster learning

11.4.4 Results and Evaluation

IC was implemented for Levels 1 and 2 of RAIL and tested on several high resolution aerial images. In this section the results obtained on the images are presented in the following ways:

(i) rules learned at each stage of the IC pipeline for Levels 1 and 2 of RAIL
(ii) leave-one-out cross validation results at each stage of the IC pipeline for Levels 1 and 2 of RAIL
(iii) test results illustrating IC in practice.

This is then followed by an evaluation of results. Fifteen grey-scale images from a rural area in Canberra, Australia were used, named from A to O. The size of each image is 1024*1024 pixels, having been cropped from a larger image of ground resolution 1.3 meters per pixel. The image characteristics are shown in Table 11.6. Although only 15 images were used, each image contains thousands of samples at the edge pair and road pair levels. Leave-one-out cross validation was used, to learn from the largest available dataset and obtain effective test sets.

Table 11.6 Image characteristics

Image	No. of edges
A	3039
B	3769
C	12482
D	9150
E	12855
F	6055
G	6103
H	7056
I	5797
J	5231
K	4761
L	9513
M	3045
N	4014
O	5559

11.4.4.1 Inductive Learning Rules

At each stage in the Inductive Clustering framework, C4.5 learns rules that are used to automate clustering of unseen images. Some examples of the learned rules for the 3 stages of Level 1 are listed below. The rules can only be applied to images with similar properties (e.g. resolution, complexity, etc.) as the training images. Level 2 rules are similar to Level 1 except that their attributes are different, and they are not shown.

Each rule identifies a partition of data via its learning attribute and gives a classification for that partition. The percentage after the classification indicates the accuracy of the rule when applied on the training data, which might not be the same when applied to test data.

11.4.4.2 Algorithm Learning

The following rules were derived from learning on images A-N, and tested on image O. Rule 1 selects KM as the best algorithm to use regardless of the image attribute (number of edges).

Rule 1: algorithm = KM -> class good [51.5%]
Rule 2: algorithm = SOM -> class bad [90.6%]
Default class: bad

11.4.4.3 Parameter Learning

The following rules were derived from learning on images A-N, and tested on image O. The algorithm was learned at the previous stage. Rule 39 selects the number of clusters to be between 8 and 15, when the number of edges is fewer than or equal to 6103, and the clustering algorithm is KM.

Rule 39: size <= 6103, algorithm = KM, ncluster > 7, ncluster <= 15
-> class good [86.8%]
Default class: bad

Cluster Learning: The following rules were derived from learning on images A-N, and tested on image O. Rule 43 selects a cluster as a road cluster if the width, internal intensity, length centre and bearing are within the learnt ranges. An additional condition it must also satisfy is on the number of edges on the image.

Rule 43: size > 3039, size <= 9513, width > 10.13, width <= 21.93, internal_intensity > 112.08, internal_intensity <= 167.52, length_center > 23.16, bearing > 2.16, bearing <= 15. 87 -> class good [71.8%]
Default class: bad

Table 11.7 Level 1 cross validation (%)*

Learning stage	Error rate	Completeness	Correctness
Algorithm	16.7 (16.7)	66.7 (66.7)	66.7 (66.7)
Parameter	38.1 (6.8)	56.0 (92.0)	54.1 (91.9)
Cluster	2.6 (0.7)	45.0 (82.7)	52.1 (91.3)

Table 11.8 Level 2 cross validation (%)*

Learning stage	Error Rate	Completeness	Correctness
Algorithm	21.7 (21.7)	0.0 (0.0)	- (-)**
Parameter	34.3 (0.8)	72.3 (99.2)	77.0 (99.7)
Cluster	6.9 (1.0)	65.1 (86.9)	39.5 (94.2)

**The results are averaged over 15 cross validation experiments. The results for test images are given first and those for training images in brackets.*
*** Undefined since no positive classification*

11.4.4.4 Cross Validation Results

The rules learned during the training phase of IC were applied to the test images, in order to pick the appropriate clustering algorithm for the test image, set its parameters automatically and finally pick out the road cluster in the image. Tables 11.7 and 11.8 present leave-one-out cross validation results for IC at Levels 1 and 2, which are summarized from the confusion matrix calculated by C4.5. The results for the 15 images have been averaged. The rules were also applied back on the training images and the training image results are shown within the brackets.

11.4.5 IC in Practice

The system is able to automatically select the clustering algorithm and parameter values and pick the road cluster. The results are shown in Table 11.9 for Level 1. The procedure starts by using the learned algorithm to find the good parameters. Usually the learned parameter is not a single number; rather it ranges between 2 and 29, and clustering starts from the smallest parameter found. The clustered data are used as input to cluster learning and if a road cluster is selected, the process is complete and the completeness and correctness values of the road cluster are evaluated. If cluster learning does not select a road cluster, the next available parameter is used. The parameter column in Table 11.9 shows the parameter range learned, and the number after the dash is the parameter value which selected a road cluster. If a parameter cannot be found (labeled Non), IC was set up to randomly select a parameter from the given range. Only 2 selected road clusters (for Image C and M) are false positives, as confirmed by examining each cluster visually, otherwise every image output has high completeness values.

Image B has a completeness value of over 100% in Table 11.9, which is due to the buffered zones in the centerline reference that was employed. When road sections are very noisy and there are many edges near the road but within the buffer zone, they may be counted as positive objects, but this only happens infrequently as the buffer zone is relatively small (3 pixels on each side of the reference).

Table 11.9 Level 1 inductive clustering

Image	Algorithm	Parameter	Completeness	Correctness
A	KM	[8,21] - 13	98.4	27.8
B	KM	[6,29] - 12	102.0	31.6
C	KM	Non - 14	90.6	18.0
D	KM	[10,29] - 11	92.6	41.3
E	KM	Non - 25	90.4	25.1
F	KM	[6,29] - 8	96.6	41.9
G	KM	[11,13] - 11	94.8	46.3
H	KM	[6,15] - 11	85.2	25.2
I	KM	[13,25] - 13	94.8	32.3
J	KM	[6,25] - 8	94.2	35.1
K	KM	[8,14] - 9	95.2	47.3
L	KM	Non - 17	94.8	27.7
M	KM	[6,21] - 7	99.6	22.1
N	KM	[11,13] - 11	93.6	52.98
O	KM	[8,15] - 8	94.2	63.6

The Level 2 error rates in Table 11.8 are similar to the Level 1 rates, but the completeness and correctness values are higher, with the exception of algorithm learning. Table 11.10 shows the Inductive Clustering results for Level 2. Three out of the 15 images did not have classified parameters (Images E, G and L), and the parameters for these images were randomly selected. 6 out of the 15 images did not have classified road clusters for the given algorithm and parameters. For those images where road clusters were selected automatically, they have a slightly lower completeness than Level 1 but higher correctness values as expected.

Two sets of images (E and O) are shown from Fig. 11.6 to Fig. 11.13. Each set contains the reference image, the input edge image, the Level 1 output (in grey-scale and color) and Level 2 output (in grey-scale and color) from IC. The grey-scale images are provided for better viewing on print. Image E shows the performance of the system even when the learned results are poor, i.e. when parameter learning finds no solutions. Image O has been cropped since the lower portion of the image is totally black as it was near the bottom edge of a larger image.

11.4.6 Evaluation of Results

The accuracy of algorithm learning in Tables 11.7 and 11.8 is fairly low and may be attributed to several factors. Firstly and most importantly, the current image characteristic used is the number of edges, which is used to differentiate between images. This is probably inadequate to characterize an image type fully; other image attributes such as land coverage and texture measures may solve this problem. Another reason for the low accuracy of algorithm learning is that the clustering algorithms used perform almost equally well. Selecting the 'best' algorithm by their best cxc cluster which only differ by approximately 1% could be misleading at best. In Level 2 when algorithms were randomly used, more than half of the images were able to obtain good results via IC, which means that the algorithm used is not so important.

Table 11.10 Level 2 inductive clustering

Image	Algorithm	Parameter	Completeness	Correctness
A	KM	(3,18] - 3	89.6	41.4
B	FF	[2,18] - Non	-	-
C	EM	[2,4] - Non	-	-
D	SOM	[2,10] - Non	-	-
E	KM	Non - 4	87.8	30.2
F	FF	[2,9] - 4	93.3	64.7
G	EM	Non - 5	92.4	66.7
H	SOM	[2,29] - 14	70.6	32.3
I	KM	[2,29] - 2	94.8	35.1
J	FF	[2,9] - Non	-	-
K	EM	[2,17] - 5	92.0	70.2
L	SOM	Non - 6	-	-
M	KM	[2,4] - Non	-	-
N	FF	[2,18] - 3	85.9	60.1
O	SOM	[2,17] - 2	92.4	65.1

Determining the best algorithm using a single cluster might itself be problematic. For example, Algorithm A might have the highest cxc value but only very few parameters for that algorithm are good. On the other hand, Algorithm B might have a slightly lower cxc value but many good road parameters. Although Algorithm A is the 'best' algorithm in terms of cxc, Algorithm B could also be considered if the parameters learned using the algorithm are weighted. This would require feedback by cross-level learning and will determine a more realistic algorithm since parameter learning has higher error rates.

The goal of IC is to automatically learn the algorithm-parameter combination that will produce a good road cluster, and parameter learning addresses this problem. In the experiments, the cross validation error rate for parameter learning was low and completeness and correctness were high in the training phase, but they were poor when applied to test images (see Tables 11.7 and 11.8). The problem stems from the paucity of training images and the fact that the current image characterization is probably inadequate. The inductive approach typically classifies false positives when it determines good parameters that are greater than 18 clusters. This is due to the nature of clustering algorithms. When the data is represented by fewer clusters, fine detail is lost, but simplification is achieved in that there is one cluster which contains all the road objects, plus other similar non-road objects. These road clusters typically have higher completeness but lower correctness compared to road clusters using a higher parameter value. If the data were represented by more and more clusters, the cluster details can be learned but road objects may be split into several clusters, with each being a different type of road. This is the reason why IC begins with the lowest parameter value learned and then works its way up the parameter range as it looks for a good road cluster. From Tables 11.7 and 11.8, cluster learning has the lowest error rate among the 3 learning stages, but its completeness and correctness values are still relatively low.

The problems that cause the false positive segmentations are due to the limited level attributes (cluster centroids) that are used for cluster learning. When the number of clusters is higher than 18, there may be multiple clusters that contain different types of road objects. Therefore the assumption that all the road objects are contained in one cluster may need to be relaxed. The randomness of cluster centroids is dependent on the initial centroid position and the ordering of data presented to the algorithm. Typically, when IC erroneously classifies a road cluster, there is usually another cluster with very similar attributes except for one attribute. To avoid these problems, inter-cluster attributes may need to be included as part of the cluster learning process. These could be distance to nearest cluster, relative spatial position to k-nearest clusters, shape and compactness. Furthermore, multiple road clusters may allow be permitted.

11.4.7 RAIL Levels 3 and 4

From the edge pairs identified in Level 2 of RAIL, junction information is extracted and has been described elsewhere (Teoh and Sowmya 2000). A junction contains three edge pairs. The edge pair and junction information are then fused at the feature level and the fused features then provide support to an extended fast marching level set method by providing confident road seeds and helping to refine road centerline extraction (Cai et al. 2005). Machine learning is utilized to learn the seeds selection rules and stopping criterion parameter rules. A set of seeds is selected according to the learned rules, which determines the initial location for the fast marching level set method. At the same time, a texture based parameter is used to reduce the speed function to zero. Then, a fast marching level set method is applied on the image again to extract the road contour and road centerline network. The results of fast marching and the road segments are then combined to improve road recognition. Figure 11.14 shows the extracted centerlines of one whole image on the left and its ground truth on the right.

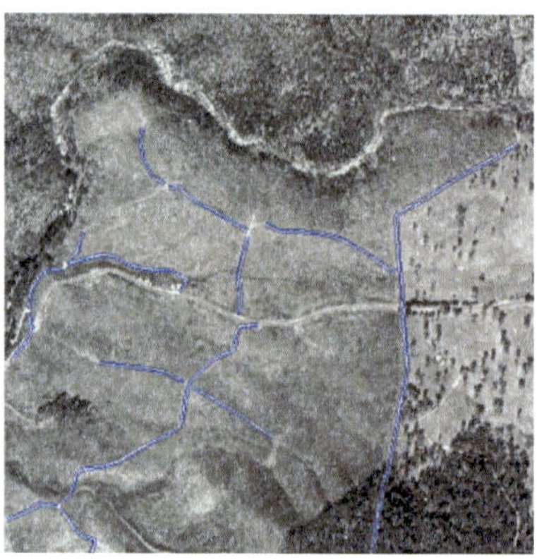

Fig. 11.6 Image E reference model

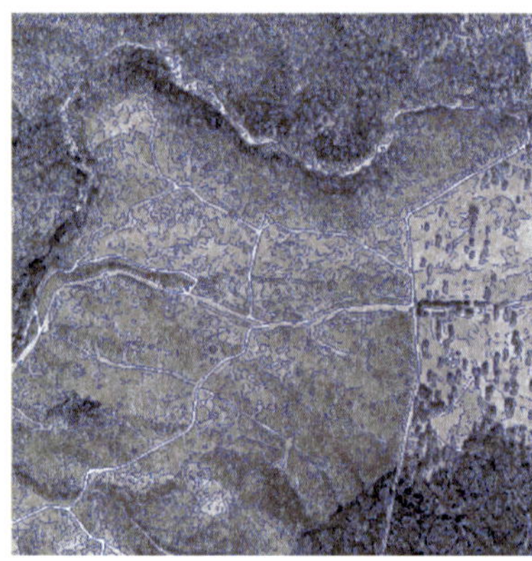

Fig. 11.7 Image E input

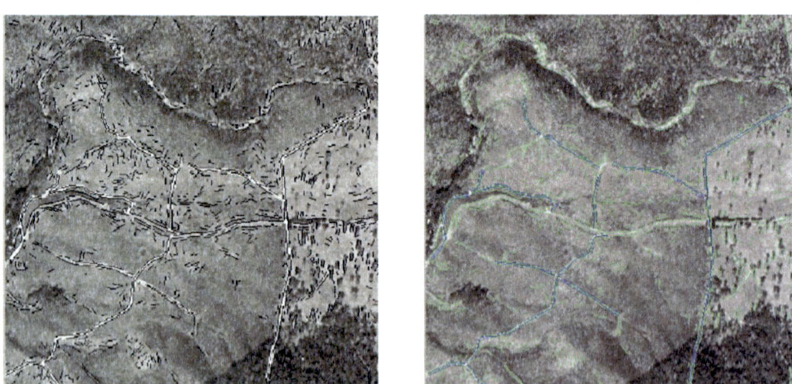

Fig. 11.8 Image E Level 1 output (left: gray scale; right: color, highlighted)

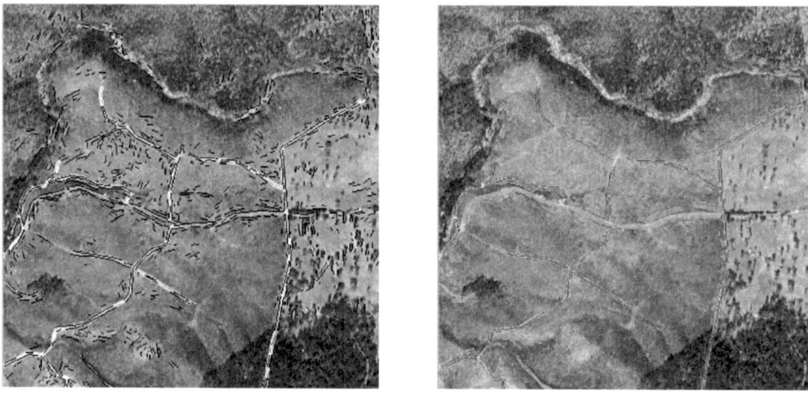

Fig. 11.9 Image E Level 2 output (left: gray scale; right: color, highlighted)

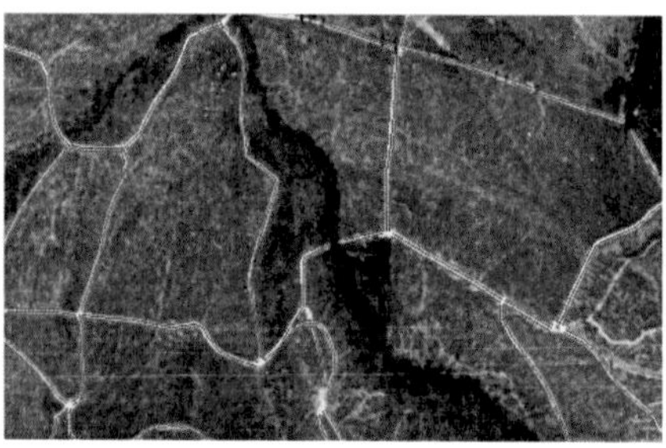

Fig. 11.10 Image O reference model

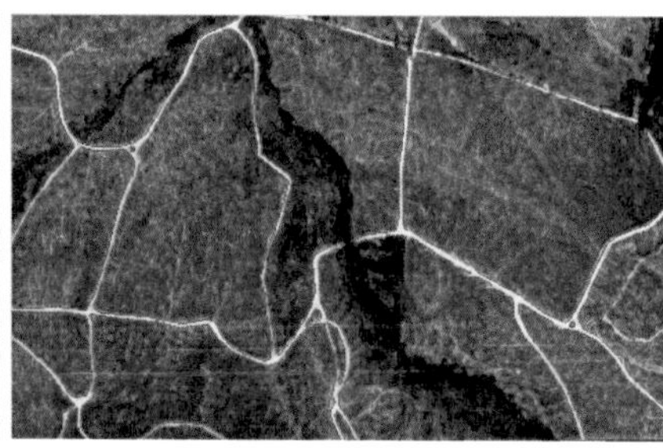

Fig. 11.11 Image O input

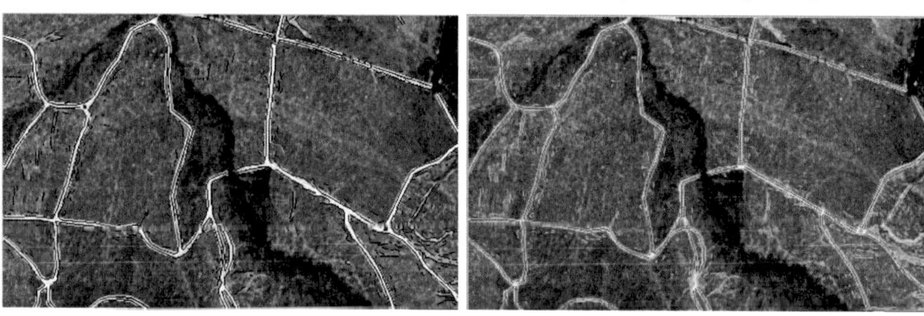

Fig. 11.12 Image O Level 1 output (left: gray scale; right: color, highlighted)

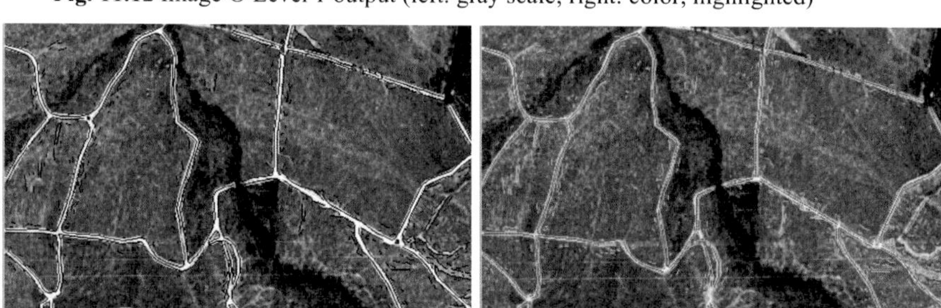

Fig. 11.13 Image O Level 2 output (left: gray scale; right: color, highlighted)

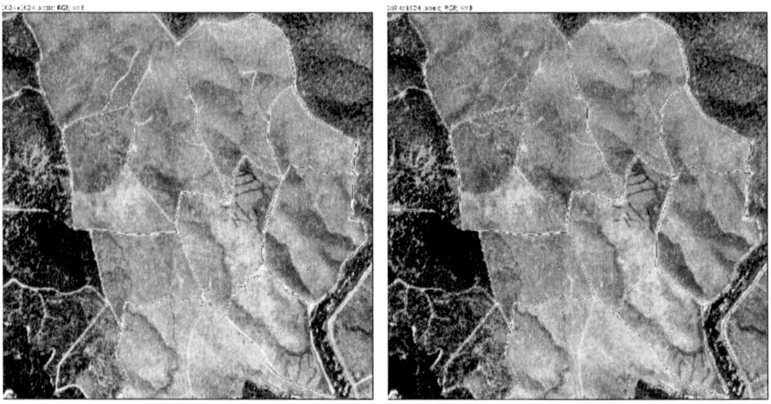

Fig. 11.14 Extracted centerlines (left) and ground truth (right)

11.5 Fusion of Multi-source Data Based on Dempster-Shafer Approach

We now turn to a problem of building extraction from multi-source data, based on a probabilistic approach of feature extraction based on the Dempster-Shafer algorithm. Two examples of this application have been undertaken by Lu et al. (2003) and Lu et al. (2006) for multispectral images and Rottensteiner et al. (2007) for images and LiDAR (Light Detection And Radar). Only the first example will be briefly described, since the latter research has been presented in Rottensteiner and Clode (2008).

Lu et al. (2003) described a system for extracting buildings and trees from digital aerial images in order to derive bare earth elevations. The system consists of three main parts: the first part derived a digital surface model (DSM) by image matching of the stereo image pair on a digital photogrammetric workstation (DPW). A multi-band classification of multispectral images using K-Means unsupervised clustering of the area and a NDVI (Normalized Difference Vegetation Index) was then used to transform the multispectral data into a single image band representing vegetation. While multispectral images supply abundant information for land cover classification, the NDVI and DSM are two key parameters, which define the difference between vegetated and non-vegetated objects. Simplistically, the areas which have heights above a certain limit are likely to be either trees or buildings. Areas with low NDVI and above the general terrain surface are likely to be buildings, whereas areas with high NDVI and are above that surface are likely to be trees. Areas with high NDVI, with heights similar to the terrain surface are likely to be grassland or cultivated areas. Therefore, based on the above hypotheses, so-called 'building interest areas', i.e. areas likely to be buildings, were derived in Part 1, and processed further in Parts

2 and 3. Part 2 defines building outlines based on the level set formulation of curve and surface motion, driven by an image-dependent speed function. Part 3 uses the Dempster-Shafer fusion theory, to combine three different data sources, namely the classified image, referred to hereinafter as the 'clustered image', DSM, and the building outlines, to extract the estimated correct building areas.

11.5.1 Shape Modeling and Image Segmentation with Level Set Method

The level set method for curve propagating interfaces was introduced by Osher and Sethian (1988), and Sethian (1999). It is based on mathematical and numerical definitions of curve and surface motion by Sethian (1985), and offers a highly robust and accurate method for tracking interfaces moving under complex motions.

Consider a closed curve moving in a plane. Let $\gamma(0)$ be a smooth, closed initial curve in a Euclidean plane R^2, and let $\gamma(t)$ be the one-parameter family of curves generated by moving $\gamma(0)$ along its normal vector field with speed $F(K)$, K is a given scalar function of the curvature. Let $\mathrm{x}(s,t)$ be the position vector which parameterizes $\gamma(t)$ by s, $0 \le s \le S$.

The level set method represents the front $\gamma(t)$ enclosing curve, as the level set $\{\Phi = 0\}$ of a function Φ. Thus, given a moving closed hypersurface $\gamma(0)$, we wish to produce a formulation for the motion of the hypersurface propagating along its normal direction with speed F. F can be a function of various arguments, including the curvature, normal direction, etc. The main idea is to embed this propagating interface as the zero level set of a higher dimensional function Φ. Let $\Phi(\boldsymbol{x}, t=0)$, where $\boldsymbol{x} \in R^N$ in N dimensional space, is defined by

$$\Phi(\boldsymbol{x}, t = t_1) = \pm d \tag{11.5}$$

where d is the distance from x to $\gamma(t=0)$, and the plus sign is chosen if the point $\boldsymbol{x}$ is outside the initial hypersurface $\gamma(t=0)$, the minus sign is chosen if the point $\boldsymbol{x}$ is inside the initial hypersurface. Thus, the initial function $\Phi(\boldsymbol{x}, t=0): \boldsymbol{x} \in R^N$ can be defined as follows:

$$\gamma(t=0) = (\boldsymbol{x} \mid \Phi(\boldsymbol{x}, t=0) = 0) \tag{11.6}$$

Now, we need to produce an equation for the evolving function $\Phi(\boldsymbol{x}, t)$, which contains the embedded motion of $\gamma(t)$ as the level set $\{\Phi = 0\}$.

Using the chain rule in Sethian (1995, 1999), the evolution equation for Φ can be a type of Hamilton-Jacobi equation. Based on the advantages of the Hamilton-Jacobi equation, in two space dimensions, a numerical approximation for the evolving function can be obtained. Using the forward and backward difference approximations in Φ, the evolving function can be described as Equation (11.7) and n defines iterations.

$$\begin{aligned}\Phi_{ij}^{n+1} &= \Phi_{ij}^{n} - F_0\Delta t((\max(D_x^-\Phi_{ij},0))^2 + (\min(D_x^+\Phi_{ij},0))^2 + \\ &+ (\max(D_y^-\Phi_{ij},0))^2 + (\min(D_y^+\Phi_{ij},0))^2)^{1/2} - \Delta t F_1 |\nabla\Phi|\end{aligned} \tag{11.7}$$

where

D_x^+ computes the new values at j using information at j and j+1;

$D_x^+\Phi_{ij} = \Phi_{i(j+1)} - \Phi_{ij}$

similarly

D_x^- computes the new values at j using information at j and j-1;

D_y^+ computes the new values at i using information at i and i+1;

D_y^- computes the new values at i using information at i and i-1;

The speed function can be as follows:

$$\begin{aligned}F &= F_0 + F_1(k) = k_I(-1.0 - 0.025K) \\ k_I(x,y) &= \frac{1}{1 + |LOG * I(x,y)|}\end{aligned} \tag{11.8}$$

where K is the curvature of level set. LOG is Laplacian of Gaussian operator.

The intrinsic geometric properties of the front curve may be determined from the level function Φ because F_1 is related to curvature K. The above level set approach can be used in high spatial dimensions.

Based on the 'building interest areas' derived from image analysis step, the level set algorithm is used to process all 'building interest areas' to delineate their boundaries. Figure 11.15 gives an example to show how the level set method works. The example demonstrates that when the evolving curve reaches the boundaries of the building, all the points on the curve stop evolving further and the computation is ended.

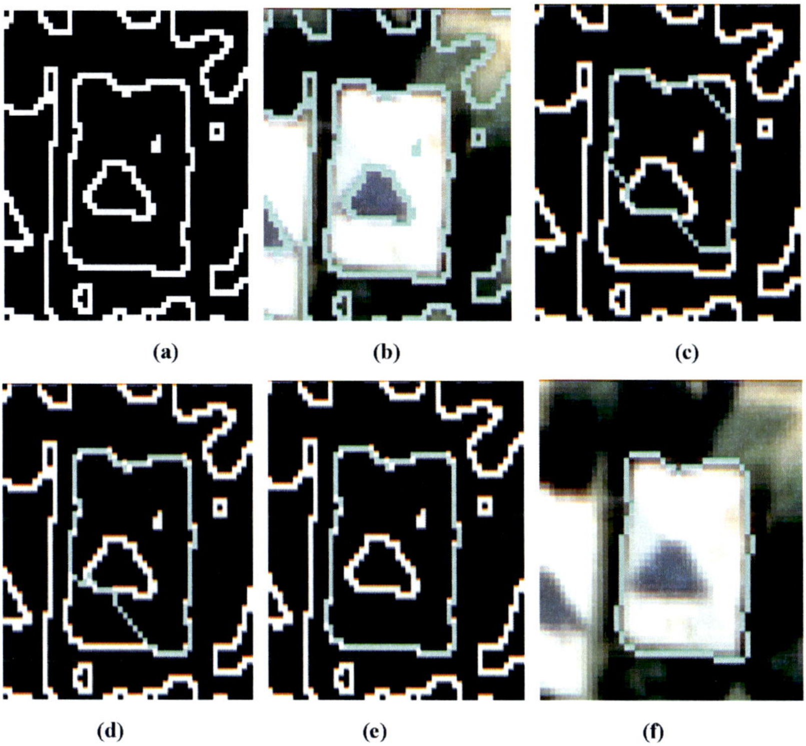

Fig. 11.15 Example of delineation the boundary of the 'building interest areas'

Figure 11.15(a) shows the zero crossings over the building areas, which are obtained by convolving the image with a LOG operator. Figure 11.15(b) shows the zero crossings overlaid on the original image. Figure 11.15(c) and 11.15(d) are the intermediate curves of the level set modeling. Figure 11.15(e) shows the final curve from the level set modeling. Only the building boundary shown in blue is extracted because the points on the boundary stop the curve from evolving further. The final boundary of the building overlaid on the original image is shown in Figure 11.15(f).

11.5.2 Data Fusion

In this step several sets of data are combined to provide a more reliable estimate of the buildings and trees, by the process of data fusion using the Dempster-Shafer algorithm. In this case there are 3 sources of data, the 'clustered image', the 'DSM', and the 'building outlines' derived by the level set.

As described in (Lu et al. 2006), the Dempster-Shafer approach is a statistically based classification algorithm which can be used to combine the knowledge from several data sources of the same region, to find the intersection of propositions on extracted information, derived from these datasets, together with the associated probabilities. This theory allows the representation of both imprecision and uncertainty and appears as a more flexible and general approach.

Assume a set of n propositions making up the hypothesis space as denoted by Θ. 2^{Θ} are the subsets of Θ. Based on the information from the data sources, a probability mass m can be assigned to any proposition or union of propositions. For $\forall \mathrm{A} \in 2^{\Theta}$. **m** is defined for every element A and the mass value $\mathbf{m}(A)$ is in the interval [0,1]. The following mass equations can be obtained:

$$\mathbf{m}(\varnothing) = 0 \tag{11.9}$$

$$\mathbf{m}(\Theta) = \sum_{A \subset 2^{\Theta}} \mathrm{m}(A) = 1 \tag{11.10}$$

where $\varnothing$ is the empty set.

In image classification, Θ is the set of hypotheses about a pixel class. The Dempster-Shafer theory permits the consideration of any subset of Θ. Applied to image classification problems, it means that single classes, as well as any union of classes can be represented. The number of classes (including all possible unions, but excluding the null set) is called the power set and is equal to $2^n - 1$. For example, if n=3, $2^3 - 1 = 7$, classes are given by C1, C2, C3, C1 $\cup$ C2, C1 $\cup$ C3, C2 $\cup$ C3, and C1 $\cup$ C2 $\cup$ C3 (Shafer 1976), Hegarat-Mascle et al. (1997), Klein (1999).

The Dempster-Shafer theory provides a representation of both imprecision and uncertainty through the definition of two parameters: support (Sup) and plausibility (Pls), which are obtained from the probability mass function m. Support for a given proposition means that all masses assigned directly by the data sources are summed. Plausibility for a given proposition means all masses not assigned to its negation are summed. For $\forall \mathrm{A_i} \in 2^{\Theta}$, the two parameters are defined respectively as follows:

$$\begin{aligned} Sup(A) &= \sum_{A_i \subseteq A} \mathrm{m}(A_i) \\ Pls(A) &= \sum_{A_i \cap A \neq \varnothing} \mathrm{m}(A_i) \end{aligned} \tag{11.11}$$

An uncertainty interval is defined by [Sup(A), Pls(A)] where

$$Sup(A) \le Pls(A)$$
$$Pls(A) = 1 - Sup(\overline{A}) \tag{11.12}$$

with $A \cup \overline{A} = \Theta$, $A \cap \overline{A} = \varnothing$.

The support value of hypothesis A may be interpreted as the minimum uncertainty value about A. Its plausibility may be interpreted as the maximum uncertainty value of A. The uncertainty interval gives a measurement of the imprecision about the uncertainty values. For several data sources, the Dempster-Shafer method allows compatible propositions to combine the probability masses from these sources to obtain a single value for the probability of the intersection (union) of the propositions.

As shown in Lu et al (2006) for multiple data sources, $\mathrm{m_t}$ is the basic probability mass provided by source t ($1 \le t \le p, p \ge 3$), the combination of all the data sources is defined as follows:

$$\mathbf{m}(A) = \frac{\sum\limits_{A_1 \cap A_2 ...... \cap A_p = A} \prod\limits_{1 \le t \le p} \mathbf{m}_t(A_i)}{1-K}, \quad K \ne 1 \tag{11.13}$$

where

$$K = \sum_{A_1 \cap A_2 ...... \cap A_p = \varnothing} \prod_{1 \le t \le p} \mathbf{m}_t(A_i)$$

In the Dempster-Shafter theory, the hypotheses about single classes and hypotheses about unions of classes are called simple hypotheses and compound hypotheses, respectively. When the probability mass of simple hypotheses are not null, a decision rule must be determined that best suits the application, such as the maximum support over simple hypotheses. The formula is as follows:

$$\max(Sup(A))$$
$$Sup(A) \ge Sup(\overline{A}) \tag{11.14}$$

The evaluation procedure by Dempster-Shafer approach to data fusion is based on spatial features. This means that the determination of probability masses and their combination is based on the frequencies of the features. For each 'building outline', there are corresponding areas obtained from the 'DSM' and 'clustered image'. The 'DSM', the 'clustered image', and the 'building outlines' are therefore assigned as data sets 1,2, and 3 respectively (Lu et al 2006). Classes C1, C2 and C3 represent trees, buildings and ground respectively.

Lu et al. (2006) have described the derivation of Table 11.11, which shows the origin of the computed probability mass, plausibility and support values for each simple and compound hypothesis based on three data sources. For the 'clustered image', the probabilities for all the simple and compound classes are represented by probability t, as an example. Similarly, the probabilities for 'building outlines' and 'DSM' are represented by u and s respectively.

For the data fusion process, the probability masses are based on information provided by each image. For the clustered image, $m_1(A)$ there are no null values assigned to C1, C2, C3 and C2 $\cup$ C3 as shown in Table 11.11. Since C2 for buildings and C3 for the ground may have the same texture in the image, there are ambiguities between these two classes in 'clustered image'. In Table 11.11, C2, C3 and C2 $\cup$ C3 are assigned the same probability, t.

Table 11.11 Dempster-Shafer calculation from three data sources ($1-k=4(ut+st+us-9uts)$)

A	m_1 (A) *clustered image*	m_2 (A) *building outlines*	m_3 (A) *DSM*	m (A)	Sup(A)	Sup($\overline{A}$)
C1 trees	$1-3t$	u	s	$\frac{4us(1-3t)}{1-k}$	$\frac{4us(1-3t)}{1-k}$	$\frac{4t(s+u-6su)}{1-k}$
C2 buildings	t	$1-3u$	s	$\frac{4st(1-3u)}{1-k}$	$\frac{4st(1-3u)}{1-k}$	$\frac{4u(s+t-6st)}{1-k}$
C3 ground	t	u	$1-3s$	$\frac{4ut(1-3s)}{1-k}$	$\frac{4ut(1-3s)}{1-k}$	$\frac{4s(u+t-6ut)}{1-k}$
C1 $\cup$ C2	0	0	s	0	$\frac{4s(u+t-6ut)}{1-k}$	$\frac{4ut(1-3s)}{1-k}$
C1 $\cup$ C3	0	u	0	0	$\frac{4u(s+t-6st)}{1-k}$	$\frac{4st(1-3u)}{1-k}$
C2 $\cup$ C3	t	0	0	0	$\frac{4t(s+u-6su)}{1-k}$	$\frac{4us(1-3t)}{1-k}$
Θ	0	0	0	0	1	0

For each extracted region defined by the 'building outlines', the numbers of pixels representing as trees, ground, glasses and building can be calculated respectively. The probability u can be defined based on the number of pixels assigned to each building region. For data set 2, 'building outlines', since they are extracted from low-level image analysis and interpretation and the level set modeling based image segmentation, they are assigned higher probabilities.

For data set 3, 'DSM', since buildings and trees are above the ground in the DSM, if they have about the same height, these two classes of areas are easily confused. As shown in the Table 11.11, class C1 representing trees, C2 of buildings and $C1 \cup C2$ are therefore assigned the same probabilities, s.

In Rottensteiner et al. (2007), which involved both images and LiDAR a different approach was taken for the definition of mass probabilities for each of the classes. In this case, emphasis was placed on the union of classes, and mass probabilities were modeled by a cubic parabola between 2 extreme values and constant values outside these 2 extremes.

11.5.3 Results of Experiments

While a number of test areas have been investigated with this approach, the results[4] of processing an area shown in (Lu et al. 2006) will be given in the following. Figure 11.16 illustrates a left image of a pair of color aerial images with 522×584 pixels in the row and column directions respectively, with GSD of 0.3 meter. The flying height is 3070 meters and the original scale of the images was 1:20,000. The scale of the images is smaller than desired, but larger scale images of the area were not available. The majority of buildings have white or red roofs, but there are some dark roof buildings as well. Based on the classification, a segmented image can be created to show the areas classified. Most of the building areas have been detected, but the dark roofed buildings are completely missed and will not be recovered. Some red roofed buildings are partly detected. Also, one car is assigned as a building in the top of image.

Fig. 11.16 Left image of stereo

Fig. 11.17 Building interest areas after clustering image analysis

[4] The results of these tests have been presented with the permission of the American Society for Photogrammetry and Remote Sensing (ASPRS).

Fig. 11.18 Building outlines overlaid on ortho image

Fig. 11.19 Result of Dempster-Shafer based on 3 datasets overlaid on ortho image

Since only the color image was available, the results of processing by Visible Vegetation Index (VVI) were obtained using only the 3 color images. These areas with high VVI represent the vegetation and the areas with low VVI represent the ground and building areas. The four information layers of DSM, classification, post classification segmentation and VVI are input into the ArcView®, using the Map Queries operation, and the possible building areas are extracted. Using the region growing algorithm, small spots which do not belong to buildings can be deleted from the Map Queries result. The derived 'building interest areas' are overlaid on the ortho image as shown in Figure 11.17. Some road areas wrongly assigned as buildings have been deleted and the correct building areas derived from classification have been successfully maintained. For every small building area, shape modeling with level set method was then implemented. Some regions in the left of the image, which belong to road areas, are still assigned as building areas after this step as shown in Fig. 11.18. Thus, it is necessary to use Dempster-Shafer data fusion method to evaluate the regions.

The Dempster-Shafer method provides a single value for the probability of the intersection (union) of the propositions. Based on the decision rule given in Equation 11.14 , the three data sources, 'clustered image', 'building outlines' and 'DSM' were combined to produce more reliable building areas shown overlaid on orthoimage in Figure 11.19.

The consequence of the data fusion is that three incorrect building areas have been detected and deleted in the final result, while the correct building areas remain unchanged. There are 50 buildings in the image and 40 building are detected. The detection rate is 80%. The DSM extraction and low-level image analysis and interpretation are important, since they supply 'building interest areas' and the level set modeling and data fusion are based on these areas. Since some dark roof buildings were missed in Part 1 of the process, they will never be recovered. The results of building extraction could be improved by refining Part 1 and using larger scale images. The improvement in building extraction by using the data fusion approach is 3 out of 50. This is important processing step because it makes the system to supply more reliable extracted buildings.

In Lu et al. (2006) 5 test areas have been presented in which 80% or more buildings have been detected in all tests as shown in Table 11.12. The Dempster-Shafer data fusion technique provides the theoretical basis for evaluating the reliability of the extracted buildings from the combination of the different data sources by a statistically-based classification. Based on the test areas shown, the results are encouraging, but further research is needed to refine these methods, in order to extract a greater number of the buildings with different textured roofs.

Table 11.12 Results of all tests in this study

Test No	Total buildings	Regions derived by level set	Detected buildings	Deleted wrong buildings	False evaluation	Detection rates
Test1	96	91	85	5	1	88.5%
Test2	32	39	31	8	0	96.8%
Test3	61	65	50	11	4	81.9%
Test4	50	43	40	3	0	80%
Test5	26	24	21	2	1	80.8%

11.6 Conclusion

This chapter has described several methods that have been developed to extract features from digital satellite and aerial images. The chapter commenced with a description of some of the issues that impact on a developer's ability to automatically extract features from images, and typical approaches taken in computer science. A review is then given of current methods that have been used to extract roads from digital images. A method based on machine learning has been described in detail for the extraction of roads at four levels: road edge pairs that enclose a segment of road; linked road edge pairs that form continuous roads; intersections to form junctions; and road networks. The approach uses a specially designed inductive learning technique called Inductive Clustering (IC) that can automatically select parameters for clustering algorithms by learning from clustering examples in the images, in three steps: algorithm learning that derives the clustering algorithm most suited to an image type; parameter learning that finds the most appropriate parameter values that give the best results for the learned algorithm and image type; and cluster learning that identifies a cluster of interest. Tests of the approach on a number of images have been presented.

A description is then given of a method of detecting buildings from multiple data sources based on the Dempster-Shafer algorithm where the evidence used to identify buildings includes a digital surface model, NDVI derived from multispectral images, image analysis based on clustering and building outlines extracted using the level set approach. The section gave an introduction to level set theory and its implementation, followed by a description of the principles of the Dempster-Shafer algorithm and the approach used to determine the mass probabilities for each data source. Examples of tests on the method have also been given which demonstrate accuracies for extracting buildings higher than 80%.

References

Agouris P, Doucette P, Stefanidis A (2001) Spatiospectral cluster analysis of elongated regions in aerial imagery. IEEE Int. Conf. Image Proc 2:789-792

Amini J, Sarahjian MR, (2000) Image map simplification by using mathematical morphology. ISPRS Int Arch Photogram Rem Sens 33(B3):36-47

Bandyopadhyay S, Sowmya A, Maulik U (2000) Genetic classifier for extracting road segments. Proc Int Conf Communications Computers and Devices (II):455-458

Barsi A, Heipke C, Willrich F (2002) Junction extraction by artificial neural network system genes, ISPRS Int Arch Photogram and Rem Sens & SIS 34(3B):18-21

Baumgartner A, Hinz S, Wiedemann C (2002) Efficient methods, and interfaces for road tracking. Int Archives Photogram and Rem Sens & SIS 34(3B):28-31

Baumgartner A, Steger C, Mayer H, Eckstein W, Ebner H (1999) Automatic Road Extraction Based on Mult-Scale, Grouping, and Context. Photogram Eng and Rem Sens 65 (7):777-785

Benjamin S, Gaydos L (1990) Spatial resolution requirements for automated cartographic road extraction, Photogram Eng. Rem Sens & SIS 56(1):93-100

Bonnefon R, Dherete P, Desachy J (2002) Geographic information system updating using remote sensing images. Pattern Recognition Letters 23(9):1073-1083

Brazdil PB, Henery RJ (1994) Analysis of results. In Michie D,

Spiegelhalter DJ, Taylor CC, Campbell J (ed), Machine Learning, Neural and Statistical Classification, Chap. 10, Ellis Horwood

Cai X, Sowmya A, Trinder JC (2005) Learning to Recognise Roads from High Resolution Remotely Sensed Images, Proc 2nd ISSNIP:307- 312

Cai X, Sowmya A, Trinder JC (2006) Learning parameter tuning for object extraction. In Lecture Notes in Computer Science, Narayanan PJ, Nayar SK, Shum H-Y (Ed) 3851:868- 877

Caruana R, Niculescu-Mizil A, Crew G, Ksikes A (2004) Ensemble selection from libraries of models. Proc. 21st Int Conf Machine Learning, ACM International Conference Proceeding Series; Vol. 69, ACM Press New York:137-144

Chanussot J, Lambert P (1998) An application of mathematical morphology to road network extraction on SAR images. Int Sym Math Morphol, Amsterdam:399-406

Chen A, Donovan G, Sowmya A, Trinder JC (2002) Inductive Clustering: automating low-level segmentation in high resolution images. ISPRS Int Arch Photogram and Rem Sens & SIS 34 (3A):73-78.

Chiang TY, Hsieh TH, Lau W (2001) Automatic road extraction from aerial images, Stanford Education – on-line and unpublished document

Cohen LD (1991) On active contour models and balloons. CVGIP Image Understanding 53:211-218

Dal Poz AP, Gyftakis S, Agouris P (2000) Semiautomatic road extraction: Comparison of Methodologies and experiments. ASPRS Ann Conf Washington DC USA

Dell'Acqua F, Gamba P, Lisini G (2003) Road map extraction by multiple detectors in fine spatial resolution SAR data. Can J Rem Sens 29(4):481-490.

Dial G, Gibson L, Poulsen R, (2001) Ikonos imagery and its use in automated road extraction. In Automatic extraction of Man-made Objects from Aerial and Space Images (III), Zurich (III), Baltsavias et al.(ed) 2001 Swets S Zeitlinger, Lisse: 357-366

Doucette P, Agouris P, Musavi M, Stefanidis A (1999) Automated extraction of linear features from aerial imagery using Kohonen learning and GIS data. Lecture Notes in Computer Science, 1737:20-33

Dzeroski S, Zenko B (2002) Is combining classifiers better than selecting the best one. Proc. 19 Int. Conf. Mach. Learn:123-130

Faber A, Förstner W (1999) Scale characteristics of local autocovariances for texture segmentation, ISPRS Int Arch Photogram Rem Sens & SIS 32(7-4-3W6): 74-78

Ferraro M, Boccignone G, Caelli T (1999) On the representation of image structures via scale space entropy conditions. IEEE Trans. Patt Anal Mach Intell 21:1199-1203

Goldberg DE (1989) Genetic algorithms in search, optimization and machine learning. Addison-Wesley, Reading, MA

Gruen A, Li H (1996) Linear Feature Extraction with LSB-Snakes from Multiple Images. Int Arch Photogram Rem Sens & SIS 31(B3):266-272

Gruen A, Li H (1997) Semi-automatic linear feature extraction by dynamic programming and LSB-Snakes. Photogram Eng Rem Sens 63:985-995

Haverkamp D, Poulsen R (2002) Complementary methods for extracting road centrelines from Ikonos imagery. Proc SPIE Image and Signal Processing for Remote Sensing VIII:501-511

Hegarat-Mascle S, Bloch I, Vidal-Madjar D (1997) Application of Dempster-Shafer Evidence Theory to Unsupervised Classification in Multisource Remote Sensing. IEEE Trans Geosci Rem Sens 35 (4):1018-1031

Hinz S, Baumgartner A (2003) Automatic road extraction of urban road networks from multi-view aerial imagery. ISPRS J Photogram Rem Sens 58(1-2):83-98

Jeon B K, Jang J, Hong K (2000) Map based road detection in spaceborne synthetic aperture radar images based on curvilinear structure extraction. Opt Eng 39(9):2413-2421

Jin X, Davis CH (2005) An integrated system for automatic road mapping from high-resolution multi-spectral satellite imagery by information fusion, Information Fusion 6(4):257-273

Kass M, Witkin A, Terzopoulos D (1988) Snakes: Active Contour Models. Int J Computer Vision:321-331

Katartzis A, Sahi H, Pizurica V, Cornelis J (2001) A model based approach to the automatic extraction of linear features from airborne images, IEEE Trans Geosci Rem Sens 39(9): 2073-2079

Keaton T, Brokish J (2003) Evolving roads in Ikonos Multispectral Imagery. Proc Int Conf Image Processing (ICIP)3: III-1001-4 vol 2

Kessler O, Askin K, Beck N, Lynch J, White F, Buede D, Hall D, Llinas I, (1992) Functional description of the data fusion process, Office of Naval Technology, Naval Air Development Center, Warminster, PA, 1992.

Kittler J, Hatef M, Duin RPW Matas J (1998) On combining classifiers. IEEE Trans PAMI 20(3):226-239

Klein LA (1999) Sensor and Data Fusion Concepts and Applications. SPIE Optical Engineering Press, Bellingham, WA

Lai JY, Sowmya A, Trinder JC, (2005) Support Vector Machine experiments for road recognition in high resolution images. Proc MLDM in Perner P and Imiya A (Ed) Machine Learning and Data Mining in Pattern Recognition, Springer Verlag, LNAI 3587, MLDM 2005 Chapter 42:426-436.

Laptev I, Mayer H, Lindeberg T, Eckstein W, Steger C, Baumgartner A (2000) Automatic extraction of roads from aerial images based on scale space and snakes. Machine Vision Applicat 12(1):23-31

Lu Y H, Trinder JC Kubik K (2003) Automatic Building Extraction for 3D Terrain Reconstruction Using Interpretation Techniques. Proceedings ISPRS Workshop on High Resolution Mapping from Space, Hannover Germany (on CD):1-9 pages

Lu YH, Trinder JC, Kubik K (2006) Automatic Building Detection Using The Dempster-Shafer Algorithm, Photogramm Eng & Rem Sens 72(4):395-404

Mayer H, Laptev I Baumgartner A (1998) Multi-scale and snakes for automatic road extraction.

5th Europ Conf on Comput. Vision:720-733

Mayer H, Hinz S, Bacher U, Baltsavias E (2006) A Test of Automatic Road Extraction Approaches. Int. Archives Photogram Rem Sens & SIS 34(3) paper 0_15:1-6

McKeown DM Denlinger JL (1988) Cooperative methods for road tracking in aerial imagery. Workshop Comput Vision Pattern Recognition:662-672

Mena JB (2003) State of the art on automatic road extraction for GIS update: a novel classification. Pattern Recognition Letters 24(16):3037-3058

Mena JB Malpica JA (2004) An automatic method for road extraction in rural and semi-urban areas starting from high resolution satellite imagery. Pattern Recognition Letters 26: 1201–1220

Nevatia R Babu KR (1980) Linear feature extraction and description. Comp. Graph Image Process 13:257-269

Osher S, Sethian JA (1988) Fronts propagating with curvature dependent speed: Algorithms based on Hamilton-Jacobi formulation. J Computational Physics 79:12-49.

Quam LH (1978) Road tracking and anomaly detection in aerial imagery, Image Understanding Workshop:51-55

Quinlan JR (1993) C4.5: Programs For Machine Learning. Morgan Kaufmann San Mateo, CA

Ravanbakhsh M, Heipke C, Pakzad K (2008) Extraction of Road Junction Islands from High Resolution Aerial Imagery Using Level Sets, ISPRS Int. Archives Photogram Rem Sens and SIS 37(3A):209-214.

Roggero M (2002) Object segmentation with region growing and principal component analysis, ISPRS Int Arch Photogram Rem Sens & SIS 34(3A):289-294

Rottensteiner F, Trinder JC, Clode S, Kubik K (2007) Building detection by fusion of airborne laser scanner data and multi-spectral images: Performance evaluation and sensitivity analysis, ISPRS Journal Photogram Rem Sens 62:135-149

Rottensteiner F, Clode S, (2008) Building and Road Extraction by LiDAR and Imagery, in Topographic Laser Ranging and Scanning, Shan J, Toth C (Ed) CRC Press, Oxford UK.

Sakoda W, Hu J, Pavlidis T (1993) Computer Assisted Tracking of Faint Roads in Satellite Imagery. ACSM/ASPRS Convention, New Orleans 9 (2):311-323

Sethian JA (1985) Curvature and the evolution of fronts. Comm In Math Phys 54:487-499

Sethian JA (1995) Shape Modelling with front Propagation: A Level Set approach, IEEE Trans PAMI 17(2):158-175

Sethian JA (1999) Level set methods and fast marching methods. Cambridge University Press

Shafer G (1976) A Mathematical Theory of Evidence. Princeton University Press, Princeton, NJ

Shanahan J, Thomas B, Mirmehdi M, Martin T, Campbell N Baldwin J (2000), A soft computing approach to road classification. J. Intell. Robot. System 29(4):349-387

Singh S, Sowmya A (1998) RAIL: Road Recognition from Aerial Images Using Inductive Learning. ISPRS Int Arch Photogram Rem Sens SIS 32(3/1):367-378

Strat TM (1995) Using Context to Control Computer Vision Algorithms in Automatic Extraction of Man-made Objects from Aerial and space Images, Birkhäuser Verlag Basel:3-12

Teoh C, Sowmya A (2000) Junction Extraction from high resolution images by composite learning, ISPRS Int Arch Photogram Rem Sens & SIS 33(B3):882-888

Tesser H, Pavlidis T (2000) Road finder front end: an automated road extraction system, IEEE Trans. Geosci. Rem Sens 38:338-341

Trinder JC Li H (1995) Semi-Automatic Feature Extraction by Snakes, In Automatic Extraction of Man-Made Objects from Aerial and Space Images Gruen A, Kuebler O, Agouris P (Ed) Birkhäuser Verlag, Basel:95-102

Trinder JC. Maulik U, Bandyopadhyay S (2000) Semi-Automatic Feature Extraction Using Simulated Annealing, Int Arch Photogram Rem Sens & SIS 33(3/2):905-909

Tupin F, Maitre H (1998) Detection of linear features in SAR images: application to road network extraction. IEEE Trans. Geosci. Rem Sens 36(2):434-453

Vosselman G, Knecht D (1995) Road tracking by profile matching and Kalman filtering. In Automatic Extraction of Man-Made Objects from Aerial and Space Images Gruen A, Kuebler

O, Agouris P (Ed) Birkhäuser Verlag, Basel:265-274

Wiedemann CC, Heipke H, Mayer. Jamet O (1998) Empirical Evaluation of Automatically Extracted Road Axes. CVPR Workshop on Empirical Evaluation Methods in Computer Vision:172-187

Wufeng C, Qiming Q (1998) A knowledge based research for road extraction from digital satellite images. Acta Scientiarum Naturalium Universitatis Pekinensis 34(2-3):54-263

Yoon T, Park W, Kim T (2002) Normalized gradient vector diffusion and image segmentation. Proc. SPIE Remote Sensing for Environmental Monitoring, GIS Application:320-328

Zafiropoulos P, Schenk TF (1998) Extraction of road structures with color energy models, Videmoe-TRICS VI Proc SPIE:276-290

Zhang C, Murai S, Baltsavias E (1999) Road network detection by mathematical morphology, ISPRS Workshop on 3D Geospatial Data Production: Meeting Applicat, Requirements, Paris:85-200

Zhao H, Kumagai J, Nakagawa M, Shibasaki R (2002) Semi automatic road extraction from high resolution satellite image, ISPRS Int Arch Photogram Rem Sens & SIS 34(3A):406-411

Ziems M, Gerke M, Heipke C (2007) Automatic Road Extraction from Remotely Sensed Imagery Incorporating Prior Information and Colour Segmentation. Int. Arch Photogram Rem Sens & SIS 36(3/W49A):141-147

Zlotnick A, Carnine PD (1993) Finding roads seeds in aerial images. CVGIP Image Understanding 57(2):243-260

Chapter 12

EFFICIENT GEOSPATIAL ANALYSIS OF REMOTELY SENSED IMAGES BY MEANS OF LINEAR FEATURE EXTRACTION AND COMBINATION

Gianni Lisini, Fabio Dell'Acqua, Paolo Gamba

12.1 Introduction

An efficient and valid interpretation of very high resolution images, either SAR or optical, must take into account the spatial details of the scene. As a result, there are many options to the very simple pixel-by-pixel classification scheme which applies to coarse resolution remote sensing data sets. One option is the exploitation of spatial features, like for example textures, that connect each pixel with its neighborhood (Dekker 2003). Another option is multi-scale analysis, where scales, and thus contexts with different size, are jointly considered to capture details at various levels (Benediktsson *et al.* 2003). A third option, explored in this chapter, is to extract significant yet simple geometrical features and use them (possibly in combination with spectral features) to improve the understanding of the scene (Xin *et al.* 2007). It is a separate approach, other than standard classification, but it can be used also for classification and change detection (which is essentially multitemporal classification).

One of the simplest yet efficient geometrical features that one can exploit in human settlements or, generally speaking, when artificial structures are of concern are linear features. They can be used as a basis for higher-level feature extraction, such as parallel pairs or junctions, and combined to obtain even more complex elements, like geometric shapes. They are very flexible elements of the scene, and can be used for a variety of applications, provided an efficient methodology for their extraction is considered, capable of dealing with different VHR images, such as SAR and optical ones. After this step, higher-level feature extraction needs to be at the same time fast and reliable, building over the feature extraction in an application-driven manner. Finally, the ability to manage these spatial features in conjunction with spectral ones must be taken into account, and an eye must be kept on pixel-based measures related to spatial information.

D. Li et al. (eds.), *Geospatial Technology for Earth Observation*,
DOI 10.1007/978-1-4419-0050-0_12,

The choice of focusing on "segments" -as we will call the linear features extracted from a VHR scene- in our research work is a result of balancing between advantages and disadvantages. The most important advantage of segments is the (relative) simplicity of the extraction routine, as well as the high flexibility in combining them into various and very different scene elements. The disadvantages are related to problems in discriminating segments belonging to geometrically very similar yet semantically completely different elements of the scene, and the high level of false positives and/or false negatives that any segment extractor produces.

Although the single steps of this approach are not novel, this chapter attempts to describe the methodology in a framework that allows recognizing many different applications as instances of the same idea. To this aim, the structure of this chapter is devoted first to describe the state of the art in technical literature for approaches based on linear feature extraction and combination for VHR image analysis. Then, a general framework is described and proposed. Following this general assessment of the problem, the following section of this chapter is conceived as the presentation of a simple yet flexible routine for segment extraction in VHR remotely sensed imagery. Section 5 explains how segments can be combined to extract higher-level geometrical elements of the scene, such as junctions and parallel pairs. Eventually, applications of the framework and the specific steps proposed in this chapter are discussed in the final section, which introduces a few applicative examples, showing how it is possible to exploit linear and higher-level features for very different purposes.

12.2 Linear Feature Extraction in VHR Image Analysis

Although there are many ways to interpret an image, remotely sensed data interpretation approaches are constrained by the nature of the data and its geographical meaning. As a result, the general trend in this specific area of image processing is on the one hand toward more refined classification algorithms, exploiting at their best the measurement vectors coming from multi- or hyperspectral sensors or from multi-scale analysis of the scene. On the other hand, the research is moving towards more and more object-based scene analysis. With respect to both these issues, more feature extraction and feature fusion methodologies are continuously seeked, tested and compared. One specific feature that has been always considered as useful is the linear one, which may be considered, according to the approach, an edge, a portion of a road, a texture element.

Although not limited to VHR images, the use of linear features has been often associated with the finest spatial resolution, because linear features are increasingly visible and meaningful when more and more scene details are available. Examples date back from the use of the first panchromatic aerial images, where linear features have been considered as the basis for the interpretation of the structures in the scene, mainly because they are mostly associated with artificial manufacts. In Nevatia and Babu (1980), for instance, the general system for extracting edges by means of image filtering, followed by aggregation of the elements into higher level feature based on specific rules has been introduced. The same ideas has been used by the authors in many paper devoted to object extraction and image interpretation (Lee *et al.* 2002), considering also multiple images to solve from complex topologies for the edges in building rooftops (Kim *et al.* 2002). Similarly, in Krishnamachari and Chellappa (1996), Markov Random Fields instead of Perceptual Grouping rules are used to provide a system able to detect and delineate buildings starting from edges. Finally, and without any intention to provide a complete list but just to give an idea of the numerous ideas developed in this field, another methodology, based on graphs to connect edges and eventually recognize buildings have been presented in Kim and Muller (1999).

On a different subject, linear feature are naturally part of any road extraction approach, and they are also instrumental for many road element detection techniques (e.g. Tavakoli and Rosenfeld 1982). As a result, basic road element extraction methods are usually considering linear features as components or seeds for the characterization of the whole network (Netenyahu *et al.* 1997, Netanyahu *et al.* 1996, Merlet and Zerubia 1996, Agouris *et al.* 2001, Kozaitis and Cofer 2005, Shi and Zhou 2002). To enhance the suitability of "standard" image processing routines to different linear element exploitation techniques, extraction algorithms suited to optical (Groch 1982) or SAR (Onana *et al.* 2001) remotely sensed data have been developed, each one with its own peculiar advantages. Fusion of the results by these extraction routines has been also proposed (Dell''Acqua *et al.* 2002) in order to improve and refine the output product of the interpretation process. Besides extracting, recognizing and joining linear elements, more advanced references in technical literature propose to combine linear features extracted from VHR images in order to detect roadsides (Amini *et al.* 2002), road intersections (Mackaness and Mackchnie 1999), zebra crossings and other similar scene elements (Jung and Paparoditis 2003). For instance in Zlotnick and Carnine (1993) antiparallel edge pairs, are used as clues to road hypotheses, and the longer the pair, the more reliable the seed for further road network processing.

Finally, a different way to exploit the information provided by linear features, more related to their being a texture component, has been used in HR and VHR images for classification purposes. The very recent example discussed in Huang *et al.* (2009) shows that linear element density and orientation can be used to discriminate among different blocks in urban areas, or as a hint to different land uses. Another interesting reference is Price (1999), where a model-based approach is used to detect urban grid patterns and classify them.

12.3 A General Framework

The examples discussed in the previous section highlight the large amount of work proposed with respect to linear feature extraction and their use for VHR scene interpretation. In order to try and provide a general framework where these examples could fit in, we introduce the procedure whose building blocks are shown in Fig. 12.1.

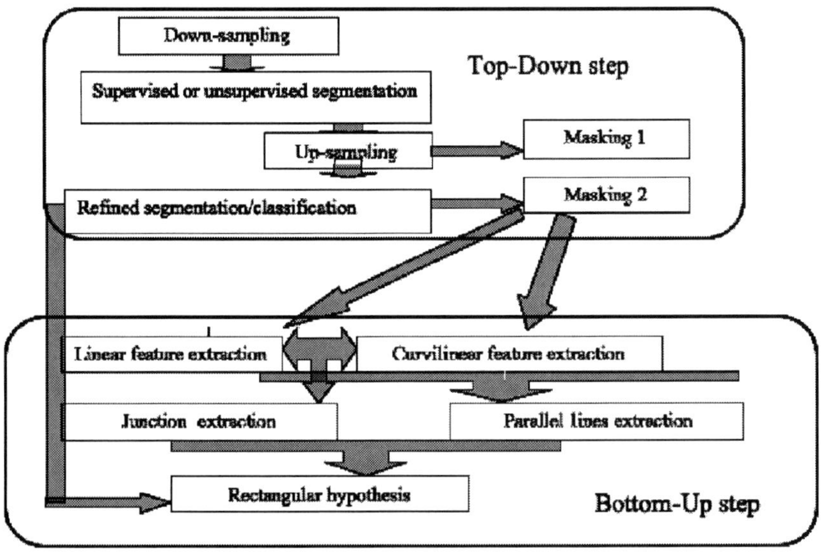

Fig. 12.1 General framework for VHR scene interpretation procedure based on feature extraction and combination, and specifically, for what concerns this chapter, on linear features

The procedure is aimed at VHR scene interpretation and includes two steps, running sequentially and able to provide eventually a full scene interpretation product: the "top-down" and the "bottom-up" steps.

The top-down step is aimed at extracting via basic scene description the most important areas (hotspots) in the scene. One way to achieve this is by means of an object-based segmentation or by a supervise/unsupervised classification. Usually this step does not require the full spatial resolution of the VHR data and should consider textural information, ancillary data and other non remotely sensed data (e.g. in situ measurements).

The bottom-up step is instead devoted to focus within the hotspots extracted during the first part of the procedure and extract basic geometrical elements of the scene, e.g. the linear segments as in this chapter. The extraction, combination and/or joint analysis of these geometrical features and the information they carry will provide in the end a refined description of the scene.

The advantage of this framework is that on the one hand it generalizes usual scene interpretation methodologies base on geometric feature alone. On the other hand, it exploits as much as possible the spectral and textural characteristics of the remote sensing data, by means of the top-down segmentation and hotspot recognition. The main idea behind this framework is therefore that the combination of these two steps is able to improve the final interpretation results of each of them by means of an "information fusion" approach. For instance, it is likely that the joint use of a map of the objects in the scene (with knowledge of their meaning) together with the geometrical features extracted from the same scene will reduce false detections (e.g. linear elements within the boundaries of areas recognized as water bodies due to their spectral signature) and it will also provide a measure of the reliability of the geometric feature extraction results (e.g. a high density of straight lines is more likely within a human settlement than in open rural regions).

According to this framework, the linear feature extraction and combination techniques described in the following two sections can be understood as a part of a procedure whose practical applications are the examples proposed and discussed in Section 6.

12.4 Linear Feature Extraction

One of the fundamental elements of the second part of the framework discussed in the previous section is therefore the extraction of all the linear features in the scene under consideration. Among them, some will represent portions of road and railway networks, river banks, others portions of building perimeters (houses, walls, embankments), as well as vegetation contours. These objects will be recognized starting from these linear elements, based on their relationship and patterns, i.e. using geometrical characteristics together with spectral characteristics. For the moment, however, the focus is on their extraction without any attempt of interpreting them.

In technical literature several examples are found of techniques designed to extract linear features (Henderson and Xia 1997, Dell'Acqua and Gamba 2001, Gruen and Li 1997, Airailu *et al.* 1994, Barzohar and Cooper 1996, German and Jedynak 1996, Park and Kim, 2001) from EO data. Each one has its own advantages and drawbacks, and no all-purpose approach has been designed so far. Moreover, the extraction procedure is highly dependent on the data and specific attention shall be given to the spatial resolution and the noise level of the image under test. As a matter of fact, the sensor (SAR, optical, multispectral, etc.) plays a role, binding on the choice of the extraction parameters as well as of the extractors.

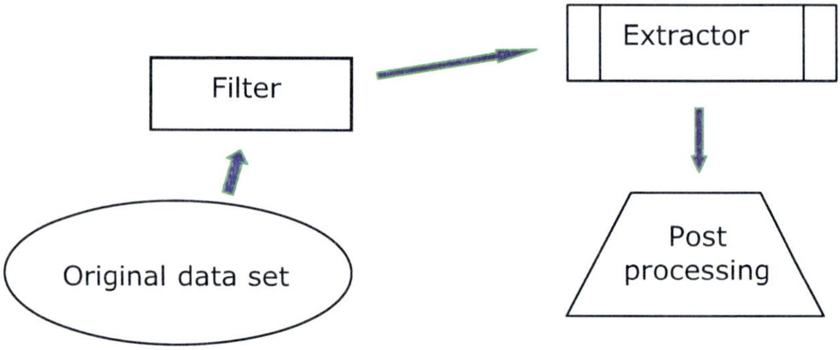

Fig. 12.2 General processing chain for linear feature extraction from EO data

As a general statement, extraction routines depend on the characteristics of the input data, and need some pre-processing before ingesting data from different sensors. The general processing chain for linear feature extraction is thus composed of a filter (or a set of filters, as further specified in the following section), the proper linear feature extraction routine, and a post-extraction refinement. The general structure of the linear feature processing chain is thus as in Fig. 12.2.

12.4.1 Adaptive Filtering

As mentioned, the image to be analyzed often requires an enhancement step, and almost always an adaptation step. The first step is aimed at emphasizing as much as possible the features to be extracted. The second step is aimed at reducing noise, at pixel level or in a more sophisticated manner considering spatial patterns. In any case, a filtering chain is needed, composed of different elements aimed at reducing the useless part of the image information content.

Fig. 12. 3 Simulated SAR urban scene

The most common implementation of this chain for linear feature detection is a set of directional filters. For example, Figure 12.3 depicts a simulated COSMO/SkyMed VHR SAR urban scene, and it is clear that some kind of filtering is required to extract, for instance, the road network. Even more filtering is required for building extraction via their footprint identification. In this type of image a simple extraction after thresholding would not identify anything interesting. This effect is due to the huge number of possible linear features on the one side and to the large amount of noise affecting the image on the other side.

Performances can be greatly improved by applying a suitable sequence of filters, optimized according to the resolution of the data and the linear feature directions. In particular, with simulated SAR images at a resolution of 3 m/pixel and 1 m/pixel respectively, the filter sequence shown in Table 1 has proved to work best (Dell'Acqua *et al.* 2003).

Table 1 Optimized directional filter sequence for the simulated SAR image in Fig. 12.3 according to Dell'Acqua *et al.* (2003).

	Horizontal filtering		**Vertical filtering**	
	Window Resol. 1m/p	Window Resol. 3m/p	Window Resol. 1m/p	Window Resol. 3m/p
Closing	1 x 7	1 x 3	7 x 1	3 x 1
Directional Filtering	5 x 5 at 90°	5 x 5 at 90°	5 x 5 at 0°	5 x 5 at 0°
Opening	11 x 1	5 x 1	1 x 11	1 x 5
Opening	1 x 3	1 x 2	3 x 1	2 x 1

The chain uses morphological filters (opening and closing) designed to emphasize structures with horizontal or vertical orientation. Corresponding results are shown in Fig. 12.4.

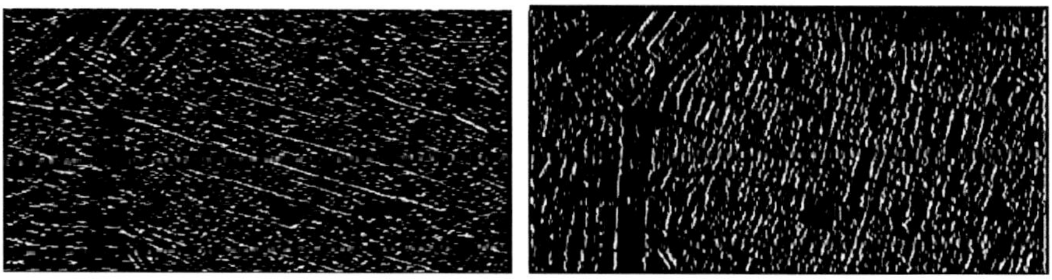

Fig. 12.4 Results of the directional filtering applied to the image SAR image in Fig. 12.3

As noted, in Fig. 12.3 roads appear with only two orientations (horizontal and vertical), but this can't be considered as a general statement. There could be portions of a city where the predominant orientations differ greatly by moving a few kilometers away and then change again in another area and so on. Generally speaking, this calls for a further adaptive filtering procedure. As a matter of fact, we don't need only a flexible filtering chain capable of adapting to different sensors and spatial resolutions, as suggested above, but also a chain capable of taking into account different orientations for the features to be extracted. The technique we briefly outline in the following paragraph is better explained in Dell'Acqua *et al.* (2005).

The first step is simply the extraction of the linear features using a coarse version of the image and no filtering at all. In case of a road network, that would for instance result into the extraction of the major roads leaving out the rest of the details. Then, for each histogram of feature orientations the two dominant directions are extracted. Please note that only two directions are considered on the basis of the assumption that in urban areas roads are locally oriented following two main directions, because of the building outlines, which tends to be rectangular. The whole approach for the adaptive directional filtering is outlined in Fig. 12.5.

One critical parameter of the above procedure is the size of the window used to compute orientation histograms, which naturally depends in a critical manner on the resolution. The rule of thumb is that it must be large enough to accommodate a significant number of segments, but not too much to be still able to discern the two predominant directions.

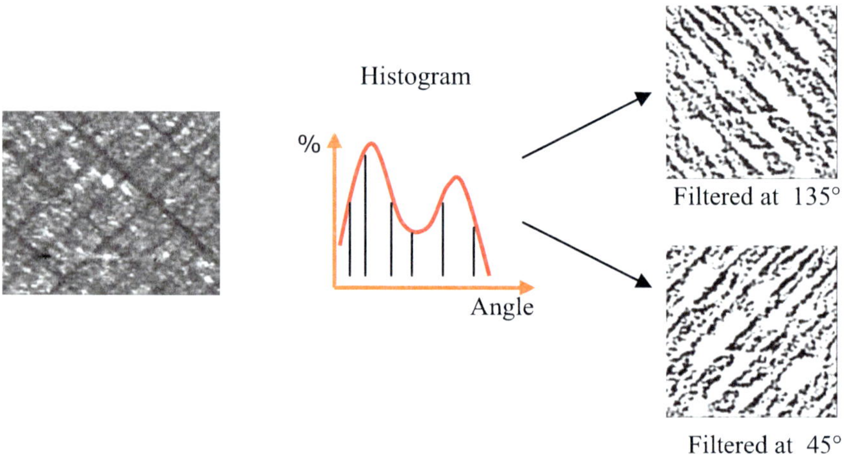

Fig. 12.5 An example of input and outputs to the adaptive directional filtering routine

Once the map of the dominant directions is computed, the entire image can be filtered, locally adapting the directional filter according to this map. Of course, since two directions are considered, two separate images are obtained, whose further analysis results still have to be somehow fused and combined.

12.4.2 Extraction Routine

The step following filtering is the linear feature extraction. In this work we rely on a simple yet efficient extraction method proposed in Lisini (2002) and used in Dell'Acqua *et al.* (2003), Dell'Acqua *et al.* (2005), Dell'Acqua *et al.* (2002),

Dell'Acqua *et al.* (2003b), Dell'Acqua *et al.* (2004). Again, in the following we summarize very briefly the approach, pointing the interested readers to the above-mentioned references.

This method was originally designed to provide a spatially pre-filtered version of the original data, where groups of pixels that can represent (portion of) roads were selected, while those with very different geometrical characteristics were discarded. The aim was to increase the reliability of a subsequent skeletonization routine and at the same time to reduce significantly the computational time. But beyond this role, this routine can be also considered as an extractor of linear features by itself. Basically, the approach compares groups of interconnected pixels with a set of linear prototypes. To avoid memory storage problems, the algorithm scales the area of analysis according to the pattern to be analyzed, choosing an optimal window (both in width and location) to characterize the part of the data set under test. The overall structure of the routine is proposed in Fig. 12.6, while a more specific explanation of the method is proposed in next paragraphs.

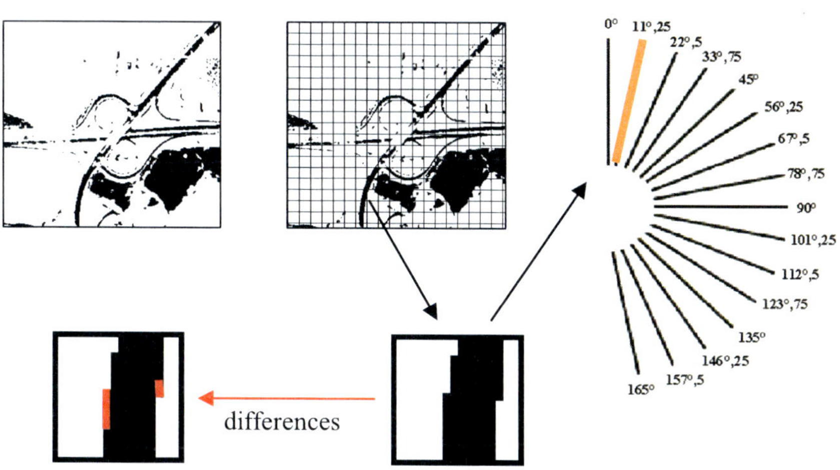

Fig. 12.6 Graphical representation of the main processing steps of the linear feature extraction procedure

More specifically, the first step of the routine is aimed at the identification of the regions of interest, where linear features (segments) may be present. To this end, the image is partitioned into squared zones, and the initial size of these zones is chosen in an arbitrary manner. Each window is then moved to achieve a higher concentration of black (segment) pixel at the centre and less on the edges so as to place the hypothetical linear feature as close as possible to the centre of the floating window. Subsequently, each of these regions is widened to reach the best scale, which is when there is a low percentage of black pixels within the window. The idea is that linear features are elongated and narrow, and are surrounded by many non-segment pixels. If in the area under test it is not possible to lower the segment pixel percentage below a certain threshold, the window is discarded along with the region contained therein.

To further reduce the set of regions to be considered for the subsequent analysis, we recall that a segment is made by a group of pixels surrounded by both sides by non-segment pixels. Regions that do not match this definition are discarded. Such verification consists of a double diagonal window scan while counting the number of switch between segment and non-segment areas.

Finally, when a candidate segment area has been identified, it is compared with a set of 16 basic prototypes (see again Fig. 12.6). The prototype is shifted from left to right and the maximum correlation value is used to obtain the best prototype and its location within the window under test.

Then, for each extracted segment the routine tries and explores if the segment is longer than what has been extracted, by moving the analysis window in the direction identified by the previous analysis. This movement continues until the end of the linear feature is reached or when the shape changes in a way that it does not fulfill the requirements of the above mentioned test. As a consequence, curvilinear features are tracked as a series of linear elements, and the tolerance to which the curve is tracked is a function of the parameters of the extraction routine (Fig. 12.7).

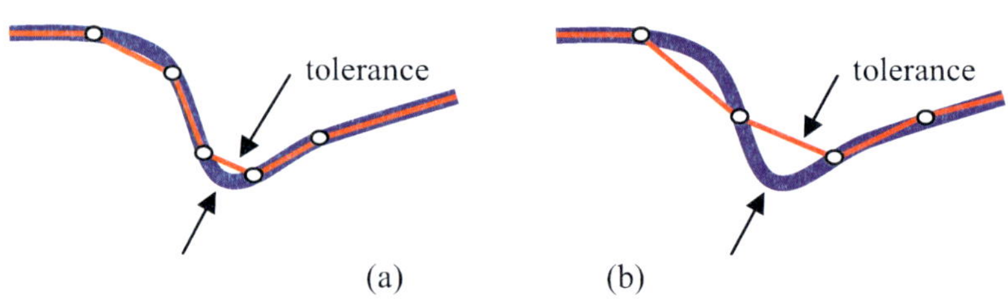

Fig. 12.7 Linear approximation of a curvilinear feature

12.4.3 Post-Extraction Refinement

The last step of the generalized linear feature extraction procedure described in Fig. 12.2 is a post-extraction refinement, which can be based for instance on techniques known as perceptual grouping. The procedure we recall here, proposed and more detailed in Dell'Acqua *et al.* (2005), has two purposes: on the one hand it removes partially or fully overlapping features, replacing them with a single segment; on the other hand it tries and connect segments whose extremes are close to each other. There are six steps to the final result, exemplified in the following figures.

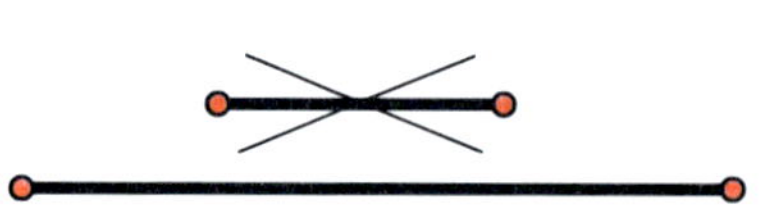

The first step performs a search on pairs of segments that are side by side. If their distance is shorter than a given percentage of the longer segment the shorter one is discarded.

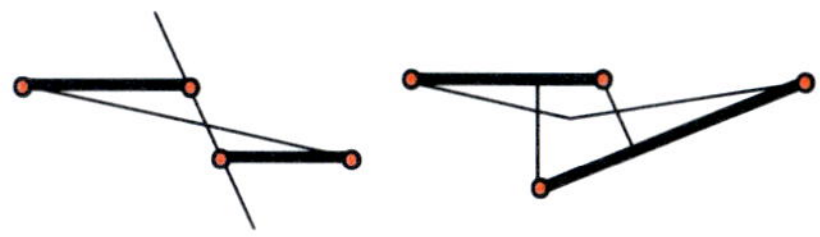

If a pair of segments has similar orientations and the segments lie close enough, they are joined together to form the chain in green.

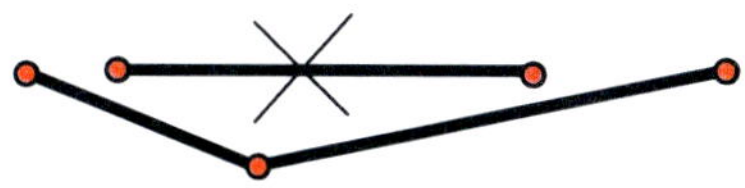

The third step requires a search involving isolated segments located side-by-side with a segment chain as the one created by the previous step. If the distance of an isolated segment from the chain is small enough, the former is removed.

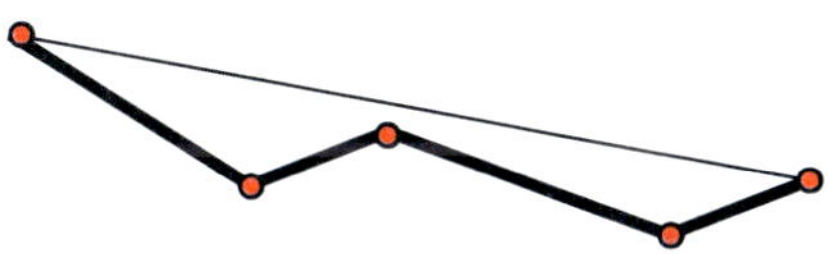

Chains of segments can then be approximated by a single segment where they meet a standard tolerance set by the user. The maximum distance between chain segments and the new segment should not exceed that tolerance.

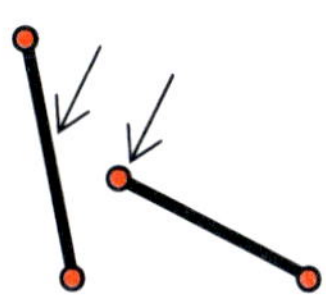

This step tests the proximity between one extreme of a segment and the point of intersection with a second one. If this distance is below a threshold, then the two segments are connected.

As a final step, small isolated segments, which do not exceed a predetermined length are removed.

In summary, it is useful to note that the parameters involved in this process are:

- maximum distance between two extremes of segments that can be interconnected;
- gap length between the closest extreme of a segment and the point of intersection with a second one;
- maximum difference in orientation between segments candidates to be interconnected;
- maximum distance between a segment in a chain of segments and the segment candidate to replace the chain;
- minimum length below which segments are removed.

Fig. 12.8 Results of the linear feature extraction chain before and after the post-extraction alignment processing step

In Fig. 12.8 two sets of segments before and after the post-extraction alignment are showed. It is clear that this procedure has a significant effect on the visual clarity of the results.

12.5 Higher-Level Feature Extraction

In the process of visual interpretation of aerial or satellite imagery, the usual approach is a top-bottom procedure. Starting from the scene, the operator partitions it into different subscenes, and focuses his/her attention on the elements of these subscenes, for a more reliable and faster data analysis. Instead, automatic and semi-automatic approaches, such as usual classification algorithms, are based on the extraction of elementary features that they use to achieve, in a bottom-up approach, the interpretation of the scene.

Linear feature extraction somehow pertains to this second class, since it is assumed that these elements of the scene are extracted earlier than any attempt to interpret them or to assign them to a class. Linear features are just elements of the scene, and their spatial relationships or clustering might indicate something about the different parts of the scene, but this is not captured by the linear features alone. As a consequence, a data interpretation approach based on linear features must consider other processing steps after the segment extraction procedure which has been proposed in the previous section. It is mandatory to try and combine these segments into more meaningful features, that we call here "higher-level" features on the basis of the assumption that this is a bottom-up approach. Usually, these new features still do not provide enough information to interpret the data, and further steps will be required. They are however a very interesting intermediate result, and they will be therefore discussed in this section, while applications will be the topic for section 4 of this chapter.

12.5.1 Junction Extractor

According to the above discussion, a first algorithm for the extraction of more significant spatial features may be one devoted to characterizing segment junctions, i.e. groups of two or more segments with specific spatial relationships based on proximity and geometrical constraints.

The first idea to look for segment junction is to start from any pair of linear features and check their actual or possible intersection points. Then, points falling within or close enough to the extracted segments are retained. Of course, in order to correctly evaluate the junctions, it is mandatory to define a number of thresholds and values, as follows:

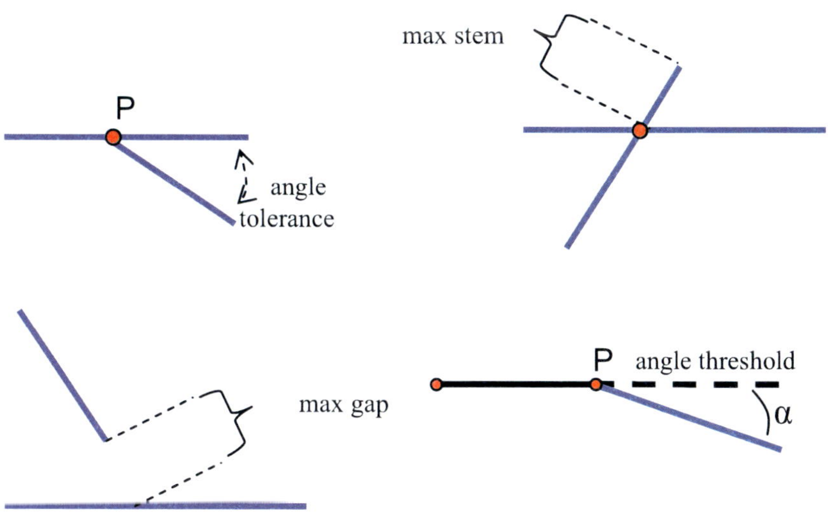

Fig. 12.9 Graphical definition of the parameters introduced in the text

1. ***angle threshold***: the minimum angle between two roads to consider them as separate and not belonging to a single segment. If the angle α (see Fig. 12.9) is smaller than this threshold, the two lines are considered as a single one. If it is greater than or equal to the threshold, the point P is a junction.
2. ***minimum stem***: maximum possible length of the part of a segment that "passes beyond" the junction, i.e. the distance between the junction point and the extreme of a segment that after crossing the junction continues for a while.

3. ***maximum gap***: maximum possible distance between the end of a segment not intersecting the junction and computed junction point, i.e. the distance between the junction point and the closest extreme of a segment that does not "hit" the junction.
4. ***minimum junction distance***: minimum distance between two junctions;
5. ***angle tolerance***: minimum possible angle between two non-overlapping segments (see again Fig. 12.9). If this angle is less than the threshold the segments are considered as overlapping, if the angle is greater than or equal to the threshold P is considered a junction.
6. ***max # of segments***: maximum number of segments forming a junction.

According to these definitions, a few classes of junction may be considered. For instance, L-shaped junctions are made up of two connected (within the max gap tolerance) segments. The angle formed by these segments is 90 degrees plus or minus the angle tolerance. The outer portion of the intersecting segments should be less than the max stem parameter. Similarly, and limiting ourselves to constraints on segment relationships, X-shaped junctions are those where the intersection point falls within both segments, and the outer portions are longer than the max stem for both segments; T-shaped junctions are formed by two segments and only for one of them the outer portion is less then the max stem threshold; finally, Y-shaped junctions are formed by three segments. Some synthetic examples of different junctions which can be recognized and classified by this approach are shown in Fig. 12.10.

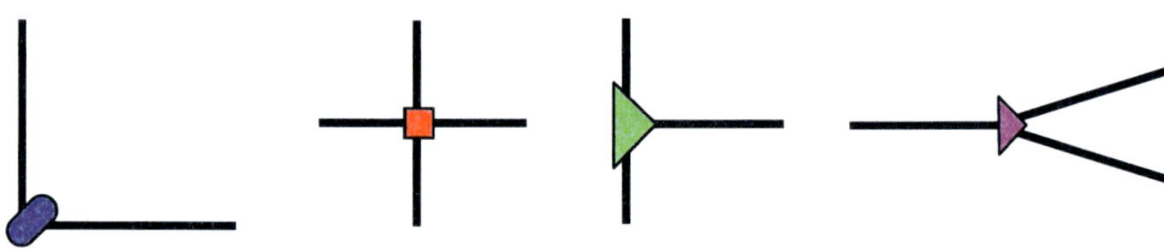

Fig. 12.10 Examples of L-, X-, T- and Y-shaped junctions.

12.5.2 Pairs of Parallel Segments

Another particularly significant combination of linear features is a pair of parallel segments. A very simple algorithm to detect these pairs involves again the search for candidate pairs that, to some extent, resemble perfectly parallel pairs of equal length segments.

As a matter of fact, to match this ideal situation with real segment pair there should be a tolerance on the orientation of the two segments and a minimum superposition between them to allow calling them a "parallel pair". Figure 12.11 graphically depicts these quantities and how they are computed and compared with the tolerances, i.e. d/D as the percentage of the shorter segment that the longer one overlaps, and the angle between the two segment orientations, that must be equal to zero within the above mentioned angle tolerance.

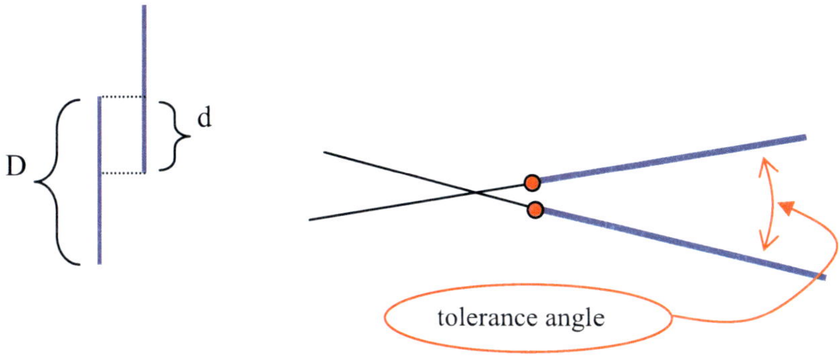

Fig. 12.11 Graphical definition of the parameters required for parallel segment pairs' identification

12.6 Application Examples

Once the higher-level features have been extracted, the only remaining step is truly image interpretation. Of course, a priori information about the scene (or the portion of the scene) under test is required at this point, and should be injected into the procedure. In the following, three different examples of smart use of the higher level features are proposed. They show a very small sample of the possibilities offered by these techniques, but can hopefully provide some sort of guidelines to apply the same idea to different problems and interpretation tasks.

12.6.1 Example 1: Building Characterization

A first example of higher-level feature extraction starting from linear features is the recognition of man-made (artificial) objects in urban areas. To this aim, the idea is to design an algorithm to find, within a segment set, specific combinations suggesting the existence of rectangular structures. The routine (presented in more detail in Lisini *et al.* 2005) is divided into two parts: the first one performs a search for rectangles as combinations of parallel segment pairs, the second one performs the same search looking for combinations of "L-shaped" junctions.

The parameters of the first part of the routine are basically only two: a tolerance on the orthogonality and the proximity between the two parallel segment pairs. The first one is computed as the difference between 90° and the angles obtained by assembling using two parallel pairs a quadrilateral figure, while the tolerance about distances is computed as the maximum distance between all the extremes of the four segments involved. Fig. 12.12 explains some of these concepts, and serves as an example for the description of the procedure, which is as follows:

- for each pair of parallel segments the mean direction is computed;
- then, according to the required tolerances, candidate couples of parallel pairs are analyzed to check for those with higher similarity to orthogonal pairs;
- for each combination that matches the orthogonality requirements, the distance tolerance is checked and, in case, the set of pairs is validated.

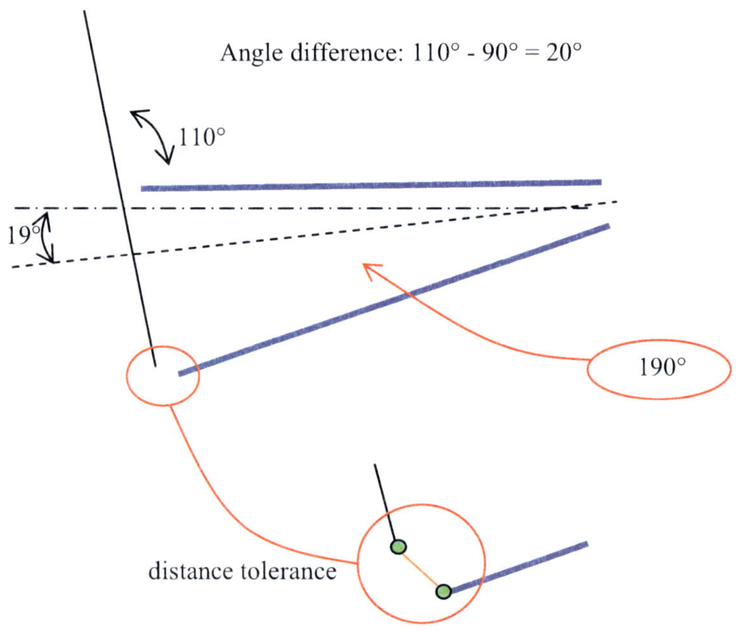

Fig. 12.12 Tolerances used for the evaluation of parallelogram instances starting from pairs of (quasi)parallel segments

The second procedure for identifying rectangles is based on junctions. First of all, each junction that is not an "L-shaped" one is transformed into one or more of them, simply by separating the contributions as exemplified in Fig. 12.13.

The complete procedure is as follows:

- for each L-shaped junction the edge orientations is analyzed;
- each junction is considered as a "promoter" of a group of 2, 3 or possibly 4 corners, but only some specific patterns are considered valid, as shown in Fig. 12.14;
- only corners that verify the alignment between their edges are considered as useful;
- each group is finally checked for proximity and an overall proximity value for each group is computed;

- finally, each junction is assigned to the group (if any) with the lowest proximity value.

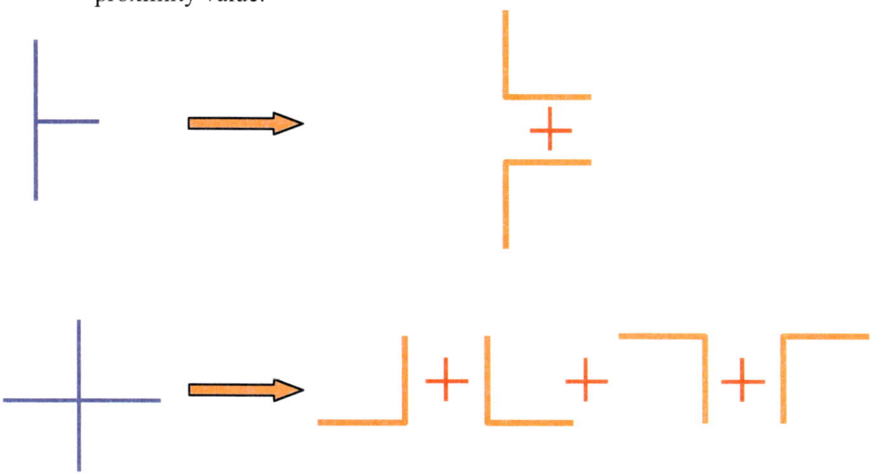

Fig. 12.13 Separation of more complex junctions into "L-shaped" ones

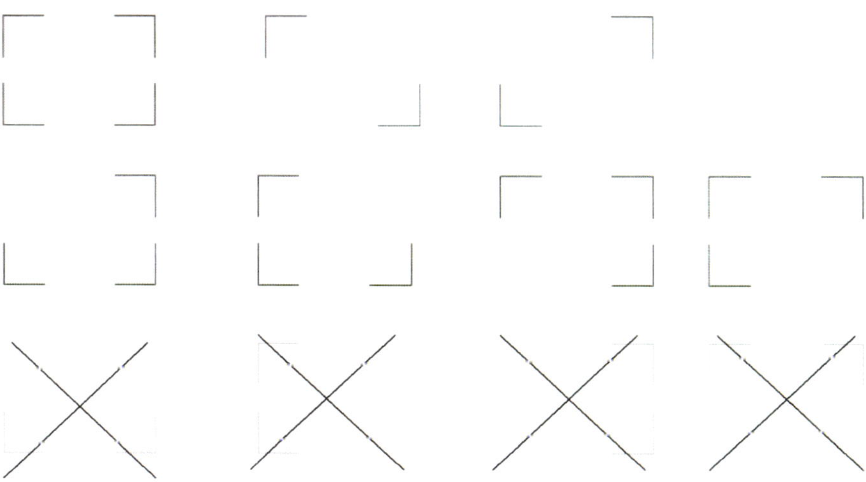

Fig. 12.14 L-shaped junctions' patterns accepted or discarded for rectangle recognition

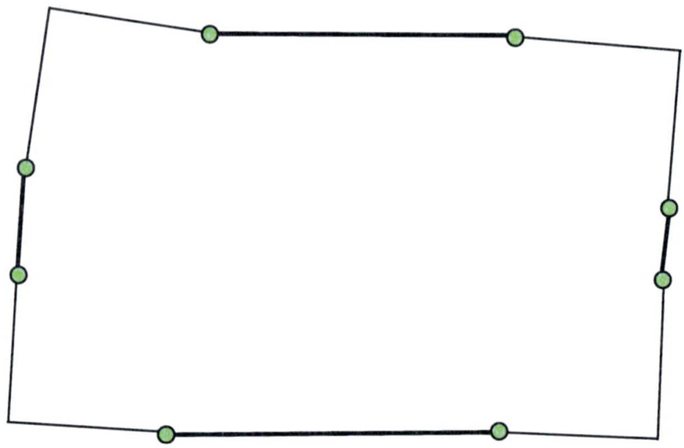

Fig. 12.15 Example of successful rectangle extraction from synthetic data

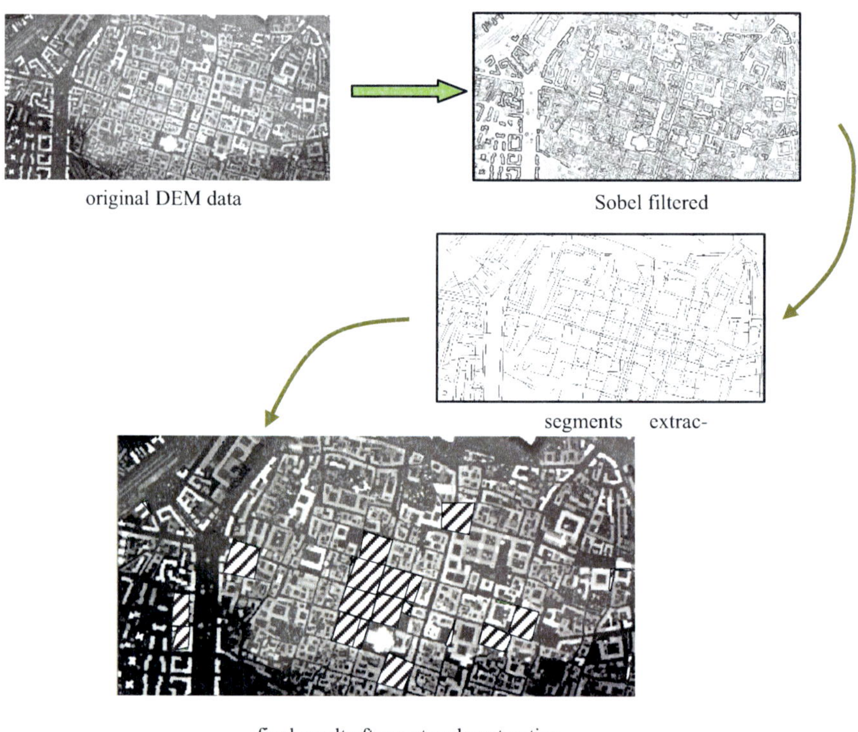

Fig. 12.16 Examples of building extraction from LIDAR imagery based on L-shaped junction combination.

In Fig. 12.15 a synthetic data example is provided, while in Fig. 12.16 an example of rectangle extraction from a LIDAR data set is shown. From the example it is clear that only a few of the building shapes can be obtained using this method, because it greatly relies on the previous feature extraction and corner recognition steps. Indeed, as mentioned above, the quantity and reliability of extracted rectangles is largely due to the inputs, and these in turn depends on the linear features provided by the extractors. The proposed procedure is robust to orientation problems and partially also to mismatches, but needs a certain extracted information to elaborate on.

12.6.2 Example 2: Automatic Coregistration of SAR Images

A further application of the geometric structures explored in the preceding paragraphs is automatic coregistration of multiple SAR images of urban area. It is well-known that one of the biggest problems encountered in SAR images of urban areas is determined by a strong layover effect that significantly changes distances between buildings in data sets acquired with different viewing angles. To compare these data, it is therefore needed a coregistration approach for these images which takes into account the huge distortions due to the fine spatial resolution. The basic methodologies used for optical images, i.e. correlation techniques, even when improved to be robust to misregistration noise (Bruzzone and Fernandez-Prieto 2000, Billimgsley 1982, Townshend *et al.* 1992, Dai and Khorram 1998), cannot be used to identify corresponding points since the same object on the scene may give rise to different responses in terms of backscattered field.

The approach explored in this section is therefore to derive junction from all images to be coregistered (see next figure). It will be through the comparison of these features in different images of the same area that the routine will try and find the points (ground control points, GCPs) to be used for coregistration.

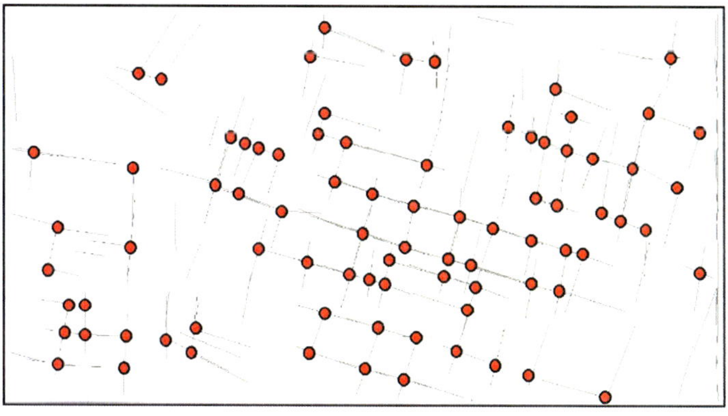

Fig. 12.17 Example of junctions extracted from the VHR SAR image in Fig. 12.3 and useful as input to the coregistration procedure

To this aim, after applying the procedure proposed in the previous sections and extracting all the junctions in the images to be coregistered, a methodology to find matches between junctions in different images has to be developed. One possibility is the approach proposed in Dell'Acqua *et al.* (2004), which is briefly recalled here. The first step in that algorithm is to look into the same location of a junction extracted from one image to seek if one or more junctions are present in the second one. This search is performed in a circle of radius R around the original position in order to take into account misalignments, in VHR SAR images largely due to the layover effects. Once all the junctions in this area have been identified, their orientation and type (L-, X-, T- shaped, ...) is considered. This is done in order to reduce the number of candidates. In fact, it is to be noted that the mutual segment orientations in a junction are invariant with respect to the point of view and so is the type of the junction.

To better identify corresponding junctions, they are identified by means of the list of angles between segments, reordered from the largest to the smallest one. Lists representing different junctions are easily compared, immediately discarding junctions with different number of segments (and therefore angles) and very different angle distributions. The degree of similarity between two junctions is obtained by comparing one by one the elements of the lists and increasing the similarity value if they are found equal or very similar according to the angle tolerance in use. In case of more junctions with the same similarity a further refinement is computed, looking for the lowest absolute difference between corresponding angles in the list. Another element for discrimination is the closeness of the candidate junction to the position of the one we are trying to match.

Fig. 12.18 shows a graphical example of the circle of validity, the ordering procedure for the list of angles corresponding to a junction and the final result of the matching procedure.

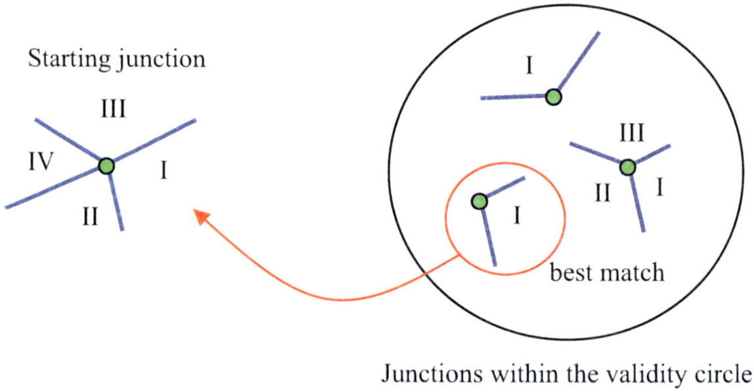

Fig. 12.18 Graphical representation of the junction matching procedure

As stated above, this approach has been discussed in Dell'Acqua *et al.* (2004) and applied to multiple SAR image coregistration with satisfying results. It has been proven that the RMS positional error using the GCPs established by this algorithm has decreased to 6.51 pixels with respect to manual extraction of GCPs through visual interpretation of an expert operator, i.e. slightly more than 8 pixels.

12.6.3 Example 3: Change Detection

A different scenario for using linear features as well as higher-level features is related to change detection, i.e. the interpretation of multitemporal data, as proposed in Dell'Acqua *et al.* (2006). According to this reference, linear features may be used to focus the analysis on the relevant changes occurred in the area under test. In Fig. 12.19 the general outline of a possible procedure is shown. This technique is designed to validate a pixel-based map through the use of the linear features that are in the image sequence, composed at least by two images.

The routine starts form the extraction of linear features from these images and then continues with a comparison of the features extracted in order to match them or to detect changes. A change map based on linear feature is thus obtained, and used to validate and improve the one achieved by more common pixel-based analysis.

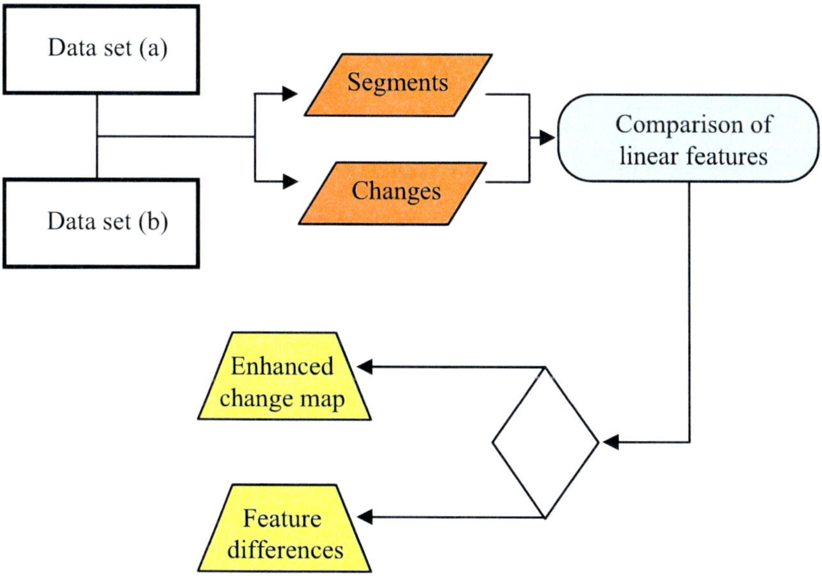

Fig. 12.19 Overall scheme of a joint pixel-based and feature-based change detection algorithm

Two main problems are encountered in this process. First of all, changes in linear features should reflect real changes, which in turn demands high reliability of the linear feature extraction routine.

The second problem is the location mismatch between linear features, due to misregistration between pre and post images and/or to different viewing points. Linear features have to be matched using a robust approach, which can be implemented, for instance, by means of spatial buffers. More specifically, once a segment has been extracted, the corresponding one in the second (or third) image is searched in a buffer around its position, in order to accommodate for the above-mentioned misregistrations. Moreover, since slight differences in orientations may be present due to residual rotational mismatches and extraction errors, the match is considered valid if the percentage of the second segment falling within the buffer around the first one is higher than a given threshold (e.g. 80%). Using segment match, a segment-based change detection can be set up. Although it refers only to the areas where reliable segments were extracted (which might correspond only to a small part of the whole scene), in these areas there is a possibility to fuse the changes obtained by directly comparing linear features and those, independently obtained, coming from a per-pixel analysis of the pre and post imagery. As a matter of fact, the above-mentioned linear feature matching procedure can output an image formed by all the unmatched segments, i.e. those corresponding to a change. The feature-based change map will be therefore the area covered by these segments plus the buffer around them, already exploited to find if a match was possible.

Finally, in order to merge together the two change maps a logical AND is used, which basically means that the final change map will present areas of change only if both inputs (pixels and linear features) detect such change. The approach is graphically shown in Fig. 12.20.

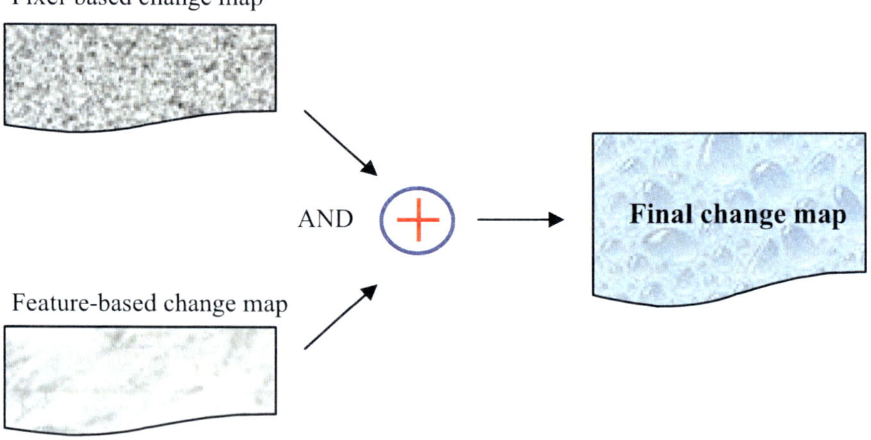

Fig. 12.20 Fusion procedure for the pixel-based and feature-based change maps.

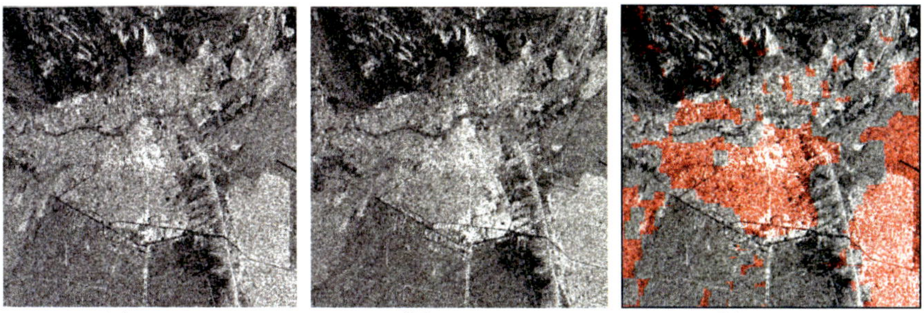

(a) (b) (c)

Fig. 12.21 Example of pre- and post-event SAR images over the Bam (Iran) area and the extracted change area validated using linear features (from Tsin *et al.*, 2006.).

As an example of the results of the procedure, Fig. 12.21 shows two SAR images of the area surrounding the city of Bam (Iran) before (a) and after (b) the 2003 earthquake. The images have been processed in order to extract relevant linear features and match those appearing on both images. Then, a very simple pixel-based change detection has been performed by computing pixel-wise the ratio of the two images, followed by a threshold to detect areas where strong changes took place. The feature-based and the pixel-based change map have been then combined to obtain the improved change map shown in Fig. 12.21(c).

12.7 Conclusions

Linear features are a very simple but really interesting way of interpreting a scene. They have been used for long time in photogrammetry and in automatic scene analysis for VHR optical images. They have however a more general usefulness in the context of remotely sensed data, even in situations where they have not been considered useful, like for SAR images. From the researches summarized in this chapter it can derived a certain ability to exploit SAR data using tools traditionally considered for optical images by means of improved extraction routines and modified radar data characterization.

However, the main aim of this chapter, after a brief survey of the technical literature on this subject, was to provide the interested reader with a general framework, with a wide applicability to different input data sets and enough flexibility to offer (partial) solutions to different problems, from classification to co-registration to change detection. None of the techniques shown in the previous sections is considered as “the best one” in absolute meaning. They all offer, however, some insight on the usability of linear features in many different application fields. Moreover, they all fit into the general framework introduced and discussed in Section 3.

Of course each of them can be improved, and the framework can be more precisely specified. For example, most of the processing steps proposed in sections 4 and 5 can be replaced by other, possibly improved, ones, and we do hope that someone in the research community will be able to do so. In that case, the procedures introduced to solve practical situations in section 6 might be more effective and provide even better results.

References

P., Agouris, P. Doucette, A. Stefanidis, "Spatiospectral Cluster Analysis of Elongated Regions in Aerial Imagery", Proc. of 2001 Int. Conf. on Image Processing, vol. II, pp. 789-792.

S. Airault, R. Ruskone, O. Jamet, "Road detection from aerial images: a cooperation between local and global methods", Image and Signal Processing for Remote Sensing, Satellite Remote Sensing I, SPIE, vol. 2315, pp. 508- 18, Rome, Italy, 1994.

J. Amini, M.R. Saradjian, J.A.R. Blais, C. Lucas, and A. Azizi, "Automatic road-side extraction from large scale imagemaps", International Journal of Applied Earth Observation and Geoinformation, Vol. 4, no. 2, pp. 95-107, 2002.

D. Barzohar, B. Cooper, "Automatic finding of main roads in aerial images by using Geometric Stochastic models and estimation", IEEE trans. PAMI, vol.18, no.7, pp. 707-721, July, 1996.

J.A. Benediktsson, M. Pesaresi, K. Amason, "Classification and feature extraction for remote sensing images from urban areas based on morphological transformations", IEEE Transactions on Geoscience and Remote Sensing, Vol. 41, no. 9, 1940–1949, Sept. 2003.

F.C. Billingsley, "Modeling misregistration and related effects on multispectral classification," Photogramm. Eng. Remote Sensing, vol. 48, pp. 421-430, 1982.

L. Bruzzone, D. Fernandez Prieto, "Automatic Analysis of the Difference Image for Unsupervised Change Detection," IEEE Trans. Geoscience and Remote Sensing, vol. 38, no. 3, pp. 1171-1182, May 2000.

R.J. Dekker, "Texture analysis and classification of ERS SAR images for map updating of urban areas in The Netherlands", IEEE Transactions on Geoscience and Remote Sensing Vol. 41, no. 9, pp. 1950–1958, 2003.

X. Dai and S. Khorram, "The Effects of Image Misregistration on the Accuracy of Remotely Sensed Change Detection," IEEE Trans. Geoscience and Remote Sensing, vol. 36, n. 5, pp. 1566-1577, Sept. 1998.

F. Dell'Acqua, P. Gamba: "Evaluation of COSMO/Skymed SAR data for urban characterisation," Proc. of the IEEE/ISPRS Joint workshop on Remote Sensing and Data Fusion over Urban Areas, Rome, Italy, 8-9 Nov. 2001, pp. 141-145.

F. Dell'Acqua, P. Gamba, G. Lisini, "Extraction and fusion of street networks from fine resolution SAR data", Geoscience and Remote Sensing Symposium, 2002. IGARSS '02. 2002 IEEE International , Vol. 1, 24-28 June 2002 pp. 89-91.

F. Dell'Acqua, P. Gamba, A. Iodice, G. Lisini, D. Riccio, G. Ruello, "Simulation and analysis of fine resolution SAR images in urban areas", 2nd GRSS/ISPRS Joint Workshop on Remote Sensing and Data Fusion over Urban Areas, Berlin (Germany), 22-23 May 2003, pp. 133-136.

F. Dell'Acqua, P. Gamba, G. Lisini, "Road map extraction by multiple detectors in fine spatial resolution SAR data", Canadian Journal of Remote Sensing, Vol. 29, n. 4, pp. 481-490, Aug. 2003.

F. Dell'Acqua, P. Gamba, G. Lisini, "Improvements to urban area characterization using multitemporal and multiangle SAR images", IEEE Transactions on Geoscience and Remote Sensing, Vol. 41 no. 9, pp. 1996-2004, Sept. 2003.

F. Dell'Acqua, P. Gamba, G. Lisini, "Co-registration of multi-angle fine spatial resolution SAR images", IEEE Geoscience and Remote Sensing Letters, Vol. 1, n. 4, pp. 237-241, Oct. 2004.

F. Dell'Acqua, P. Gamba, G. Lisini, "Road extraction aided by adaptive directional filtering and template matching", Proc. of URBAN 2005, Tempe AZ, USA 14-16 March 2005.

F. Dell'Acqua, P. Gamba, G. Lisini: "Change Detection of Multi-Temporal SAR Data in Urban Areas Combining Feature-Based and Pixel-Based Techniques", IEEE Trans. on Geoscience and Remote Sensing, vol. 44, n.10, pp. 2820-2827, Oct. 2006.

D. German, B. Jedynak, "An active testing model for tracking roads in satellite images", IEEE Trans. PAMI, vol.18, no.1,pp 1-14, January 1996.

W.-D. Groch, "Extraction of line shaped objects from aerial images using a special operator to analyze the profiles of functions", Computer Graphics and Image Processing, Vol. 18, no. 4, pp. 347-358, 1982.

A.W. Greun, H, Li, "Linear Feature Extraction with 3-D LSB Snakes", Proc. SPIE, Vol. 3072, pp. 119-130, 1997.

F.M. Henderson, and Z.G. Xia: "SAR applications in human settlement detection, population estimation and urban land use pattern analysis: a status report," IEEE Trans. on Geoscience and Remote Sensing, Vol. 35, No. 1, pp. 79-85, 1997.

X. Huang, L. Zhang, and P. Li, "Classification of Very High Spatial Resolution Imagery Based on the Fusion of Edge and Multispectral Information", Photogrammetric Eng. Remote Sens., Vol. 74, no. 11, pp. 1585-1596, 2009.

F. Jung, N. Paparoditis, "Extracting 3D free-form surface boundaries of man-made objects from multiple calibrated images: a robust, accurate and high resolving power edgel matching and chaining approach", ISPRS Archives, Vol. XXXIV, Part 3/W8, Munich, 17.-19. Sept. 2003, pp. 39-44.

T.J.Kim, J.P.A Muller,. "Development of a graph-based approach for building detection", Image and Vision Computing , vol. 17, no. 1, pp. 3-14, 1999.

Z. Kim, A. Huertas, R. Nevatia, "Automatic Description of Buildings with Complex Rooftops from Multiple Images", Proc. of 2001 Conference on Computer Vision Pattern Recog, vol. II, pp. 272-279.

S.P. Kozaitis, R.H. Cofer, "Linear Feature Detection Using Multiresolution Wavelet Filters", Photogrammetric Eng. Remote Sens, Vol. 71, no. 6, pp. 689-698, 2005.

S. Krishnamachari, R. Chellappa, "Delineating Buildings by Grouping Lines with MRFs", IEEE Trans. Image Procesing, vol. 5, no. 1, pp. 164-168, 1996.

S.C. Lee, S.K. Jung, R. Nevatia, "Integrating ground and aerial views for urban site modeling", Proc. of 2002 Intern. Conf. on Pattern Recognition vol. IV, pp. 107-112.

G. Lisini, "Tecniche di estrazione e fusione di reti stradali su immagini telerilevate (in Italian)", Tesi di Laurea, Mar. 2002, Laboratorio di Telecomunicazioni, Università di Pavia.

G. Lisini, P. Gamba, F. Dell'Acqua, G. Trianni, W. Tompkinson, "Image interpretation through problem segmentation for very high resolution data", Proc of IGARSS'05, Seoul (Korea), 25-29 July.

W.A. Mackaness, G.A. Mackechnie, "Automating the Detection and Simplification of Junctions in Road Networks", GeoInformatica, vol. 3, no. 2, pp. 185-200, 1999.

N. Merlet, J.B. Zerubia, "New Prospects in Line Detection by Dynamic-Programming", IEEE Trans. Pattern Anal. Machine Intell., Vol. 18, no. 4, pp. 426-431, 1996.

N.S. Netanyahu, V. Philomin, A.J. Stromberg, "Robust Detection of Road Segments in Noisy Aerial Images", Proc. of 1996 Int. Conference on Pattern Recognition, vol. II, pp. 151-155.

N.S. Netanyahu, V. Philomin, A. Rosenfeld, A.J. Stromberg, "Robust Detection of Straight and Circular Road Segments in Noisy Aerial Images", Pattern Reconition, vol. 30, no. 10, pp. 1673-1686, 1997.

R. Nevatia, K.R. Babu, "Linear Feature Extraction and Description", Computer Graphics Image Processing, vol. 13, no. 3, pp. 257-269, 1980.

V.P. Onana, E. Trouve, G. Mauris, J.P. Rudant, E. Tonye, "Thin linear features extraction in SAR images by fusion of amplitude and coherence information", Proc. of the IEEE 2001 International Geoscience and Remote Sensing Symposium. Vol. 7, pp. 3012–3014.

S. R. Park, T. Kim, "Semi-Automatic road extraction algorithm from IKONOS images using template matching", Proc. of the 22nd Asian Conference on Remote Sensing, pp 1209-1213, 2001.

K.E. Price, "Road Grid Extraction and Verification", Proc. of 1999 ISPRS Congress, pp. 101-106.

W.Z. Shi, C.Q. Zhu, "The line segment match method for extracting road network from high-resolution satellite images", IEEE Transactions on Geoscience and Remote Sensing , Vol. 40, no. 2, pp. 511-514, 2002.

M. Tavakoli, A. Rosenfeld, "Building and Road Extraction from Aerial Photographs", IEEE Trans. Systems, man, Cybernetics, vol. 12, no. 1, pp. 84-91, 1982.

J.R.G. Townshend, C.O. Justice, C. Gurney, and J. McManus, "The impact of misregistration on change detection," IEEE Trans. Geoscience and Remote Sensing, vol. 30, pp. 1054-1060, Sept. 1992.

Yanghai Tsin, Sing Bing Kang, R. Szeliski, "Stereo matching with linear superposition of layers", IEEE Transactions on Pattern Analysis and Machine Intelligence, Volume 28, no. 2, pp. 290-301, Feb. 2006.

H. Xin, Z. Liangpei, L. Pingxiang, "Classification and Extraction of Spatial Features in Urban Areas Using High-Resolution Multispectral Imagery", IEEE Geoscience and Remote Sensing Letters, Vo. 4, no. 2, pp. 260-264, April 2007.

A. Zlotnick, P.D. Carnine, "Finding Road Seeds in Aerial Images", Computer Viion Graphics Image Processing, vol. 57, no. 2, pp. 243-260, 1993.

Chapter 13

GEOSPATIAL SERVICE WEB

Jianya Gong, Huayi Wu, Wenxiu Gao, Peng Yue, Xinyan Zhu

13.1 Introduction

13.1.1 Web and Services

The Web has become a platform where data, information and knowledge can be published, discovered and retrieved. The growth of the Web has resulted in the Web-based sharing of distributed resources. The trend to enhance the collaboration, creativity and rich user experiences has led the evolution of Web to the so-called Web 2.0, which is represented by a set of typical applications including blog, wiki, social networking (O'Reilly 2005). Another trend to increase the intelligence of the Web is the effort to move the Web to a universal medium for data, information and knowledge exchange, i.e. the Semantic Web, in which the semantics of information on the Web are defined, making it possible for the Web to understand (Berners et al. 2001). Grid computing, a blueprint for the computing infrastructure (Foster and Kesselman 2004), provides a set of middleware to back up the processing capabilities of Web.

New information architecture, such as Service-Oriented Architecture (SOA), affects the way that those trends of Web involvement are implemented. The information tools and applications are encapsulated as services which are accessible through standard interfaces and protocols. Through the standards for Web Service, multiple services can be discovered and integrated automatically on the Web. The process and results of integration can be even accessible as new services. The service-oriented applications and their potentials have been claimed to increase individual and collective scientific productivity by making powerful information tools available to scientists, and thus enables the widespread automation of data analysis and computation (Forster 2005).

D. Li et al. (eds.), *Geospatial Technology for Earth Observation*,
DOI 10.1007/978-1-4419-0050-0_13,

13.1.2 Geospatial Domain

With the advancement of Earth Observation (EO) and information technologies, the capabilities for collecting, transferring, processing and sharing geospatial data has dramatically increased in recent years. For example, NASA's Earth Observing System (EOS) alone is generating about 3.5 terabytes of data each day (Yue et al. 2007). Geospatial data are collected by different organizations and archived at globally distributed locations. The exploding number of distributed sensors can also provide real-time or near real-time data. An integrated and distributed Space-Earth information system, therefore, is needed to cover the lifecycle of geospatial application, i.e. from the collection of geospatial data to the end user-oriented applications, thus significantly enhancing the efficient and effective sharing of geospatial data, information and knowledge.

The distinguished features of the Web such as distribution and openness make the Web a promising platform for supporting an integrated and distributed Space-Earth system. There has been lots of Web-based geospatial information sharing and applications. Web Service technologies are already being widely used in geospatial domain for distributed geospatial data sharing, such as the U.S. Department of Energy's Earth System Grid (ESG) (Bernholdt et al. 2005), the U.S. National Science Foundation (NSF) funded GEONGrid (GEON 2003) and the UK e-science program (Hey and Trefethen 2005). The geospatial data obtained from the diverse sources are often incompatible in terms of the temporal and spatial coverage, resolution, format, and map projections. Processing distributed geospatial data for information and knowledge requires interoperability among diverse data resources. This necessity led to the development of a set of standard interfaces for geospatial Web services, and a number of interoperable services have been available in the geospatial community. Most of the services are compliant with the Open Geospatial Consortium (OGC) standards.

13.1.3 Objective of this Chapter

This Chapter is to propose a new concept, Geospatial Service Web (GSW), to umbrella and envision the ultimately seamless integration of all types of distributed geospatial resources, which are currently or will be in the future Web discoverable, accessible, integratible, and "plug-and-play". It provides a framework towards an integrated Space-Earth system through the Web and service technologies. From this perspective, it will also contribute to the implementation of Global Earth Observation System of Systems (GEOSS). The following sections describe the concept and framework of GSW. Major service components in GSW are also introduced. Finally, we address the implementation of GSW.

13.2 Concepts of Geospatial Service Web

Geospatial Service Web (GSW) is a virtual geospatial infrastructure which integrates various geospatial-related resources. GSW unifies the functions of geospatial acquisition system, data transformation system, distributed spatial data collection, high-capability server system, large volume storage system, remote sensing and GIS system. These functions are implemented by web services and communicated through the standardized protocols of the Internet. Fig. 13.1 shows the logical components and structures of GSW. With GSW, it is able to effectively describes, organizes, manages, manipulates, interchanges, searches and releases the geospatial-related resources.

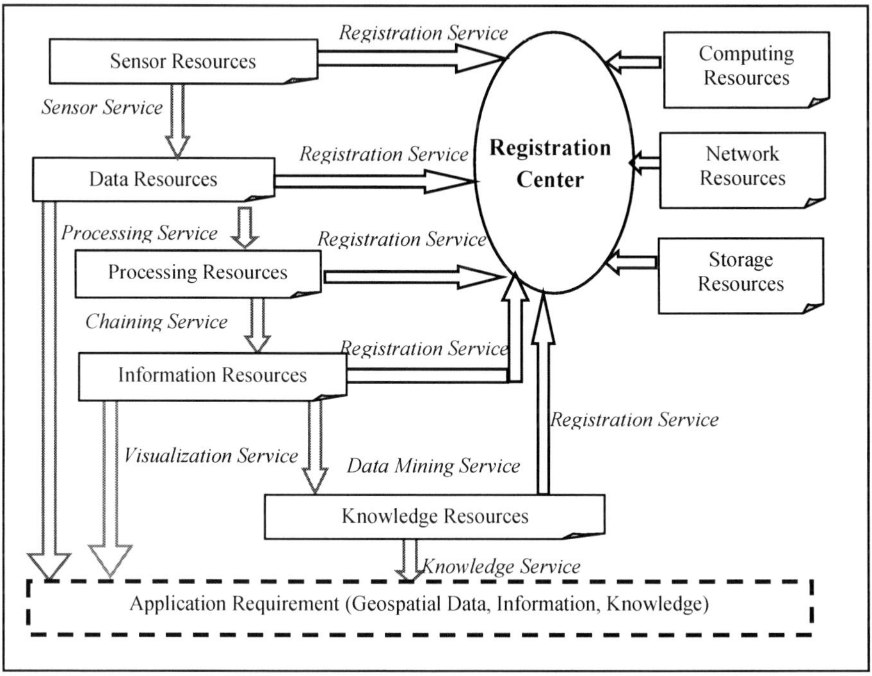

Fig. 13.1 The procedure of geospatial service from sensor to knowledge

A registration center is the core of the virtual infrastructure, which accepts and archives the registration information of all resources. Thus, all of the resources can be acquired and accessed through the center. Sensor resources cover all kind of sensors for data acquisition including space-borne sensors, air-borne sensors and handheld sensors. The original Earth Observation data acquired by sensor resources constructs huge geospatial data resources. Processing resources are the collections of theoretical models, process models and behavior models which are responsible to pre-process, transform, compact, project, generalize, visualize geospatial data for a specific application context. In many cases, these models can be composed orderly into an integrated model to implement a complex function and further to derive potential information and knowledge in datasets. Finally, all of the resources are combined by associated web services and provide services as a whole for geospatial-related applications. In addition to the resources of geospatial domain, the general resources, including computing resources, network resources and storage resources, are also indispensable for GSW, but they are beyond the theme of this chapter.

In conclusion, the mission of GSW is to:

- acquire global geospatial data for all season, all day and all directions by all kinds of sensors on satellite, aircraft and surface.
- chain the whole process seamlessly from sensors to application services by unified information networks, including satellite communicate, data relay network and wired or wireless computer communication networks.
- register sensors, computing resources, storage resources, internet resources, geospatial data and manipulate software, geospatial knowledge on the Internet, and process geospatial data online quantitatively, automatically, intelligently and real-timely.
- provide geospatial services, compose virtual service chains and transmit user-required information with the most effective and efficient ways.

13.3 Framework of Geospatial Service Web

Based on the concept of geospatial service web described in the previous section, Fig. 13.2 illustrates the corresponding framework including five basic components: geospatial resource component, geospatial service component, geospatial service application component, geospatial service security component, and geospatial service standard component.

Geospatial resource components are the cornerstone which involves almost all applicable resources in digital environment. Geospatial service component is the bridge between geospatial resources and geospatial service applications, which provides the functions with service like accessing, processing, transporting, and visualizing data. Some geospatial services can be combined into a specific application system, or be developed into individual application tools, or be used to build a visualization environment for geospatial data. Geospatial standards are fundamental supports for communications and data sharing between heterogeneous components in the framework. Geospatial service security component protects geospatial data and services from illegal usage or attack.

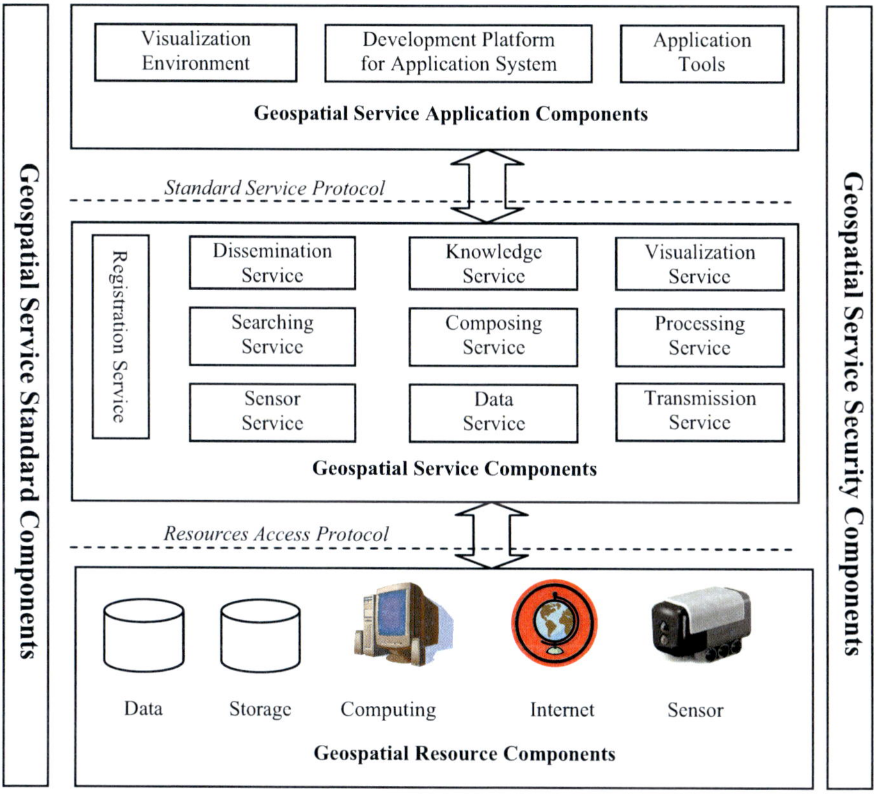

Fig. 13.2 Framework of Geospatial Service Web

13.3.1 Geospatial Resource

The framework of Geospatial Service Web involves various geospatial resources, such as geospatial data, storage, computing, internet and sensor resources. These resources play fundamental roles in the whole framework.

Geospatial data resources include basic geographic, remote sensing, global positioning, application-based thematic, social statistics data, and field observation data. The efficient management of these data resources is a prerequisite of applications, which requires high efficient data model with multiple dimensions including temporal dimension and multiple scales. Moreover, concurrent control and access on the Internet are also key issues to the efficiency of the data resources application.

Storage resources provide the space for storing the data. Computing resources support data processing and managing. Internet resources are responsible for communication among different components deployed on the Internet.

Sensor resources include satellite sensors, laser scanners, CCD cameras, biologic detectors, radiation sensors, geodetic sensors etc. Their responsibilities are to collect real-time geospatial data.

13.3.2 Geospatial Service

Geospatial service is the core component in the framework. The objective of the framework is to decompose the whole geospatial processes into appropriate single steps or functions, and further wrap these steps and functions into geospatial services, and finally the geospatial services will be invoked to implement the behaviors required by clients.

Geospatial data service and geospatial processing service are two basic and core types of geospatial service. The former is used to access different types of geospatial data, while the latter provides various functions for data manipulation. Besides the two basic geospatial services, more services are included to provide extensive geospatial functions like resources registration, composing, searching, knowledge, visualization, and data transmission. These services can be implemented in various platforms and by technologies based on standards. Therefore, the evaluation of the quality of geospatial services is indispensable before they are invoked.

13.3.3 Geospatial Service Applications

Geospatial service application is the top layer of the framework, which directly communicates with clients of geospatial services. According to its purpose, geospatial application can be created at one of these different levels: visualization of geospatial data, geospatial application tool or application system.

For visualization of geospatial data, geospatial services are just invoked to display geospatial data provided by data resources, which is the most general and simplest application of geospatial services.

For the second type, clients compose the geospatial services into a scalable web-based geospatial application tool according to the requirements of their application. This tool can be further deployed in this framework as a usual service, which provides a way of self-growing of Geospatial Service Web.

The third type, development of an application system, is an advanced integration of geospatial services. By combining existing geospatial services, clients can develop application system for their specific application domain.

13.3.4 Geospatial Service Standard

Acquisition, transmission, storage, representation, management and sharing of geospatial data are main tasks of geospatial services. By following geospatial service standards, all of the components involved in the framework of Geospatial Service Web can work together in harmony. Geospatial service standard is a collection of the standards related to geospatial data, data services, data processing services and other services in the framework. More details will be introduced in section 5.

13.3.5 Geospatial Service Security

As a result of the development of geospatial service, huge amount of geospatial information are published over the Internet. The information should not only be utilized but also be protected from illegal or wrong usage. Basically geospatial service security needs to consider geospatial information regarding privacy, confidentiality of nondisclosure geospatial information, integrity and authenticity of geospatial information, the access privilege of geospatial information, copyright of geospatial information and availability of geospatial information service. In order to make geospatial services on the Internet secure, security policy, analysis of potential threats and risks, and a reasonable and logical security system are essential.

13.4 Geospatial Services

The geospatial service components are functional building blocks in GSW. These components cover a large extent of geospatial services. This section discusses some main geospatial services.

13.4.1 Geospatial Sensor Service

Geospatial sensor service is a set of services focusing on discovery and access of sensor observations for all sensors including remote, in-situ, fixed and mobile sensors, as well as receiving alerts and planning sensors tasks to acquire observations of interest. Since various sensors exist around different communities, a standard common conceptual model for representing observation results is necessary to easily discover, access and utilize their observations without the need to support sensor-specific data formats. The Observation and Measurements (O&M) specification (Cox, 2007) from OGC has addressed this issue and provided GML encodings for this model. Depending on the requirements, the observation result may be delivered to the user in its elemental form (i.e. the initial measurement), or value-added data and its derived form after a series of processes. The latter one is commonly available in the intelligent applications of on-board processing of observations. The capabilities of sensors and sensor systems can be modeled through the functional descriptions of these processes, including the measurement and post-measurement processing. OGC has developed the Sensor Model Language (SensorML) specification to define and encode these processes. Therefore, these standards for modeling and encoding sensor observations, sensors and sensor systems are the first step to provide the geospatial sensor service.

In the context of GSW, geospatial sensor service will provide the standard Web Service interface to access sensors and sensor observations. Based on evaluation of the capability descriptions for sensors discovered from a sensor registry (e.g., an OGC Catalogue Service implementation based on the ebRIM profile), geospatial sensor service can determine the feasibility of collecting data from those sensors, decompose the data request for managing the planning, deploy mobile sensors where applicable, make observation schedules, assign tasks to sensors, collect observation data, process the data if necessary, archive them and disseminate them to requestors. In case of the accidental failure of a sensor, geospatial sensor service can support the in-orbit repairing and recovery, or rescheduling the plan with the dynamic constellation. In emergency response applications, through the subscription and publish mechanism, geospatial sensor service can also deliver alerts from sensors or sensor systems to notification receivers. As such, a set of standard interface specifications, such as OGC Sensor Observation Service (SOS), Sensor Planning Service (SPS), Sensor Alert Service (SAS), Web Notification Service (WNS), can work together to discover, access and control sensors.

13.4.2 Geospatial Transmission Service

Geospatial transmission service implements protocols used to transfer geospatial data between distributed information systems through Space-Earth network. The protocols can support the transmission of large volume of data provided by space-based high-resolution Earth observation sensors. Together with the improvement in the data compression rate and algorithm, geospatial transmission service will deliver the coordinated transmission service through combining multiple communication channels and optimized encodings. Different wire or wireless networks, including fiber optic, microwave and satellite-based networks are integrated, equipped with standard transfer protocols to provide a seamless communication network.

13.4.3 Geospatial Data Service

Geospatial data service provides a common data environment supported by a set of standard interfaces for finding and accessing distributed data. It allows geospatial services and value-added applications to access diverse data provided by various providers in a uniform manner. The interface specifications can follow the OGC and ISO standards that are widely used by geospatial communities for sharing data and resources. The most significant types of data concerned in geospatial domain are feature, coverage and observation. Existing data collections, including different formats of high-resolution EOS data and various sensor data can be supported, and when provided on the fly, may often go through necessary query processing and data transformation and reduction, such as georectification, subsetting or resampling. Transaction operation can also be supported when needed, including create, update, and delete operations on geographic features.

13.4.4 Geospatial Processing Service

Generally speaking, a geospatial processing service is a wrap of a specific geoprocessing algorithm. The data required by a geospatial processing service and its output can be delivered over the Web. The geo-processing algorithm provided by a geospatial service may handle only a tiny part of the overall geo-processing (such as coordinate transformation) or a large aggregated processing (e.g., Earth system prediction models). In both situations, the service should be well defined externally, have clear input and output requirements, and can be independently executable. Therefore, taxonomy of geospatial processing services should be defined scientifically, as well as the interfaces are standardized. As such, services developed by different organizations can be chained to fulfill users' requests. Internally, geospatial processing service acquires the data across the Web, initiate the execution of the algorithm, and manage the output for being accessed over the Web. The algorithms can be executed using high performance computing utilities, thus achieve a good efficiency through parallel execution.

Both OGC and ISO standards define taxonomies to categorize geospatial services. Geospatial processing service is one of the six categories in the geospatial service taxonomy. The establishment of sub-categories in geospatial processing service, i.e. geospatial processing service - spatial, geospatial processing service - thematic, geospatial processing service - temporal and geospatial processing service - metadata, is based on the General Feature Model specified in ISO 19109 (ISO19119:2005). The interface for all these geospatial processing services needs to be standardized. There is a promising standard in OGC, namely the Web Processing Service (WPS), which specifies how to describe, advertise, discover and bind Web-enabled geo-processing processes, as well as the standardization of data inputs and outputs for these processes. Thus it provides a general mechanism to allow client to perform distributed geospatial processing in GSW.

13.4.5 Geospatial Chaining Service

Individual services can be reused to construct different geo-processing workflows for geospatial knowledge discovery, geospatial information visualization and value-added data production. In a distributed data and information environment, such as the World Wide Web, there are many independent data and service providers. A complex geo-processing workflow may include services come from multiple service providers. Therefore, service chaining is important for geospatial processing and services integration. Service chaining is defined as a sequence of services where, for each adjacent pair of services, occurrence of the first action is necessary for the occurrence of the second action (ISO/TC 211, 2005). When services are chained, they are combined in a dependent series to achieve larger tasks. The WfMC Terminology and Glossary document (WfMC, 1999) defines the Workflow as "the automation of a business process, in whole or part, during which documents, information or tasks are passed from one participant to another for action, according to a set of procedural rules". Workflow is a key technology for automating business processes that involve access to several applications. From the point of this view, the geospatial service chaining is indeed a construction of workflows, and it specifically deals with Web-based applications.

Three types of chaining are defined in ISO 19119 (ISO/TC 211, 2005): user-defined (transparent) – a human user defines and manages the chain; workflow-managed (translucent) – a human user invokes a service that manages and controls the chain and is aware of the individual services in the chain; aggregate (opaque) – a human user invokes a service that carries out the chain, and has no awareness of the individual services in the chain. The transparent chaining and the translucent chaining involve human interactions. The opaque chaining performs without human interactions, thus Artificial Intelligence (AI) approaches are needed. The logical and semantic representation of geospatial services plays an important role when applying AI approaches. Geospatial chaining service in GSW will support all three types of service chaining through providing approaches from manual construction of service chain to automatic or semi-automatic chaining of geospatial services.

13.4.6 Geospatial Searching Service

Geospatial searching service provides functionalities of search, discovery and organization of geospatial data, service and even knowledge over the Web. There are two prominent ways to support the distributed geospatial information search: one is to search through a Web search engine, and the other is to query a metadata catalogue service which works as a central registry of Web-based geospatial resources. For the previous one, search functions that can effectively dig out information from unorganized and unstructured Web data are the basis to the extraction and classification of geospatial information. The latter one is more related to the current effort of OGC on CSW, a specification for Web-based geospatial catalogue service. The search can also be more than the keyword match. Geospatial ontologies can be incorporated into the search so that semantic information of keywords can be used (Yang 2008). These two approaches can also be combined together, e.g., the search result from the Web search engine can be catalogued into the registry service for later cost-effective discovery. No matter which approach is selected, recall and precision are important evaluation criteria for geospatial searching services.

13.4.7 Geospatial Knowledge Service

The Web can be a platform for exchanging not only data and information but also knowledge. The solid geospatial data and information foundation on the Web is now supporting the emerging applications of knowledge management that analyzes and integrates diverse information sources to support geospatial knowledge discovery and reuse. GSW can support the sharing of geospatial knowledge in the distributed environment through the geospatial knowledge service. A geospatial knowledge service can provide access to a knowledge base in a distributed environment. Its interface specification can support the publishing, querying and updating of geospatial knowledge. Geospatial knowledge will be formalized, stored in a knowledge base, retrieved by a query language, and applied into geospatial problem solving and decision making. The approaches for formal representation of geospatial knowledge such as ontologies and rules are important towards the sharing of geospatial knowledge. Typical geospatial knowledge includes spectral library for featured crops, geo-processing models, and geospatial taxonomies. The knowledge would allow geospatial resources to be annotated, discovered, and integrated intelligently within GSW.

13.4.8 Geospatial Visualization Service

Geospatial visualization service presents geospatial information in a human perceptual format such as 2D maps and 3D globes. Portrayal services such as feature portrayal services and coverage portrayal services can produce rendered outputs for the visualization of 2D map (e.g., cartographically portrayed maps or annotated images). Distributing and visualizing 3D geospatial information over the Web requires a set of technologies and integrated tools, which can enable interoperable fusion of multi-source data, network-based on-the-fly 3D visualization, and interactive local-to-global multi-resolution visualization. Heterogeneous data including terrain, image, 3D models and vector can be delivered to a wide variety of multi-user clients over the Web.

13.4.9 Geospatial Registration Service

Geospatial registration service is the interface of a network-based meta-information repository (e.g., a registry, catalog or clearinghouse). This service contains information about information (meta-information) available over the Internet or in the holdings of digital libraries but not the information itself. The registration services help requestor find the right geospatial information with well-organized clues. Geospatial data and services are cataloged in a registry with their properties and capabilities. The underlying registry information model should also support the registration of sensor and knowledge resources. Thus, it is necessary to provide a unifying interface in registration service to find different information sources according to their organization structures. In case of multiple registration services, they can be federated to provide a comprehensive query for geospatial information (Yang et al 2008).

13.4.10 Geospatial Dissemination Service

Geospatial dissemination service provides personalized, on-demand geospatial data dissemination within GSW. It is supported by distributed geospatial data archives and data pools. From an engineering perspective, it is an application which integrates geospatial data service, geospatial registration service, and geospatial transmission service to deliver rich geospatial content effectively and efficiently over the Web. In addition, custom or semi-custom spatial clients are parts of the dissemination service. Therefore, geospatial dissemination service is a solution and service to meet the geospatial needs of customers.

13.4.11 Quality of Geospatial Service

A complete description of the quality of geospatial service (QoGS) could encourage the sharing and using of geospatial services. The contents, elements and measurements of QoGS should be established to facilitate the evaluation of geospatial services. A QoGS-aware GSW application schema is needed and incorporated into various aspects of GSW applications, such as geospatial service description, discovery and composition.

13.5 Geospatial Service Standards

13.5.1 Overview

Geospatial data is the core ofthe GSW. The acquisition, transmission, storage, representation, management and share of geospatial data are main tasks of geospatial services. In order to integrate physically distributed geospatial services provided by different developers, it is important to standardize general characteristics and behaviors related to geospatial services. Many organizations, such as ISO/TC211, OGC and FGDC, have proposed a series of geographic information standards.

Fig. 13.3 shows the standards in five groups, general standards for geospatial data, general standards for geospatial data acquisition, standards for geospatial data service, standards for geospatial processing service and standards for geospatial applications. The standards in each group are not exhaustive enumerations but a few examples. ISO/TC211 and OGC have published many standards for geospatial data, data services, geospatial processing services and geospatial data acquisition. However, there still remains very large space for developing necessary standards for geospatial applications such as knowledge service, service composition, and visualization service.

This section mainly introduces three groups of standards: the general standards for geospatial data, standards for geospatial data service and standards for geospatial processing service issued by ISO/TC211 and OGC. The first group includes the standards defining the basic elements of geospatial data. This group plays a fundamental role in the whole geospatial service web. The standards in the second group are used in geospatial data access and sharing in distributed environment. The third group focuses on geospatial processing.

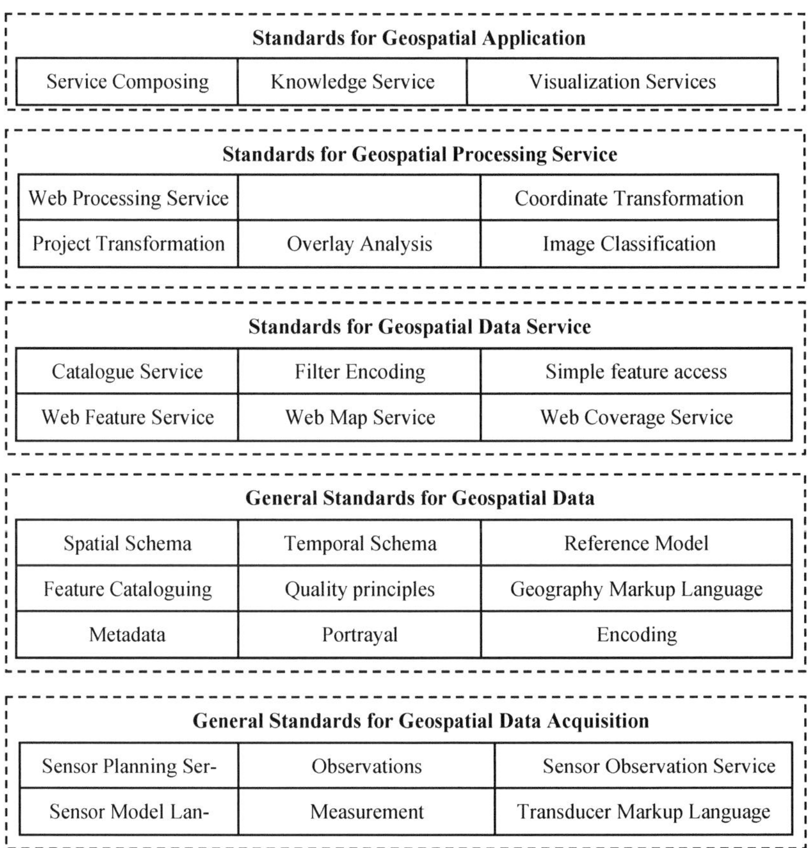

Fig. 13.3 Geospatial service standards

13.5.2 General Standards for Geospatial Data

General standards for geospatial data define the basic elements related to geospatial features and phenomena like spatial and temporal schema, portrayal and encoding rules, reference model, and metadata. This standard group plays a fundamental role in the whole geospatial service web.

ISO 19107:2003 - Spatial Schema specifies conceptual schemas for describing the spatial characteristics of geographic features, and a set of spatial operations consistent with these schemas. It treats vector geometry and topology up to three dimensions. It defines standard spatial operations for use in access, query, management, processing, and data exchange of geographic information for spatial (geometric and topological) objects of up to three topological dimensions embedded in coordinate spaces of up to three axes.

ISO 19108:2002 - Temporal Schema defines concepts for describing temporal characteristics of geographic information. It depends upon existing information technology standards for the interchange of temporal information. It provides a basis for defining temporal feature attributes, feature operations, and feature associations, and for defining the temporal aspects of metadata about geographic information. Since this International Standard is concerned with the temporal characteristics of geographic information as they are abstracted from the real world, it emphasizes valid time rather than transaction time.

ISO 19110:2005 - Methodology for Feature Cataloguing (Nebert and Whiteside 2005) defines the methodology for cataloguing feature types and specifies how the classification of feature types is organized into a feature catalogue and presented to the users of a set of geographic data. ISO 19910:2005 may be used as a basis for defining the universe of discourse being modeled in a particular application, or to standardize general aspects of real world features being modeled in more than one application.

ISO 19111:2007 - Spatial Referencing by Coordinates defines the conceptual schema for the description of spatial referencing by coordinates, optionally extended to spatio-temporal referencing. It describes the minimum data required to define one-, two- and three-dimensional spatial coordinate reference systems with an extension to merged spatial-temporal reference systems. It allows additional descriptive information to be provided. It also describes the information required to change coordinates from one coordinate reference system to another.

ISO 19113:2002 - Quality Principles establishes the principles for describing the quality of geographic data and specifies components for reporting quality information. It also provides an approach to organizing information about data quality.

ISO 19115:2003 - Metadata defines the schema required for describing geographic information and services. It provides information about the identification, the extent, the quality, the spatial and temporal schema, spatial reference, and distribution of digital geographic data.

ISO 19117:2005 - Portrayal defines a schema describing the portrayal of geographic information in a form understandable by humans. It includes the methodology for describing symbols and mapping of the schema to an application schema. It does not include standardization of cartographic symbols, and their geometric and functional description.

ISO 19118:2005 - Encoding specifies the requirements for defining encoding rules to be used for interchange of geographic data within the ISO 19100 series of International Standards.

ISO 19131:2007 - Data product specifications specifies requirements for the specification of geographic data products, based upon the concepts of other ISO 19100 International Standards. It also provides help in the creation of data product specifications, so that they are easily understood and fit for their intended purpose.

ISO 19134:2007 -Location-based services - Multimodal Routing and Navigation specifies the data types and their associated operations for the implementation of multimodal location-based services for routing and navigation. It is designed to specify web services that may be made available to wireless devices through web-resident proxy applications, but is not limited to that environment.

ISO 19136:2007 - Geography Markup Language (GML, Cox 2004) is an XML encoding in compliance with ISO 19118 for the transport and storage of geographic information modeled in accordance with the conceptual modeling framework used in the ISO 19100 series of International Standards and including both the spatial and non-spatial properties of geographic features.

13.5.3 Standards for Geospatial Data Service

The standards for geospatial data service are used in geospatial data access and sharing in distributed environment.

ISO 19119:2005 - Services identifies and defines the architecture patterns for service interfaces used for geographic information, defines its relationship to the Open Systems Environment model, presents a geographic service taxonomy and a list of example geographic services placed in the services taxonomy. It also prescribes how to create a platform-neutral service specification, how to derive conformant platform-specific service specifications, and provides guidelines for the selection and specification of geographic services from both platform-neutral and platform-specific perspectives.

ISO 19125-1:2004 - Simple Feature Access -- Part 1: Common Architecture establishes a common architecture for geographic information and defines terms to use within the architecture. It also standardizes names and geometric definitions for types for geometry. *Part 2: SQL Option* specifies an SQL schema that supports storage, retrieval, query and update of simple geospatial feature collections via the SQL Call Level Interface (SQL/CLI) and establishes architecture for the implementation of feature tables.

ISO 19116:2004 - Positioning Service specifies the data structure and content of an interface that permits communication between position-providing device(s) and position-using device(s) so that the position-using device(s) can obtain and unambiguously interpret position information and determine whether the results meet the requirements of the use. A standardized interface of geographic information with position allows the integration of positional information from a variety of positioning technologies into a variety of geographic information applications, such as surveying, navigation and intelligent transportation systems. ISO 19116:2004 will benefit a wide range of applications for which positional information is important.

ISO 19128:2005 - Web Map Server Interface (WMS, DLB 2004) specifies the behavior of a service that produces spatially referenced maps dynamically from geographic information. It specifies operations to retrieve a description of the maps offered by a server, to retrieve a map, and to query a server about features displayed on a map. ISO 19128:2005 is applicable to pictorial renderings of maps in a graphical format; it is not applicable to retrieval of actual feature data or coverage data values.

The OpenGIS Web Coverage Service Interface Standard (WCS) defines a standard interface and operations that enables interoperable access to geospatial "coverages" [http://www.opengeospatial.org/ogc/glossary/c]. The term "grid coverages" typically refers to content such as satellite images, digital aerial photos, digital elevation data, and other phenomena represented by values at each measurement point.

The OpenGIS Web Feature Service Interface Standard (WFS, Vretanos 2002) defines an interface [http://www.opengeospatial.org/ogc/glossary/i] for specifying requests for retrieving geographic features [http://www.opengeospatial.org/ogc/glossary/g] across the Web using platform-independent calls. The WFS standard defines interfaces and operations for data access and manipulation on a set of geographic features.

13.5.4 Standards for Geospatial Processing Service

Enabling geospatial processing on the Internet requires the development of a wide variety web services to support atomic geospatial operations as well as sophisticated modeling capabilities. It is important to standardize the way that these processes are called, in order to reduce amount of programming required, and to facilitate the implementation and adoption of new services (OGC 05007r7).

Therefore, OGC issued the OpenGIS® Web Processing Service (WPS) to provide rules for standardizing how inputs and outputs (requests and responses) for geospatial processing services, such as polygon overlay. The standard also defines how a client can request the execution of a process, and how the output from the process is handled. It defines an interface that facilitates the publishing of geospatial processes and clients' discovery of and binding to those processes.

The OpenGIS® Coordinate Transformation Service Standard (CT) provides a standard way for software to specify and access coordinate transformation services for use on specified spatial data. This standard addresses a key requirement for overlaying views of geodata ("maps") from diverse sources: the ability to perform coordinate transformation in such a way that all spatial data are defined relative to the same spatial reference system.

Besides the two formal standards, OGC also proposed some discussion papers as candidate implementation specification about web processing, like Web Coverage Processing Service (WCPS) and Web Image Classification Service (WICS).

The Web Coverage Processing Service (WCPS, Lansing 2002) supports retrieval and processing of geo-spatial coverage data. WCPS grounds on the coverage model of the OGC Web Coverage Service (WCS, Evans 2003) Implementation Specification where coverages are defined as "digital geospatial information representing space-varying phenomena", currently constrained to equally spaced grids.

The Web Image Classification Service (WICS) supports classification of digital images. A digital image is composed of pixel values organized into one or more two-dimensional arrays. The two dimensions of an image represent two axes in space based on a spatial coordinate reference system. The dimensions of the different 2-D arrays comprising an image must be the same and represent exactly the same spatial locations.

In addition to the above standards and discussion, other kinds of web processing service also need the support of associate standards. Moreover, according to the framework of GSW proposed in Section 3, more standards and specifications are expected to be developed to support geospatial knowledge service, dissemination service and searching service.

13.6. Implementation of Geospatial Service Web

13.6.1 Overview

A framework of GSW is designed to enable the sharing and processing of geospatial information over the Web. More generally defined Web Service can support many capabilities of GSW. The implementation of a GSW can use existing Web Service technologies. Geospatial standards will ensure the interoperability of GSW implementation. For standards that are not available at geospatial domain, the implementation will adopt W3C Web Services standards which are widely used in general information domain. The standard-based interoperable architecture allows the "plug-and-play" of community-developed, standard-compliant Web service. The development of both consensus-based standards and the standard-compliant, interoperable, distributed service components will ensure the openness, growth and evolution of GSW.

The implementation will also leverage the new and ongoing development in the Semantic Web, Artificial Intelligence (AI), Grid Computing technology and Sensor Web etc. Semantic Web technologies, which give machine-processable meanings to the documents, allow the semantics of data and services machine-understandable and thus can be processed by machines (reasoning) for more effective discovery, automation, integration and reuse of geospatial data and services. The Semantic Web community works closely with AI community. The Semantic Web community applies ontology ideas developed in the AI community to various aspects of Web Services and Web information search and manipulation. Thus, these technologies show considerable promise, and would allow representation and sharing of geospatial knowledge in a Web-based distributed environment.

Grid Computing technology, a rapid developing technology focusing on distributed resources sharing and coordinated problem solving among dynamic virtual organizations, provides a scalable secured sharing of distributed computational resources. The distributed environment supported by the Grid technology, i.e. Grid environment, carries with it various functional components including security, scheduling, data transfer and monitoring. Therefore, it provides a promising prospect to the effective sharing of distributed geospatial resources (e.g. geospatial data, geospatial analysis functions) and supports the integrated analysis of geospatial data.

Sensor Web refers to Web accessible sensor networks and archives sensor data that can be discovered and accessed using standard protocols and application program interfaces (APIs) (Botts et al. 2007). The goal of the Sensor Web is to provide mechanisms to integrate space-airborne and in-situ sensors and enable the Web-based sharing, discovery, exchange and processing of sensor observations, as well as the task planning of sensor systems. So, the implementation of GSW should incorporate the achievement of the Sensor Web. The approach to the Sensor Web is to establish the standards foundation for plug-in-sensor-and-play Web-based sensor networks through harmonizing the existing standards for geospatial data sharing and processing. OGC has developed several encoding and service interface specifications for the enablement of Sensor Web.

13.6.2 System Architecture for the Implementation of GSW

Fig. 13.4 shows the system architecture towards the implementation of GSW. The implementation of GSW will integrate and communicate different types of space-earth data acquired by using various earth observation technologies such as satellite, airplane and in-situ observation. The application areas of GSW are diverse, such as meteorology, agriculture, forestry, transportation and digital city. The GSW is built upon the open, consensus-based standards (i.e. specifications for geospatial information resources in Fig. 13.4) that will allow the "plug-and-play" of community-developed, standard-compliant components and services. The following paragraphs provide detail descriptions of the major components for the implementation of GSW.

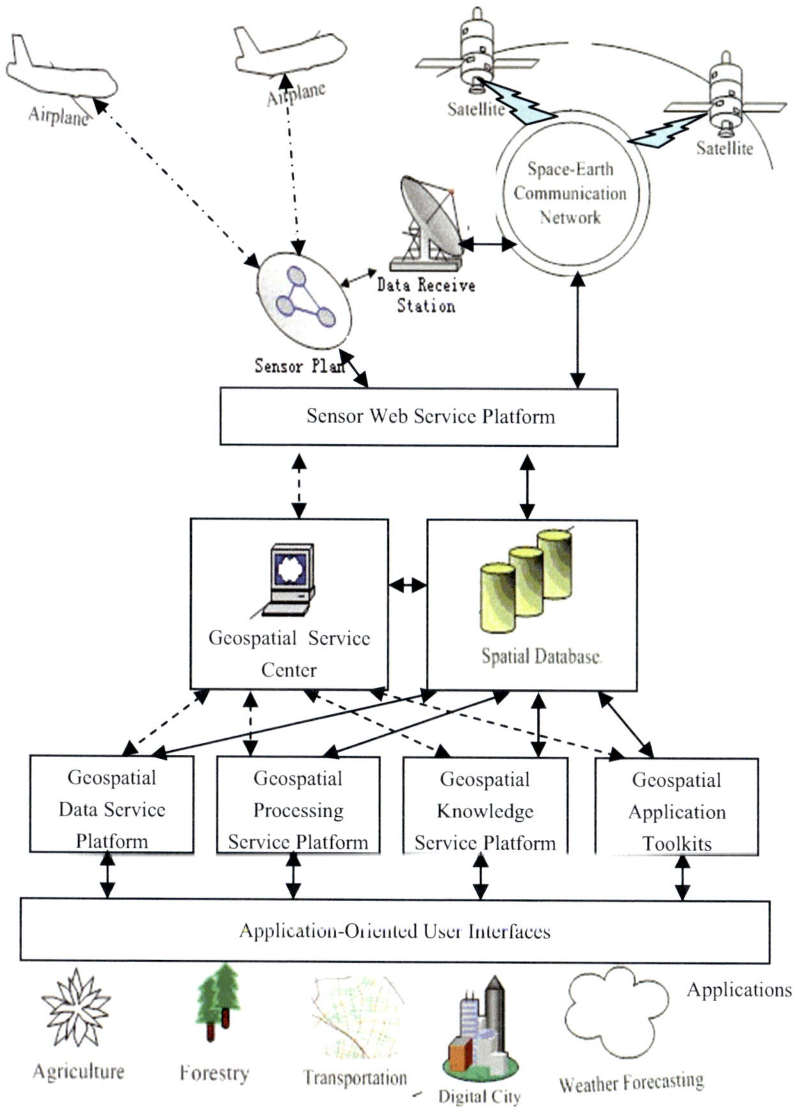

Fig. 13.4 System architecture towards the implementation of GSW

Sensor Web Service Platform provides the interfaces in GSW to register, plan and monitor various space borne, airborne and ground sensor. Sensor Web technologies will play an important role towards the implementation of this platform.

The sensor planning mechanism enables the dynamic and live operations on the sensor resources.

Geospatial Service Center is an application software designed to register, search, locate, bind, schedule, and allocate geospatial resources over the Web. The users' requests are generally decomposed into programmatic requests with filtering conditions to Web-based metadata catalogue services or Internet search engines to answer users' request. The decomposition requires the semantic information in the requests to be correctly understood and processed. Technologies and tools for natural languages understanding by machines and semantic information processing including Semantic Web technologies will be used. In addition, the responses from metadata catalogue services or internet search engines will be processed and integrated based on the correct understanding of users' requests.

Spatial Database Management Platform provides the interfaces to access and manipulate various geospatial data. The data management components in a grid environment help to index the data replicas and manage the reliable file transfer. The information services and execution scheduler help to employ high-performance computing (HPC) resources in a coordinated manner to achieve a computationally hard geospatial task, thus moving the grid-computing from concept to operation. Spatial database management technology is mature. However, some new technologies such as P2P and distributed database management will be used.

Geospatial Data Service Platform provides the discovery and access of heterogeneous data in diverse data archives, ranging from small data providers to multiple-petabyte Earth Observation System (EOS) data pools. With the enablement of Sensor Web, it can further acquire real-time or near real-time data in all-weather, all-wave, all-time conditions. The most promising approach for this platform implementation is to use OGC Web Data Services Specifications including WFS, WCS, WFS, SOS and CSW, since these specifications are widely accepted and allow seamless access to geospatial data in a distributed environment, regardless of the format, projection, resolution, and the archive location.

Geospatial Processing Service Platform includes two major components: Geospatial Processing Service Middleware and Services Integration Environment.

- GeoSpatial Processing Service Middleware is a type of software warehouse. Various geospatial processing components are developed according to the interface specification of Web service and registered in Geospatial Service Center. Some geo-processing tasks are complex and time-consuming. Geospatial Processing Service Middleware will provide HPC power for intensive geo-processing tasks while at the same time hide the underlying complexity as much as possible from the users. Geo-processing tasks, such as geospatial buffer operation, overlay analysis, network analysis, terrain analysis, statistical analysis, image processing and data mining, involve large volumes of data and require high performance computational resources. Thus the middleware needs to work with the Geospatial Service Center to support data-intensive and computer-intensive geospatial applications, for example, running a large number of parallel jobs on a computer cluster.

- Services Integration Environment in GSW follows the publish-find-bind paradigm in the SOA. Individual geospatial Web Services are published though external registries so that they also can be found through the registries. In a Web-based distributed environment, the sites and applications are changing frequently. This paradigm can allow the dynamic binding between service providers and requestors, thus adapting the changing Web environment well. Individual services also act as the building blocks for dynamically constructing complex geo-processing models which can perform larger tasks. The construction of geo-processing models is a service chaining process. The result of service chaining can be encoded in a workflow language. A promising candidate is the Web Services Business Process Execution Language (WSBPEL), shortly known as BPEL, an industry-wide standard provided by Organization for the Advancement of Structured Information Standards (OASIS).

Geospatial Knowledge Service Platform provides some toolkits for describing and modeling geospatial knowledge, managing and accessing knowledge base, and registering and searching geospatial knowledge in a Web service environment. The design of geospatial knowledge services is highly related with the current progress of knowledge management, AI and Semantic Web technologies. For example, geospatial ontology represented using the Web Ontology Language (OWL) from the Semantic Web are organized in the knowledge base using elements following the Resource Description Framework (RDF) triple form: subject-predicate-object, because RDF, the basis of OWL, provides a flexible model for describing Web resources and relations among these resources. Different ontology can be aggregated into distributed triple store, backed up by inference engines, shared and accessed using SPARQL Protocol And RDF Query Language (SPARQL), a standard query language for RDF.

Geospatial Application Toolkits provide tools, library and APIs that can be used to develop GSW architecture and applications. It brings the power of distributed data and other computational resources to end users' desktops, manipulates the resources as if they are local resources, and allows users to conveniently put the resources together. The users can plug-in any geospatial data resources and processing functionalities as they want. For example, a geospatial service module development environment is included in the toolkit which includes a set of libraries for handling the interface protocols, data encoding and decoding, and general utility functions. By using those libraries, Web Service modules developed by users will be standard-complaint, interoperable and easily plugged into GSW.

Application-Oriented User Interfaces are a set of user interfaces to perform various tasks, such as a client to allow users to develop and test the service chain visually, or to "drill-down" into subchains, to animate service chain execution, to check intermediate results, etc. Since geospatial services in GSW are standard-compliant and different standards exist, e.g., WFS, WCS, WMS, an integrated, multiple-protocols client is also needed to provide access to all geospatial data and other resources in GSW.

13.6.3 Prototype for Geospatial Service Web

The above section designs a comprehensive framework of GSW which involves various components. Geospatial Service Center is one of the key components, which acts as a bridge to connect the other components of GSW. Fig.13.5 shows the user interface of a prototype for Geospatial Service Center of the architecture described in the above section. The left tree lists the available services registered in the center, e.g. geospatial data services, processing services, and map portrayal services. This center accepts the registration of data type, data instance, service type, service instance and map symbols. Users can view the information of a service by clicking it on the tree.

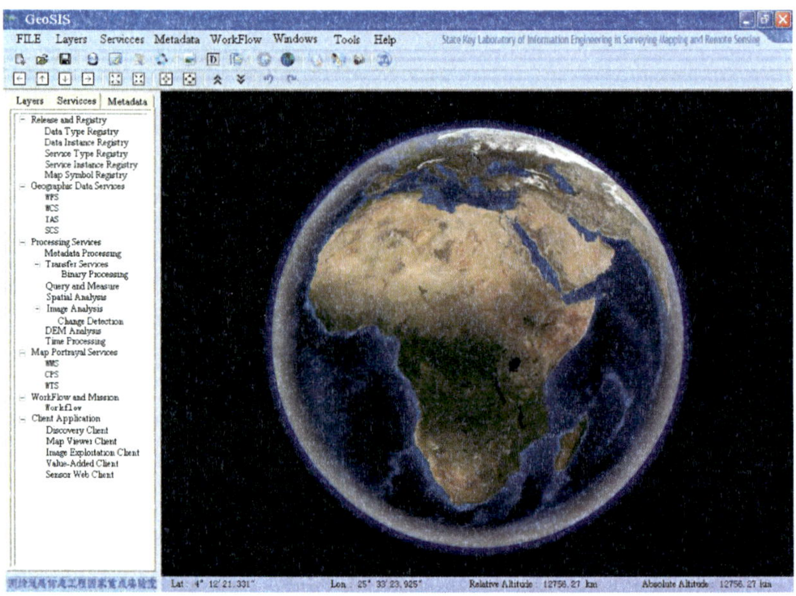

Fig.13.5 The user interface of a prototype for Geospatial Service Center

The prototype provides an environment for composing service chains as application-specific workflows. Fig. 13.6 illustrates an example of a service chain composed for flood submergence analysis. From the palette on the left column, the useful services are chosen and dragged to the right area, then these services are chained with a logical order according to the specific requirements of flood submergence analysis. Finally, an abstract chain will be built and stored as an expert workflow. The abstract chain will be transformed into a BPEL service chain and executed by the engine of service chain. This kind of abstract chains can be reused and adjusted for different applications. Fig. 13.7 presents the result of the service chain for flood submergence analysis. The dark part in the center of the map is the area submerged by flood.

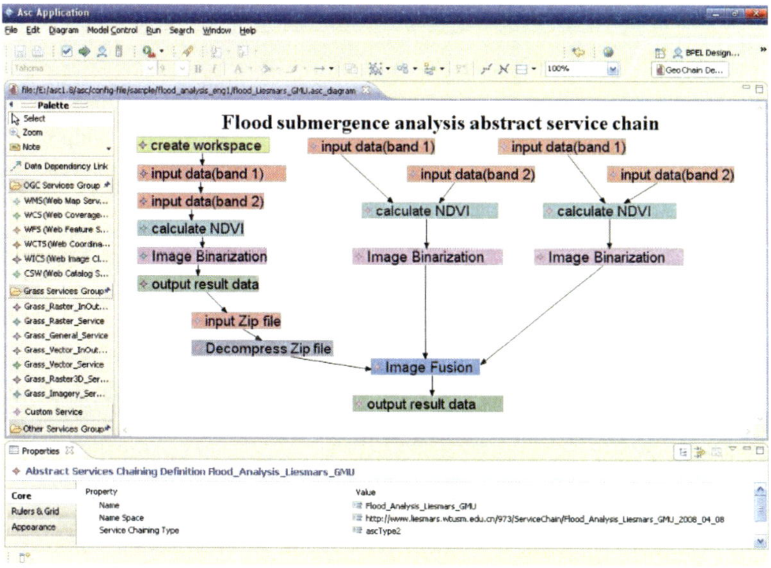

Fig. 13.6 A service chain for flood submergence analysis

Fig. 13.7 The result of the service chain for flood submergence analysis

13.7 Conclusion

With the Web Service technology advancing, the services deployed and distributed in the Internet are not only services of data, implied information and static knowledge, but also geospatial processing services, other related geospatial services and their combinations, which are generating user-induced information and dynamically growing knowledge. Meanwhile, semantic tools are helping the interaction between service and service, and between users and services. Thus, a new concept, Geospatial Service Web (GSW), to umbrella the basic framework of the future of geospatial information technology, is proposed in this chapter. While data, information and knowledge services are still essential bricks of this web, the focus of this web is on atom processing and processing combination services that collaborate to simulate, deduce and predict geographic phenomena, processes and results. This new concept will also expand the reach of geospatial connection from both spatial and temporal dimensions. In the data source rim, this web extends its antenna from static database to all data collecting sensors from satellite-based, airborne to ground and mobile. In the application end, the web supports from visualization until real-time automatic decision support application. What is more, this web will have a mechanism of rule-based self-growing. New combinations of geospatial processing services can be deployed, registered and included in the repository geographic models. As a pioneer effort, this chapter systematically preaches this new thought, outlines the concepts, framework, technologies and standards of GSW.

Acknowledgements

This study was supported by the National Key Basic Research and Development Program of China (973 Program) (Grant: 2006CB701304) and the Program for New Century Excellent Talents in University (Grant: NCET-06-0619). Thanks are due to Ms. Yumin Chen, Mr. Jing Chen, Mr. Longgang Xiang, Mr. Shengyu Shen and Mr. Zhipeng Gui for creating the prototype.

References

Berners-Lee T, Hendler J, and Lassila O (2001). The Semantic Web. Scientific American, 284(5):34-43.

Bernholdt D, Bharathi S, Brown D, Chancio K, Chen M, Chervenak A, Cinquini L, Drach B, Foster I, Fox P, Garcia J, Kesselman C, Markel R, Middleton D, Nefedova V, Pouchard L, Shoshani A, Sim A, Strand G, Williams D (2005). The Earth System Grid: Supporting the Next Generation of Climate Modeling Research. Proceedings of the IEEE, Vol. 93, No.3, 485-495.

Botts M, Percivall G, Reed C, and Davidson J, eds (2007). OGC Sensor Web Enablement: Overview and High Level Architecture, Version 3, OGC 07-165. Open GIS Consortium Inc.

Cox S (Ed.) (2004). OGC Geography Markup Language, implementation specification, http://www.opengis.net/gml/.

Cox S (Ed.) (2007). Observations and Measurements - Part 1 - Observation schema, Version 1.0, OGC 07-022r1. Open Geospatial Consortium, Inc., 85 pp.

de La Beaujardiere J (2004). Web Map Service Ver.1.3, OGC implementation specification, http://portal.opengis.org/files/?artifact_id=5316.

Evans J (Ed.) (2003). Web Coverage Service Ver.1.0, OGC implementation specification, http://portal.opengeospatial.org/files/?artifact_id=3837&version=2.

Foster I and Kesselman C (Eds.) (2004). The Grid 2: Blueprint for a New Computing Infrastructure, Morgan-Kaufman.

Forster I (2005). Service-Oriented Science. Science. 308(5723):814-817.

GEON (2003). A Research Project to Create Cyberinfrastructure for the Geosciences. NSF/ITR. www.geongrid.org .

Hey T and Trefethen A E (2005). Cyberinfrastructure for e-Science. Science. 308(5723): 817-821.

ISO 19108:2002 Geographic Information - Temporal Schema

ISO 19113:2002 Geographic Information - Quality principles

ISO 19107:2003 Geographic Information - Spatial Schema

ISO 19115:2003 Geographic Information - Metadata

ISO 19125-1:2004 Geographic Information - Simple feature access -- Part 1: Common architecture

ISO 19116:2004 Geographic Information - Positioning Service

ISO 19110:2005 Geographic Information - Methodology for Feature Cataloguing

ISO 19117:2005 Geographic Information - Portrayal

ISO 19118:2005 Geographic Information - Encoding

ISO19119:2005 Geographic Information - Services.

ISO 19111:2007 Geographic Information - Spatial referencing by coordinates

ISO 19131:2007 Geographic Information - Data product specifications

ISO 19134:2007 Geographic Information - Location-based services

Lansing J (Ed.) (2002). Web Coverage Portrayal Service Ver.0.0.2, http://portal.opengeospatial.org/files/?artifact_id=1121.

Nebert D and Whiteside A (Eds.) (2005). Catalog Services Ver.2, OGC implementation specification, http://portal.opengis.org/files/?artifact_id=5929.

O'Reilly T (2005), What is Web 2.0, http://oreilly.com/pub/a/oreilly/tim/news/2005/09/30/what-is-web-20.html

Vretanos PA(Ed.) (2002). Web Feature Service Ver. 1.0, http://portal.opengeospatial.org/files/?artifact_id=7176.

Yang C, Li W, Xie J, and Zhou B (2008). Distributed geospatial information processing: sharing earth science information to support Digital Earth, International Journal of Digital Earth. 1(3): 259-278.

Yue P, Di L, Yang W, Yu G and Zhao P (2007). Semantics-based automatic composition of geospatial Web services chains. Computers & Geosciences, 33(5): 649-665.

Chapter 14

OPTIMAL COMPOSITION ALGORITHM CONCERNED WITH RESPONSE TIME FOR REMOTELY SENSED IMAGE PROCESSING SERVICES

Qing Zhu, Xiaoxia Yang, Haifeng Li

14.1 Introduction

As remote sensing technologies have become ever more powerful due to the introduction of multi-platforms and multi-sensors, hundreds of terabytes of image data can be made available daily. But in many cases, raw remotely sensed images are not directly useful without further processing. There are more and more needs to aggregate remotely sensed image processing to satisfy the increasing demands of various applications. Remotely sensed image processing services are modular components that are self-contained, self-describing and can be published, located, and invoked across a network to access and process remote sensing data (Onchaga 2004). Remotely sensed image processing services encapsulate all processing functions into services and combine them into a service chain to provide a value-added service. The various requirements of users can be achieved by combining different existing data and services into a value-added service chain.

For remotely sensed image processing service chaining, an important problem is finding suitable services and to select the most suitable one according the task requirements. Quality of service(QoS) is an important factor that should be considered in the process of service composition. In general, there would be many individual processing services offering similar functionality but with different qualities. Different QoS metrics, such as time, cost, reputation and availability, and so on, have been discussed in (Zeng *et al.* 2004). Several QoS metrics to geospatial information service have been given, including performance, reliability, availability, security and reputation (Onchaga *et al.* 2008). Among these QoS metrics, response time is always the most important factor for two reasons because response time is the key, and prevalent, problem in the design and management of a service chain (Eder *et al.* 1999, Gillmann *et al.* 2002). As to remotely sensed image processing services, it is difficult to estimate and manage response time for two addi-

D. Li et al. (eds.), *Geospatial Technology for Earth Observation*,
DOI 10.1007/978-1-4419-0050-0_14,

tional reasons: in most remotely sensed image applications such as in case of emergency, the requirement for response time of the service chain is very rigorous; to allow for a dynamic network environment and the uncertainty of processing service QoS, response time will vary within a range rather than being a specific value.

In current QoS-based service research, the response time is commonly considered as a certain value (Zeng *et al.* 2004). However, it is better to estimate response time in a probability manner for remotely sensed image processing services. The response time of a service chain contains two factors: expected value and variance. The variance represents the stability of the response time of a service, which is an important consideration in the service selection. Unfortunately, the variance is always ignored by researchers (Pozewaunig and Eder 1997). Based on the probability theory, this paper constructs the constraint between the expected value, the variance, and the user's requirement of response time. By using the critical path method (CPM), some critical services which have direct and crucial effects on the response time guarantee are picked out. To satisfy the constraint, an optimal algorithm of service composition is proposed to reselect appropriate remotely sensed image processing services. Thus, the optimized service chain can meet the response time requirement of the user with higher probability than before.

The rest of this chapter is arranged as follows. Some related works are introduced in the next section. The third section describes an example of remotely sensed image processing service chain. The probability response time estimation model is introduced in section 4. The fifth section describes the optimal composition algorithm of remotely sensed image processing services with response time. The experimental results are illustrated in section 6. Finally, a few concluding remarks are presented.

14.2 Related Work

There are already several methods to specify and estimate the QoS of a service chain based on the aggregations of the QoS of the component services (Zeng *et al.* 2004, Cardoso *et al.* 2004). A quality-aware service chain was presented and a QoS model was defined to describe non-functional quality characteristics of spatial information services and service chains (Onchaga 2004, Onchaga 2006). The QoS model mainly includes response time, cost, reliability, availability, security, reputation, interactivity support, location, and health of the GI service. But almost all the existing researches have focused only on the static case by assuming that each service has deterministic QoS metrics.

Crossflow Project uses the Continue Time Markov Chain (CTMC) to calculate and forecast the execute time of workflow (Klingemann and Wäsch 1999, Damen and Derks 2000, Grefen 2000). Continue Time Markov Chain and Markov reward models are used to estimate the transmission time and resource waiting time (Gillmann and Weikum 2000, Gillmann *et al.* 2002), which is similar to Crossflow's. The ADEPT Project assigns the least continuous time and the longest continuous time to the tasks in the workflow, and model the longest continuous time constraint (Dadam *et al.* 2000). However, the system only supports the management and monitoring of the longest continuous time constraint. A temporal model for workflow was proposed and describes two concepts, the duration space and the instantiation space, which are used to model the absolute and relative deadline constraints and dynamic verification of their temporal consistency (Marjanovic *et al.* 2000). However, temporal model is only useful to manage the change of workflow rather than the analysis, estimation, and management of response time. The prominent impact of slow services in parallel on the overall response time was addressed, and the performance can be improved significantly by reducing the time spent at the slow services (Menasce 2004).

14.3 Remotely Sensed Image Processing Service Chain

Remotely sensed image processing service chain is defined as a composition of services, the remote sensing data and the precedence relation between them. In this chapter, a directed acyclic graph (DAG) is used to represent a remotely sensed image processing service chain. In DAG, nodes represent services, and an arrow from service node S_i to service node S_j if and only if S_j is a direct successor to S_i. And S_i is a direct predecessor of S_j. For each S_j, there is a set of predecessor services that must be completed before S_i begins.

Typical remotely sensed image-based change detection examples are employed throughout this paper. Change detection is the process of identifying differences in the state of an object or phenomenon by images in a certain time interval. Remotely sensed image has been the most important data source for change detection, and the large collection of past and present remotely sensed images makes it possible to analyze spatio-temporal pattern of environmental elements and impact of human activities. As a key element for many applications of earth observation, such as disaster relief, loss estimate, and forest fire monitoring, change detection technique is of great potential and urgent demands. Figure 14.1 illustrates the process of change detection with remotely sensed image, which generally consists of such steps as image acquirement, pre-processing, image registration, and change detection. Node Start represents the beginning of the service chain, and node End represents the terminal of the service chain. This example defines service S1 as the first service. Its successor services, S2 and S5, are parallel services, which can be executed concurrently. Services S4 and S7 are direct predecessors of S8, only if services S4 and S7 are both finished, is service S8 allowed to execute. After completion of S8, the service S9 can be executed sequentially.

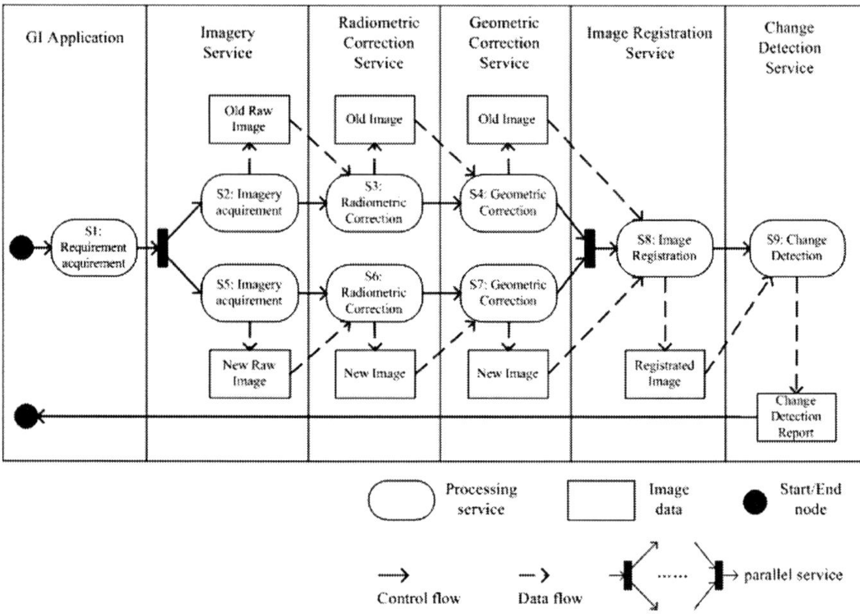

Fig. 14.1 Change detection service chain

14.4 The Probability Response Time Estimation Models

Given a remotely sensed image processing services, the response time t measures the expected delay between the moment when a request is sent and the moment when the results are received. The total response time can be computed as follow:

$$t = t_{wait} + t_{process} + t_{transmission} \tag{14.1}$$

Equation 4.1 means that the total response time is the sum of the service waiting time t_{wait}, the processing time $t_{process}$ and the transmission time $t_{transmission}$.

14.4.1 The Estimation for a Single Service's Response Time

Before the service chain is executed, the response time of a service is unknown exactly. From previous research, it is assumed that the duration of each service is known with certainty. Because of its rich data types, huge data volumes, and complex data processing, remotely sensed image processing usually involves large and heterogeneous data and multiple computation steps and service providers. There usually is considerable uncertainty about what the response time will be. But almost all the existing researches assumed that service has a deterministic response time. In fact, considering only the single or mean response time value is not accurate enough when uncertainties are taken into account. To correct this shortcoming, three different types of estimation of response time are used to assess its probability distribution in this paper. The classic three time estimation is adapted to estimate the response time (Pozewaunig and Eder 1997):

The optimistic time a – the shortest time that the service can be completed.
The pessimistic time b – the longest time that a service might need.
The most likely time m – the completion time having the highest probability.

Two equations (Pozewaunig and Eder 1997) are made to convert a, b and m into estimations of the expected value T and variance σ^2 of the response time required by the service.

$$T = \frac{1}{6}(a + 4m + b) \tag{14.2}$$

$$\sigma^2 = \frac{1}{36}(b - a)^2 \tag{14.3}$$

Two assumptions are required to enable the calculation of the probability of the remotely sensed image processing service response time, namely:

1. The response times of all remotely sensed image processing services are statistically independent;
2. The response time has normal distribution (Hwang *et al.* 2007).

14.4.2 Service Path

A *service path* is a service sequence from the Start node to the End node in which the head node of each arrow is identical to the tail node of the next arrow. In Fig.14.1, there are two service paths in the service chain, Start→ S1→ S2→ S3→ S4→ S8→ S9→ End and Start→ S1→ S5→ S6→ S7→ S8→ S9→ End. For each path p_i in the service chain, the total response time is decided by the time of its component services at each step. Then, the service chain is considered to be completed when all its service paths have been completed. The response time expected value $T(p_i)$ and variance $\sigma^2(p_i)$ of p_i are given by:

$$T(p_i) = \frac{1}{6}\sum_{i=1}^{m}(a_i + 4m_i + b_i) \tag{14.4}$$

$$\sigma^2(p_i) = \frac{1}{36}\sum_{i=1}^{m}(b_i - a_i)^2 \tag{14.5}$$

The variance of service path completion time can be calculated by summing the variances of the completion times of the services in the path. Given this variance, one can calculate the probability that the service path will be completed within a certain time assuming a normal probability distribution for the service path. The normal distribution assumption holds if the number of services in the path is large enough for the central limit theorem to be applied. Applying the assumption that the service chain is normally distributed, questions such as the following can be answered: What is the probability that the path will be completed within the required time? Equations 14.6 and 14.7 can be used to estimate the probability *Pr(*μ*)* that the service chain will be completed by a given time T_k. Equation 14.6 gives the relation between the expected value T_E, variance σ_E, requirement time T_k , and probability coefficient μ of response time. The probability *Pr(*μ*)* is given by 14.7, which is function of μ:

$$\mu = \frac{T_K - T_E}{\sigma_E} \tag{14.6}$$

$$\Pr(\mu) = \frac{1}{\sqrt{2\pi}}\int_{-\infty}^{\mu} e^{-x^2/2}\mathrm{dx} \tag{14.7}$$

The uncertain response time of services for Fig. 14.1 are given in Table 14.1.

Table 14.1 *a*, *b* and *m* of services in Fig 14.1

si	a	b	m	*T*	σ [2]
s1	272	310	300	297	44.44
s2	15	100	30	39	200.69
s3	16	62	30	33	56.25
s4	30	50	40	40	11.11
s5	10	75	30	34	117.36
s6	10	70	35	37	100
s7	15	45	25	27	25
s8	20	200	60	77	900
s9	30	300	100	122	2025

According to the standard normal distribution, the probability of the service chain completing within [T_E- σ_E, T_E+ σ_E] is 68%; the probability of completing within [T_E-2 σ_E, T_E+2 σ_E] is 95%; and the probability of completing within [T_E-3

σ_E, $T_E+3\sigma_E$] is 99%. There is a map from μ to $Pr(\mu)$, which is shown in Fig. 14.2. According to Table 14.1, the expected value T_E, variance σ_E, probability coefficient μ, and probability $Pr(\mu)$ can be calculated, which are shown in Table 14.2. Since the minimal probability of the service path decides the probability of the total service chain, the probability of the service chain completing within 800 seconds is less than 99.97%, within 700 seconds is less than 94.74%, and within 600 seconds is less than 44.43% respectively.

Table 14.2 Service paths with T_k=800, T_k=700 and T_k=600

path	component services	T_E	σ_E^2	T_k=800		T_k=700		T_k=600	
				μ	$Pr(\mu)$	μ	$Pr(\mu)$	μ	$Pr(\mu)$
P1	Start→ S1→ S2→ S3→ S4→ S8→ S9→ End	608	3237	3.37	0.9997	1.62	0.9474	-0.14	0.4443
P2	Start→ S1→ S5→ S6→ S7→ S8→ S9→ End	594	3213	3.63	0.9998	1.87	0.9693	0.11	0.5438

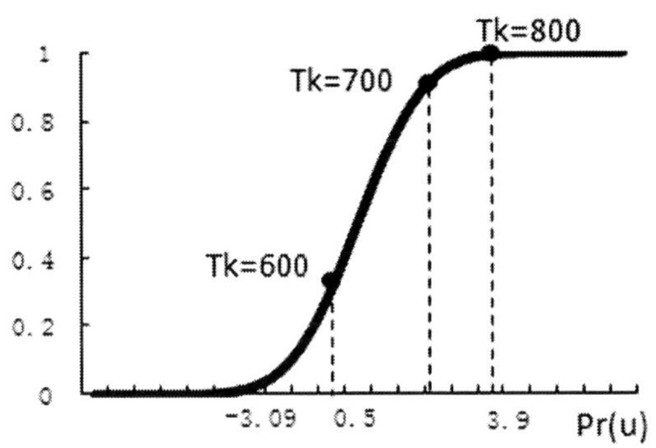

Fig. 14.2 The probability of function about μ of Path 1 in Fig. 14.1

14.4.3 The Critical Path of Service Chain

CPM is a well known project planning method that is frequently used in project management software (Philipose 1986). It can be used to manage the length of time required to complete the service chain.

The response time of a remotely sensed image processing service chain is decided by the longest service path, which is called the *critical path*. Note that the longest services path does not mean the one including most services, but the one that has the longest response time. The critical path of a service chain always requires longer total response time than any other path. The critical path determines the shortest possible execution time required for the service chain.

All service chain structures do have a critical path. If the critical path is not immediately obvious, it may be helpful to determine the following four quantities for each service.

The *earliest start time*, represented by $T_{ES}(s_i)$, is the best time point at which the service s_i can start to ensure minimal response time of the service chain.

$$T_{ES}(s_i) = \begin{cases} 0, & p_i^B = \varnothing \\ \max\left\{T_{ES}(s_j) + t_i, \forall s_j \in P_i^B\right\}, & p_i^B \neq \varnothing \end{cases} \tag{14.8}$$

where p_i^B is the set of direct predecessors of s_i, and t_i is the response time of s_i.

The *earliest finish time*, represented by $T_{EF}(s_i)$, is the best time point at which the service s_i can finish to ensure minimal response time of the service chain.

$$T_{EF}(s_i) = T_{ES}(s_i) + t_i \tag{14.9}$$

The *latest start time*, represented by $T_{LS}(s_i)$, is the worst time point at which the service s_i must start without delaying the response time of the service chain.

$$T_{LS}(s_i) = T_{LF}(s_i) - t_i \tag{14.10}$$

The *latest finish time*, represented by $T_{LF}(s_i)$, is the worst time point at which the service s_i must finish without delaying the response time of the service chain.

$$T_{LF}(s_i) = \begin{cases} T_{LF}(s_N), & p_i^A = \varnothing \\ \min\left\{T_{LF}(s_j) - t_j, \forall s_j \in P_i^A\right\}, & p_i^A \neq \varnothing \end{cases} \tag{14.11}$$

where $T_{LF}(s_N)$ is the response time of service chain, and P_i^A is the set of direct successors of s_i.

For service s_i, the *slack* represented by $TF(s_i)$ is the amount by which the starting time of service s_i could be delayed beyond its earliest possible starting time without delaying the completion of the service chain (assuming no other services are delayed). $TF(s_i)$ can easily be expressed in terms of $T_{ES}(s_i)$ and $T_{EF}(s_i)$, or $T_{LS}(s_i)$ and $T_{LF}(s_i)$ as follow:

$$TF(s_i) = T_{LS}(s_i) - T_{ES}(s_i) \text{ or } TF(s_i) = T_{LF}(s_i) - T_{EF}(s_i) \tag{14.12}$$

The slack can be used as a measure of how important it is to keep each service's response time from greatly exceeding our estimate for it. The slack information can also be used to control the remotely sensed image processing services' start times and execution durations, to avoid time failures at run time and to use time information for optimizing service chain execution.

Any service with a total slack of zero is critical service. The critical path is the path in which none of the services have slack. Every delay of a critical service also causes a delay of the overall remotely sensed image processing service chain. If the non-critical services speed up or slow down within slacks, the response time of the service chain does not change. The amount of time that a non-critical service can be delayed without delaying the service chain is referred to as slack time.

Fig. 14.4 is essentially the same as the service chain in Fig. 14.1. However, each service node is labelled with three time values that represent the response time and the earliest and the latest finish time for service executions as show in Fig. 14.3.

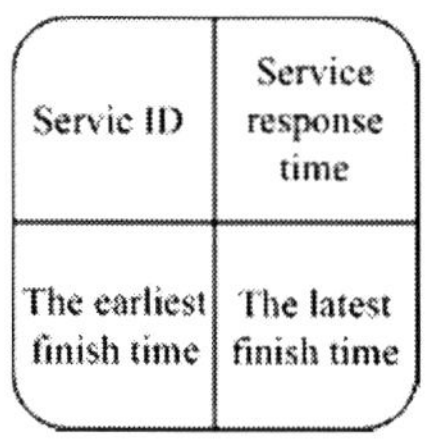

Fig. 14.3 Service node with 3 time values

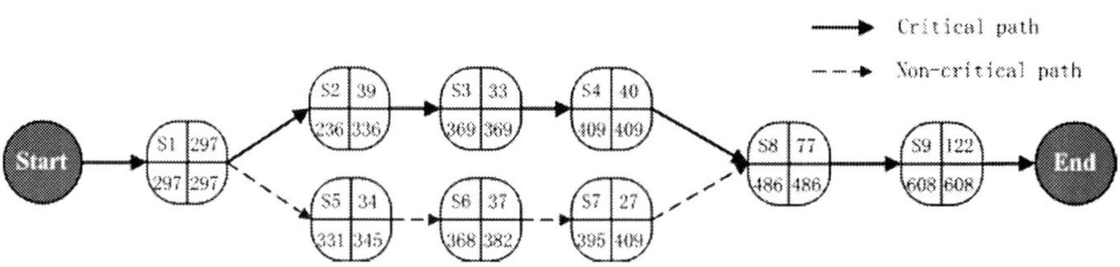

Fig. 14.4 Remotely sensed image-based change detection service chain with time information

Since the critical path determines the response time of the service chain, the service chain can be accelerated by configuring the quicker remotely sensed image processing services to decrease the response time of the remotely sensed image processing service chain.

14.5 Design of an Optimal Composition Algorithm

Since remotely sensed image processing is data intensive and computing intensive, its processing time depends on the skill and the hardware resource heavily. In general, there always exists more than one service to perform the similar function but with different qualities. These services are functionally equivalent and thus can be replaceable by each other, which are called candidate services. Considered a service path composed of n steps $S_1,\ldots,S_n$. Each step S_i can be bound to m candidate services $s_{i1},\ldots,s_{im}$ which are functionally equivalent. So, it is an essential problem of remotely sensed image processing service application to optimize the selection of suitable service from these candidates.

The optimal strategy of our algorithm is that if the response time of the critical path cannot meet the requirements (both time and probability), a faster candidate service is picked out to replace the current one. Given the response time threshold, *Time,* and probability threshold, *Probability*, the flow chart of the algorithm is as follows.

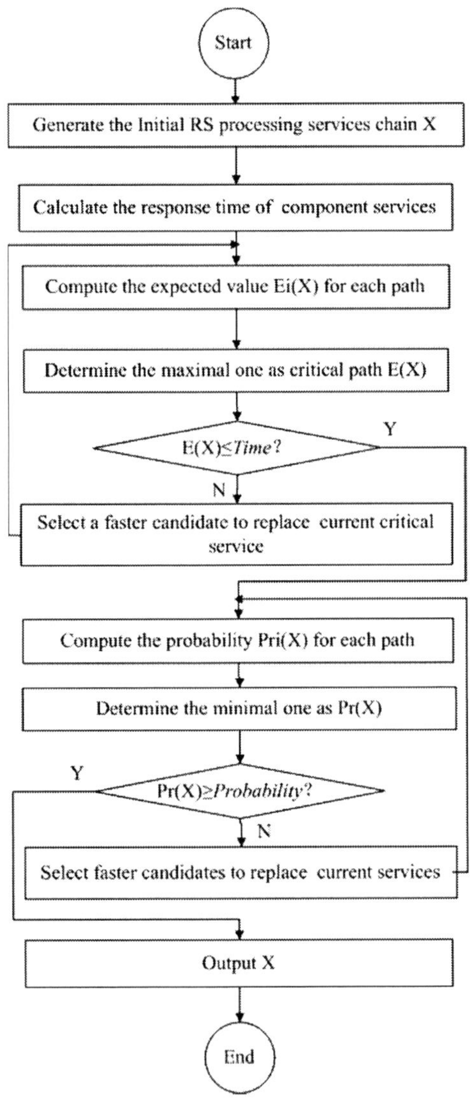

Fig. 14.5 The flow chart of optimal composition algorithm of remotely sensed image processing service with response time

Table 14.3 Candidate services for the service chain in Fig. 14.1

si		si1	si2	si3	si4	si5	si6	si7	si8	si9	si10
s1	T	297	200	310	254	196	405	360	228	400	305
	σ^2	44	64	576	225	361	484	289	117	484	256
s2	T	39	15	20	27	45	40	38	25	36	35
	σ^2	201	144	56	117	49	36	11	16	84	100
s3	T	33	30	34	20	15	18	40	37	30	35
	σ^2	56	49	44	121	36	81	169	117	100	324
s4	T	40	35	30	45	40	37	20	33	46	30
	σ^2	11	36	81	225	256	201	144	169	81	56
s5	T	34	21	18	23	35	30	26	17	35	40
	σ^2	117	289	121	225	117	361	201	44	36	36
s6	T	37	32	19	25	30	42	30	22	20	28
	σ^2	100	81	49	56	64	529	400	225	256	44
s7	T	27	20	18	30	32	24	15	31	25	34
	σ^2	25	100	49	64	121	100	11	36	225	289
s8	T	77	180	270	60	35	70	105	300	190	100
	σ^2	900	289	529	324	121	441	256	576	400	529
s9	T	122	180	300	80	60	50	120	200	240	155
	σ^2	2025	576	900	117	361	121	117	289	324	441

This algorithm involves the following steps:

(1). According to the remotely sensed image process, generate an initial service chain $X=(s_1, s_2, \ldots, s_n)$ randomly.

(2). Using the three time estimation, calculate the response time required for each service.

(3). Compute the expected value of response time of each services path.

(4). Determine the longest path of the remotely sensed image processing service chain as critical path, which has the expected value of $E(X)$.

(5). If $E(X) \leq Time$, then go to step 7, else go to step 6.

(6). For every critical service s_i, t_i is the expected response time of s_i, S_i* is the set of candidates of s_i. And the services in S_i* have the equivalent function but a faster response time. If S_i* is not empty, select the slowest candidate s_i* to replace s_i, else keep s_i. Turn to step 3.

(7). Compute the probability $Pr_i(X)$ of completion of each service path within *Time*.

(8). Determine the minimal probability as $Pr(X)$.

(9). If $Pr(X) \geq Probability$, then turn to step 11, else turn to step 10.

(10). For every service s_i in the above path, t_i is the expected response time of s_i, v_i is the variance of s_i, S_i' is the set of candidates of s_i. And the services in S_i' have the equivalent function but a faster response time or a smaller variance. If S_i' is not empty, select the slowest candidate s_i' to replace s_i, else keep s_i. Turn to step 7.

(11). Output X=(s_1, s_2,…, s_n).

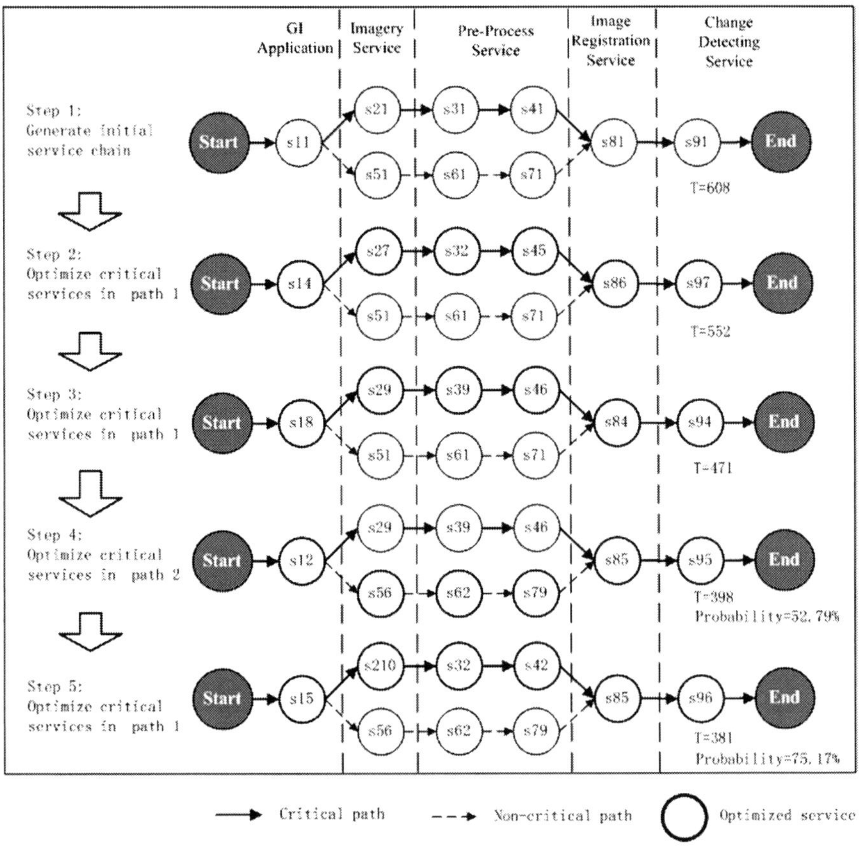

Fig. 14.6 The process of optimal composition of remotely sensed image-based change detection service chain

14.6 Experimental Results

In order to evaluate the proposed probability model and optimal composition algorithm, we developed the remotely sensed image-based change detection application with experiments, and the service chain is shown in Fig. 14.1. Five scenarios were designed to test the effect and efficiency of our probability model and optimal composition algorithm under different scales.

- Scenario 1: each step has 10 candidate services.
- Scenario 2: each step has 20 candidate services.
- Scenario 3: each step has 30 candidate services.
- Scenario 4: each step has 40 candidate services.
- Scenario5: each step has 50 candidate services.

Exhaustive search and local optimization were developed to compare with our algorithm in experiments. The time of composing services was recorded to measure the efficiency of different composition algorithms. For the purpose of soundness, the optimal composition is executed 100 times in each scenario to get average computing time. The comparison of average computation time between these three algorithms is shown in Table 14.4.

Table 14.4 Comparison results of computing time

Scenario		1	2	3	4	5
Number of candidate services		10	20	30	40	50
Local optimization	mean computing time (second)	0.17	0.17	0.17	0.17	0.17
Exhaustive search	time	0.38	1.23	3.83	7.70	16.38
Our algorithm	time	0.23	0.38	0.52	0.82	1.39

The comparison of computing times among these three algorithms is shown in Fig. 14.7. In all algorithms, the computation time increased when the number of candidate services increased. The computation time of exhaustive searching is the highest of these three algorithms. Although the computation time of our algorithm is higher than that of local optimization, it is still acceptable when the number of candidate services is large.

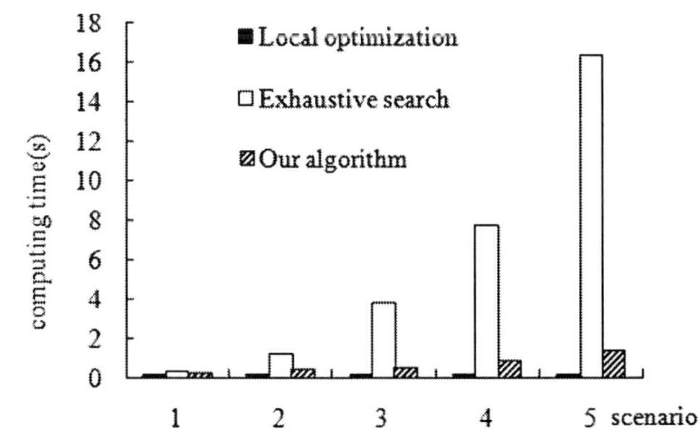

Fig. 14.7 The computing time of experience

A *successful* remotely sensed image processing service chain satisfies the response time requirement with a certain probability. The service chain has been composed N times in each scenario, among which M service chains are successful service chains. In order to evaluate the effect of our algorithm, let Response Time Success Rate($RTSR$) be the function of M and N (Equation 14.13). The comparison of $RTSR$ among these three algorithms is show in Table 14.5 and Fig. 14.8.

$$RTSR = M / N \tag{14.13}$$

Table 14.5 Comparison results of RTSR

Scenario		1	2	3	4	5
Number of candidate services		10	20	30	40	50
Local optimization	*RTSR* (successful service chains / total service chain)	66/100	78/100	80/100	85/100	89/100
Exhaustive search	*RTSR*	100/100	100/100	100/100	100/100	100/100
Our algorithm	*RTSR*	94/100	98/100	100/100	100/100	100/100

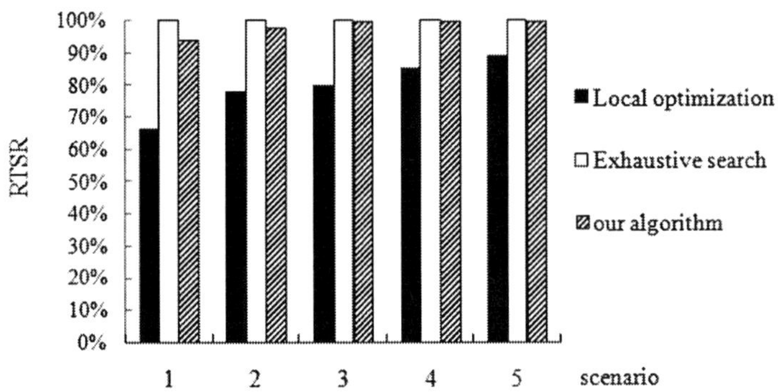

Fig. 14.8 The *RTSR* of experience results

RTSR is the radio of successful service chains to total service chains in scenario. The higher the *RTSR* is, the more effective the composition algorithm is. Since exhaustive search tries all possible composition, it will get a successful service chain when the user requirement is satisfied. Therefore, exhaustive search has a highest *RTSR* value. The *RTSR* of our algorithm indicates that our algorithm is effective, since it is similar to that of exhaustive search and better than local optimization.

14.7 Conclusion

The selection and optimization of services with a response time guarantee are essential problems in QoS-aware remotely sensed image processing service composition. Considering dynamic network environment and the uncertainty in QoS of service, response time usually varies within a range rather than being a specific value. In this paper, the probability response time estimation model is established, which can be used to estimate the probability that the service chain will be completed by a given time. In general, there always exists more than one service to perform the similar function but with different qualities. In order to ensure that the response time of service chains meet the users' requirements, an optimal composition algorithm with CPM response time guarantee is proposed to optimize the service chain. A distinguishing feature of our algorithm is its ability to deal with uncertainty in service response time, which makes the remotely sensed image processing service chain more reliable. Experimental results show that our algorithm play well in complex remotely sensed image processing service chains with multi-path structure in dynamic network environment.

Acknowledgments

The work described in this paper is supported by the National Key Basic Research and Development Program of China (No.2004CB318206), the National Natural Science Foundation of China（40721001）and the Leading Academic Discipline Project of Shanghai Educational Committee, Project No.J50104.

References

Alonso G, Casati F, Kuno H et al (2004) Web Services - Concepts, Architectures and Applications. Springer, Heidelberg

Canfora G, Penta M Di, Esposito R et al (2005) An approach for QoS-aware service composition based on genetic algorithms. Proceedings of the 2005 conference on Genetic and evolutionary computation

Cardoso J, Sheth A, Miller J et al (2004) Quality of service for workflows and Web service processes. J. Web Semantics 13: 281-308

Dadam P, Reichert M, Kuhn K (2000) Clinical Workflows–The Killer Application for Process-oriented Information Systems? 4th Int'l Conf. on Business Information Systems

Damen Z, Derks W (2000) Business-to-business E-Commerce in a Logistics Domain. The CAiSE*00 Workshop on Infrastructures for Dynamic Business-to-Business Service Outsourcing

Eder J, Panagos E, Rabinovich M (1999) Time constraints in workflow systems. Proceedings of the 11th International Conference

Gillmann M, Weikum G (2000) Performance and Availability Assessment for the Configuration of Distributed Workflow Management Systems. International Conference on Extending Database Technology

Gillmann M, Weikum G, Wonner W (2002) Workflow management with service quality guarantees. Proceedings of ACM SIGMOD International Conference on Management of Data

Grefen P (2000) CrossFlow: Cross-organizational Workflow Management in Dynamic Virtual Enterprises. International Journal of Computer Systems Science & Engineering, 15(5): 277–290

Hwang S, Wang H, Tang J et al. (2007) A probabilistic approach to modeling and estimating the QoS of web services-based workflows. Int. J. Inf. Sci. 177: 5484-5503

Klingemann J, Wäsch J (1999) Deriving Service Models in Cross-Organizational Workflows. Proceedings of RIDE - Information Technology for Virtual Enterprises

Marjanovic O (2000) Dynamic Verification of Temporal Constraints in Production Workflows. Proceedings of the 11th Australian Database Conference

Menasce D (2004) Response-time analysis of composite Web services. IEEE Internet Comput. 8(1):90–92

Onchaga R (2004) Modelling for quality of services in distributed geoprocessing. 20th Congress of the ISPRS

Onchaga R (2006). Quality of service management framework for dynamic chaining of geographic information services. Int. J. Appl. Earth Obs. Geoinf. 2006(8): 137-148

Onchaga R, Morales J, Lambert J (2008). An ontology framework for quality of geographical information services. Proceedings of the 16th ACM SIGSPATIAL international conference on Advances in geographic information systems

Philipose S (1986) Operations Research - A Practical Approach. Tata McGraw-Hill, New York

Pozewaunig H, Eder J (1997) ePERT: Extending PERT for workflow management systems. First European Symposium in Advances in Databases and Information Systems

Zeng L, Benatallah B, Ngu A et al (2004) QoS-aware middleware for Web services composition IEEE Trans. Software Eng. 30(5):311–327

Chapter 15

CADASTRAL MAPPING WITH EARTH OBSERVATION TECHNOLOGY

Gottfried Konecny

15.1 Introduction

The International Federation of Surveyors (FIG) has recently stated that about 30 to 50 countries in the world have cadastral systems in operation. Another 50 countries are in the process of establishing one. The remainder of 90 countries do not have a land registry system and they do not have the funds to establish one.

Hernando de Soto has been a primary promoter of land registration systems. (de Soto 2000). His philosophy was adopted by the World Bank and by international donors in support of land registration because secure rights to land not only avoid conflicts, but they also permit to use land as a collateral in a widened capital market.

During the last 10 years US $1.2 billions have been spent on the introduction and the renewal of land registration and cadastral systems. Most of the problem areas are institutional and political. These issues are not the topic of this chapter. But after legal and institutional issues have been solved, technological issues are important to be in a position to rapidly and cost-effectively implement cadastral registration systems. Purpose of this work is therefore to discuss the technical alternatives which have been brought about by the recent rapid advances of technology.

15.2 Definitions

A cadastre in general is a systematic collection of (spatial) data, which can be queried and maintained. A (land) cadastre in particular is a systematic collection of data on land on a land parcel basis. Such a cadastre consists of two parts: "the book", which contains non-spatial data associated with the parcel and "the map" describing the parcel spatially.

D. Li et al. (eds.), *Geospatial Technology for Earth Observation*,
DOI 10.1007/978-1-4419-0050-0_15,

15.3 History and Diverse Approaches

The "cadastral book" and the "cadastral map" has been introduced by Napoleon in the early 19th century in Europe as a "tax cadastre". The recording of property rights was obligatory and the description of the property was assured by the map geometry with 1 m ground accuracy.

Starting from the tax cadastre an "ownership protection" cadastre has evolved by about 1900 rendering a service to land owners, in which parcel boundaries were monumented and surveyed to cm accuracy. This became practice in Switzerland and Germany.

During the 20th century it was realized that the cm accuracy was only desirable for private purposes, since the public only had an interest in a "spatial data infrastructure" for integrating cadastral with topographic and infrastructure data in form of a "multipurpose cadastre".

The original establishment of a cadastre required an adjudication process, in which the owners of land rights had to state or prove their claims to legal and the survey authorities for a land holding. The maintenance of the cadastral system required obligatory registration procedures for the transfer of rights on land, which had to be recorded in books, with copies of the book entries in form of land titles.

The ability to maintain a cadastral register depends on the legal practices in use in different countries of the world. "Private Conveyancing" is used in parts of North America, when land transactions are privately arranged by sales contracts with no security to the purchaser unless he is able to purchase an expensive "title insurance". Somewhat more secure is a "deed registration", also practiced in North America, where the transfer of land in form of a sales contract is registered at the courthouse and attested by a lawyer or a notary public. "Title registration" requires the existence of an obligatory registration. The proof of the registration is the title granted to the owner. In Central European countries there is no need to issue a title, since all valid property rights are registered in the "cadastral book" with the guarantee of the state. The owner can, if needed, request a copy of the book entry.

15.4 Requirements for Cadastral Registration

- There must be a unique description of the right (e.g. ownership, lease, encumbrance)
- The parcel must be characterized by a unique parcel identifier
- The person holding the land right must also be uniquely identified (e.g. birthplace and date, or social security number)
- Every land transaction must be updated in "near real time" (e.g. at the end of a working day).

15.5 Cadastral Maps

In some countries cadastral maps may be useful, but they are not required. In these countries the maps are helpful in identifying the general location of a parcel, but the relative location of the boundaries is described in relative field survey measurements without the need for georeferencing. The map then only serves to identify the parcel, but not its accuracy. Such a map is called an "Index Cadastral Map". Nevertheless, an accurate cadastral map is an asset to a multipurpose Geographic Information Systems (GIS), in which cadastre, topography and utilities can be overlaid.

Some cadastral maps also have the disadvantage that they only show the boundaries and the parcels of private owners, but not of the public lands (roads etc.) which lie between them. A multipurpose cadastre must include all land parcels, including those of street sections.

15.6 Cadastre 2014

In 1994 the FIG Commission 7 on the Cadastre has drafted a "road map" where the development of the cadastre is developing, under the title "Cadastre 2014" (20 years from then, Kaufmann et al. 1998). Six requirements for the creation and maintenance of a modern cadastre were stated:

- The cadastre should cover all lands (public and private) including all rights and restrictions. Geocoding and a topological boundary structure is a requirement to assure, that parcels do not overlap.
- In a data system there should be no separation between (analogue) maps and (analogue) registers.
- The use of paper and pen shall be replaced by computer systems to permit automation.
- Cadastral mapping as a standalone activity shall be abolished.
- The public and the private sectors will closely work together in establishing and in maintaining a cadastral system.
- The cadastre will be cost recovering.

15. 7 Cadastral Surveys

15.7.1 Geocoding of Parcels

With today's capabilities geocoding of images, digital maps and survey data should solely be done based on Global Navigation Satellite Systems (GNSS) technology, such as Global Positioning System (GPS) and Differential Global Positioning System (DGPS). Simple code receivers can only assure accuracies of about 5 m; with corrections transmitted by a service to 0.5 to 2 m. This is sufficient for attribute data collection for an area already mapped for high resolution satellite image data or for orthophoto data.

The geocoding may simply consist of determining the coordinates of a point within a land parcel (a pointer, a "centroid", or a "TOID" as it is called in the Ordnance Survey of Great Britain). The point permits to specify a geo-location defined on a datum within the bounds of a parcel. This permits to relate alphanumeric attribute data contained in a land parcel register in form of a relational database. Such an attribute can be the name of the owner or right holder for the parcel. Fig. 15.1 depicts the use of pointers shown as dots on a digital map. To these dots attributes are geographically attached in a relational database.

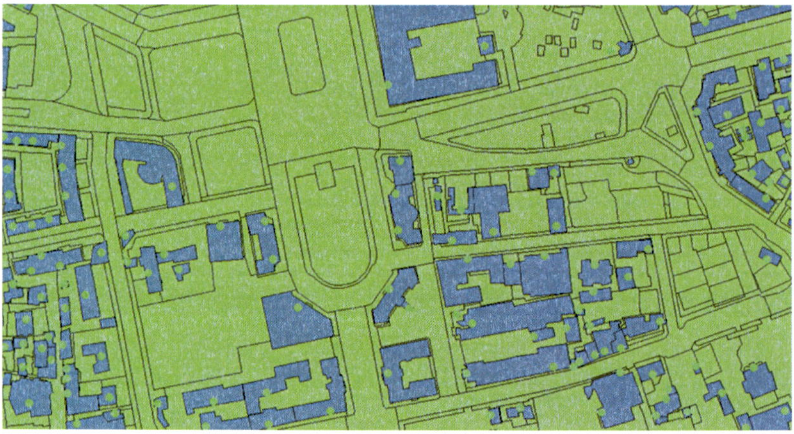

Fig. 15.1 Pointers located in the digital map or image as dots to attach attributes in the database

The process of adjudication, to be discussed later, then merely consists of identifying the rightful owner or right holder of the land. He is asked to provide documentation certified by a notary (a deed), that the rights belong to the identified person. This can be done without the description of the property boundaries. Even during this simple process the use of aerial photos or satellite images, as well as of updated topographic maps, may help to identify the existence of a parcel in a unique way.

15.7.2 Survey of Parcel Boundaries

Obviously it is better, if the interrelationship of all existing parcels in the neighbourhood is described by the tracing of the parcel boundaries from boundary point to boundary point, determined by its coordinates. In this way the topology of the boundary points and boundary lines determines the parcel fabric, and assures, that the land area is totally covered without gaps, particularly if the parcels are not only restricted to land parcels but also to adjacent road sections. A parcel may then be treated as a uniquely identifiable object, to which sub-objects, such as buildings erected on the parcel may be linked. Objects and sub-objects may have their own attributes attached.

The topology of the parcel fabric describes the geometric neighbourhood relationship between adjacent parcels, even though the coordinates of the boundary points may not be of the highest accuracy. While for a more precise geocoding to dm or even cm level with a survey on the ground by more expensive phase receivers used in Real Time Kinematic (RTK) mode or in conjunction with Continuously Operating Reference Stations (CORS) systems are required may be possible. In areas where GNSS surveys suffer from disturbances (urban canyons and power lines) such surveys are supplemented by local total station surveys. In any case such surveys are the most time consuming and costly.

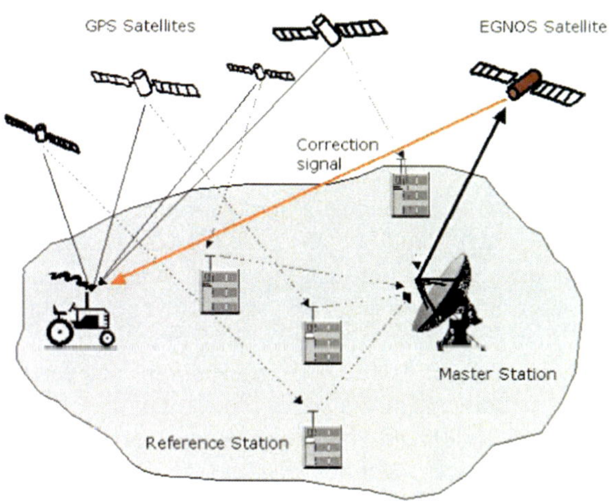

Fig. 15.2 GPS Corrections

Only few countries, such as Germany, Austria and Switzerland have been able to maintain cadastral data bases, which contain object oriented cadastral information with coordinates of boundary points at cm accuracy and the topology formed by boundary lines for all parcels and buildings. The accuracy of boundary points is reached by GPS CORS at distances of about 50km, which correct the GPS-GNSS phase signals for the atmospheric disturbances at cm level. But other areas in the world may benefit from a similar technology that corrects inexpensive GPS code receiver signals by worldwide networks, such as Fugro-Omnistar to dm level. The corrections are transmitted by special communication satellites, such as European Geostationary Navigation Overlay Service (EGNOS) shown in Fig. 15.2.

15.7.3 Use of Images for Cadastral Mapping

A much more rapid and cost effective procedure is to use aerial photos, satellite images and even updated topographic maps for the purpose of interpreting parcel boundaries from the grey level contrasts of the photographic or satellite images, or the mapped topographic objects. There is no guarantee, however, that the images or maps permit to identify the actual legal boundaries unless they are legally established as so called "un-sharp boundaries", or unless they are accepted by the adjacent owners in a "photo-adjudication process".

In many countries index cadastral maps are used (e.g. Sweden and Australia) in which the approximate boundaries are described on a map containing the unique parcel numbers and the graphic location of the boundaries. This can be done at a rather small scale, e.g. 1:5000 or 1:10 000. The accurate survey measurements are then separately contained in survey plans at larger scale (e.g. 1:1000) for each parcel by relative survey measurements or by a plan showing the coordinates of the boundary points. These survey plans may be drawn in analogue form and raster scanned to be part of the parcel database, or they may be issued as computer graphic outputs.

15.7.4 Image Scale

The ground resolution of an analogue photograph or an orthophoto depends on the image scale. The image should be observable by the human eye at a resolution of 5 lp/mm. At that resolution a physical object on the ground should be recognizable (e.g. field corner, house corner). This ground resolution naturally depends on the physical parcel size. In rural areas with field size dimensions of greater than 100m an orthophoto scale of 1:10 000 is appropriate, and for the settlements a scale of 1:2000 is recommended. For example the Ukraine is being covered entirely by orthophotos 1:10 000 and the villages and cities with orthophotos 1:2000. In more densely settled areas, such as Thailand the orthophoto scales should be 1:5000 for rural areas and 1:1000 for settlements. But as the production of orthophotos from analogue images has become an automated process, which depends on generating digital images by a scanning process, the required resolution is also determined by the required pixel size. This orthophoto generation process is at least 5 times faster and more cost effective than manual plotting operations.

The resolution of digital images is determined by pixel size. It should be related to ground sampling distance (GSD) and not to image scale. When compared to image scale 5lp/mm corresponds to 10 lines/mm. This corresponds to a GSD of 1m for the scale 1:10 000 and of 20cm for the scale 1:2000. In the Ukraine GSD´s of 40 to 60cm are being used for the smaller orthophoto scale covering the entire country and of 10 to 15 cm for the larger scale orthophoto coverage of the settlements.

The satellite images, orthophotos (or maps) to be geocoded need to be contained in digital raster (or vector) form and they need to be geometrically transformed by shifts, affine, polynomial or least square fitting models to control in a digital workstation.

The images or maps can be loaded in sections (tiles) into a tablet PC or a PDA (Personal Digital Assistant) for use in conjunction with GPS/DGPS) or total station ground surveys. The image or map helps in identification of a point, the survey in the precise coordinate determination. This is not only of advantage in surveys for the first establishment of a cadastre (as it was done in Georgia or Cambodia). But is particularly recommended in update surveys, which have to be done on a transaction basis, when a parcel boundary is to change.

If a cadastral database consists of a correct topologically closed parcel fabric, which has been established at lower accuracies of 0.5m by photo adjudication or of 1 to 5m by satellite image adjudication, then there is the possibility, that sporadic surveys done on a transaction bases will gradually improve the accuracy of the parcel fabric for a cadastral registration system.

Fig. 15.3 illustrates the relationship between resolution of images from satellites and aircraft versus the repeatability to be able to obtain images. Only meteorological satellites with 1 to 5 km resolution systematically permit imaging several times per day, an aerial photographic coverage at dm to m resolution is practically not affordable except at intervals of a few years, and the original survey at cm to dm accuracy took historically in Europe more than 50 years on the ground.

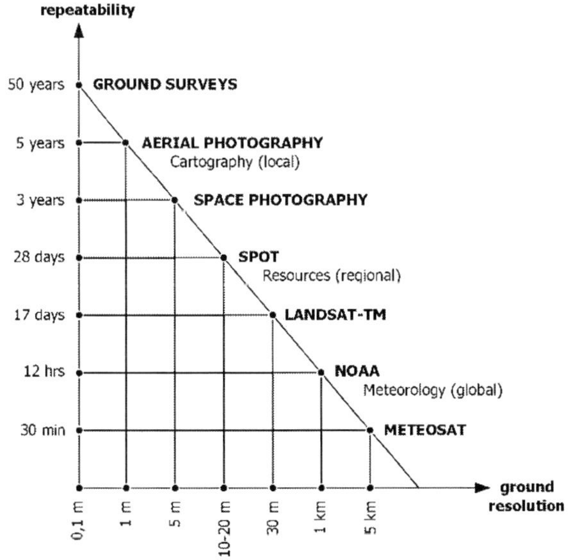

Fig. 15.3 Resolution versus repeatability

15.7.5 Adjudication

The adjudication is a process in which the owners of a land parcel present their contracts, deeds or titles to the adjudication team. These documents, if not otherwise available to the authorities, are photographed on site. Then owners and neighbours identify the boundaries of their parcel. They confirm by signature the location of eventually marked and DGPS surveyed boundaries or boundary monuments.

Modern technology permits to send the signed and photographed documents via mobile technology to the field office. Such an adjudication is possible in case of actual field surveys on the ground, when the neighbouring owners agree to the surveyed and eventually even monumented boundary points. The involved owners confirm this in an adjudication protocol.

When the adjudication is carried out on the basis of images, the neighbouring parties agree to the identified locations in the images and attest this by signature,

15.7.6 Monumentation

While monumentation of parcel boundaries was an issue around 1900 (in Germany it was made obligatory at that time), coordinate surveys to control via total stations or by DGPS have made reconstruction of a point inexpensive and fast. Monuments are often lost due to construction after a short time. Therefore, monumentation now becomes optional, if the client wishes to pay for it.

15.7.7 Surveyed Objects

The basic objects of a cadastral survey are boundary points. Some countries (Germany) do not accept curved boundaries. Curves are realized as chords with the line connecting the boundary points as the actual property limit.

Important are furthermore “permanent” buildings, which in most countries are considered part of the cadastre (e.g. Germany and the Netherlands). In some countries building outlines even refer to separate parcels (e.g. Serbia).

Topographic features are, however, not part of the cadastre. In Germany, cadastral maps are at the scale 1:1000. At that scale topography is a value added survey, which is not part of the base data provided by the cadastral administration. Municipalities may contract inclusion of topographic features as value added data to the private sector. This is done for street furniture (parking lots), utilities and trees (tree cadastre). The basic German topographic map is not at the scale 1:1000 but at the scale 1:5000. The AAA concept (AFIS = control monuments; ATKIS = topographic dataset 1:5000; ALKIS = Cadastral basemap 1:1000) attempts to integrate topography with respect to the cadastral base map.

In Britain, the opposite approach is taken. There a basic topographic map is maintained at the scale 1:1200 by the Ordnance Survey, and the Land Register only adds a cadastral layer for its own use. Maps at smaller scales should use the base map via generalization. The Ordnance Survey in Britain has committed itself to have an update goal of topography of not more than 6 months.

In Germany the updating of the cadastral basemap is in near real time with the update of the 1:5000 topographic maps every 5 years. Due to changes of topography (new construction) this is not fully acceptable, but a remedy is possible by producing digital orthophotos every 2 to 2 ½ years at the scale 1:5000. Orthophotos may thus be considered as a map substitute which only cost 1/5 of a line map and are producible in 1/5 of the time. An added advantage is that such orthophotos may be inserted into browsers, such as Google Earth for urban areas, for which Google has an interest and will purchase the products.

15.8 Cadastral Data Bases

A modern cadastral system only makes sense, if it is organized in a digital data base.

15.8.1 Data Models

The basic cadastral data model shall be simple; a land object (parcel) is linked to a certain type of right, such as

- ownership for a private parcel
- encumbrance for access of an object
- building located on a parcel
- land use for a parcel or a group of parcels
- environmental restrictions for a defined area
- informal status of an object (tribal land).

All objects must be continuously maintained.

The object is geometrically defined in a "map". Preferably this is done in form of shapefiles with topology. The attributes are linked either directly to the object or they are attached to the defined area via label point. Otherwise, it is also possible to use CAD geometry without topology. The label point is again the link to the attributes.

Attributes for land objects are stored in a relational database. They contain:

a) - an application ID for the transaction
- the application status
- the status of the workflow (start)
- the date of registration (end)

b) – the unique parcel ID

c) – the type of right (ownership, encumbrance, lease, building, land use, environmental zone)

d) – the claimant (name, personal ID, % of ownership)

15.8.2 Software Needs

Relational Database

Depending on the size of the database may be used
- Access (personal database)
- SQL Server (enterprise database)
- Oracle (enterprise database)
- Open Source databases

GIS tools

The most widespread use is with ESRI tools, such as ArcGIS (with ArcView for personal database), ArcEditor or ArcInfo via ArcSDE or ArcServer for enterprise database.

Other usable tools

- Oracle Spatial
- Intergraph Geomedia
- Bentley's Microstation and Geographics
- Autodesk's AutoCAD

15.8.3 Specialized Cadastral Software

A number of specialized cadastral software products have been developed:
- Arc Cadastre (Swedesurvey) which is not entirely up to date technology
- LM by AED SICAD for European workflows
- ESRI Cadastral Parcel Manager for US and Australian workflows.

Another option is the customization to ArcGIS or other products using local expertise. This approach has been followed in
- Serbia
- the UK
- Germany

In Germany this customization is being done by dedicated software companies (ConTerra) for the Government.

15. 9 Fast, Low Cost Realization of Cadastral Systems

It has been the experience, that large operating cadastral systems in Central and Northern Europe place their emphasis on maintenance of the records. In the transformation countries some old cadastral information may exist from the 1940's. In former Soviet areas such information is not available. In Serbia, Bulgaria, Poland, Albania the task is therefore to superimpose the existing situation contained in satellite images or orthophotos with outdated, not properly geodetically referenced and perhaps not yet digitized or scanned cadastral maps for an updating process.

Fig. 15.4 illustrates, how in Tirana, Albania at least the buildings of a not maintained cadastral database could be digitized to 2m accuracy from a geocoded QuickBird satellite image with 60cm ground sample distance. The orthophoto and buildings could be visualized by ArcView from an ArcGIS database.

Fig. 15.5 demonstrates, how the ArcGIS capabilities are also able to display attributes derived from the pointers, such as land use.

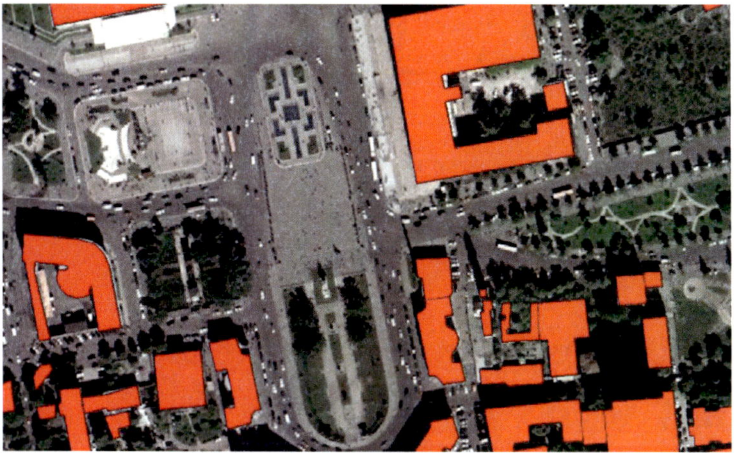

Fig. 15.4 Digitized buildings

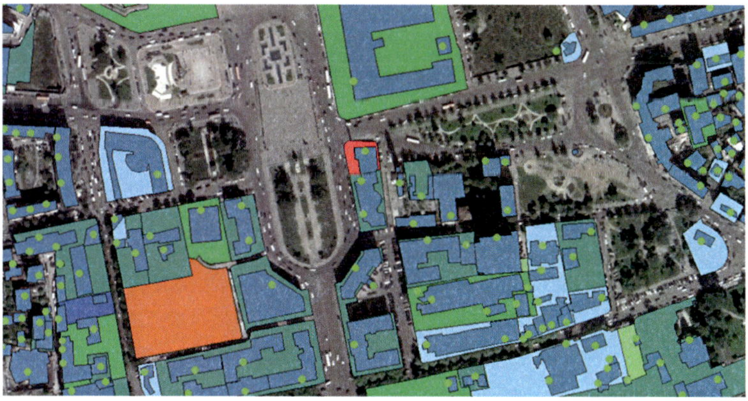

Fig. 15.5 GIS use of parcels; the attributes from the relational database permit to display the parcels in different colors (dark blue = building; red = shops; orange = commercial building; dark green = garden; light green = meadow; light blue = gravel surface)

In former Soviet controlled areas a new cadastral fabric needs to be created. Particularly in these areas it can be done without the existence of a survey profession, if a group of talented young individuals are trained on new technology, as was done in Georgia.

The trained individuals were able to use the KFW (Kreditanstalt füt Wiederaufbau = German International Development Bank) donated equipment forming competing companies managing to establish a cadastral survey system by high-tech adjudication in a 5 year period. In the first year the survey and adjudication cost per parcel was around US $10. It diminished in the 5th year of operation to US $2 per parcel. This shows that the establishment of a cadastral system is possible in the shortest possible time at low cost. It is now the government's task to show that maintenance of the system can be done in a sporadic manner.

Fig. 15.6 illustrates the components of the technology used in the project: a GPS base station, a DGPS rover, a bluetooth connected field pen computer.

Fig. 15.7 shows the adjudication process in the field combined with the ground survey (images of the orthophoto are used in the field pen computer for better orientation in the field).

Fig. 15.8 shows how legal documents of the owners are copied during the adjudication process. The copies may directly be transmitted via bluetooth to the field office.

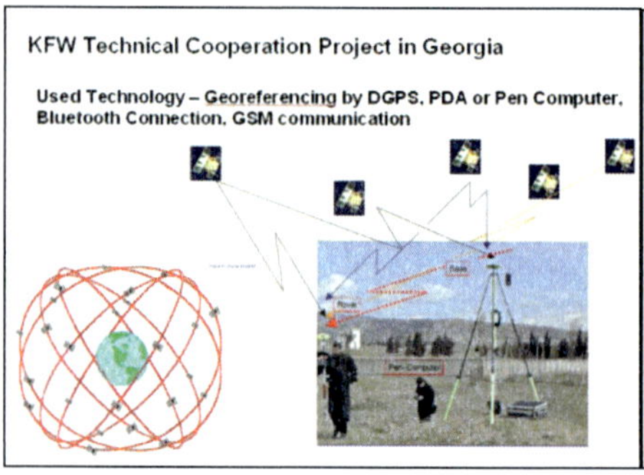

Fig. 15.6 Georgia KFW project

Fig. 15.7 Field survey and field adjudication

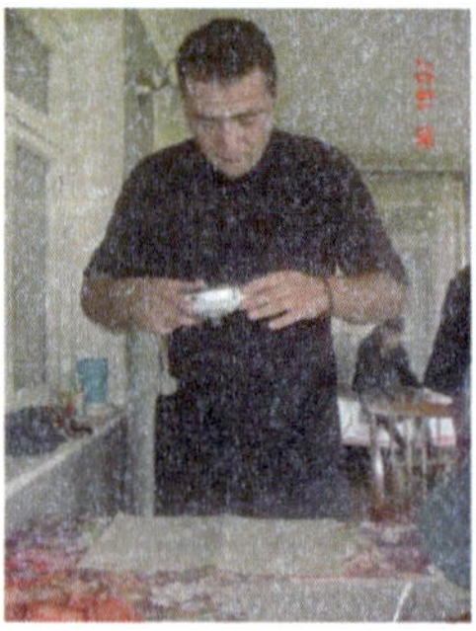

Fig. 15.8 Copying of legal documents of the owners by digital photography

There are a number of possibilities to reduce the cost of a cadastral system:

1. If there is only one administration in charge of land registration and cadastral mapping (as this is the case in Serbia) duplications of data and mismatches of information can be avoided reducing administrative costs.
2. The adjudication does not have to be done by survey measurements, as long as the parcel boundaries are identified in satellite images or orthophotos on which they are marked. If there is a need for a more precise survey due to construction activities (e.g. urban renewal or rural reallocation of parcels) then this could be done on sporadic ad hoc basis.
3. For the maintenance of the cadastre there are basically 3 scenarios:

   Already build-up areas, for which very little changes are anticipated. In these areas sporadic surveys would be used. Modern technology for survey and data base design permit the possibility to improve the originally entered coarse accuracy of 2 m from satellite images or 0.5 m from orthophotos for boundary points to DGPS accuracy of a few centimetres in a sporadic way.

   Areas in which new settlements are planned and computer designed on geocoded maps or orthoimages. The computer planned boundary points can directly be staked out by survey instruments in the field using DGPS and total stations. The record of these plans and stakeouts can immediately be used for the content of the cadastral geodatabase.

   Rural areas, in which accurate surveys are lacking, and in which combined adjudication – survey procedures are recommended.

4. No cadastral adjudication and surveys should be undertaken, unless there is a firm commitment by the authorities to maintain the system.
5. This commitment will also generated trust in the cadastral system, so that it will be used for mortgaging. A secure cadastral system will allow to use up to 90 % of the value of the land as mortgage collateral, while unsecure systems may only lend 10 % of the value.
6. The main requirement is to create a rapid and low cost cadastral system, which should be designed in such a way, that it can be improved by subsequent transactions.

References

de Soto, Hernando, 2000, "The Mystery of Capital", Bantam Press, ISBN 3 87134 449 4

Kaufmann, Jörg and Steudler, Daniel, 1998, "Cadastre 2014", FIG Congress (Brighton), Commission 7

Chapter 16

SPATIO-TEMPORAL PATTERN ANALYSIS OF LAND COVER CHANGE: A CASE STUDY IN ARIDZONE

Qiming Zhou

16.1 Introduction

Remotely sensed data have been widely used for environment change study in the past decades and large collections of remote sensing imagery have made it possible to analyze long-term change of environmental elements and impact of human activities. Research has been widely reported on methodology of remote sensing change detection and monitoring (e.g. Singh 1989, MacLeod and Congalton 1998, Mas 1999, Lu *et al.* 2004).

Change detection approaches can be characterized into two broad groups, namely, bi-temporal change detection and temporal trajectory analysis (Coppin *et al.* 2004). The former measures land cover changes based on a 'two-epoch' timescale, i.e. the comparison between two dates. Even if land cover information sometimes is acquired for more than two epochs, the changes are still measured on the basis of pairs of dates. The latter analyses the changes based on a 'continuous' timescale, i.e. the focus of the analysis is not only on what has changed between dates, but also on the progress of the change over the period. At present, most change detection methods belong to bi-temporal change detection approach including, for example, image differencing (Weismiller *et al.* 1977, Maktav and Erbek 2005), vegetation index differencing (Muttitanon and Tripathi 2005), change vector analysis (CVA) (Malila, 1980, Lunetta *et al.* 2004), principal component analysis (PCA) (Byrne *et al.* 1980, Liu *et al.* 2004), post-classification comparison (Weismiller *et al.* 1977, Dewidar 2004), multi-temporal composite and classification (Zhao *et al.* 2004), and artificial neural network (ANN) (Dai and Khorram 1999, Liu and Lathrop Jr. 2002).

In general, the aim of bi-temporal change detection is to obtain details of 'change/no change' or 'from-to' information in between the detection dates. These methods are often based on medium and high resolution remotely sensed images (e.g. Landsat TM, ETM, SPOT, IRS and AVIRIS), as its applications often require more accurate measurements on, for example, the area ratio changes (Harris 2003, Weber *et al.* 2005), the conversion matrix of land cover change (Zhao *et al.* 2004,

D. Li et al. (eds.), *Geospatial Technology for Earth Observation*,

DOI 10.1007/978-1-4419-0050-0_16,

Muttitanon and Tripathi 2005), and the spatial pattern changes characterized by land cover metrics changes (Narumalani *et al.* 2004, McConnell *et al.* 2004).

The temporal trajectory analysis is to discover the trend of the land cover change by constructing the 'curves' or 'profiles' of multi-temporal data. In contrast to bi-temporal change detection, the temporal trajectory analysis is typically based on low spatial resolution images such as AVHRR and MODIS, which are characterized as low-cost with a high temporal resolution. The trade-off of using these images, however, is the lost of spatial details that makes auto-classification very difficult with low accuracy, so that the temporal trajectory analysis is commonly restricted in, for example, vegetation dynamics in large areas (Lambin and Ehrlich 1997, Myneni *et al.* 1997, Kawabata *et al.* 2001, Dessay 2004, Herrmann *et al.* 2005, Olsson *et al.* 2005), or change trajectories of individual land cover classes (Mertens and Lambin 2000, Petit *et al.* 2001, Crews-Meyer 2001, Southworth *et al.* 2002). Quantitative parameters such as normalized difference vegetation index (NDVI) or area of given land cover class are often used as the dependent variables for the establishment of change trajectories.

To study the human impact on the natural environment, it is often required to recover the history of land cover change and relate the spatio-temporal pattern of such change to other environmental and human factors, rather than merely relying on the change of areas or indices. With the accumulation of remotely sensed images over the past 40 years, it is now possible to analyze the categorical changes of land cover types using higher resolution imagery. Attempts have been made and reported on applications of bi-temporal change detection methods to multiple epochs (Masek *et al.* 2000, Herold *et al.* 2003, Liu and Zhou 2004, 2005, Lunetta *et al.* 2004, Tang *et al.* 2005, Zhou *et al.* 2008a).

In a landscape perspective, land cover change will cause the landscape modified. To measure and distinguish landscape patterns (e.g. fragmentation and connectivity) various landscape metrics have been proposed to describe landscapes at levels of patch, class and landscape (McGarigal *et al.* 2002, Leitao *et al.* 2006). For modelling land cover change in space and time, the trajectory analysis by remote sensing and pattern metrics have been integrated to quantify patterns of landscape changes (Croissant 2004, Crews-Meyer 2006). However, to better understand the spatio-temporal pattern of land cover change, the landscape metrics on change trajectories need to be further investigated and interpreted (Zhou *et al.* 2008b). The connection between the change of landscape patterns and the impacts by human activities and natural forces can then be further explored (Zhou *et al.* 2008c, Zhou and Sun 2008).

This study seeks an efficient and practical methodology to integrate multi-temporal and multi-scale remotely sensed data from various sources with a time frame of 13 years. Based on this, the spatio-temporal pattern of environmental change, which is largely represented by the land cover (e.g. vegetation and water) change, has been analyzed. The history of land cover change for every location in the study area is traced, and the specific nature of such change related to human activities can also be identified.

A practical approach has been proposed to quantify spatial pattern of land cover change for the establishment of a model on change dynamics in an aridzone of western China. The major factors (or driving forces) of these changes and their spatio-temporal patterns are also addressed in order to seek a balance between natural resources and human activities. To simplify the research issues, the spatio-temporal change of the farmland cover type is used as the indicator of the human impacts on the landscape.

16.2 Methodology

The methodology of this study is based on the classifications of multi-temporal images. Firstly, the multi-temporal images were classified to extract the farmland land cover type on different observation dates. Farmland change trajectories, or 'pixel history', were then established to create categorical shift patterns between farmland and other land cover types in the dimension of time. Finally, selected landscape metrics of the farmland cover type and trajectories were derived and analysed.

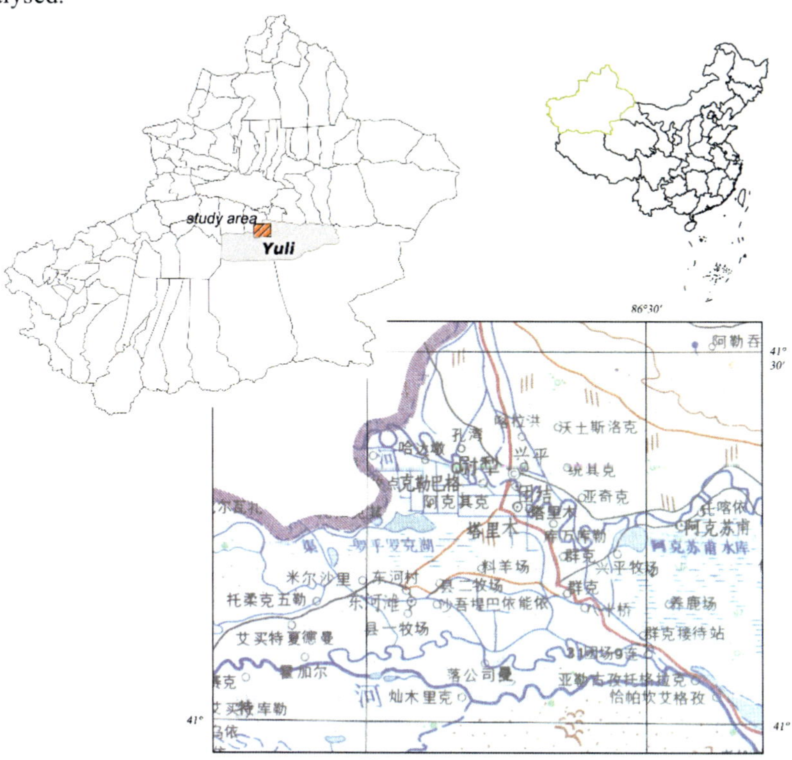

Fig.16.1 The study area

16.2.1 Study Area and Data

The study area is centered at Yuli County in Xinjiang Uygur Autonomous Region of China (Figure 16.1). It is located at the lower reach of the Tarim River – the longest inland river of China. At the fringe of the Taklimakan Desert, the oases along the 'green corridor' of the Tarim River and other rivers are considered to be the most important habitation areas in the arid zone of China with a population density greater than that of cities in the eastern part of the nation. The environment is generally dry with a harsh condition and the ecosystem is fragile by keeping the delicate balance among various environmental factors. As the evaporation is far greater than rainfall, the human survival and vegetation growth heavily rely on water supplement from the rivers.

In this fragile environment, the land cover has shown frequent changes, especially in the last decade. This seems largely due to the human impacts corresponding to the government policies that encouraged large-scale commercial operations for agriculture, particularly the cotton cultivation. According to the local bureau of statistics, more than 97% of the local farmland was for the production of cotton in 2005 and this ratio continued to grow in the following two years (BSY 2006, 2007). Natural factors, however, also played an important role, particularly the change of water resources caused by natural processes such as flood and drought.

To investigate the farmland change in the past decade, we have used five mid-resolution multi-temporal remote sensing images, including multispectral images from Landsat-5, Landsat-7, China-Brazil Environment and Resource Satellite (CBERS-02), and Beijing-1 micro-satellite (BJ-1) (Table 16.1). The 2005 CBERS-2 image was registered and geo-referenced to a topographic map at a scale of 1:50,000. The other images were then geometrically rectified and registered using image-to-image registration based on this master image. Efforts were made to control registration errors within half a pixel of the correspondent image. In addition, the images were subset to fit the study area, so as to be comparable to detect land cover changes.

Table 16.1 Multi-temporal satellite images used in this study

Satellite	*Sensor*	*Spatial-resolution (m)*	*Acquisition date*
Landsat 5	TM	30	Sept. 25, 1994
Landsat 7	ETM+	30	Sept. 17, 2000
CBERS-02	CCD	19.5	Sept. 15, 2005
BJ-1	CCD	32	Aug. 10, 2006
BJ-1	CCD	32	Aug. 31, 2007

16.2.2 Land Cover Classification

In order to minimize the seasonal impacts of multi-temporal and multi-sensor images, the post-classification comparison method is employed. This method is considered to be effective in minimizing the problem of radiometric calibrations between dates and caused by different sensors (Coppin *et al.* 2004). A supervised classification approach using the Maximum Likelihood Classifier (MLC) is adopted to classify images. Five or six land cover types were classified for each image and then they were merged into only two targeted classes, namely, "farmland" and the "others".

The spatial scale which represents the level of landscape is an important issue in ecological spatial pattern analysis (McGarigal *et al.* 2002, Leitao *et al.* 2006). A large scale may provide a detailed land cover spatial pattern. However, sometimes when a broader perspective in landscape is considered, a smaller scale is more appropriate. In this study, the analytical scale depends on the spatial resolution of images as pixel-based classifications were undertaken. The multi-resolution images were accordingly resampled into 30m (approximately the lowest spatial resolution of all used images) using the nearest neighbour method to preserve the original spectral information (Lillesand *et al.* 2004) and to minimize the impacts of resampling on classified images. After classification, a 3×3 majority filter was applied to remove isolated patches.

Due to the difficulty of acquiring historical ground data over a long period of time, collecting reference data for accuracy assessment on multi-temporal images always presents a serious constraint (Zhou *et al.* 2008b). In this study, the accuracy assessment on classified images was based on a random sampling scheme and image interpretation. For each classified image, over 200 sample points were randomly selected on the original images with the reference to some supplementary data sets such as historical aerial photos, thematic maps and ground GPS samples.

16.2.3 Establishment of Land Cover Change Trajectories

Temporal trajectory analysis based on a time series of imagery has proven to be a good way to understand the spatio-temporal pattern of ecosystem dynamics (Coppin *et al.* 2004, Crews-Meyer 2004, Kennedy *et al.* 2004). Although numerous quantitative indictors (such as vegetation index and temperature) can be used for trajectory analysis, methods have also been proposed to create cover type shift patterns or categorical trajectories (Zhou *et al.* 2008a). The focus of this study is the change of farmland in the study area, so that the change between different crops and between other cover types is not concerned. Thus, the land cover trajectory in this research is simply defined as the situation of changes between farmland and the other land cover types. A trajectory may be specified as "farmland – others – farmland – others – others", meaning that the land was once periodically cultivated, but finally abandoned. For the five-epoch, two-class scenario, the total number of possible trajectories is 32 (2^5). Figure 16.2 shows all possible land cover change trajectories and the meaning of changes based on this classification scheme.

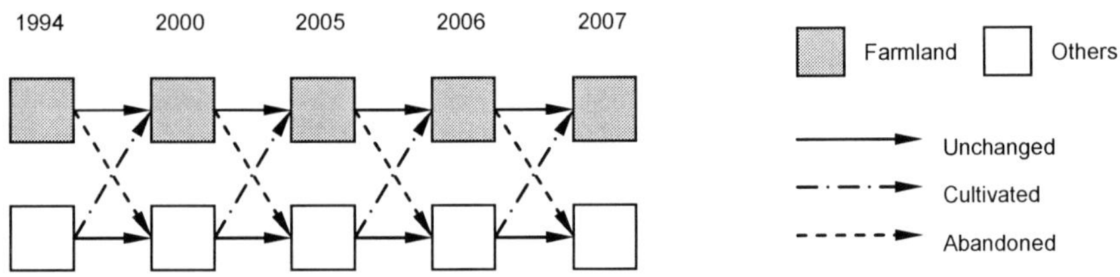

Fig. 16.2 Trajectories of changes between farmland and the other land cover types

It should be noted that a farmland is categorised as "abandoned" when it shifted to and remained as the "others" cover type. However, some of such changes might be the result of construction such as roads and built-up areas. To minimize this kind of uncertainty, the road network and built-up areas were interpreted, mapped and integrated into GIS to create a mask, which was then overlaid with the classified images to exclude built-up areas from the trajectory computation.

To establish the change trajectories, all classified images are integrated into a GIS with the raster format. After assigning value "1" (for farmland) or "0" (for others) to a corresponding bit position in a binary number for each pixel of each classified image, the multi-temporal images were merged together to identify every possible change trajectory with a unique number. Figure 16.3 demonstrates the process.

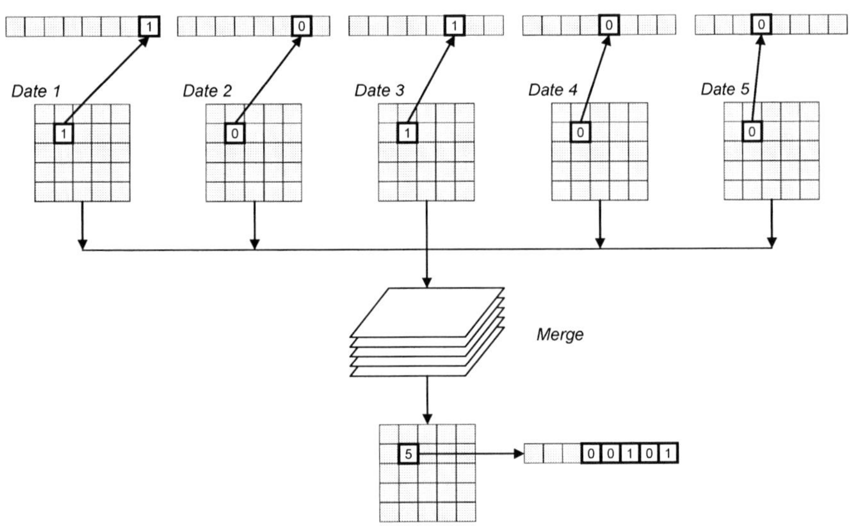

Fig. 16.3 Establishing trajectories in GIS

16.2.4 Landscape Metrics Analysis of Change Pattern

Metrics calculation for quantifying land cover change trajectories are conducted using FRAGSTATS 3.3, a program that can compute most metric indices for landscape pattern analysis (McGarigal *et al.* 2002). In this study, the farmland change trajectories are reclassified into classes according to the change scenarios. The primary focus of this study is on the spatial pattern of trajectories instead of each patch (or pixel), thus only class-level metrics are meaningful and can be used to explain change patterns. What is more, since the extent and fragmentation of a particular class is the principal concern (Leitao *et al.* 2006), the class-level characteristics are common in various studies so that comparison between results of different works is possible. In this study, four class-level metrics supported by FRAGSTATS have been selected, namely, Percentage of Landscape (PLAND), Normalized Landscape Shape Index (NLSI), Interspersion and Juxtaposition Index (IJI) and Area Weighted Fractal Dimension Index (AWFDI).

Table 16.2 Selected metrics for spatial pattern analysis of change trajectories (retrieved and modified from McGarigal et al. (2002) and McGarigal and Marks (1995))

Metrics	*Equation*		*Interpretation*
PLAND	$PLAND = P_i = \frac{\sum_{j=1}^{n} a_{ij}}{A} \times 100$	(16.1)	Quantifies the proportional abundance of each patch type in the landscape.
NLSI	$NLSI = \frac{e_i - \min e_i}{\max e_i - \min e_i}$	(16.2)	NLSI = 0 when the landscape consists of a single square or most compact patch. NLSI → 1 when the patch class is maximally disaggregated.
IJI	$IJI = \frac{-\sum_{k=1}^{m}\left[\left(\frac{e_{ik}}{\sum_{k=1}^{m} e_{ik}}\right)\ln\left(\frac{e_{ik}}{\sum_{k=1}^{m} e_{ik}}\right)\right]}{\ln(m-1)} \times 100$	(16.3)	IJI → 0 when the patch adjacencies becomes uneven, and IJI → 100 when all patch classes are equally adjacent.
AWFDI	$AWFDI = \sum_{j=1}^{n}\left[\left(\frac{2\ln(0.25 p_{ij})}{\ln(a_{ij})}\right)\left(\frac{a_{ij}}{\sum_{j=1}^{n} a_{ij}}\right)\right]$	(16.4)	AWFDI → 1 for shapes with very simple perimeters such as squares, and AWFDI → 2 for shapes with highly convoluted, plane-filling perimeters.

(1) P_i = proportion of the landscape occupied by class i; a_{ij} = area (m^2) of patch ij; A = total landscape area (m^2).

(2) e_i = total length of edge of class i in terms of number of cell surface including all landscape boundary and background edge segments involving class i.

(3) e_{ik} = total length (m) of edge in landscape between classes i and k; m = number of classes.

(4) p_{ij} = perimeter (m) of patch ij; a_{ij} = area (m^2) of patch ij.

PLAND is selected to show the domination of farmland in the landscape so that the overall impact of farmland can be assessed. NLSI specifies the degree of aggregation of farmland. The greater degree of aggregation usually implies a larger scale of agricultural operations. IJI illustrates the state of adjacency between farmland trajectories. This to some extent explains how the new farmland expansion or abandonment occurred in relation to the existing farmland. AWFDI measures the complexity of farmland patches. A complex patch shape is often regarded as the indicator of poorly planned (random) or uncontrolled farmland expansion. The definitions and interpretations of these selected metrics are listed in Table 16.2.

16.3 Results

16.3.1 Classification Accuracy

The spatial registration accuracy for the images is adequate for this study with RMSE less than 0.2 pixels. On the basis of the assessment on only two combined classes, the classifications have shown high accuracy between farmland and the others. The overall accuracies of classifications range from 88.9% to 95.2%, with Kappa coefficients ranging from 0.762 to 0.896 (Table 16.3).

Table 16.3 The overall accuracies and kappa of the classifications of individual images

Images	*1994*	*2000*	*2005*	*2006*	*2007*
Overall accuracy (%)	93.8	95.2	88.9	90.7	92.5
Kappa coefficient	0.877	0.896	0.762	0.799	0.837

16.3.2 Farmland Change Trajectories

Table 16.4 lists trajectories of farmland expansion and abandonment. Table 16.5 shows the change of farmland area and the proportion of permanent, ephemeral and abandoned farmlands. The dominant trajectories are the expansion of stable farmland (i.e. the land has remained as farmland since cultivation and never been abandoned). In terms of area, the trajectories of permanently abandoned farmland (i.e. farmland that was abandoned and has never been re-cultivated since) are not significant. It should also be noted that there have been a large area of ephemeral farmlands accounting for 22.7% of total farmland area in 2007. These ephemeral farmlands are only periodically cultivated and usually do not have adequate infrastructure and water resource to support sustainable agricultural landuse. Therefore, they are prone to the farmland abandonment and most likely to be where land degradation occurs.

Table 16.4 Trajectories of farmland expansion and abandonment

*Trajectory**	*Description*	*Area (kha)*	*% Stable Farmland*	*Annual Growth (%)*
XXXXX	Old farmland	6.76	20.1	-
OXXXX	Cultivated since 2000	7.17	21.3	12.8
OOXXX	Cultivated since 2005	8.55	25.4	10.0
OOOXX	Cultivated since 2006	3.27	9.7	14.5
OOOOX	Cultivated since 2007	7.88	23.4	30.6
XOOOO	Abandoned since 2000	0.32	2.6	-
XXOOO	Abandoned since 2005	0.20	0.8	10.1
XXXOO	Abandoned since 2006	0.13	0.4	25.6
XXXXO	Abandoned since 2007	1.08	3.0	165.8
OXOXO XXOOX ...	Ephemeral farmland	-	-	-

* the cover type "X" stands for the "farmland" class, "O" stands for the "others" class (ordered as 1994-2000-2005-2006-2007).

Table 16.5 Change of farmland area and the proportion of permanent, ephemeral and abandoned farmlands

	Farmland (kha)	*Stable (kha)*	*%*	*Abandoned (kha)*	*%*	*Ephemeral (kha)*	*%*
1994	12.2	6.76	55.4	-	-	5.44	44.6
2000	23.5	13.93	59.3	0.32	2.6	9.57	40.7
2005	34.4	22.48	65.3	0.52	2.2	11.92	34.7
2006	36.1	25.75	71.3	0.65	1.9	10.35	28.7
2007	43.5	33.63	77.3	1.73	4.8	9.87	22.7

Figure 16.4 illustrates the farmland change trajectories. Since the land cover changes are mainly caused by increased farmland in the past decade, the farmland expansion trajectories are highlighted, representing the old farmland since 1994 and the expansions since 2000, 2005, 2006 and 2007.

Figure 16.5 highlights the spatial distribution of the trajectories of abandoned and ephemeral farmlands. Generally speaking, the abandoned and ephemeral farmlands were mostly distributed in the marginal area, particularly at the fringe of water bodies and wetlands. This suggests that the flood and salinity caused by the change of water level and inappropriate irrigation could be the main reason for farmland abandonment.

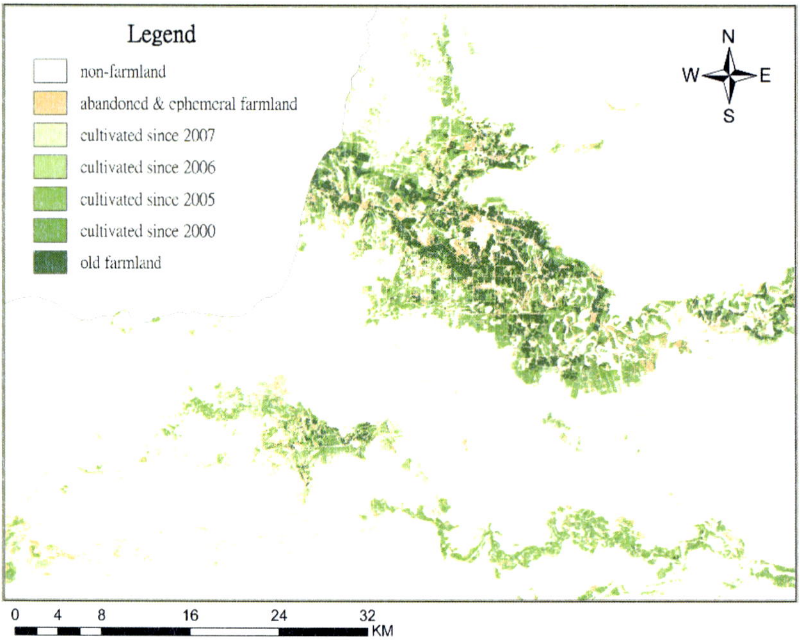

Fig. 16.4 Farmland expansion trajectories in Yuli County

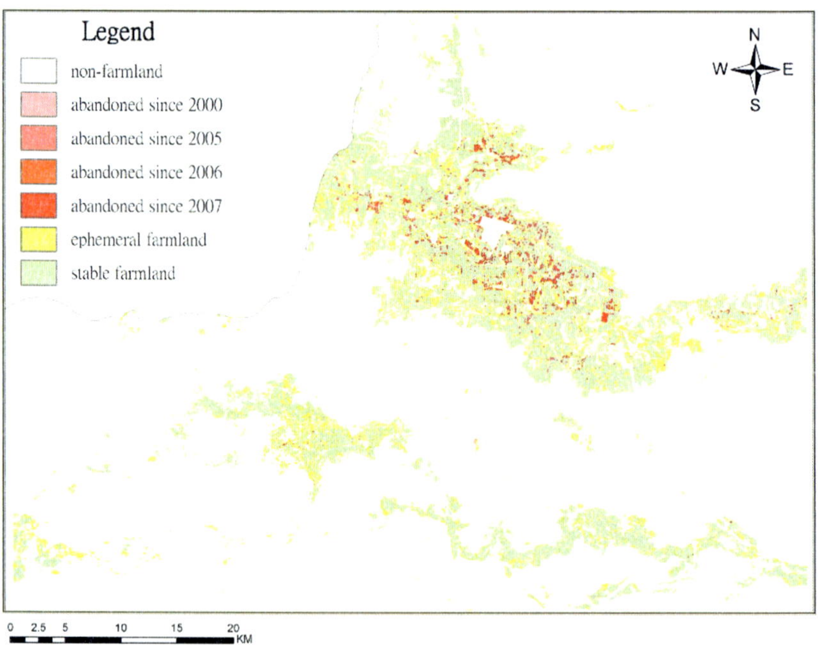

Fig. 16.5 Abandoned and ephemeral farmland trajectories in Yuli County

16.3.3 Metrics of Farmland Trajectories

The selected metrics have been calculated for the trajectory classes. Table 16.6 shows PLAND and IJI of the farmland expansion/abandonment trajectories.

Table 16.6 The class-level landscape metrics of farmland expansion/abandonment trajectories

Trajectory	*Description*	*PLAND (%)*	*IJI*
XXXXX	Old farmland	2.02	62.70
OXXXX	Cultivated since 2000	2.14	61.10
OOXXX	Cultivated since 2005	2.55	59.30
OOOXX	Cultivated since 2006	0.98	53.20
OOOOX	Cultivated since 2007	2.35	51.10
XOOOO	Abandoned since 2000	0.10	49.20
XXOOO	Abandoned since 2005	0.07	72.70
XXXOO	Abandoned since 2006	0.05	71.00
XXXXO	Abandoned since 2007	0.33	54.30
OXOXO			
XXOOX	Ephemeral	3.26	85.03
…			

While considering the spatial patterns of farmlands that were kept cultivated or abandoned at a given time point, the farmland expansion/abandonment trajectories have been merged. For example, the stable farmlands in 2005 include trajectories of "old farmland", "cultivated since 2000" and "cultivated since 2005", while permanently abandoned farmlands in 2005 are composed of trajectories of "abandoned since 2000" and "abandoned since 2005". Table 16.7 summarises the class-level metrics on the merged trajectories classes.

Table 16.7 The class-level landscape metrics of permanently cultivated/abandoned farmland trajectories

Year	*Description*	*Area (kha)*	*PLAND (%)*	*NLSI*	*AWFDI*
1994		6.76	2.02	0.172	1.170
2000		13.93	4.16	0.161	1.200
2005	Stable farmland	22.48	6.71	0.145	1.224
2006		25.75	7.69	0.134	1.240
2007		33.63	10.04	0.145	1.305
2000		0.32	0.10	0.545	1.066
2005	Permanently	0.52	0.18	0.484	1.091
2006	abandoned	0.65	0.23	0.454	1.099
2007		1.73	0.56	0.376	1.132
	Ephemeral	9.87	3.26	0.421	1.157

16.4 Discussion

16.4.1 Spatio-Temporal Patterns Shown by Trajectory Metrics

Usually, landscape metrics are based on land cover types, which are used to describe spatial patterns of landscapes. In this study, we have employed the landscape metrics to describe the nature of change trajectories. Through this approach, attempts have been made to link the trajectory metrics to the environmental impacts of human activities in the study area.

PLAND is a widely used parameter which can determinate the majority land cover type in a landscape. In this study, since the farmland is the only land cover type of interest, PLAND on farmland expansion/abandonment is shown as the indicator of the farmland change rate.

The results show that during the 13-year period from 1994 to 2007, the area of stable farmland has expanded from 6.76 to 33.63 kha (Table 16.7), increasing by nearly 400%. More significantly, the growth has been accelerated recently. During the period of 11 years from 1994 to 2005, the annual growth of stable farmland was about 10-13%. This annual growth rate increased to 14.5% in 2006 and reached an alarming level of 30.6% in 2007. Meanwhile, the area of permanently abandoned farmland has increased from 0.32 to 1.73 kha, accounting for about 4.8% of total farmland area in 2007. Although in terms of area the farmland abandonment has not been significant compared to the expansion of stable farmland, its annual increasing rate has surged from 10.1% before 2005 to 25.6% in 2006 and 165.8% in 2007. The change of farmland expansion/abandonment can be well illustrated by corresponding change in PLAND of the farmland cover type, as shown in Figure 16.6.

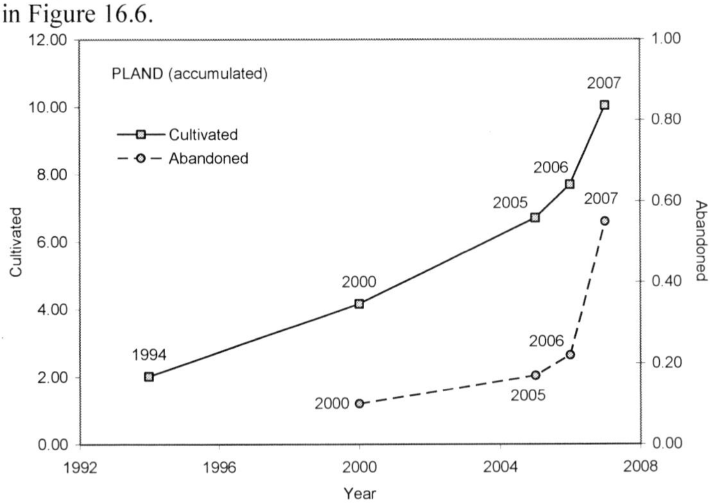

Fig. 16.6 PLAND of permanent farmland expansion/abandonment from 1994 to 2007

IJI measures the adjacency of a patch type to the others. A higher value means the patch type is well interspersed among the others (i.e., equally adjacent to each other), whereas lower value shows a poorly interspersed situation (e.g., a circular distribution of patch types). In this study, the IJI of expanding farmland tends to decrease over time (Table 16.6 and Figure 16.7), while it shows generally higher values for abandoned farmlands. The decreasing IJI indicates that the expansion of farmland was largely located at the outskirt of the existing farmland, thus the new farmland was more likely to intersperse with those just cultivated. This is particularly the case in the recent development since 2006, suggesting that the spatial pattern of farmland expansion was largely dependent upon the agricultural infrastructure (e.g. the irrigation system) in the study area, where the old farmland started in close proximity to water resources while spreading away with the development of infrastructure and increasing investment. The higher IJI values for abandoned farmland suggest that the farmland abandonment seems to occur randomly and does not yet follow a specific spatial pattern.

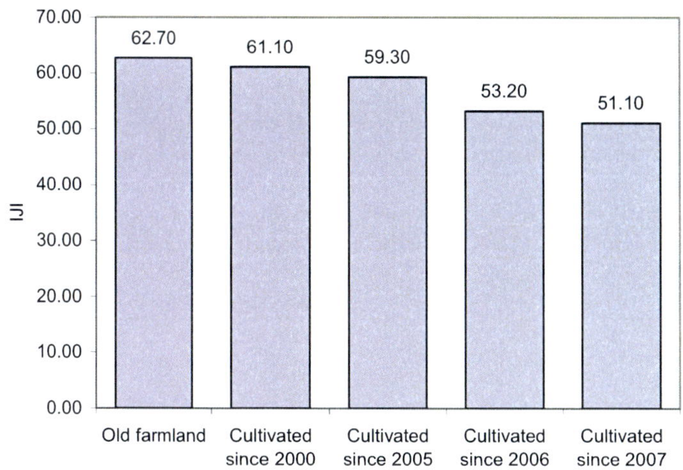

Fig. 16.7 IJI of farmland expansion from 1994 to 2007

NLSI is a measure of the aggregation of a class. As shown in Table 16.7 and Figure 16.8, both stable and permanently abandoned farmlands present the tendency of decreasing NLSI, illustrating the increasing degree of patch aggregation. This implies that the previously sparse patches of stable farmlands tend to be merged into larger farmland patches, suggesting that a larger scale of agricultural operation is in place. Correspondingly, the farmland abandonment also becomes more aggregated considering the increasing abandoned farmland area as indicated by PLAND. It should be noted that abandoned farmlands still have much higher NLSI values than those of stable farmlands, suggesting that farmland abandonment was still rare and random. However, the rapidly decreasing rate of NLSI of permanently abandoned farmlands in recently years certainly shows a warning trend of larger, more aggregated farmland abandonment.

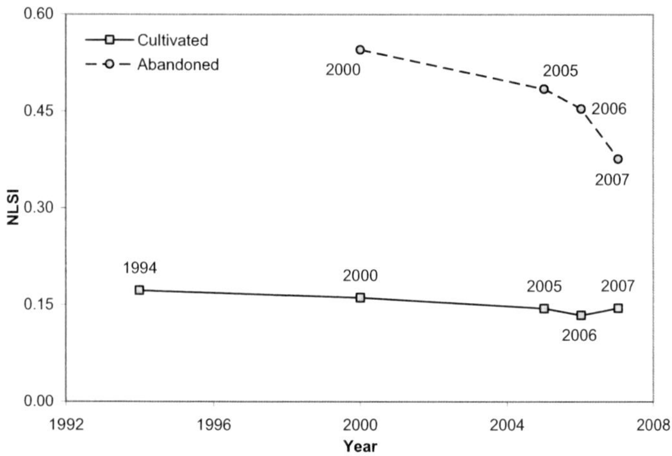

Fig. 16.8 NLSI of permanent farmland expansion/abandonment from 1994 to 2007

AWFDI is the measurement of the complexity of a shape. While the stable or permanently abandoned farmlands have merged together, the patch sizes increased with more complicated shapes. This trend is shown in Table 16.7 and Figure 16.9 where both stable and permanently abandoned farmlands show increasing AWFDI values over time. Similarly to NLSI, the permanently abandoned farmland has shown less complication in shape compared to the stable farmland, but also a rapid increase of complexity over time in the similar way as shown by NLSI.

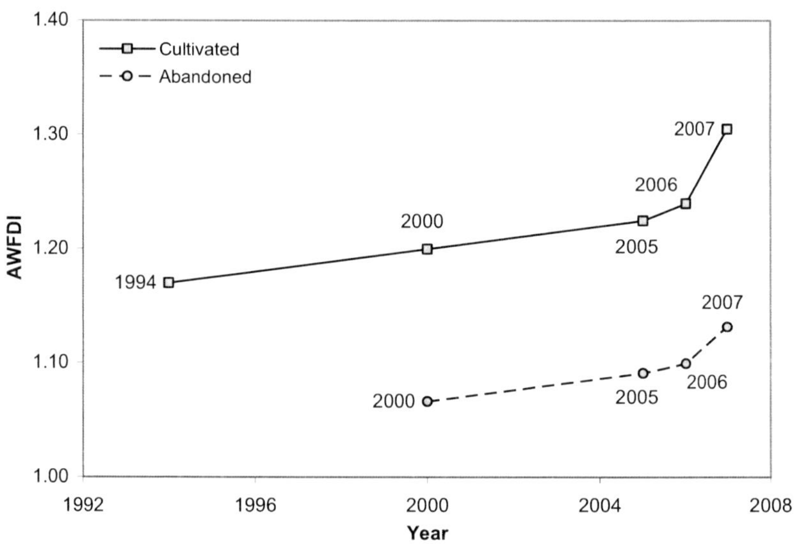

Fig. 16.9 AWFDI of permanent farmland expansion/abandonment from 1994 to 2007

16.4.2 Limitation of the Study

With the constraint of image data acquisition, it is impossible to acquire multi-temporal images of same kind over the 13 year study period. Post-classification comparison method has therefore been employed to cope with the multi-sensor multi-temporal images. The principal advantage of the post-classification comparison method is that the images acquired on different dates by different sensors are independently classified, so that the problems of radiometric calibration among images of different dates by different sensors are minimized. However, it has been argued that the post-classification comparison method may overestimate land cover change due to uncertainties in data (Vanoort 2005), including those by errors in image registration and classification, and misunderstanding of trajectories. The final accuracy of the post-classification comparison and trajectory analysis is largely dependent upon the accuracy of the initial classifications (Coppin *et al.* 2004).

In this study, farmland cover types have been delineated through image classification. When assessing the accuracy of classification, reference data sets have been acquired through visual interpretation of images with the assistance of field data. Ideally, error assessment when using post-classification comparison method should follow the traditional error assessment method using simultaneous ground reference data (Liu and Zhou 2004). Nevertheless it is argued that sampling on typical visual interpretation keys for general classes such as farmland and others would be sufficient for obtaining high-quality reference data sets for classification accuracy assessment.

16.4.3 Comparison of Metrics-based Methods

Most metrics analyses are based on land cover types and focused on the pattern or structure of landscapes. A change detection based on that was only a numeric comparison. The results cannot reveal the spatial pattern of the change. This study takes the approach that the class-level landscape metrics are applied to the land cover change trajectories, rather than land cover types at a given point of time. The results reflect not only the pattern of land cover type distributions but also the spatial pattern of the change. The latter can further be analysed to disclose the reason why a change happened and the trend of how the change will develop.

16.5 Conclusion

This study has demonstrated a method to quantify spatial pattern for analyzing land cover changes effectively by calculating class-level landscape metrics of change trajectories. Different from the traditional landscape metrics on land cover types, the metrics based on change trajectories that can better reveal spatial pattern of changes. In the environment, the supply of arable land has been considered to be unlimited but use of such land is severely limited by the supply of water. Our approach does not only consider the spatial pattern of expanding farmland, but also takes care of the situation of the process of land cover change (or temporal patterns of farmland expansion). Through using multi-temporal images, the causes and consequences of farmland expansion can be better monitored and analyzed, so as to provide better clues for the solution of sustainable development in such fragile ecosystem.

The results show that farmland has rapidly increased in the study area during the 13 year study period. In contrast, signs have shown that the lack of water supply has resulted in farmland abandonment that could cause severe land degradation due to salinity and wind erosion. Large-scale commercial agricultural practice has obviously increased demand on natural resources, particularly the water resource. When the shortage of water supply happens, newly cultivated farmland is more likely to be abandoned. In the year of 2007, the highest expansion rate of new cultivation was observed, while the area of abandoned farmland also reached the largest. It is therefore fearful that the rapid expansion of agricultural land may not be sustainable and may lead to the rapid farmland abandonment in unfavourable environmental conditions.

Although challenges remain with numerous research questions unanswered, the study nevertheless encourages further effort to develop more advanced methodology for establishing more robust and reliable spatial pattern indicators, so that the spatio-temporal pattern of land cover change can be better understood. Attempts will be made to establish a regression model between water resources and spatio-temporal pattern of farmland change trajectories. The aim of further studies is also on the regional balance of water resources, corresponding to the dynamic change of landuse and crop cover.

Acknowledgements

The research is supported by National Key Basic Research and Development Program (2006CB701304), Research Grants Council Competitive Earmarked Research Grant (HKBU 2029/07P), and Hong Kong Baptist University Faculty Research Grant (FRG/06-07/II-76). The author would like to acknowledge the contributions from Dr. Li Baolin and Mr. Alishir Kurban of Chinese Academy of Sciences in early field investigation of this study. The author would also like to thank Mr. Sun Bo of Hong Kong Baptist University for his contribution in data processing.

References

Bureau of Statistics of Yuli County (BSY) (2006), Statistics News. Retrieved June 21, 2008 from Bureau of Statistics of Xinjiang Bayingolin Mongol Autonomous Prefecture website: http://tjj.xjbz.gov.cn/html/xstj/652823/2007918163639.htm.

Bureau of Statistics of Yuli County (BSY) (2007), Statistics News. Retrieved June 21, 2008 from Bureau of Statistics of Xinjiang Bayingolin Mongol Autonomous Prefecture website: http://tjj.xjbz.gov.cn/html/xstj/652823/20061025182742.htm.

Byrne, G.F., Crapper, P.F. and Mayo, K.K. (1980), Monitoring land cover change by principal component analysis of multitemporal Landsat data, *Remote Sensing of Environment*, 10(3): 175-184.

Coppin, P., Jonckheere, I., Nackaerts, K., Muys, B. and Lambin (2004), Digital change detection methods in ecosystem monitoring: a review. *International Journal of Remote Sensing*, 25(9): 1565–1596.

Crews-Meyer, K.A. (2001), Assessing landscape change and population-environment interactions via panel analysis. *Geocarto International*, 16(4): 69-79.

Crews-Meyer, K.A. (2004), Agricultural landscape change and stability in northeast Thailand: historical patch-level analysis, *Agriculture Ecosystems and Environment*, 101: 155-169.

Crews-Meyer, K.A. (2006), Temporal extensions of landscape ecology theory and practice: Examples from the Peruvian Amazon, *The Professional Geographer*, 58(4): 421-435.

Croissant, C. (2004), Landscape patterns and parcel boundaries: an analysis of composition and configuration of land use and land cover in south-central Indiana, *Agriculture Ecosystems and Environment*, 101: 219-232.

Dai, X.L. and Khorram, S. (1999), Remotely sensed change detection based on artificial neural networks. *Photogrammetric Engineering and Remote Sensing*, 65(10): 1187-1194.

Dessay, N., Laurent, H., Machado, L.A.T., Shimabukuro, Y.E., Batista, G.T. and Diedhiou, A. (2004), Comparative study of the 1982–1983 and 1997–1998 El Niño events over different types of vegetation in South America. *International Journal of Remote Sensing*, 25(20): 4063–4077.

Dewidar, K.H.M. (2004), Detection of land use/land cover changes for the northern part of the Nile delta (Burullus region), Egypt. *International Journal of Remote Sensing*, 25(20): 4079–4089.

Harris, R. (2003), Remote sensing of agriculture change in Oman. *International Journal of Remote Sensing*, 24(23): 4835–4852.

Herold, M., Goldstein, N.C. and Clarke, K.C. (2003), The spatiotemporal form of urban growth: Measurement, analysis and modelling. *Remote Sensing of Environment*, 86: 286-302.

Herrmann, S.M., Anyamba, A. and Tucker, C.J. (2005), Recent trends in vegetation dynamics in the African Sahel and their relationship to climate. *Global Environmental Change*, 5: 394–404.

Kawabata, A., Ichii, K. and Yamaguchi, Y. (2001), Global monitoring of inter-annual changes in vegetation activities using NDVI and its relationships to temperature and precipitation. *International Journal of Remote Sensing*, 22: 1377–1382.

Kennedy, R. E., Cohen, W. B. and Schroeder, T.A. (2007), Trajectory-based change detection for automated characterization of forest disturbance dynamics, *Remote Sensing of Environment*, 110: 370-386.

Lambin, E.F. and Ehrlich, D. (1997), Land-cover changes in sub-Saharan Africa, 1982–1991): application of a change index based on remotely sensed surface temperature and vegetation indices at a continental scale. *Remote Sensing of Environment*, 61: 181–200.

Leitao, A.B., Miller, J., Ahern, J. and McGarigal, K. (2006), *Measuring Landscapes: A Planner's Handbook*, Island Press, Washington, D.C.

Lillesand, T.M., Kiefer, R.W. and Chipman, J.W. (2004), *Remote Sensing and Image Interpretation* (Fifth Edition), Wiley, New Jersey.

Liu, H. and Zhou, Q. (2004), Accuracy analysis of remote sensing change detection by rule-based rationality evaluation with post-classification comparison. *International Journal of Remote Sensing*, 25(5): 1037–1050.

Liu, H. and Zhou, Q. (2005). Establishing a multivariate spatial model for urban growth prediction using multi-temporal images, *Computers, Environment and Urban Systems*, 29(5): 580-594.

Liu, X. and Lathrop Jr., R.G. (2002), Urban change detection based on an artificial neural network. *International Journal of Remote Sensing*, 23: 2513–2518.

Liu, Y., Nishida, S. and Yano, T (2004), Analysis of four change detection algorithms in bi-temporal space with a case study. *International Journal of Remote Sensing*, 25(11): 2121–2139.

Lu, D., Mausel, P., Brondízio, E. and Moran, E. (2004), Change detection techniques. *International Journal of Remote Sensing*, 24(12): 2365–2407.

Lunetta, R.S., Johnson, D.M., Lyon, J.G. and Crotwell, J. (2004), Impacts of imagery temporal frequency on land-cover change detection monitoring. *Remote Sensing of Environment*, 89(4), 444-454.

MacLeod, R.D. and Congalton, R.G. (1998), A quantitative comparison of change-detection algorithms for monitoring eelgrass from remotely sensed data. *Photogrammetric Engineering and Remote Sensing*, 64(3): 207-216.

Maktav, D. and Erbek, F.S. (2005), Analysis of urban growth using multi-temporal satellite data in Istanbul, Turkey. *International Journal of Remote Sensing*, 26(4): 797–810.

Malila, W.A. (1980), Change vector analysis: an approach for detection forest changes with Landsat, *Proceedings of the 6th Annual Symposium on Machine Processing of Remotely Sensed Data*, Purdue University, West Lafayette, IN: 326-335.

Mas, J.F., 1999, Monitoring land-cover changes: a comparison of change detection techniques. *International Journal of Remote Sensing*, 20(1): 139-152.

Masek, J.G., Lindsay, F.E. and Goward, S.N. (2000), Dynamics of urban growth in the Washington DC metropolitan area, 1973–1996, from Landsat observations. *International Journal of Remote Sensing*, 21(18): 3473-3486.

McConnell, W., Sweeney, S.P. and Mulley, B. (2004), Physical and social access to land: spatio-temporal patterns of agricultural expansion in Madagascar. *Agriculture, Ecosystems and Environment*, 101: 171–184.

McGarigal, K. and Marks, B.J. (1995), FRAGSTATS: spatial pattern analysis program for quantifying landscape structure, USDA Forestry Service General Technical Report PNW-351.

McGarigal, K., Cushman, S.A., Neel, M.C. and Ene, E. (2002), FRAGSTATS: Spatial Pattern Analysis Program for Categorical Maps, Computer software program produced by the authors at the University of Massachusetts, Amherst. Available at the following web site: www.umass.edu/landeco/research/fragstats/fragstats.html.

Mertens, B. and Lambin, E.F. (2000), Land-cover-change trajectories in southern Cameroon. *Annals of the Association of American Geographers*, 90(3): 467-494.

Muttitanon, W. and Tripathi, N.K. (2005), Land use/land cover changes in the coastal zone of Ban Don Bay. *International Journal of Remote Sensing*, 26(11): 2311–2323.

Myneni, R.B., Kefling, C.D., Tucker, C.J., Asrar, G. and Nemani, R.R. (1997), Increased plant growth in the northern high latitudes from 1981 to 1991. *Nature*, 386: 698–702.

Narumalani, S., Mishra, D.R. and Rothwell-R.G. (2004), Change detection and landscape metrics for inferring anthropogenic processes in the greater EFMO area. *Remote Sensing of Environment*, 91: 478-489.

Olsson, L., Eklundh, L. and Ardö, J. (2005), A recent greening of the Sahel—trends, patterns and potential causes. *Journal of Arid Environments*, 63: 556–566.

Petit, C., Scudder, T. and Lambin, E. (2001), Quantifying processes of land-cover change by remote sensing: resettlement and rapid land-cover changes in south-eastern Zambia. *International Journal of Remote Sensing*, 22(17): 3435-3456.

Singh, A. (1989), Digital change detection techniques using remotely-sensed data. *International Journal of Remote Sensing*, 10(6): 989-1003.

Southworth, J., Nagendra, H. and Tucker, C. (2002), Fragmentation of a landscape: Incorporating landscape metrics into satellite analyses of land-cover change. Landscape Research, 27(3): 253-269.

Tang, J., Wang, L. and Zhang, S. (2005), Investigating landscape pattern and its dynamics in Daqing, China. *International Journal of Remote Sensing*, 26(11): 2259–2280.

Vanoort, P.A.J. (2005), Improving land cover change estimates by accounting for classification errors, *International Journal of Remote Sensing*, 26(14): 3009-3024.

Weber, C., Petropoulou, C. and Hirsch, J. (2005). Urban development in the Athens metropolitan area using remote sensing data with supervised analysis and GIS. *International Journal of Remote Sensing*, 26(4): 785–796.

Weismiller, R.A., Kristoof, S.J., Scholz, D.K., Anuta, P.E. and Momen, S.A. (1977), Change detection in coastal zone environment, *Photogrammetric Engineering and Remote Sensing*, 43(12): 1533-1539.

Zhao, G.X., Lin, G. and Warner, T. (2004), Using Thematic Mapper data for change detection and sustainable use of cultivated land: a case study in the Yellow River delta, China. *International Journal of Remote Sensing*, 25(13): 2509–2522.

Zhou, Q. and Sun, B. (2008), Analysing the spatio-temporal pattern of changing farmland in China's aridzone, in *Remote Sensing for Agriculture, Ecosystems, and Hydrology X*, Neale, C.M.U., Owe, M. and D'Urso, G. (eds.), 15-18 September, Cardiff, Proceedings of SPIE, Vol. 7104: 71040U.

Zhou, Q., Li, B. and Kurban, A. (2008a), Trajectory analysis of land cover change in arid environment of China, *International Journal of Remote Sensing*, 29(4), 1093 – 1107.

Zhou, Q., Li, B. and Kurban, A. (2008b), Spatial pattern analysis of land cover change trajectories in Tarim Basin, northwest China, *International Journal of Remote Sensing*, 29(19): 5495-5509.

Zhou, Q., Li, B. and Sun, B. (2008c), Modelling spatio-temporal pattern of landuse change using multi-temporal remotely sensed imagery, *The International Archives of The Photogrammetry, Remote Sensing and Spatial Information Sciences*, Chen, J., Jiang, J. and van Genderen, J. (eds.), 37(B7): 729-734.

Chapter 17

REMOTE SENSING OF WATER ENVIRONMENT

Xiaoling Chen, Zhifeng Yu

17.1 Introduction

Water, the hydrosphere, covers approximately 71% percent of the earth. It consists of ocean, river, lake, marsh, glacier, snow, groundwater, air moisture, and so on. Water environment, closely linked with human being's life, is facing serious problems of pollution and eutrophication. Water resource's protection and management has become more and more important in the world.

Water has been traditionally monitored by *in situ* measurement, to take point samples at regular intervals. But point samples are not adequate to observe spatial and temporal variations in a large area. Remote sensing has provided a new way to obtain water quality data over large areas simultaneously. Various kinds of remotely sensed images, including air-borne and space-borne optical (passive visible and infrared, laser), and passive and active microwave (e.g., Synthetic Aperture Radar, SAR) images, have become important information source for monitoring and detecting water quality. Satellite sensors such as CZCS (Coastal Zone Color Scanner), SeaWiFS (Sea-viewing Wide Field-of-view Sensor), MODIS (Moderate Resolution Imaging Spectroradiometer), MERIS (Medium Resolution Imaging Spectrometer) and Landsat series with various spatio-temporal and spectral resolutions can provide more timely synoptic water quality data (Chen *et al.* 2004). Therefore, remote sensing could be used as an independent measurement tool by water management authorities (Dekker *et al.* 2001, 2002).

When remote sensing became the irreplaceable advanced approach in the field of Earth observation, water remote sensing has also been adopted as one of the key technologies in ocean science in the late 20th century. In recent years, along with the aggravation of the ecological environment problem, coastal and inland water quality has deteriorated in varying degrees, giving the remote sensing technique an opportunity in water monitoring, study and exploitation. With the help of the remote sensing technique, we can realize oil spill detection, red tide detection and so on. Further more, as an important branch of water remote sensing, water color remote sensing from airborne or space-borne sensors can observe the concerned parameters related to chlorophyll, Suspended Sediments (SS), Colored Dissolved Organic Matter (CDOM), etc. based on the absorbing and scattering

D. Li et al. (eds.), *Geospatial Technology for Earth Observation*,
DOI 10.1007/978-1-4419-0050-0_17,

properties of water body in the range of visible wavelength. This makes water color remote sensing a useful tool to retrieve information about the water environment. Two issues are very important for water color remote sensing: atmospheric correction and the quantitative retrieval of water quality parameters. This chapter will discuss how remote sensing technique is used in water environment analysis.

17.2 Optical Characteristics of Water Body

17.2.1 Constituents of Water Body

From an optical perspective, in addition to pure water itself, the optical properties of natural bodies of water are mainly influenced by three constituents: phytoplankton, suspended sediment, and colored dissolved organic matter (CDOM) that is also called yellow substance or gelbstoff. Pure water is defined as a chemical pure substance composed of a mixture of several water isotopes, each of different molecular mass (Pozdnyakov *et al.* 2003).

The principal drifting organic matter in natural water consists of phytoplankton and other microscopic organisms. For convenience, it is often called the "phytoplankton" component, in recognition of its major influence on optical properties. The phytoplankton component is the single-celled plants that form the base of the aquatic food web, and plays an important role in the global carbon cycle. Chlorophyll mainly exists in phytoplankton and other animalcule. The concentration of Chlorophyll-a, the main phytoplankton pigment, is often used to estimate phytoplankton biomass (IOCCG, 2000).

Suspended sediment refers to the suspended microscopic solid particular matter that is not included in the phytoplankton component and commonly has a diameter of less than 2 millimeters. The presence of terrigenous suspended particles is a result of coastal erosion, catchment runoff, river discharge, and long- and short-range transport of atmospheric particulates followed by dry and wet deposition (Pozdnyakov *et al.* 2003). As a result, suspended particles brought into suspension can significantly modify the color of oceans and play an important role in determining the optical properties of the water.

CDOM is a collective term for a vast array of chemically complex organic compounds dissolved in natural water, consisting of humic and fulvic acids. CDOM usually has two origins: one is degradation of biological organism on the spot, and the other is direct inputs of terrestrial matter. It can absorb not only blue–green visible light but also light in the UV-A and UV-B. This absorption by CDOM can significantly influence the light field in the water and subsequently affect marine ecosystems (Chen *et al.* 2004).

17.2.2 Impact of Water Body on Electromagnetic Radiation

When the sun light arrives at water surface, the solar radiance can be divided into following parts:

1) Radiance reflected by water surface: a small proportion of incident light (about 3.5%) is reflected directly to the sky, and form the reflected radiance;

2) Solar radiance absorbed by the water: except for specular reflection of sun light, the maximal part of radiance is absorbed by water;

3) Scattered radiance: a part of incident light is scattered by the suspended material and organism, forming the scattering light below water surface, the light of which goes out of water surface and form back scattered light.

4) Radiance reflected from the bottom: In shallow and clear waters, a significant part of the light from the sun may get to the bottom, and be reflected from it. Some of the scattered and reflected photons eventually find their way to the remote sensor.

Because the effect of atmospheric scattering, water constituents absorbance and some other influence factors, the water-leaving radiance received by the sensor at the top of atmosphere has suffered from attenuation and departure (Fig. 17.1).

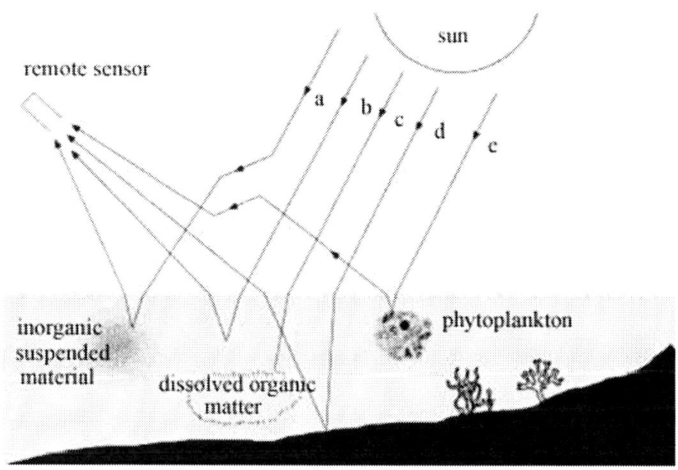

Fig. 17.1 Factors that influence upwelling light leaving the sea surface: (a) Upward scattering by inorganic suspended material; (b) Upward scattering from water molecules; (c) Absorption by the yellow-substances component; (d) Reflection off the bottom; and (e) Upward scattering from the phytoplankton component (Reprinted with permission from IOCCG).

Below water surface, the attenuation of water-leaving radiance is caused by the absorbance and scattering of the suspended microscopic particles and water itself. The water-leaving radiance detected by the sensor has a function relationship with the concentration of each material. The absorbance and scattering coefficient can be expressed as the function of the concentration of each material below water surface.

In order to detect the concentration of every constituent of waters, the effect of absorbance and scattering that every constituent has on light and the properties of absorbance and scattering of every constituent must be known firstly, and then the relationship between water-leaving radiance and the concentration can be built.

As the basic component of waters, the strong reflection of pure water lies in blue and green wavelength. Both the chlorophyll and suspended sediment have prominent effect of absorbance and scattering on visible light that penetrate into the waters. The impact of CDOM on the optical properties of water is more easily retrieved than chlorophyll and suspended sediment, because only the absorbance contribution of CDOM towards water will be taken into account, ignoring its scattering properties.

17.2.3 Optical Classification

Waters can be divided into inland waters, costal waters, and open sea according to their geographical locations and optical properties.

According to the influence of water constituents on the optical properties, waters can be partitioned into Case I waters and Case II waters using the bipartite classification scheme (Morel and Prieur 1977, Gordon and Morel 1983). The water classification is based on the relative contributions of the three types of substance: phytoplankton, suspended material, and yellow substances. Classification does not depend on the magnitude of individual contribution (IOCCG, 2000). International Ocean Color Coordinating Group (IOCCG) has given a pictorial representation in Fig. 17.2 (IOCCG, 2000).

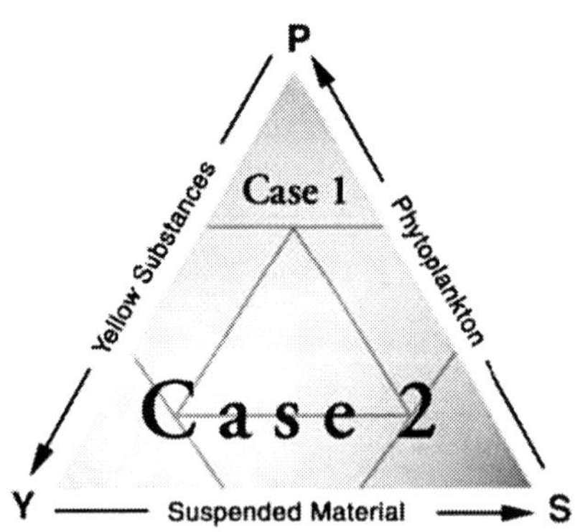

Fig. 17.2 Diagrammatic representation of Case I and Case II waters (Reprinted with permission from IOCCG)

Case II waters mainly distribute in coastal area and estuary regions which are affected by the inputs of terrestrial matter while the open sea is usually Case I waters. The optical properties of Case I waters are mainly influenced by phytoplankton and their accompanying and co-varying retinue of material of biological origin, and can be modeled as a function of phytoplankton concentration. The variations in optical properties of Case II waters rely on not only phytoplankton and related particles, but also inorganic suspended sediment and colored dissolved organic matter that vary independently of phytoplankton (IOCCG, 2000).

17.2.4 Inherent Optical Properties and Apparent Optical Properties of Waters

Remote sensing is an important way to observe the optical properties of waters and the research of optical properties of waters is especially important. Two reasons can account for it. On one hand, the contribution made by the water-leaving signals to the total signal detected by the sensor is very small; on the other hand, the retrieval algorithm of water color remote sensing is sensitive to the error of water-leaving radiance.

The optical properties of waters can be divided into apparent optical properties (AOPs) and inherent optical properties (IOPs). Apparent optical properties, mainly consisting of upwelling irradiance, downwelling irradiance, water-leaving irradiance, remote-sensing reflectance, ratio of irradiance, diffuse attenuation coefficient for downwelling irradiance (K_d), and diffuse attenuation coefficient for upwelling irradiance (K_u) are optical properties that are influenced by the angular distribution of the light field, as well as by the nature and quality of substances present in the medium (IOCCG, 2000). Since apparent optical properties will be changed with the incident light field, the measurements at different time and place must be normalized so as to compare with each other.

Inherent optical properties, mainly consisting of absorbance coefficient, scattering coefficient, attenuation coefficient, and volume scattering function are independent of variations in the angular distribution of incident light field and are solely determined by the type and concentration of substance present in the medium. Since inherent optical properties are not influenced by the illumination conditions in a multi-component medium, the total inherent optical properties can be derived by a simple addition of the contributions of the individual components, which may not hold for apparent optical properties (IOCCG, 2000).

17.3 Data Acquisition

There are two major ways, *in situ* / field observation and space / air borne remote sensing, to acquire data for water environment analysis. In order to retrieve quantitative values of water quality parameters from remote sensed image, in situ data should be measured to validate the retrieval algorithms or models.

17.3.1 In Situ Measurement

At the time of overpass of satellites, simultaneously measured in situ data will be desired to support the validation of algorithms for atmospheric correction, retrieval of water quality parameters, including chlorophyll, suspended sediments, and colored dissolved organic matter. We can directly obtain absorption coefficient (a), beam attenuation coefficient (c), remote-sensing reflectance, atmospheric parameters and water quality parameters by in situ measurement and laboratory analysis. The total scattering coefficient (b) can be indirectly derived by subtracting a from c in the same waveband: $b = c - a$.

In order to retrieve water quality parameters from remote sensed image, we need not only to measure absorption coefficient, but also to determine how the absorption is distributed among different components, CDOM, phytoplankton, and inorganic particulates. There are three ways to achieve these measurements. The first way is combination of the filter technique for absorption measurements for the particulate absorption and a spectrophotometer method for measuring absorption by yellow substances. The second way is to use an in situ absorption meter. The third way is to use an integrating cavity meter. Each way has their own advantages and disadvantages for particular waters (IOCCG, 2000).

The measurement of 'remote-sensing reflectance' just above the water surface (spectral upwelling radiance divided by downwelling irradiance, i.e. $L_u(\lambda)/E_d(\lambda)$) is an essential work that must be carried out by any water color remote sensing algorithm. The standard method for estimating this value is to measure profiles of $L_u(\lambda)$ and $E_d(\lambda)$ and use values of K_u and K_d calculated from these profiles to extrapolate the profiles back up to the surface. It is a difficult measurement and calculation because of wind-roughened water surfaces, partly cloudy skies, ship shadows, and other common difficulties. The second approach is to measure remote-sensing reflectance using a hand-held spectrometer pointed down at the water surface. The third approach to estimating remote-sensing reflectance at the surface is to use the measured values of absorption a and back-scattering coefficient b_b and a radiative transfer model to compute remote-sensing reflectance from the inherent optical properties of water (IOCCG, 2000).

There are some atmospheric and hydrological parameters, including wind speed, relative humidity, barometric pressure, water depth, and water temperature, which can be obtained by *in situ* observation.

17.3.2 Remote Sensing Measurement

The development of remote sensing technique has provided us a new way to obtain earth observation data. Some launched satellites have special sensors for water color. The Nimbus-7 satellite containing the Coastal Zone Color Scanner (CZCS), the first generation of ocean color sensors, was launched in America in 1978. This satellite ran on-orbit for more than 8 years, and colleted much useful oceanic data for investigation. CZCS is a scanning radiometer viewing the ocean in six co-registered spectral bands, five in the visible and near IR (443, 520, 550, 670, and 750 nm, labeled Bands 1, 2, 3, 4, and 5, respectively), and one in the thermal IR band (10.5-12.5 μm, labeled Band 6). The goal of the CZCS is to provide evaluation of the near-surface concentration of phytoplankton pigments (the sum of the concentrations of chlorophyll a plus phaeophytin a) and total suspended particles by measuring the spectral radiance backscattered out of the ocean (Gordon *et al.* 1987).

The Seastar satellite loaded with Sea-viewing Wide Field-of-view Sensor (SeaWiFS) launched in America in August 1997. As a second generation of ocean color scanners, SeaWiFS is superior to its predecessor CZCS in terms of radiometric sensitivity, Signal-to-Noise Ratio (SNR), and spectral resolution. In addition, SeaWiFS has an additional 4 spectral bands for bio-optical applications (band 1 and band 3 centered at 412 nm and 490 nm respectively) and atmospheric correction (band 7 and band 8 centered at 765 nm and 865 nm respectively). The goal of SeaWiFS is to help the oceanographic society to get better understand to the biological and biogeochemical processes of the ocean (Chen *et al.* 2003).

As the forerunners of water color satellite sensors, CZCS and SeaWiFS have led to a series of increasingly-sophisticated sensors, such as MODIS, MISR, OCM, GLI, OCI, OSMI, MERIS, and so on. These new sensors have better radiometric performances and an increased number of spectral channels (from 5 for CZCS up to 36 for MODIS and GLI). In addition to the sensors mentioned previously, some other optical (passive visible and infrared, active laser), and air-borne and space-borne passive and active microwave (e.g., Synthetic Aperture Radar, SAR) sensors can also be used to serve for water environment analysis.

17.4 Atmospheric Correction

More than 90% of the signal measured by an ocean color satellite sensor is due to the confounding influence of the atmosphere. The atmospheric and ocean surface effects must be removed before ocean radiance signals can be analyzed for the purposes of understanding the ocean biosphere. This step of processing satellite ocean color imagery is referred to as the atmospheric correction procedure (Antoine *et al.* 2003, Gordon 1997). In this part, the atmospheric radiative transfer model and atmospheric correction model of Case Ⅰ and Case Ⅱ waters in the field of water color remote sensing will be introduced.

17.4.1 Ocean-Atmospheric Radiative Transfer Model

Radiative transfer theory is distinguished by the fact that it is one of the branches of physics that can be made to rest on a single principle from which all the salient structures of the theory can be systematically deduced (Mobley 1994). Before any interpretation of the marine signals can be made, the crucial problem in detecting ocean color from space is to make an accurate "atmospheric correction" (IOCCG, 1998). The backscattered radiation by air molecules and aerosols is predominant in forming the radiance detected by the sensors at the top of the atmosphere (TOA) in the visible part of the spectrum (IOCCG, 1998). The incident radiant energy arriving at ocean surface can be mainly divided into two parts, one part entries into the water, and the other part is reflected into the sky. As shown in Fig. 17.3, (a) water-leaving radiance, (b) attenuating water-leaving radiance, (c) water-leaving radiance scattered out of the instantaneous field of view (IFOV), (d) specular reflection of direct sunlight at the sea surface, (e) specular reflection of scattered sky light at the sea surface, (f) reflected radiance scattered out of the instantaneous field of view (IFOV), (g) attenuating reflected radiance, (h) scattered solar light, (i) scattered sky light by atmosphere, (j) water-leaving radiance scattered out of the instantaneous filed of view (IFOV) of the sensor and traveling in the direction of the sensor, (k) specular reflection of direct sunlight scattered out of the instantaneous field of view (IFOV) and traveling in the direction of the sensor, (L_w) total water-leaving radiance, (L_r) total specular reflection at water surface within the instantaneous filed of view (IFOV) of the sensor, (L_p) atmospheric radiance, the radiant energy received by a remote sensed sensor includes: b, d, e, h, i, j, k. It is only the upwelling light from the sea surface (about 10% of the total signal at the top of atmosphere) that carries useful information of the water body. The atmospheric contributions and specular reflections at the sea surface constitute noise that has to be removed. Thereby, atmospheric correction of water color remote sensing is a course to extract a small signal from a large signal, and it is a crucial point desiderating resolution of water color remote sensing.

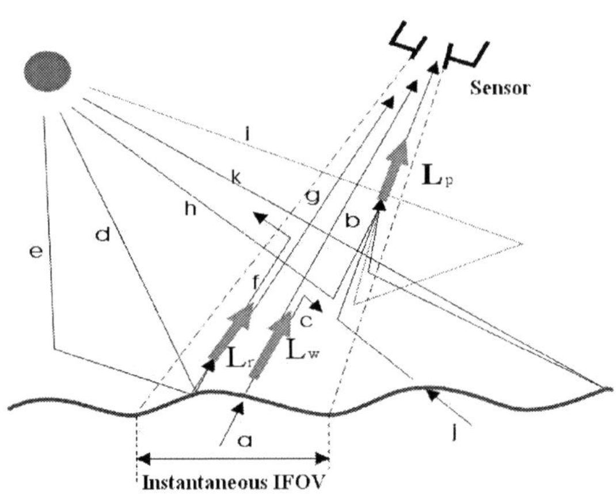

Fig. 17.3 Radiative transfer process of ocean-atmosphere

According to the typical ocean-atmosphere radiative transfer model, the radiance received by the water color sensor consists of three parts: water-leaving radiance from water surface, specular reflection of radiance from water surface, and scattered radiance by the atmosphere. The atmospheric correction equation can be depicted as follows:

$$L_t(\lambda) = L_r(\lambda) + L_a(\lambda) + L_{ra}(\lambda) + \mathrm{T}(\lambda)L_g(\lambda) + L_b(\lambda) + t(\lambda)L_f(\lambda) \\ + t(\lambda)(1-w)L_w(\lambda) \tag{17.1}$$

where $L_t(\lambda)$ is the total radiance received by the sensor of water color remote sensing; $L_r(\lambda)$ is the Rayleigh scattering radiance; $L_a(\lambda)$ is the aerosol scattering radiance; $L_{ra}(\lambda)$ is the multiple scattering between Rayleigh and aerosol radiances; $L_g(\lambda)$ is the contribution arising from specular reflection of direct sunlight from the sea surface (sun glitter); $T(\lambda)$ is the direct transmittance of the atmosphere or Beam Transmittance; $L_b(\lambda)$ is the radiance from the bottom of the waters; $L_f(\lambda)$ is the contribution arising from sunlight and skylight reflecting from individual whitecaps on the sea surface; $t(\lambda)$ is diffuse transmittance, that is the attenuation coefficient of the atmospheric transmission between satellite and water surface; w is covering rate of whitecaps; $L_w(\lambda)$ is the desired water-leaving radiance.

According to the different weights and influences of all the radiance, the factors having impact on atmospheric correction can be divided into three grades: Rayleigh scattering has the largest impact on atmospheric correction, occupying about 80% of the radiance generated along the optical path by scattering in the atmosphere and by specular reflection of atmospherically scattered light (skylight) from the sea surface; aerosol scattering is in the next place; and small impact coming from other factors including multiple scattering, polarization, roughness of the water surface, reflectance of whitecaps, water-leaving radiance, air pressure and spatio-temporal diversification of ozonic concentration is the last. In the process of dealing with the data of water color remote sensing, some parameters with little impact on atmospheric correction are usually ignored.

17.4.2 Atmospheric Correction Algorithms

In term of atmospheric correction, several techniques exist for the ocean color bands, such as Coastal Zone Color Scanner (CZCS), Sea-Viewing Wide Field-of-View Sensor (SeaWiFS), and Moderate-resolution Imaging Spectroradiometer (MODIS), based on their ability to gain information about the spectral dependency of aerosol scattering through measurements at near-infrared bands for extrapolating the atmospheric radiance into the visible bands (Gordon and Wang 1994).

For Case I waters (clear water), based on the black pixel assumption, the values of water-leaving radiance in the near-infrared (NIR) bands are negligible, so in the atmospheric correction of ocean color data, the aerosol radiative properties and water-leaving radiance values at various bands are to be easily determined (Gordon and Wang 1994).

For Case II waters (turbid water), the assumption that the water-leaving radiance is zero at near-infrared bands does not hold, because turbid water constituents (suspended sediments, bubbles, etc) can contribute significant amounts of radiance to the atmospheric correction bands (765nm and 865nm for SeaWiFS, 748nm and 869nm for MODIS ocean color data), which make it difficult to separate the aerosol information from the total recorded signal exactly. As a result, the current standard atmospheric correction algorithms frequently overestimates the radiance from aerosol scattering and aerosol-Rayleigh interaction, and underestimates water-leaving radiance at some visible bands, even yielding negative values at the green and blue spectra.

17.4.2.1 Atmospheric Correction Algorithm over Case I Waters

Gordon CZCS Algorithm for Single Scattering Effects over Clear Waters

The purpose of atmospheric correction is to obtain water-leaving radiance L_W that contains useful information about water constituent. The radiance backscattered from the atmosphere and sea surface should be removed from the total radiance L_t. In the process of atmospheric correction of CZCS imagery, Gordon firstly put up with approximate atmosphere correction algorithm suitable for Case I waters assuming single scattering over clear water (Gordon 1978, Gordon *et al.* 1981, Gordon *et al.* 1987).

On the basis of assuming single scattering theory, ignoring direct sun glint and assuming that the sea surface is flat, the total sensor radiance $L_t(\lambda)$ consists of three components: $L_r(\lambda)$, Rayleigh scattering; $L_a(\lambda)$, aerosol scattering; and $t(\lambda)L_\omega(\lambda)$, the water-leaving radiance diffusely transmitted to the top of the atmosphere. The atmospheric correction equation can be simplified as follows:

$$L_t(\lambda) = L_r(\lambda) + L_a(\lambda) + t(\lambda)L_w(\lambda) \tag{17.2}$$

where $t(\lambda)$ is the diffuse transmittance of the atmosphere between the sea surface and the sensor.

Atmospheric Correction Algorithm for the Second-generation Ocean Color Scanners

The atmospheric correction algorithm of Gordon CZCS algorithm for Single Scattering effects over clear waters is not suitable for the second generation of ocean color scanners. In 1994, based on the research of CZCS atmospheric correction algorithm, Gordon and Menghua Wang put up with a new atmospheric correction algorithm by use of two near infrared bands of SeaWiFS sensors (Gordon and Wang 1994), which was launched in August 1997. The new algorithm takes the assumption that the water-leaving radiance of atmospheric-correction bands at 765 nm and 865 nm approaches zero, according to which the radiance by aerosol scattering can be estimated. And then the aerosol scattering radiance at near infrared band can be extrapolated into visible bands. After that, water-leaving radiance can be ultimately extracted. Because this algorithm needs the least in situ measurement data and can be regularly operated, it has been applied in the SeaWiFS Data Analysis System (SeaDAS) software as a standard atmospheric-correction algorithm for SeaWiFS and MODIS imageries by NASA.

The critical point of the atmospheric correction algorithm of the second generation ocean color scanner is to achieve optical characteristic of all the aerosol existing in the study area. Because of the high spatio-temporal variation of aerosol property and limitations of aerosol understanding, it is not possible to thoroughly resolve atmospheric correction of ocean color remote sensing by use of aerosol models embedded in SeaDAS software, especially for Case II waters.

17.4.2.2 Atmospheric Correction Algorithm over Case II Waters

For Case II waters, a number of improvements were introduced to correct the effect of the near-Infrared (NIR) contribution on water-leaving radiance and retrieve the aerosol scattering and aerosol-Rayleigh interaction term properly in atmospheric correction procedure (Arnone *et al.* 1998, Chen *et al.* 2003, He *et al.* 2004, Hu *et al.* 2000, Land and Haigh 1996, Lavender and Nagur 2002, Liao *et al.* 2005, Pan and Mao 2001, Ruddick *et al.* 2000, Siegel *et al.* 2000, Shanmugam and Ahn 2007, Shutler *et al.* 2007). There are several approaches commonly used: (1) assuming that the backscattering or absorption coefficients of the water constituents are available, take an iterative or spectral optimization approach to obtain both the oceanic and atmospheric parameters simultaneously (Arnone *et al.* 1998, Chomko *et al.* 2003, Gao *et al.* 2007, Hoyningen-Huene *et al.* 2006, Land and Haigh 1996, Zhao and Nakajima 1997); (2) assuming that the type of aerosol or some water properties does not vary much over a relatively small spatial scale (50-100km), and there are some relationships can be expressed by an approximate model, apply the optical properties observed over adjacent areas to the turbid water pixels (Hu *et al.* 1998, 2000, Ruddick *et al.* 2000, Sturm and Zibordi 2002, Zagolski *et al.* 2007); (3) use the multi-platform vicarious inter-calibration approach to correct the satellite data for ocean color remote sensing (Hu *et al.* 2001, Wang and Franz 2000). The first method generally works well, but some parameters should be known prior to use and generally, it is very difficult to use. The second one uses the hypothesis of homogenous aerosol type or water optical properties to correct satellite imagery over turbid waters, however, the assumption always does not hold well (Hu *et al.* 2000). The third approach may improve the results of atmospheric correction, but the high variation of the atmospheric condition have not been taken into account during the images obtained by multi-sensors, furthermore, all the passive sensors have limit abilities to separate the water-leaving signals from the total radiance measured by satellite at the NIR bands over turbid waters. In addition, many ground-based aerosol observations have been taken over or near coastal water area (Behnert *et al.* 2007, Bryant *et al.* 2006, Knobelspiesse *et al.* 2004, Sakerin and Kabanov 1997, Vergaz *et al.* 2005), but there is just a little improvement in ocean color remote sensing over turbid waters, because the aerosol data and satellite images generally are not simultaneous in spatial and temporal space. In order to find a better algorithm over Case II waters, we have tried the atmospheric correction method using active and passive remote sensing over turbid coastal waters, which will be depicted in Section 17.4.3.2 (Tian *et al.* 2009).

17.4.3 Case Studies of Atmospheric Correction over South China Coastal Turbid Waters (Case II Waters)

In China, some studies were done with atmospheric correction over Case II waters along the China Seas. The issue of the atmospheric correction results having negative values in some complicated water areas using standard atmospheric correction algorithm has been discussed. Two case studies related to atmospheric correction over the turbid coastal waters of South China will be discussed.

17.4.3.1 Atmospheric correction of SeaWiFS imagery for turbid waters in South China coastal waters

In 2003, a practical approach for atmospheric correction of SeaWiFS imagery for turbid waters in the Pearl River Estuary and its adjacent areas locating in Southern China coastal areas was proposed (Chen *et al.* 2003).

We define:

$$L_{t-r}(\lambda) = L_t - L \qquad (17.3)$$

In this approach, it has been found that there is a prominent linear relationship between $L_{t-r}(765)$ and $L_{t-r}(865)$ from normal coastal water area to very turbid coastal water area. According to this approach, a result of atmospheric correction of SeaWIFS imagery acquired on March 1st, 2001 was achieved, as shown in Figure 17.4 (band: 490nm and 555nm).

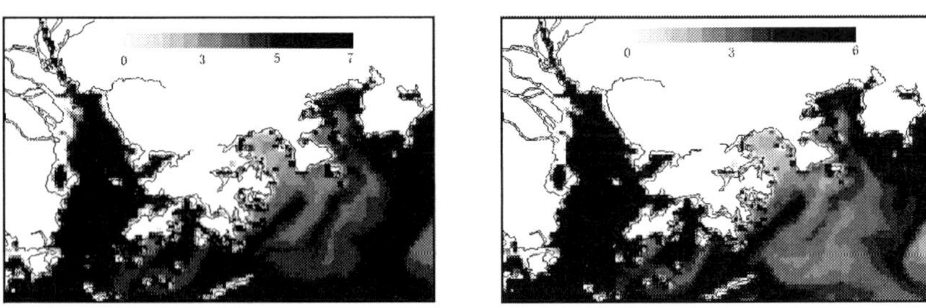

Fig. 17.4 Water-leaving radiance: (left) $L_w(490)$; (right) $L_w(555)$

17.4.3.2 Atmospheric Correction of MODIS/Aqua Ocean Color Imagery Based on CALIOP Aerosol Data

In 2009, we proposed an atmospheric correction approach for ocean color imagery over turbid coastal waters using active and passive remote sensing (Tian *et al.* 2009).

As is known, Aqua, CloudSat, CALIPSO, PARASOL and Aura satellites form the "A-Train" constellation (Anderson *et al.* 2005). In the constellation, CALIPSO (Cloud-Aerosol LIDAR, Infrared Pathfinder Satellite Observation) satellite lags Aqua by only 1 to 2 minutes in the afternoon, and CALIOP (Cloud-Aerosol Lidar with Orthogonal Polarization) is the primary instrument onboard CALIPSO satellite, so CALIOP data and MODIS/Aqua imagery can be approximately considered to be obtained simultaneously. Furthermore, CALIOP monitors aerosol information by an active way. The signals from sea surface have little effects on CALIOP measurements, so CALIOP aerosol data can be easily adopted to avoid the turbid water signal disturbance in the retrieval of the atmospheric effects at NIR bands in ocean color remote sensing atmospheric correction study.

The method to retrieve the water leaving radiance from MODIS/Aqua imagery over Chinese turbid costal waters using CALIOP aerosol data is described in detail in Figure 17.5.

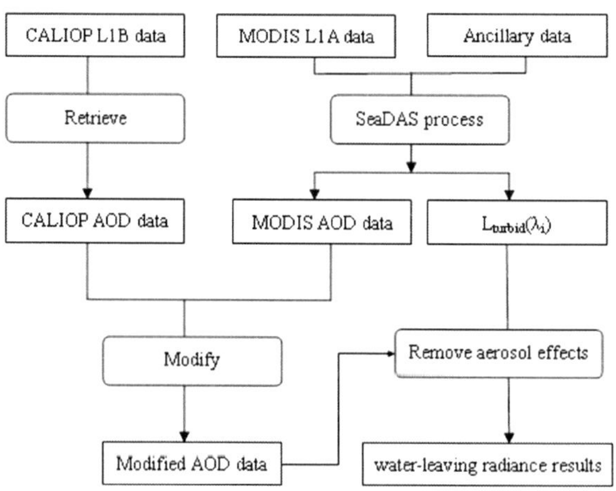

Fig. 17.5 Framework of the atmospheric correction method by combining MODIS and CALIOP data. $L_{turbid}(\lambda_i)$ is the preliminary atmospheric correction results that the effects of Rayleigh scattering, sunglitter, whitecaps have been removed.

To test the modified atmospheric correction method, we have applied the proposed scheme to the MODIS/Aqua scene acquired on 5 October 2006 over the South China Sea and its coastal waters, including the Pearl River Estuary and its adjacent turbid waters. Fig. 17.6 shows the AOD distributions at 531nm before and after the adjustment for the MODIS/Aqua imagery on 5 October 2006. From Fig. 17.6a, we can find that there are many extremely high values over turbid waters. The modified AOD distribution in Fig. 17.6b seems to be reasonable, and the modified AOD values are lower than those by SeaDAS over turbid waters (in the rectangular box). From the comparison with the water-leaving radiance products at 412nm processed by SeaDAS (see Fig. 17.7a), we can find that the extremely high AOD values in Fig. 17.6a are corresponding to the Pearl River Estuary and its adjacent turbid coastal water, where the standard atmospheric correction can not give satisfied results. The overestimation of aerosol optical properties should partially result in the underestimation of water-leaving radiance and the failure of atmospheric correction.

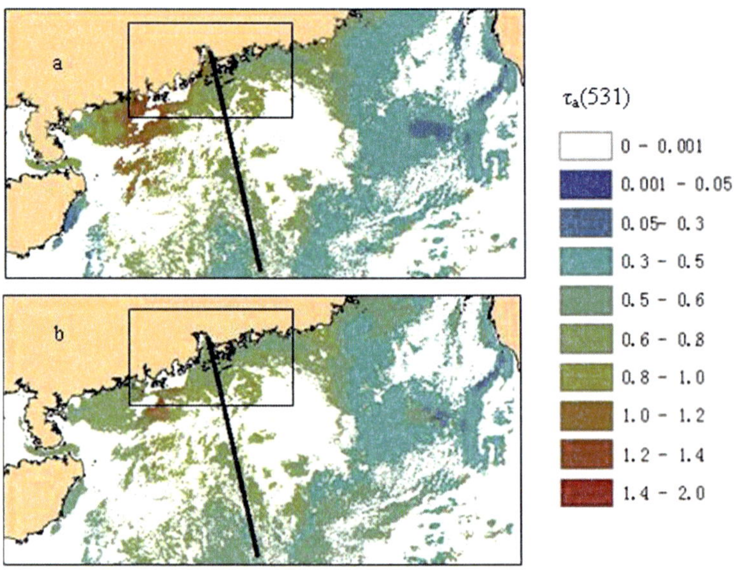

Fig. 17.6 MODIS/Aqua AOD distributions at 531nm ($\tau_a(531)$) on 5 October 2006 before and after incorporating CALIOP data (straight black line): (a) $\tau_a(531)$ cal-culated by SeaDAS. Values in red area are abnormal (note the large values in the rectangular box over the Pearl River Estuary); (b) $\tau_a(531)$ modified by CALIOP data (note the large values in the rectangular box over the Pearl River Estuary decrease significantly).

The water-leaving reflectance derived from MODIS image at 412nm by SeaDAS and our proposed atmospheric correction are shown in Fig. 17.7a and Fig. 17.7b, respectively. In the yellow color area shown in Fig. 17.7a, there are hundreds of pixels with "negative water-leaving reflectance" flags. The number of negative $\rho_w^*(412)$ pixels drops significantly by using our improved atmospheric correction method based on CALIOP aerosol data (Fig. 17.7b). The result is obviously improved with our proposed atmospheric correction method. Similar improved results can be found for MODIS bands at 443nm, 488nm, 531nm, 551nm and 667nm.

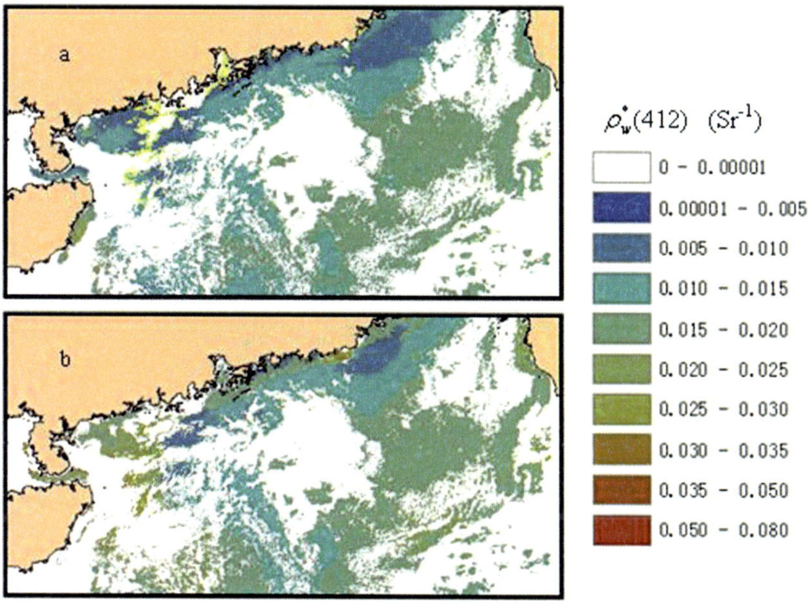

Fig. 17.7 Comparison of the water-leaving reflectance at MODIS/Aqua 412nm band on 5 October 2006 (unit: Sr^{-1}). (a) Result derived with the MODIS operational atmospheric correction algorithm. The yellow area is caused by negative water-leaving reflectance. (b) Result derived with our proposed atmospheric correction algorithm. Land is marked in brown color. White color refers to failure areas caused by cloud or other reasons.

17.5 Water Environment Information Retrieval and Analysis

17.5.1 Introduction

Because of its prominent advantage of synchronous observation in a large area coverage compared with the traditional water quality measurements, the remote sensing technique has currently been widely used for water environment analysis. The remote sensing imageries can be used for Information extraction and change detection, such as chlorophyll-a (Chl-a) and suspended sediment concentration retrieval, oil spill detection, and red tide detection.

Some water quality parameters, Chl-a, suspended sediment (SS), colored dissolved organic matter (CDOM), water depth, and water temperature can be quantitatively retrieved to understand the spatio-temporal variation of water environment after atmospheric correction is implemented over a remote sensing image to remove the effect of atmosphere and water surface from total radiance received by the sensor at the top of atmosphere. Nowadays, the retrieval of water constituents is relatively successful in Case I waters (open sea), because Case I waters is mainly influenced by phytoplankton and related particles. Although a lot of approaches including theoretical, semi-analytical, and empirical models have been proposed, it is very difficult to retrieve the concentration of the surface constituents in the Case II waters. The main reason is that the optical properties of the constituents (Chl-a, SS, and CDOM) are complicated in the Case II complex waters.

Recent years, harmful algal blooms have become a serious environmental problem in coastal areas on a worldwide scale (Anderson and Garrison 1997). Hence, it is of great importance to mitigate the impacts of such harmful algal blooms and therefore there is a need to monitor the blooms and to forecast their development and movement (Stumpf 2001). Because phytoplankton blooms are sporadic in time and isolated in space, they are hard to monitor by conventional methods (Gower *et al.* 2005). Thus, because of the synoptic and repeat coverage, it is an effective way to monitor such blooms through remote sensing technology.

As a major environmental threat, ocean oil pollution has been a subject of increasing public concern. In this background, particular attention is paid to the enormous environmental damage caused by oil-tanker accidents or by the illegal cleaning of tankers (Fiscella *et al.* 2000). According to the European Space Agency (1998), operative discharges from ships account for 45% of the oil pollution, and observed oil spills correlate very well with the major shipping routes (Lu 2003, Lu *et al.* 1999), such a large amount of oil is deliberately pumped into the marine environment every day. Since oil spills can cause substantial damage to the marine environment, monitoring illegal ship discharges is thus an important component in ensuring compliance with the marine protection legislation and the general protection of coastal environments.

17.5.2 Oil Spill Detection

The traditional survey of oil spills can not provide sufficient spatial or temporal coverage to assess the magnitude, area coverage, or duration of the spills (Hu *et al.* 2003), however, the techniques for remote sensing of aquatic oil spills including optical (passive visible and infrared, and laser fluorosensors), microwave, and radar (e.g., Synthetic Aperture Radar) approaches using aircraft or satellites are considered to be an effective way to monitor oceanic oil spill.

With the all-weather and all-day capabilities, microwaves are commonly used for ocean oil pollution monitoring by remote sensing. As is known, space-borne synthetic aperture radar (SAR) has particularly become a very popular tool for ocean oil slick monitoring due to its wide area coverage, and cloud-free, day and night operation (Lu *et al.* 2003). Under certain conditions, an oil slick will dampen capillaries and short gravity waves, reduce the radar backscatter coefficient, and appear as a dark slick in a SAR or SLAR (Side-Looking Airborne Radar) image of the ocean surface (Espedal *et al.* 1999). A major part of the oil spill detection problem is distinguishing oil spills from other natural phenomena (look-alikes) that dampen the short waves and create dark patches on the surface, such as natural films/slicks, algae, grease ice, threshold wind speed areas, wind sheltering by land, rain cells, shear zones, internal waves, etc. (Solberg *et al.* 2007). In the beginning, an oil slick in a SAR image is usually visually detected by a trained human interpreter based on experience and other relevant information. In order to speed up the detection process, algorithms for automatic detection that can help in screening the images and prioritizing the alarms have been gradually proposed. Solberg *et al.* (1999) developed a semi-automatic oil slick detection system, in which high probability oil slicks were first automatically identified, but with other possible slicks to be manually inspected. Solberg *et al.* (2007) proposed algorithms for automatic detection of oil spills in synthetic aperture radar (SAR) images, which consist of three main parts, namely: 1) detection of dark spots; 2) feature extraction from the dark spot candidates; and 3) classification of dark spots as oil spills or look-alikes. Although a SAR sensor is the most efficient satellite sensor for oil spill detection (Solberg *et al.* 2007), it also has some limitations (Hu *et al.* 2003). Firstly, it does not have capabilities for oil spill thickness estimation and oil type recognition. Secondly, SAR data are expensive, are not available daily, and are limited by sea state (Liu *et al.* 2000). In addition, the optimal wind speed to effectively detect oil spill slicks by SAR is about 1.5 to 6 m/s. Other features, such as phytoplankton slicks and fresh water slicks, may present interference and cause false detection (Lin *et al.* 2002).

In addition to SAR, other space-borne remote sensing devices also have some potential for oil spill monitoring. An example from Lake Maracaibo, Venezuela, Hu *et al.* (2003) proved the possibility of oil spill monitoring by MODIS instrument. What's more, hyperspectral sensors used for oil spill monitoring have a potential for detailed identification of materials (e.g crude oil or light oil) and better estimation of their abundance (Brekke *et al.* 2005). Since oil absorbs solar radiation and re-emits a portion of this energy as thermal energy, IR (infrared) sensors observe thick oil slicks as hot, intermediate thickness of oil as cool, and thin oil is not possible to be detected (Fingas and Brown 1997). UV technology can be used to detect oil spills as the spill displays high reflectivity of UV radiation even at thin layers, but the UV instrument is not usable at night and wind slicks, sun glints, and biogenic material can cause false alarms in the UV data (Brekke *et al.* 2005).

17.5.3 Retrieval of Water Depth

It is very useful to obtain bathymetric information in many remote areas. For example, Coral reefs, by their nature, strongly influence the physical structure of their environment, and water depth information is fundamental to discriminate and characterize the coral reef habitat, such as patch reef, spur-and-groove, and seagrass beds (Stumpf *et al.* 2003). In addition, knowledge of water depth also allows estimation of bottom albedo, which can improve habitat mapping (Mumby *et al.* 1998). However, traditional *in situ* surveys are time consuming and labor intensive, generally lack the spatial resolution and precision required to detect subtle changes before they become catastrophic, and can be difficult to maintain from year to year (Orth and Moore 1983, Peterson and Fourqurean 2001). With the development of geospatial technology, the feasibility of mapping bathymetry from satellite or aircraft imagery has been demonstrated by several researchers, with techniques ranging from photointerpretation and photogrammetry using black-and-white photography to computer analysis of multispectral satellite imagery (Philpot 1989). The interpretation of this photography is impeded by the fact that water depth variations are not easily distinguished from bottom color differences, but the use of digitally recorded multispectral scanner data permits corrections to be made for surface reflection effects and also allows the possibility of automatic recognition of bottom features and water depth using radiometric techniques (Lyzenga 1978).

Legleiter *et al.* (2004) summed up two popular models developed for bathymetric mapping in shallow coastal waters: (1) the linear transform introduced by Lyzenga (1978) and extended by Philpot (1989) and (2) a ratio-based technique used more recently by Dierssen *et al.* (2003) and Stumpf *et al.* (2003).

17.5.4 Retrieval of Water Temperature

The water surface temperature is a very important parameter of the water environment. Jessup *et al.* (1997) pointed out that applications of thermal infrared (TIR) remote sensing have been important in obtaining sea surface temperature measurements for oceanographic and meteorological applications. Applications of TIR remote sensing have been conducted to map surface temperature and circulation patterns in lakes, since water density is a complex function of temperature, and any investigation of lake dynamics is incomplete without measurements of the thermal behavior of the water body (Anderson *et al.* 1995). In addition, water temperature in rivers and streams has been identified as a critical element in the restoration of freshwater aquatic ecosystems and fishery managements. Torgersen *et al.* (2001) developed an airborne remote sensing method to measure spatial continuous patterns of stream temperature and evaluated the physical factors that influence the accuracy of thermal remote sensing of flowing waters.

Remote measuring water temperature is made with a sensor that detects thermal radiation (3~5 and 8~14 μm wavebands) emitted from the upper 0.1 mm of the water surface (Anderson *et al.* 1984, Atwell *et al.* 1971, Robinson *et al.* 1984). The temperature of the water surface measured remotely with a thermal sensor as the radiant water temperature (T_r) is different from the kinetic water temperature (T_k) measured 10 cm below the surface with a thermometer (Torgersen *et al.* 2001). Emissivity, atmospheric absorption, TIR reflection, and surface characteristics are the factors that influence measurements of T_r (Smith *et al.* 1996). Emissivity describes the actual absorption and emission properties of the water surface and is expressed as a ratio of the emittance from the water surface at a given temperature to that from a black body at the same temperature (Avery and Berlin 1992, Avery and Berlin 1992). Water has an emissivity very close to 1.0

and a high thermal inertia, so it is relatively easy to obtain measurements of T_r, while atmospheric effects can be corrected based on water vapor content and transmission along the target sensor path (Torgersen *et al.* 2001).

17.5.5 Retrieval of Total Suspended Sediment Concentration

Using empirical statistics, the key to the quantitative retrieval of suspended sediment (SS) is the establishment of the relationship between spectral reflectance and the concentration of suspended sediment. That is:

$$SSC = f(R_{rs}) \tag{17.4}$$

where R_{rs} is spectral reflectance, and SSC stands for suspended sediment concentration.

According to much in situ measurement result, the function relationship between spectral reflectance R_{rs} and suspended sediment concentration SSC can be analyzed as follows:

1) The remote-sensing reflectance increases with the increase of the concentration of suspended sediment (SSC), and that is $dR_{rs}/dSSC > 0$;

2) The variance ratio $dR_{rs}/dSSC$ is not constant, and it decreases with the increase of the concentration of SS, that is $d^2R_{rs}/dSSC^2 < 0$;

3) When $SSC = 0$, R_{rs} is a constant that is greater than zero; R_{rs} will approach the extreme 1 with the increase of SSC.

In general, the retrieval algorithm can be portioned into the theoretical algorithm and empirical algorithm.

17.5.5.1 Theoretical Algorithms

On the basis of marine optics and the radiative transfer model, through simulation experiment, we can explore the correlation between electromagnetic radiation and SSC, and then a series of semi-analytical algorithms can be derived which combine the radiative transfer model with the empirical equation and is the approximate solution of the radiative transfer equation. The common semi-analytical retrieval models of SSC mainly include Gordon model and negative exponent model.

1) Gordon model

According to the approximate formula of quasi single scattering of reflectance proposed by Gordon *et al.*, Munday got the approximate model of remote-sensing reflectance of the water (Munday *et al.* 1979):

$$R_{rs} = f\left[\frac{b_b(\lambda)}{a(\lambda)+b_b(\lambda)}\right] \tag{17.5}$$

where a is the total absorption coefficient of water; b_b is the total back-scattering coefficient of water, f is the functional relation.

The retrieval accuracy is not high using Gordon formula in practical calculation. It has two reasons: firstly, Gordon formula is derived by approximation, while Whitlock (1976) showed that there is an obvious non-linear relation between R_{rs} and $\frac{b_b(\lambda)}{a(\lambda)+b_b(\lambda)}$. Secondly, Gordon formula is obtained on the hypothesis that the optical properties of water is homogeneous, which is not tenable for the water containing suspended sediment. In fact, the concentration of suspended sediment varies prominently in vertical direction, so does the optical properties of water.

2) Negative exponent model

In the process of simplifying the radiative transfer equation, Li (1986) took the vertical variation of optical properties of water into account and proposed a plane layered model. Provide that the optical property is the function of water depth z. The negative exponent model is got:

$$R_{rs} = A + B(1 - e^{-D \cdot SSC}) \tag{17.6}$$

where A , B , D are all dimensionless constants. Negative exponent mode overcomes the weakness of other relation expressions which are only suitable for low concentration of suspended sediment and is closer to the trend how the reflectance varies with the variety of SSC .

Semi-analytical mode, as the approximation and simplification of theoretical mode, is simple and beneficial to operational application. However, when the mode is being established, some hypothesis is used in order to reduce the unknown quantities; what's more, semi-analytical mode relies on the accuracy of in situ measurement data to some extent. All the factors will bring inevitable error in retrieval result using semi-analytical model.

17.5.5.2 Empirical Algorithms

Empirical algorithm is used to establish the correlation between the remote-sensing data and simultaneously in situ measured data. The simultaneously in situ measured data is necessary for empirical algorithm, especially in estuary or regions affected greatly by tide, weather and hydrological condition. The existing empirical algorithms mainly include linear model and logarithm model.

1) Linear model

Weisblatt firstly found that there was a linear relation between suspended sediment concentrations and irradiance (Weisblatt 1973). The general expression of linear relation is:

$$R_{rs} = A + B \cdot SSC \quad (B > 0) \tag{17.7}$$

where R_{rs} is remote-sensing reflectance at a certain wavelength; SSC is the concentration of suspended sediment; A, B are constants. This expression can not satisfy the above-mentioned relation characteristics between SSC and water-leaving radiance (See Fig. 5.1), so it can only be applied in lower concentration of suspended sediment water.

2) Logarithm model

The general expression of logarithm is:

$$R_{rs} = A + B \cdot \ln(SSC)(B > 0) \tag{17.8}$$

where A, B are constants. Klemas *et al.* (1974) found that logarithm relation model is better than linear relation model in describing the relationship between suspended sediment concentration and irradiance in term of Landsat MSS images in Delaware Bay.

For the moment, the common retrieval models just suitable for specific region are semi-analytical models or empirical models and it is time consuming to obtain sufficient simultaneously in situ measured data. Furthermore, the retrieval models can not be effectively transplanted to different water regions in the spatio-temporal scale.

17.5.6 Retrieval of CDOM Concentration

The retrieval of CDOM (colored dissolved organic matter) concentration by remote sensing can be partitioned into two aspects: firstly, it is required to remove the influence of CDOM before retrieving other water color parameters (Tassan 1988, Carder *et al.* 1989); secondly, we need to study how to retrieve the CDOM concentration.

At present, the modes of retrieval of CDOM concentration mainly include two categories: one is to extract the CDOM concentration directly; the other is to calculate the absorption coefficient at the characteristic wavebands of CDOM and use absorption coefficient to represent the CDOM concentration, which is indirect extracting mode of CDOM.

17.5.6.1 Direct Extracting Mode

CDOM is the colored fraction of the dissolved organic carbon (DOC), and commonly DOC can be used to represent CDOM. According to the in situ measured spectrum and the concentration of DOC at Pearl River Estuary of southern China, Chen *et al.* (2003) established the retrieval mode of regression analysis between DOC concentration and optimal wavebands combination as follows:

$$\lg(DOC) = 1.2419 \cdot \lg\left(\frac{R_{670}}{R_{412}}\right) - 0.2614 (n = 130, r = 0.916) \quad (17.9)$$

where R refers to simulated spectral reflectance.

According to in situ measured spectrum and the concentration of DOC of eight reservoirs in Colorado, Arenz *et al.* (1996) established the regression relational expression as follows:

$$DOC = 0.55 \cdot \left[\left(\frac{R_{716}}{R_i}\right)^{-9.6}\right]\left[\left(\frac{R_{706}}{R_{670}}\right)^{-9.6}\right] \quad (17.10)$$

where R_i refers to spectral reflectance.

17.5.6.2 Indirect Extracting Mode

According to the relevant material about Golfo di Napoli, S.Tassan proposed a model to extract the absorption coefficient of CDOM at the wavelength of 440nm using SeaWiFS data (Tassan, 1994):

$$\lg[-a_g(440)] = -3.0 - 1.93\left[\left(\frac{R_1}{R_3}\right)(R_2)^{0.5}\right] \quad (17.11)$$

where R_1, R_2, R_3 are the reflectance of 1st, 2nd, 3rd bands of SeaWiFS respectively; a_g is the absorption coefficient of CDOM.

17.5.7 Retrieval of Chlorophyll-a Concentration

Most early studies concerning the remote sensing of ocean color were directed toward the extraction of the surface chlorophyll-a concentration from the spectral radiance upwelling above the sea surface (Gordon *et al.* 1983). Chlorophyll-a, responsible for photosynthesis, is the pigment present in living plants. Accurate estimate of chlorophyll-a concentration is essential for estimating primary productivity, biomass, and so on.

Since CZCS was launched in 1978, many methods of retrieving chlorophyll-a concentration were proposed. Inverse modeling of remotely sensed radiance spectra, which has been applied to data of the Coastal Zone Color Scanner (CZCS) (Doerffer and Fischer 1994), has turned out to be a useful method for estimating concentrations of chlorophyll-a, which uses an iteration technique. O'Reilly *et al.* (1998) put forward the OC2 algorithm that was initially used by NASA in the operational processing of SeaWiFS data. OC2 estimates chlorophyll-a concentration for Case 1 waters as the function of the ratio between remote sensing reflectance at the wavebands of 490 and 555 nm, using coefficients derived by a statistical fit with SeaBASS (SeaWiFS Bio-optical Archive and Storage System) data (O'Reilly *et al.* 1998). Later on, Kahru and Mitchell (1999) showed that OC2 overestimates chlorophyll-a at high concentration in the California Current area. Using a larger data set, O'Reilly *et al.* (2000) proposed an updated version of the ocean chlorophyll-a two- (OC2v4) and four-band (OC4v4) algorithms, with a suggestion that the OC4v4 is expected to perform better than OC2v4 when applied to satellite-derived water leaving radiances both in oligotrophic and eutrophic conditions. Subsequently, NASA adopted OC4v4 algorithm to process the global SeaWiFS data. Through analysis, D'Ortenzio *et al.* (2002) found a systematic overestimation of chlorophyll-a concentration by National Aeronautics and Space Administration (NASA) global algorithms (OC2v4 and OC4v4) in the Mediterranean Sea. Differences in the performance of regional and global algorithms might be enhanced by peculiarities in the atmospheric term (Dierssen and Smith 2000, Jorgensen 1999, Kahru and Mitchell 1999, Le'on *et al.* 1999), which indeed affect radiances and the result is important in a misfit of the algorithm (D'Ortenzio *et al.* 2002). In terms of the ocean-color remote sensing, the inverse method using the NN (Neural Network) was first tested for SeaWiFS by Schiller and Doerffer (1993) and adopted as a standard algorithm for the coastal region for MERIS by Buckton *et al.* (1999), and Doerffer and Schiller (1997). Tanaka *et al.* (2004) demonstrated the inverse NN method by using ADEOS/OCTS data. For high-spatial resolution data, the inverse NN method was tested by Keiner and Yan (1998) for estimating sea-surface chlorophyll-a and sediments, and by Zhang *et al.* (2002) for estimating chlorophyll-a, suspended sediments, turbidity, and Secchi disk depth. By using ASTER (Advanced Spaceborne Thermal Emission and Reflectance Radiometer) data with 15-m spatial resolutions, Kishino *et al.* (2005) used the Neural Network (NN) method to estimate chlorophyll-a, suspended matter and CDOM.

All the algorithms for retrieving chlorophyll-a concentration can be partitioned to empirical model, theoretical model, and fluorescence line height (FLH) model. On the basis of in situ measured data, the empirical algorithm is to establish the relationship between the spectral radiance above water surface and in situ measured concentration. The common expression is as follows:

$$C = a\left(\frac{R_1}{R_2}\right)^{\beta} + r \tag{17.12}$$

or

$$\lg C = \lg a + \beta \lg \frac{R_1}{R_2} = \lg a + \beta \lg \frac{L_w(\lambda_1)}{L_w(\lambda_2)} \tag{17.13}$$

where C is the chlorophyll-a concentration; R_i is the reflectance at certain waveband; a β and r is constant, which can be induced by regression equation.

Theoretical algorithm is used to describe the relationship between the concentration of in-water constituents and water-leaving radiance or remote-sensing reflectance using the bio-optical model. In order to simplify the calculation, some assumptions are introduced, and then semi-analytical algorithms are generated, such as the algebraic method, which is the easiest algorithm. There is the following relation between inherent optical properties (IOPs) and remote-sensing reflectance:

$$R_{rs} = \frac{L_w(\lambda)}{E_d(\lambda,0^+)} \approx \sum_{i=1}^{2} g_i \left(\frac{b_b}{a+b_b}\right)^i \tag{17.14}$$

where $g_1 \approx 0.0949I$; $g_2 \approx 0.0794I$; a is the total absorption coefficient; b_b is back-scattering coefficient; $a+b_b$ is attenuation coefficient.

Fluorescence line height (FLH) method is one of the common fluorescence height methods. Its calculation formula is as follows:

$$C = a(FLH)+b \tag{17.15}$$

where the unit of FLH is $(mW \cdot cm^{-2})/(sr \cdot nm)$; C is the chlorophyll-a concentration, and its unit is $(mg \cdot m^{-3})$; a and b are the coefficients obtained by regression analysis.

Although the bio-optical algorithms have been successfully used for estimating water constituents of open sea water from remotely sensed ocean color data (Gordon *et al.* 1980, Kishino *et al.* 1998, O'Reilly *et al.* 1998), it is difficult to be applied in the turbid coastal waters because the constituents of seawater mainly consist of phytoplankton pigments, while coastal areas are optically characterized by high concentrations of phytoplankton pigments, suspended sediment, and colored dissolved organic matter (CDOM) (IOCCG, 2000). Hence, further research is needed for a better retrieval algorithm of chlorophyll-a concentration.

17.5.8 Primary Productivity Estimation

Primary productivity (PP) is the amount of carbon fixed by plants over time through photosynthesis and forms a fundamental basis for both the terrestrial and marine food webs (Smyth *et al.* 2005). The amount of carbon fixed by phytoplankton is estimated to be between 45 and 50 $Gt \cdot C \cdot yr^{-1}$ accounting for 48% of the total carbon uptake (Field *et al.* 1998). Since ocean PP represents a critical process for the oceanic carbon cycle (Antoine *et al.* 1996), it is of great sense to establish the model of estimating ocean PP. Previously, thousands of measurements of marine phytoplankton productivity have been made at discrete locations throughout the world's oceans since the introduction of the radiolabelled carbon uptake method (i.e. ^{14}C method) in 1952 (Steemann Nielsen 1952).With the development of geospatial technology, satellite water color remote sensing data provides a convenient way of estimating chlorophyll-a concentration. The mathematical models for estimating primary production from satellite data are known as primary productivity algorithms (Campbell *et al.* 2002). In the early days of the Coastal Zone Color Scanner (CZCS), simple statistical relationships, which estimate time- and depth-integrated primary production as a function of sea surface chlorophyll-a, were proposed for calculating primary production from the surface chlorophyll-a concentration (e.g., Smith and Baker 1978, Eppley *et al.* 1985). Over the past 2 decades, scientists have sought to improve algorithms by combining the satellite-derived chlorophyll-a data with other remotely sensed fields, such as sea surface temperature (SST) and photosynthetic available radiation (PAR) (Campbell *et al.* 2002). Based on this, several models of estimating marine primary productivity have been proposed.

Behrenfeld and Falkowski (1997) proposed the vertically generalized production model (VGPM) belonging to the complicated depth-integrated models (DIMs). This model combines the physiological process of phytoplankton photosynthesis with empirical relationship. All the important parameters in VGPM model can be obtained by satellite remote sensing data, and VGPM model has a high accuracy, so it is widely used to estimate marine primary productivity. In China, Li *et al.* (2003) calculated the monthly average primary productivity of China East Ocean in 1998 using the VGPM (See Table 17.1).

Table 17.1 Monthly average primary productivity ($mg \cdot m^{-2} \cdot d^{-1}$) of China East Ocean in 1998 using VGPM (Reprinted with permission from Acta Geographica Sinica)

Month	Average	Month	Average	Month	Average	Month	Average
January	440.10	April	610.54	July	607.57	October	543.30
February	519.26	May	681.84	August	534.58	November	522.46
March	583.87	June	696.06	September	516.04	December	464.70

17.5.9 Red Tide Detection

Red tide is the common name for what scientists now prefer to call "harmful algae blooms (HABs)". HABs are a phenomenon caused by marine micro-algae, bacteria, and protozoa, which bloom or aggregate and cause discoloration of the water and threaten the marine environment and human health. Some phytoplankton blooms are caused by the toxic species that can produce brevetoxins which accumulate in shellfish (e.g., oyster and clam) and cause deaths of fish, bird, and marine mammals, and can irritate the eye and respiratory systems of animals including humans (Hu *et al.* 2005). The term "red tide" is often used colloquially to describe these phenomena, which is the change of the ocean color caused by an algal bloom. It is known that not all algal blooms are harmful, not all algal blooms cause discoloration of water, and not all algal blooms are associated with tides.

As the first generation of ocean color remote sensing sensors, Coastal Zone Color Scanner (CZCS) was demonstrated to have the potential to detect a major Karenia brevis bloom in the western Florida waters (Steidinger *et al.* 1981). Similarly, the second generation ocean color remote sensing sensor, SeaWiFS (Sea viewing Wide Field of view Sensor), imagery is used to monitor the birth, development, and termination of phytoplankton blooms in the north-east Atlantic (Zeichen *et al.* 2000). Hu *et al.* (2005) used near real-time fluorescence data from the MODIS satellite sensor to detect and trace a harmful algal bloom (HAB), or red tide, in SW Florida coastal waters from October to December 2004, and found that MODIS fluorescence line height (FLH in $W \cdot m^{-2} \cdot \mu m^{-1} \cdot sr^{-1}$) data showed the highest correlation with near-concurrent in situ chlorophyll-a concentration (Chl-a in $mg \cdot m^{-3}$), which demonstrated that the MODIS FLH data provide an unprecedented tool for research and managers to study and monitor algal blooms in coastal environments.

Although much work has been done about red tide which explored some evidence that various HABs species may have distinguishable optical properties, the satellite chlorophyll concentrations suffer from uncertainties in the atmospheric correction and interference of other colored compounds and/or shallow bottom, further researches are being conducted (Hu *et al.* 2005).

17.6 Water Environment Management - A Case Study of the Pearl River Estuary and the Adjacent Coastal Waters of Hong Kong

17.6.1 Introduction

Water quality has gradually experienced degradation in many areas of the world because of pollution and heavy use resulting from expansion of human activities. The anthropogenic effects on the Pearl River estuary and the adjacent coastal waters of Hong Kong have become increasingly marked since 1980's when rapid industrialization and urbanization occurred in Guangdong Province of South China. Hong Kong has a population of over 6,000,000 people and relies heavily on its coastal waters for recreation and aquaculture, cooling water for air-conditioning plants, disposal of treated waste water, shipping, and navigation. In order to preserve the water quality and to assess the effectiveness of control measures, Hong Kong's coastal waters have been delineated into 10 Water Control Zones (WCZs) according to the locations and characteristics of water quality and a comprehensive monitoring program has been run by the Environmental Protection Department of Hong Kong Government (EPDHK) since April 1986 which covers 79 water monitoring stations in the open sea and semi-enclosed bays, but excluding the typhoon shelters (EPDHK, 1998). The earliest records could be as early as 1982 when the monitoring had been made by the agriculture, Fisheries and Conservation Department of Hong Kong Government before EPDHK was established. The monitoring stations that are near the coast or heavily influenced by human activities have been sampled monthly or bi-weekly and the monitoring stations located in the open sea or less anthropogenically affected region are monitored either in odd or even months since 1988. Twenty-four water quality parameters indicating the oceanographic conditions, the transparency and light penetration, the organic pollution, the faecal pollution, and the eutrophic condition, were collected for every monitoring station, respectively. Although Hong Kong's coastal waters have been regularly monitored since the 1980's (Chen et. al., 2002), the point data sampled at long intervals are not adequate to catch the spatial and temporal variations. It is not easy to obtain the water quality over large areas simultaneously. Various kinds of remote sensed images have become an important information source for monitoring and detecting coastal water quality. Satellite sensors such as CZCS, SeaWiFS, MODIS and Landsat series with various spatio-temporal and spectral resolutions can provide more timely synoptic data of water quality. So remote sensing could be taken as an independent measurement tool for water management authorities (Dekker *et al.* 2001, 2002). The high spatial resolution of the Landsat TM image is useful for monitoring the aquatic systems in a small turbid water body, and it has been reported to be able to effectively analyze the ocean color parameters in estuarine and coastal waters by various authors (e.g. Munday and Alföldi 1979, Lathrop and Lillesand 1986, Doerffer *et al.* 1989, Ritchie *et al.* 1990, Richard and Lathrop 1992, Lavery *et al.* 1993, Tassan 1987, 1992, 1993, Dekker *et al.* 2001, 2002).

17.6.2 Spatio-temporal Monitoring Data Organization and Analysis System Construction

All the monitoring data used in this study area can be divided into two types: spatial data with definite geographical location and measurements of monitoring water quality parameters. The combination of above two types of data provides with the spatio-temporal characteristics of water quality parameters. In this study, water quality monitoring data was organized by Geodatabase. In this case, the water quality monitoring stations can be zoned by distance, and the water quality data can be more directly analyzed in the spatial domain.

For a water quality sample, there exists 4 types of data, location, monitoring depth, monitoring date and water quality parameters, which corresponds to a point in a 4 dimensional space. It was abstracted as follows:

$$\phi = \phi(x, y, z, t) \tag{17.16}$$

where ϕ is a sample of water quality parameters, x, y, z represent 3 dimensional coordinates of monitoring location, re t is monitoring date in this study. Above equation can be expressed as follows: respectively, and

$$\phi = \phi(s, z, t) \tag{17.17}$$

where s stands for monitoring location (monitoring station here), $s = s(x, y)$, and corresponds to a pair of coordinates (x, y).

Taking the study area as a continuous region, Ω, which can be divided into n sub-regions, Ω_1, Ω_2, ……, Ω_n, through spatial analysis such as buffering with the aid of GIS. Every sub-region has a 1:1 relation with a monitoring location (monitoring station), and a number of monitoring stations will be included in each region, taking as, S_1, S_2, ……, S_n.

According to above description, a related-data model can be described as follows:

$$\begin{cases} \phi = \phi(x, y, z, t) \\ s = s(x, y) \\ \phi = \phi(s, z, t) \end{cases} \tag{17.18}$$

In this model, for a given sample, $(s, z, t, \Phi(s, z, t))$, if $s \in Sn$, then the sample can be expressed as $(s, z, t, \Phi(s, z, t), n)$, where, n stands for sub-region Ωn. It clearly shows that the spatial location and water quality parameters can be combined for an integrated analysis in this way.

Based on the proposed related-data model, a GIS software tool, ArcGIS Desktop, and statistical software tool, SPSS, were selected to manage and analyze the spatial relations, and water quality's statistical characteristics. In our spatio-temporal analysis, ArcMap and SPSS were linked by dbf file. Data can be freely communicated in the two systems by using the proposed data-related model. The spatial relation of monitoring stations could be explored by spatial analysis in the ArcMAP, and the water quality parameters related to spatial relationship could be extracted based on the proposed related-data model, and sent to SPSS through DBF for further statistical analysis.

17.6.3 Statistical Characteristics of Chlorophyll-a Level and its Impact Factors

The level of chlorophyll-a, which indicated the scale of algal biomass (Yung *et al.* 1997), is mostly correlated with the number of red tide incidents in Hong Kong's coastal waters. The result shows an obviously positive correlation between the average chlorophyll-a concentration and frequency of red tide occurrence. So the chlorophyll-a concentration which has been used as an indication of the intensity of algal growth and hence the degree of eutrophication was selected as the indicator to study the temporal and spatial variations associated with red tide. The distribution pattern identifies the impacts of eutrophication in various areas. Generally, for unpolluted offshore oceanic waters, the chlorophyll-a level is below 2 μg/l. A mean chlorophyll-a level above 10μg/l can be regarded as unacceptably high and such a level indicates eutrophication. An algal bloom was defined as an incident with the chlorophyll-a concentration above 40μg/l for a lake environment (James and Havens 1996).

The characteristics of monthly chlorophyll-a level and its correlated parameters from 1988 to 1999 in the 37 monitoring stations had two kinds of distinct differences in the temporal and spatial domains. One type is in the eastern part of Hong Kong's coastal waters mainly consisting of semi-enclosed bays, which was more influenced by the tidal flushing characteristics, and had been heavily attacked by red tides since the 1980s. Another type is in the west of Hong Kong's coastal waters. As it is influenced by the interaction between the Pearl River waters and the ebb and flood of tides, there were less red tide incidents because of the influence of the Pearl River discharge, monsoon-affected currents and the mixed and mainly semi-diurnal tidal currents that made the waters more turbid and turbulent.

Statistical analysis shows that the chlorophyll-a concentration was influenced by the nutrients, physical and aggregate properties, and organic contents of marine water in the Pearl River estuary and its adjacent Hong Kong's coastal waters. The indicator of organic contents, BOD5, was the common influencing factor. The primary significant factors of nutrients and physical and aggregate properties had different combinations from place to place. It can be classified into two types in terms of flow conditions. In the east waters where the general water circulations were very weak, the condition was favorable to the occurrence of red tides. The most significant nutrients were nitrogen and phosphorus. In the west waters, oceanographic conditions were the most significant influencing factors of chlorophyll-a level. Nutrients also played an important role, and phosphorus was the most significant factor. Parameters related to transparency and light penetration were more important to the chlorophyll-a concentration in the east waters than that in the west waters. However, the oceanographic conditions had a more significant impact in the west waters than that in the east waters.

17.6.4 Water Quality Spatial Pattern Identification

It is much more difficult for the quantitative use of optical satellite data for synoptic monitoring of surface water quality in the coastal waters because of the complex characteristics and absence of in situ field water quality data measured simultaneously with the satellite overpass. When the time difference between image acquisition and water quality sample collection is not substantial, the spatial patterns of coastal water quality could be indirectly detected by ocean color classification through establishing the relationships between spectral reflectance and water quality characteristics. The ocean color related to the reflectance spectra is affected by the nature and the quantity of the main optically active constituents. So the ocean color data will give information on the existence and concentration of optically active constituents, thus, indicating the water quality status of the coastal waters. This would then provide a very useful tool for water quality management and dynamic detection. In order to do above analyses, 5 optically active water quality parameters, namely turbidity (TURB), suspended solids (SS), total volatile solids (TVS), chlorophyll-a (Chl-a), and phaeo-pigment (PHAE) were selected. TURB is a qualitative measurement of water clarity. It measures the amount of light scattered from a sample and can be caused by organic suspended solids, inorganic particulates, and the effect of colored dissolved matter (Han and Rundquist 1998). SS and TVS both measure actual weights of material per volume of water. Generally, more solids cause greater scattering. Chl-a and PHAE are parameters related to phytoplankton. The absorption of Chl-a in the visible spectral domains makes it behave as a negative correlation with spectral reflectance.

The spatial variations of coastal water quality were studied by using the remote sensed image and water quality measurements not simultaneous with the satellite overpass in the Pearl River estuary and the adjacent coastal waters of Hong Kong. The Landsat TM bands 1-4 acquired on Dec. 22, 1998 were selected for color classification. The COST method was employed to derive the spectral reflectance from the TM image bands 1-4. After a geometric processing, the reflectance of the 88 sample sites was estimated from the TM image. A clustering analysis of the 88 samples showed that the reflectance could be divided into 5 classes. The reflectance in each band showed a consistent decrease from class 1 to class 5. The reflectance of TM bands 2 and 3 had the largest difference between classes, and the maximum reflectance clearly shifted towards shorter wavelength from class 1 to class 5. These trends showed a close relationship between reflectance and particulates in the study area, which was verified by a correlation analysis of the optically active water quality parameters and reflectance for the same sample sites.

In order to study the spatial patterns of ocean color related to water quality, 88 samples were randomly divided into two sets: a training set with 65 samples and a testing set with 23 samples. As the data sets for training and testing were small, we selected three classifiers to compare the classification results. Through a comparison of 3 classification methods, the SVM was proven the best method for the condition of small sample population. Although the three methods had different accuracies with 78.3%, 82.6% and 91.3% for MLH, NN and SVM, respectively, all the results provided similar spatial patterns. In order to examine whether the five classes classified by spectral reflectance could be well interpreted by the water quality characteristics, five optically active water quality parameters, namely TURB, SS, TVS, Chl-a, and PHAE, were selected to do the analysis of variance (ANOVA). The analysis results showed that the difference of water quality parameters in different classes was statistically significant at 0.01 at the significant level. It clearly indicated that the study area could be divided into five categories of water quality characterized by different spectral reflectance and optically active substances in the water. Class 1 appeared in the west part of the study area belonging to the Pearl River estuary, and was characterized by the highest levels of TURB, SS, TVS and PHAE and the lowest level of Chl-a concentration. Class 5 was located in the east part of the study area mostly enclosed by land and the water exchange was not strong, so the Chl-a was at the highest level and TURB and SS at the lowest level. As for the other three classes, they were in the middle part of the study area and were influenced by tidal flushing. There was a consistent decrease of TURB, SS, TVS and PHAE from classes 2 to 4 and a minor fluctuation in Chl-a. Therefore, we could identify a type of turbid water body characterized as 'sediment dominated' coastal waters in the west part of the study area, and a type of water body with the most nutrients characterized as 'phytoplankton dominated' coastal waters in the east part of study area. An intermediate type appeared between these two types of water bodies in the middle part of study area. It provides a basis to discriminate different areas in the estuary and coastal waters when a dynamic monitoring of the spatial patterns of water quality is required. The results clearly showed that the ocean color categories classified by spectral reflectance could indicate the spatial distribution of water quality. By establishing a relationship between reflectance and optically active water quality parameters, ocean color classification could be a scientific and economical tool for coastal water quality management and dynamic monitoring through integrating remote sensing observation with a regular water quality monitoring program.

17.6.5 Spatio-temporal Analysis of Coastal Water Environment based on RS/GIS Technology and Statistical Software

RS/GIS technology combined with statistical analysis is the basis for scientific management and decision making to coastal water environment, and therefore plays a key role in the coastal water environmental information system. In this paper, two water environmental parameters, suspended sediment and dissolved oxygen, were selected to analyze their spatio-temporal regime based on RS/GIS technology and statistical software.

17.6.5.1 RS based Spatio-temporal Analysis of Suspended Sediment in the Coastal Waters

Suspended sediment, as an important carrier of land source pollution, brings a large number of nutrients and pollutants to the coastal waters. Its concentration changed according to the dynamics of watershed's environment and human activities, and has become a natural tracer of pollutant diffusion. It is difficult to get enough spatio-temporal information of suspended sediment concentration using the conventional measurement. Remote sensing provides with a possibility to rapidly and periodically monitor the distribution and variation of suspended solids simultaneously in a large area (Chen 2003, Pan *et al.* 2003, Zhu *et al.* 2001). In this study, the Pearl River estuary and its adjacent coastal waters were selected as a case to study the spatio-temporal dynamics of suspended sediment, and 152 suspended sediment concentration resultant maps reversed from NOAA/AVHRR images acquired from 1995 to 20000 were used. An algorithm using an analytical model based on the difference of the NOAA/AVHRR Channels 1 and 2 reflectance data is developed for the retrieval of the suspended sediment in coastal and shelf waters. The algorithm was tested using the data from seven transects, and the retrieved results for the Pearl River Estuary were compared with the sea-truth data with good agreement (Li *et al.* 1998, Huang 2001, Deng *et al.* 2002) .

Three time scales, month, season, and hydrologic period were selected to analyze the suspended sediment concentration resultant maps, and 13 levels of suspended sediment concentration with an interval of 10 mg/l were divided in this paper (Fig. 17.12).

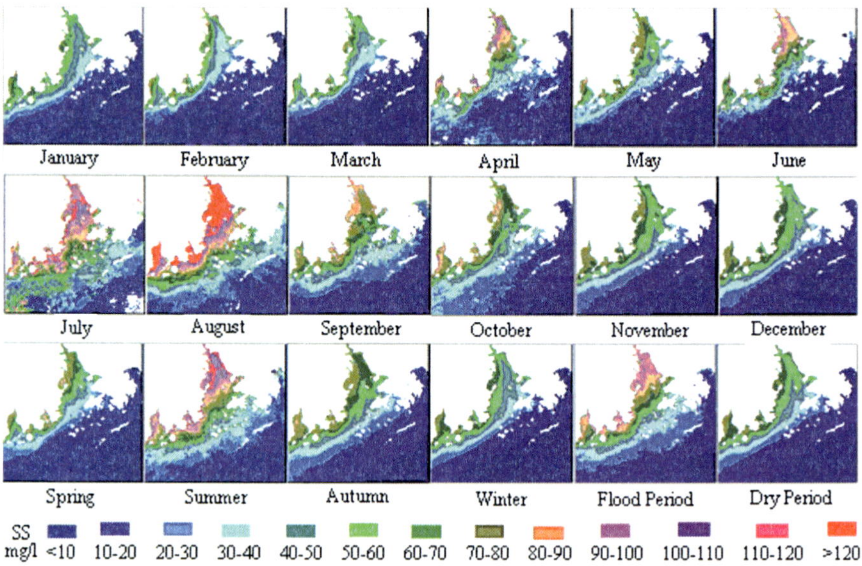

Fig. 17.8 Average distribution of suspended sediment concentration in different time scales in the Pearl River estuary and its adjacent Hong Kong's coastal waters

The results showed that the highest concentration of suspended sediment was near the five diversion mouths of Pearl River. The concentration of suspended sediment decreased towards the ocean from every mouth of the Pearl River, and the shape and gradient of concentration contour reflected the impact of land runoff on the suspended sediment. The flood period was characterized by the high suspended sediment concentration in the wholly estuary, which revealed a positive correlation between the suspended sediment concentration and the watershed's rainfall.

17.6.5.2 Spatio-temporal Analysis of Dissolved Oxygen (DO) in the Coastal Waters in the Aid of GIS and Statistical Software

As the coastal water environment is mainly influenced by land source pollution, a buffer function was used to divide the study waters into three zones from land to ocean, 0-1 km, 1-4 km and 4-20 km (Fig. 17.13). The monitoring stations were retrieved through spatial operation, and the related attribute data would be correspondently selected to write to dbf files. The dbf files were loaded in SPSS to do statistical analysis after adding regional identifications. The distribution and variation of water environment in different buffer zones were received through analyzing the sampling measurements falling in different spatial units.

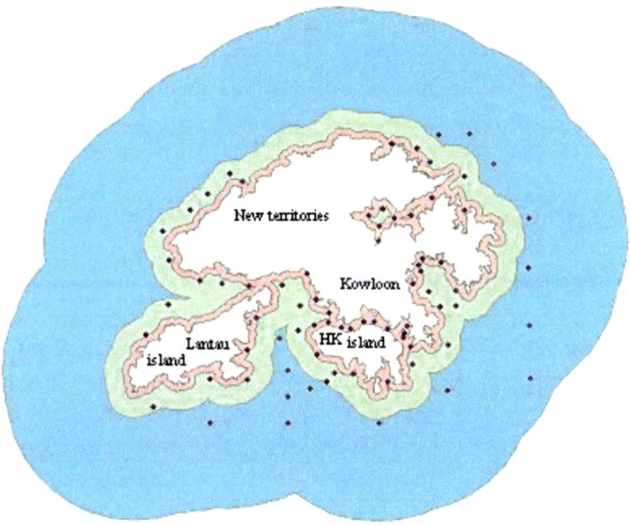

Fig. 17.9 Buffer zones (0-1 km in pink, 1-4 km in green, 4-20 km in blue) and monitoring stations in Hong Kong's coastal waters

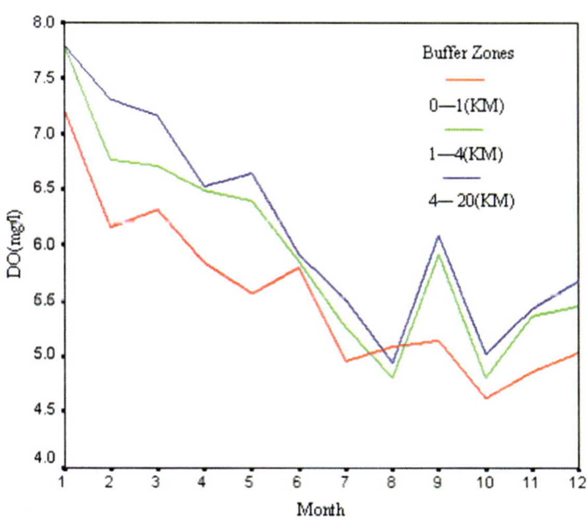

Fig. 17.10 Water quality analysis for different buffer zones in Hong Kong's coastal waters in 2001

In this study, dissolved oxygen (DO) was selected to analyze the coastal water environment. DO is an indicator of measuring the water polluted degree caused by organic and non-organic pollutants. Low DO concentration usually corresponds to serious pollution. The analytical results of DO in Hong Kong's coastal waters in 2001 in Fig. 17.14 showed that DO had an obvious increase towards ocean, and the DO levels within 1 km were obviously lower than other two spatial areas in most time periods, which reflected that the effective impact of biochemical processes related to land source pollution mainly occurred within 1 km in the coastal waters. This conclusion can provide a guide for coastal water environment management.

17.7 Conclusions

Most of streams, lakes, estuary, and costal waters all over the world have been contaminated by human activities and shown various degrees of degradation, which has seriously threatened the social development, and therefore, has been paid more attention by governments and scientists. In order to mitigate the water environmental problems, a long term water quality monitoring is needed for helping policy makers, researchers, and the public to understand the status of water quality conditions and the complexity of factors involved, identify specific existing or emerging water quality problems, and catch changes or trends in water quality over time. Many countries have developed water quality observing systems to measure water quality dynamically. In order to mine much more useful information from the monitoring data, suitable tools for analysis and result visualization are required. Water environmental management needs effective tools for scientific analysis. In this chapter, a study case in the Pearl River estuary and Hong Kong's coastal waters was taken for a spatio-temporal pattern identification and detection of water environment based on remote sensing and in situ measurements. Managing and preserving water resources is a formidable challenge given the rapid pace of change affecting water environment, so the remote sensing technique, as a fast, accurate, and quantitative tool, has been widely used for water environment analysis. The seriously polluted areas would be shown by spatial analysis, which would provide a scientific basis for water pollution control and other environmental decision making. The spatio-temporal analysis based on remote sensing would be sure to provide an effective technical tool in a coastal water environment management.

Acknowledgements

This work was funded by the 973 Program (Grant No. 2009CB723905), the 863 project (Grant No. 2007AA12Z161), NSFC (Grant Nos. 40676094, 40721001), Sino-Germany Joint Project (Grant No. 2006DFB91920), and MOST of China (Grant No. 2007BAC23B05).

References

Anderson, D. M. and D. L. Garrison (eds.) (1997): The Ecology and Oceanography of Harmful Algal Blooms. Limnol. Oceanogr., 42, 1009-1305.

Anderson, J.M., Duck, R.W. and McManus, J., 1995. Thermal radiometry: a rapid means of determining surface water temperature variations in lakes and reservoirs. Journal of Hydrology 173, pp. 131-144.

Antoine, D., and A. Morel (1996), Oceanic primary production: 1. Adaption of spectral light-photosynthesis model in view of application to satellite chlorophyll observations, Global Biogeochem. Cycles, 10, 43-55.

Arenz R F , Lewis WM, Saunders J F. Determination of chlorophyll and dissolved organic carbon from reflectance data for Colorado reservoirs[J] . Int . J . Remote Sensing ,1996 ,17(8) :1 547-1 566.

Atwell, B.H., McDonald, R.B. and Bartolucci, L.A., 1971. Thermal mapping of streams from airborne radiometric scanning. Water Resources Bulletin 7, pp. 228-243.

Avery, T.E. and Berlin, G.L., 1992. Fundamentals of remote sensing and airphoto interpretation, Macmillan, New York.

Behrenfeld, M. J., and P. G. Falkowski. Photosynthetic rates derived from satellite-based chlorophyll concentration, Limnology and Oceanography, 1997, 42 (1):1 -20.

Belknap, W. and Naiman, R.J., 1998. A GIS and TIR procedure to detect and map wall-base channels in western Washington. Journal of Environmental Management 52, pp. 147-160.

Brekke, C. and A.H.S. Solberg, Oil spill detection by satellite remote sensing. Remote Sensing of Environment, 2005. 95(1): p. 1-13.

Buckton, D., O'Mongain, E., and Danaher, S. (1999). The use of neural networks for the estimation of oceanic constituents based on the MERIS instrument. International Journal of Remote Sensing, 20, 1841-1851.

Bukata R P,J H Jerome *et al.*Optical Properties and Remote Sensing of Inland and Coastal Water.CRC Press,Boca Raton,Florida 1995:362pp.

Campbell, J., *et al.* (2002), Comparison of algorithms for estimating ocean primary production from surface chlorophyll, temperature, and irradiance, Global Biogeochem. Cycles, 16(3), 1035.

Carder KL , Steward R G, Harvey G R , *et al* . Marine Humic and fulvic acids : Their effects on remote sensing of ocean chlorophyll [J] . Limnol . Oceanogr. ,1989 ,34 (1) :68 -81.

Chen Q. C., Wei J., and Shi P.(2003). Atmospheric correction of SeaWiFS imagery for turbid waters in Southern China coastal areas. Proceedings of SPIE -The International Society for Optical Engineering, 4892, 80-86.

Chen Q. C., Pan Z.L., Shi P., Simulation of Sea Water Reflectance and Its Application in Retrieval of Yellow Subtance by Remote Sensing Data. Journal of Tropical Oceanography, 2003, 22(5):34-39.

Chen X H, Chen Y Q, Lai G Y. Modeling of the transport of suspended solids in the estuary of Zhujiang River[J]. Acta Oceanological Sinica, 2003, 25(2): 120-127.

Chen X. L., Li, Y. S., Li, Z. L.. Statistical characteristics of chlorophyll-a concentration in Hong Kong's coastal waters. Journal of Geographical Science, 2002, 12 (3): 331-342.

Chen X., Y.S. Li, Z. Liu, K. Yin, Z. Li, O.W. Wai and B. King,2004. Integration of multi-source data for water quality classification in the Pearl River estuary and its adjacent coastal waters of Hong Kong, Cont. Shelf Res. 24 (16): 1827-1843.

Chen Z Q， Li Y， Pan J M. Distributions of colored dissolved organic matter and dissolved organic carbon in the Pear River Estuary， China. Cont. Shelf Res. ， 2004， 24: 1845～1856.

Dekker, A.G., Vos, R.J., Peters, S.W.M., 2001. Comparison of remote sensing data, model results and in situ data for total suspended matter (TSM) in the southern Frisian lakes. The Science of the Total Environment 268, 197-214.

Dekker, A.G., Vos, R.J., Peters, S.W.M., 2002. Analytical algorithms for lake water TSM estimation for retrospective analysis of TM and SPOT sensor data. International Journal of Remote Sensing 23 (1), 15-35.

Deng M, Huang W, Li Y. Data collection of remote sensing derived suspended sediment concentration in Zhujiang River estuary[J]. Oceanologia Et Limnologia Sinica, 2002,33(4):341-348.

Dierssen, H. M., and Smith, R. C. (2000). Bio-optical properties and remote sensing ocean color algorithms for Antarctic Peninsula waters. Journal of Geophysical Research, 105, 26301-26312.

Dierssen, H. M., Zimmerman, R. C., Leathers, R. A., Downes, T. V., and Davis, C. O. (2003). Ocean color remote sensing of seagrass and bathymetry in the Bahamas Banks by high-resolution airborne imagery. Limnology and Oceanography, 48, 444-455.

Doerffer, R., Fischer, J., Stössel, M., Brockmann, C.. 1989. Analysis of Thematic Mapper data for studying the suspended matter distribution in the coastal area of the German Bight (North Sea). Remote Sensing of Environment 28, 61-73.

Doerffer, R., and Fischer, J. (1994). Concentrations of chlorophyll, suspended matter, and gelbstoff in case II waters derived from satellite coastal zone color scanner data with inverse modeling methods. Journal of Geophysical Research, 99(C4), 7457-7466.

Doerffer, R., and Schiller, H. (1997). Pigment index, sediment and gelbstoff retrieval from directional water leaving radiance reflectances using inverse modelling technique. Algorithm Theoretical Basis Document (ATBD) 2.12, Doc. No. PO-TN-MEL-GS-005.

D'Ortenzio, F., *et al.*, Validation of empirical SeaWiFS algorithms for chlorophyll-alpha retrieval in the Mediterranean Sea -A case study for oligotrophic seas. Remote Sensing of Environment, 2002. 82(1): p. 79-94.

EPDHK. 1998. Marine water quality in Hong Kong in 1997, Prepared by Environmental Protection Department, Hong Kong.

European Space Agency. (1998). Oil pollution monitoring. ESA brochure: ERS and its applications—Marine, BR-128, 1.

Espedal, H.A. and T. Wahl, Satellite SAR oil spill detection using wind history information. International Journal of Remote Sensing, 1999. 20(1): p. 49-65.

Falkowski,P. G. 1980. Light-shade adaption in marine phytoplankton,p. 99-119. In P. G. Falkpwski [ed.], primary productivity in the sea. plenum.

Field, C. B., M. J. Behrenfeld, J. T. Randerson, and P. Falkowski (1998), Primary production of the biosphere: Integrating terrestrial and oceanic components, Science, 281, 237-240.

Fingas, M. F., and Brown, C. E. (1997). Review of oil spill remote sensing.Spill Science and Technology Bulletin, 4, 199– 208.

Fiscella, B., *et al.*, Oil spill detection using marine SAR images. International Journal of Remote Sensing, 2000. 21(18): p. 3561-3566.

Garrett, A.J. and Hayes, D.W., 1997. Cooling lake simulations compared to thermal imagery and dye tracers. Journal of Hydraulic Engineering 123, pp. 885-894.

Gordon, H. R. Removal of atmospheric effects from satellite imagery of the ocean, Applied Optics. 1978, 17:1631-1636.

Gordon, H. R., Clark, D.K., Clear water radiances for atmospheric correction of coastal zone color scanner imagery. Applied Optics. 1981, 20: 4175 -4180.

Gordon, H. R., Castano, D. J., Coastal Zone Color Scanner atmospheric correction algorithm: multiple scattering effects. Applied Optics. 1987, 26: 2111-2122.

Gordon, H. R., Brown J. W., and Evans, R. H., Exact Rayleigh scattering calculations for use with the Nimbus-7 Coastal Zone Color Scanner. Applied Optics. 1988, 27: 862-871.

Gordon, H. R., and Clark, D. (1980). Remote sensing optical properties of a stratified ocean: an improved interpretation. Applied Optics, 19, 3428-3430.

Gordon H. R. , Clark D. K. , Brown J. W. , Brown o. B. , Evans R. H. , and Broenkow W. W., "Phytoplankton pigment concentrations in the Middle Atlantic Bight: comparison of ship determinations and CZCS estimates," Appl. Opt. 22, 20-35(1983).

Gordon, H. R. and Morel, A. (1983). Remote Assessment of Ocean Color for Interpretation of Satellite Visible Imagery. A Review, Lecture Notes on Coastal and Estuarine Studies, R. T. Barber, N. K. Mooers, M. J. Bowman and B. Zeitzschel (eds.), Springer-Verlag, New York, 114 p.

Gordon, H. R. and Wang, W., Retrieval of water-leaving radiance and aerosol optical thickness over the oceans with SeaWiFS: a preliminary algorithm, Applied Optics. 1994, 33: 443-452.

Gower, J., *et al.*, Detection of intense plankton blooms using the 709 nm band of the MERIS imaging spectrometer. International Journal of Remote Sensing, 2005. 26(9): p. 2005-2012.

Haltrin, V. I., Kattawar, G. W. and Weidemann, A. D. 1997. Modeling of elastic and inelastic scattering effects in ocean optics. In: Ocean Optics XIII. Proc. SPIE, 2963: 597-602.

Han, L., Rundquist, D.C.. 1998. The impact of a wind-roughened water surface on remote measurements of turbidity. International Journal of Remote Sensing 19 (1), 195-201.

Hu, C.M., *et al.*, Red tide detection and tracing using MODIS fluorescence data: A regional example in SW Florida coastal waters. Remote Sensing of Environment, 2005. 97(3): p. 311-321.

Hu, C.,Mqller-Krager, F. E., Taylor, C. J.,Myhre,D.,Murch, B.,Odriozola,A.L., *et al.* (2003). MODIS detects oil spills in Lake Maracaibo, Venezuela. EOS, Transactions, American Geophysical Union, 84(33), 313, 319.

Huang, W. Satellite remote sensing applications in the Pearl River estuary and Hong Kong coastal waters. M.Sc. Project Report, Hong Kong University of Science and Technology, Hong Kong, 2001

IOCCG (1998). Minimum Requirements for an Operational, Ocean-Colour Sensor for the Open Ocean, Reports of the International Ocean-Colour Coordinating Group, No. 1,IOCCG, Dartmouth, Canada.

IOCCG (1999). Status and Plans for Satellite Ocean-Colour Missions: Considerations for Complementary Missions. Yoder, J. A. (ed.), Reports of the International Ocean-Colour Coordinating Group, No. 2,IOCCG, Dartmouth, Canada.

IOCCG (2000). Remote Sensing of Ocean Colour in Coastal, and Other Optically-Complex, Waters. Sathyendranath, S. (ed.), Reports of the International Ocean-Colour Coordinating Group, No. 3, IOCCG, Dartmouth, Canada.

James R Thomas, Havens K E, 1996. Algal bloom probability in a large subtropical lake. Water Resources Bulletin, 33(5):995-1006.

Jessup, A.T., Zappa, C.J., Loewen, M.R. and Hesany, V., 1997. Infrared remote sensing of breaking waves. Nature 385, pp. 52-55.

Johnsen, G. and Sakshaug, E. (1996). Light harvesting in bloom-forming marine phytoplankton: species-specificity and photoacclimation. Scient. Mar., 60: 47-56.

Jorgensen, P. V. (1999). Standard CZCS Case 1 algorithms in Danish coastal waters. International Journal of Remote Sensing, 20, 1289-1301.

Kahru, M., and Mitchell, B. G. (1999). Empirical chlorophyll algorithm and preliminary SeaWiFS validation for the California Current. International Journal of Remote Sensing, 20, 3423-3429.

Karabashev, G. S. (1998). On concentration dependence of chlorophyll fluorescence in theoceanic waters of diverse trophicity. Oceanology, 38: 342-346.

Keiner, L. E., and Yan, X. (1998). A neural network model for estimating sea surface chlorophyll and sediments from Thematic Mapper Imagery. Remote Sensing of Environment, 66, 153-165.

Kirk, J. T. O. 1981. Monte Carlo study of the nature of the underwater light field in, and the relationships between optical properties of, turbid yellow waters. Aust. J. Freshwater Res., 32: 517-532.

Kishino, M., Ishimaru, T., Furuya, K., Oishi, T., and Kawasaki, K. (1998). Inwater algorithm for ADEOS/OCTS. Journal of Oceanography, 54, 431-436.

Kishino, M., A. Tanaka, and J. Ishizaka, Retrieval of Chlorophyll a, suspended solids, and colored dissolved organic matter in Tokyo Bay using ASTER data. Remote Sensing of Environment, 2005. 99(1-2): p. 66-74.

Klemas V, Bartlett D, Philpot W *et al.*, 1974. Coastal and estuarine studies with ERTS-1 and Skylab. Remote Sensing of Environment, 3(3): 153-174.

Lathrop Jr., R.G., Lillesand, T M.. 1986. Use of Thematic Mapper Data to Assess Water Quality in Green Bay and Central Lake Michigan. Photogrammetric Engineering and Remote Sensing 52, 671-680.

Lavery, P., Pattiaratchi, C., Wyllie, A., Hick, P., 1993. Water quality monitoring in estuarine waters using the Landsat Thematic Mapper. Remote Sensing of Environment 46 (3), 268-280.

Le'on, J. F., Chazette, P., and Dulac, F. (1999). Retrieval and monitoring of aerosol optical thickness over an urban area by spaceborne and groundbased remote sensing. Applied Optics, 38, 6918-6926.

Legleiter, C. J. , D A. Roberts, W A. Marcus, and M A. Fonstad, "Passive optical remote sensing of river channel morphology and in-stream habitat: Physical basis and feasibility" (2004). Remote Sensing of Environment. 93 (4), pp. 493-510.

Letelier, R.M. and Abbott, M.R. 1996. An analysis of chlorophyll fluorescence algorithms for the Moderate Resolution Imaging Spectrometer (MODIS). Rem. Sens. Environ., 58: 215-223.

Lee, Z., Carder, K. L., Hawes, S. K., Steward, R. G., Peacock, T. G. and Davis, C. O. 1994. Model for the interpretation of hyperspectral remote-sensing reflectance. Appl. Opt., 33: 5721-5732.

Li J., A study on determination of concentration of suspended solids in water by remote sensing.Acta Scientiae Circumstantiae, 1986,6(2):166-173.

Li G.S, Wang F., Liang Q., Li J.L.. Estimation of Ocean Primary Productivity by Remote Sensing and Introduction to Spatio-temporal Variation Mechanism for the East China Sea. Acta Geographica Sinica, 2003, 58(4):484-493.

Li, Y., Huang W., Fang M.. 1998. An algorithm for the retrieval of suspended sediment in coastal waters of China from AVHRR data. Continental Shelf Research, 1998, 18(5): 487-500.

Lin, I.-I., L. S.Wen, K.-K. Liu,and W.-T.Tsai, Evidence and quantification of the correlation between radar backscatter and ocean colour supported by simultaneously acquired in situ sea truth,Geophys.Res. Lett.29, 102.1-4, 2002.

Liu,A. K.,S.,Y.Wu,W.Y.Tseng,and W.G. Pichel,Wavelet analysis of SAR images for coastal monitoring,Can.J.Rem.Sens.,26, 494-500, 2000.

Lu, J., Marine oil spill detection, statistics and mapping with ERS SAR imagery in south-east Asia. International Journal of Remote Sensing, 2003. 24(15): p. 3013-3032.

Lyzenga, D. R. (1978). Passive remote-sensing techniques for mapping water depth and bottom features. Applied Optics, 17, 379-383.

M. Wang, "Atmospheric correction of the second generation ocean color sensors," Ph.D. dissertation (University of Miami,Coral Gables, Fla., 1991).

Mobley,Curtis D.Light and water-radiative transfer in natural waters[M].Academic Press.1994.

Morel, A. and Prieur, L. (1977). Analysis of variations in ocean color. Limnol. Oceanogr. 22:709-722.

Morel, A. and Bricaud, A., 1981. Theoretical results concerning light absorption in a discrete medium, and application to specific absorption of phytoplankton. Deep-Sea Res., (Part A), 28: 1375-1393.

Morel, A. 1988. Optical modeling of upper ocean in relation to its biogenous matter content (Case I waters). J. Geophys. Res., 93: 10,749-10,768.

Morel, A.,Ahn, Y-H., Partensky, F., Vaulot, D. and Claustre H. 1993. Prochlorococcus and Synechococcus: A comparative study of their optical properties in relation to their size and pigmentation. J. Mar. Res., 51: 617-649.

Morel A. Consequences of a Synechococcus bloom upon the optical properties of oceanic (case Ⅰ) waters. Limnol. Oceanogr. (1997) 42:1746-1754.

Mueller, J. L. and Austin, R.W., 1995. Ocean Optics protocols for SeaWiFS validation. Revision 1.NASA Tech. Memo., Vol. 25: 66pp.

MUMBY, P. J., C. D. CLARK, E. P. GREEN, AND A. J. EDWARDS. 1998. Benefits of water column correction and contextual editing for mapping coral reefs. Int. J. Remote Sens. 19: 203-210.

Munday, Jr., J. C., Alföldi, T. T.. 1979. Landsat test of diffuse reflectance models for aquatic suspended solids measurement. Remote Sensing of Environment. 8: 169-183.

O'Reilly J.E., S. Maritorena, B.G. Mitchell, D.A. Siegel, K.L. Cardel and S.A. Garver *et al.*, Ocean color chlorophyll algorithms for SeaWiFS, Journal of Geophysical Research-Oceans 103 (1998) (C11), pp. 24937-24953.

O'Reilly J.E., S. Maritorena, D. Siegel, M.C. O'Brien, D. Toole and B.G. Mitchell *et al.*, Ocean color chlorophyll a algorithms for SeaWiFS, OC2, and OC4: Version 4. In: S.B. Hooker and E.R. Firestone, Editors, SeaWiFS postlaunch technical report series, SeaWiFS postlaunch calibration and validation analyses, Part 3 vol. 11, NASA/GSFC (2000), pp. 9-23.

Orellana M.V. and M.J. Perry, An immunoprobe to measure Rubisco concentration and maximal photosynthetic rates of individual phytoplankton cells. Limnol. Oceanogr. 37 (1992), pp. 478-490.

ORTH, R., AND K. MOORE. 1983. Chesapeake Bay: An unprecedented decline in submerged aquatic vegetation. Science 222:51-53.

Pan J M, Zhou H Y, Hu C Y, Liu X Y, Dong L X, Zhang M. Nutrient profile in interstitial water and flux in water-sediment interface of Zhujiang estuary of China in summer[J]. Acta Oceanologica Sinica, 2002, 24(3): 52-59.

PETERSON, B. J., AND J. W. FOURQUREAN. 2001. Large-scale patterns in seagrass (Thalassia testudinum) demographic in south Florida. Limnol. Oceanogr. 46: 1077-1090.

Pozdnyakov, D., Grassl,H. *et al*, Colour of inland waters: a methodology for its interpretation, Chichester: Springer-Praxis, 2003.

Philpot, W. D. (1989). Bathymetric mapping with passive multispectral imagery. Applied Optics, 28, 1569-1578.

Richard G., Lathrop, Jr. (1992). Landsat Thematic Mapper Monitoring of Turbidity Inland Water Quality, Photogrammetric Engineering and Remote Sensing. 58 (4), pp465-470.

Ritchie, J. C., Cooper, C. M., Schiebe, F. R.. 1990. The relationship of MSS and TM digital data with suspended sediments, chlorophyll, and temperature in Moon Lake, Mississippi. Remote Sensing of Environment 33, 137-148.

Robinson, I.S., Wells, N.C. and Charnock, H., 1984. The sea surface thermal boundary layer and its relevance to the measurement of sea surface temperature by airborne and spaceborne radiometers. International Journal of Remote Sensing 5, pp. 19-45.

Rothman, L. S., Gamache, R. R., Goldman, A., Brown, L. R., Toth, R. A., Pickett, H. M., Poynter,R. L., Flaud, J.-M., Camy-Peyret, C., Barbe, A., Husson, N., Rinsland, C. P. and Smith, A. H.1987. The HITRAN database: 1986 edition. Appl. Optics, 26: 4058-4097.

Ruddick K.G., Ovidio F. and Rijkeboer M. , Atmospheric correction of SeaWiFS imagery for turbid coastal and inland waters, Applied Optics 39 (6) (2000), pp. 897-912.

Smith, W.L., Knuteson, R.O. and Revercomb, H.E., 1996. Observations of the infrared radiative properties of the ocean: implications for the measurement of sea surface temperature via satellite remote sensing. Bulletin of the American Meteorological Society 77, pp. 41-51.

Smyth, T. J., Tilstone, G. H. and Groom, S. B. (2005) Integration of radiative transfer into satellite models of ocean primary production. Journal of Geophysical Research -Oceans, 110, p. 10014.

Solberg, A.H.S., C. Brekke, and P.O. Husoy, Oil spill detection in Radarsat and Envisat SAR images. Ieee Transactions on Geoscience and Remote Sensing, 2007. 45(3): p. 746-755.

Solberg, A.H.S., G. Storvik, R. Solberg, and E. Volden, "Automatic detection of oil spills in ERS SAR images," IEEE Trans. Geosci. Remote Sens., vol. 37, no. 4, pp. 1916-1924, Jul. 1999.

Stavn, R. H. 1992. External factors and water Raman scattering in clear ocean waters: skylight, solar angle, and the air/water interface. In: Ocean Optics XI. Proc. SPIE, 1750: 138-148.

Stumpf, R. P., Holderied, K., and Sinclair, M. (2003). Determination of water depth with high-resolution satellite imagery over variable bottom types. Limnology and Oceanography, 48, 547-556.

Subramaniam, A. and Carpenter, E. J. 1994. An empirically derived protocol for the detection of blooms of the marine cyanobacterium Trichodesmium using CZCS imagery. Int. J. Remote Sensing, 15: 1559-1569.

Sugihara S., Kishino, M. and Okami, N. 1984. Contribution of Raman scattering to upward irradiance in the sea. J. Oceanogr. Soc. Japan, 40: 397-403.

Sukenik, A., J. Bennett and P. Falkowski, Light-saturated photosynthesis-limitation by electron transport or carbon fixation?. Biochem. Biophys. Acta 891 (1987), pp. 205-215.

Steemann Nielsen,E. 1952. The use of radiao-active carbon (C14) for measuring organic production in the sea. J. Cons. Int. Explor. Mer 18: 117-140.

Schiller, H., and Doerffer, R. (1993). Fast computational scheme for inverse modeling of multispectral radiances: Application for remote sensing of the ocean. Applied Optics, 32, 3280-3285.

Steidinger, K.A., Haddad, K.D., 1981. Biologic and hydrographic aspects of red tides. Bioscience 31, 814-819.

Stumpf, R.P., 2001. Applications of satellite ocean color sensors for monitoring and predicting harmful algal blooms. J. Human Ecol.Risk Assess. 7, 1363-1368.

Stumpf R.P. and M.A. Tyler, Satellite detection of bloom and pigment distributions in estuaries, Remote Sensing of Environment 24 (1988), pp. 385-404.

Tanaka, A., Kishino, M.., Doerffer, R., Schiller, H., Oishi, T., and Kubota, T. (2004). Development of a neural network algorithm for retrieving concentrations of chlorophyll, suspended matter and yellow substance from radiance data of the ocean color and temperature scanner. Journal of Oceanography, 60, 519-530.

Tassan, S.. 1987. Evaluation of the potential of the Thematic Mapper for Marine application. International Journal of Remote Sensing 8, 1455-1478.

Tassan S. The effect of dissolved 'yellow substance' on the quantitative retrieval of chlorophyll and total suspended sediment concentrations from remote measurements of water colour[J] . Int . J . Remote Sensing ,1988 ,9(4) :787-797.

Tassan S. Local algorithms using SeaWiFS data for the retrieval of phytoplankton , pigments , suspended sediment , and yellow substances in coastal waters[J]. Appl. Opt., 1994, 33(12):2 369-2378.

Tassan, S. (1993). An algorithm for the detection of the White-Tide ("mucilage") phenomenon in the Adriatic Sea using AVHRR data. Remote Sens. Environ., 45: 29-42.

Tassan,S.. 1992. An algorithm for the identification of benthic algae in the Venice lagoon from Thematic Mapper data. International Journal of Remote Sensing 13 (15), 2887-2909.

Tassan,S.. 1993. An Improved In-water Algorithm for the Determination of Chlorophyll and Suspended Sediment Concentration from Thematic Mapper Data in Coastal Water. International Journal of Remote Sensing 14 (6), 1221-1229.

Tassan S. and d'Alcalà M.R.. Water quality monitoring by Thematic Mapper in coastal environments. A performance analysis of local bio-optical algorithms and atmospheric correction procedures. Remote Sensing Environ. 45, 177-191 (1993).

Tian L. Q.,Chen X. L.,et al. Atmospheric correction of ocean color imagery over turbid coastal waters using active and passive remote sensing. Chinese Journal of Oceanology and Limnology, 2009, 27(1): 124-128.

Torgersen, C.E., *et al.*, Airborne thermal remote sensing for water temperature assessment in rivers and streams. Remote Sensing of Environment, 2001. 76(3): p. 386-398.

Torgersen, C.E., Price, D.M., Li, H.W. and McIntosh, B.A., 1999.Multiscale thermal refugia and stream habitat associations of chinook salmon in northeastern Oregon. Ecological Applications 9, pp. 301-319.

Viollier, M. and Sturm, B. 1984. CZCS data analysis in turbid coastal water, J. Geophys. Res., 89:4977-4985.

Weisblatt, E. A., Zaitzeff, J. B., and Reeves, C. A.: 1973,Classification of Turbidity Levels in the Texas Marine Coastal Zone, Machine Processing of Remote Sensing Data Conference Proceedings. Laboratory for Application of Remote Sensing, Purdue University, Lafayette, Indiana. October 16-18, pp. 42.

Whitlock, C.H.,NASA TMX-73906,1976.

Williams A N, Grabau W E, 1973. Sediment concentration mapping in tidal estuaries. Third Earth Resources Technology Satellite-1 Sym. NASA SP-351, 1347-1386.

Yung Y K, Wong C K, Broom M J *et al.*, 1997, Long-term changes in hydrography, nutrient and phytoplankton in Tolo Harbour, Hong Kong, Hydrobiologia, 352:107-352.

Zeichen, M.M. and I.S. Robinson. Detection and monitoring of algal blooms using SeaWiFS imagery. 2000. Venice, ITALY: Taylor and Francis Ltd.

Zhang, Y., Pulliainen, J., Koponen, S., and Hallikainen, M. (2002). Application of an empirical neural network to surface water quality estimation in the Gulf of Finland using combined optical data and microwave data. Remote Sensing of Environment, 81, 327-336.

Zhu X G, He Z J, Deng M. Monitoring of water color for Pearl River estuary over twenty years. Journal of Remote Sensing. 2001, 5(5): 396-401.

Chapter 18

FLOOD MAPPING AND DAMAGE ASSESSMENT – A CASE STUDY IN THE STATE OF INDIANA

Jie Shan, Ejaz Hussain, KyoHyouk Kim, Larry Biehl

18.1 Introduction

Flood mapping, damage assessment, and disaster remediation involve activities and efforts from a number of governmental agencies. Under the National Flood Insurance Act 1968, the Federal Emergency Management Agency (FEMA) is responsible for identifying flood hazards nationwide, publishing and updating flood hazard information in support of the National Flood Insurance Program (NFIP). Over a period of two decades, FEMA has produced over 90,000 flood hazard maps covering approximately 150,000 square miles of floodplain. Recently, about 75% of the flood hazard maps inventory became over 10 years old. In 2003, a program was initiated for flood hazard map modernization including the conversion of paper maps to digital format. Since flood hazard mapping is part of the NFIP, a variety of maps indicating various degrees of insurance risk and premium rating are produced. However, the basic hazard maps, indicating the 1 in 100 years (1%) floodplain and the 1 in 500 years flood (0.2%) outlines, are normally produced based on detailed hydraulic modeling of river reaches at the community scale. All flood maps are made available to the public through the FEMA Map Service Center. These maps can be purchased in paper or CD format and can be viewed online (http://msc.fema.gov/). Beginning on October 1, 2009, FEMA will provide only one paper flood map and the Flood Insurance Study (FIS) report to each mapped community. All other distribution of maps and Flood Insurance Study reports will be converted to digital delivery. FEMA will continue to provide free digital map products and data to federal, state, tribal, and local NFIP stakeholders. In addition to the FEMA mapping effort, which is specifically linked to the NFIP, some states have their own flood mapping programs. They produce flood "awareness" maps that simply show flood prone areas without specific depth or other flood hazard data for a particular flood event.

US Geological Survey (USGS) regularly records and ensures the provision of real time water level data from its gauging stations on rivers and streams to the National Weather Service (NWS) for flood forecast and flood warning to monitor the frequent changing flood situations. USGS teams visit the affected fields to set

D. Li et al. (eds.), *Geospatial Technology for Earth Observation*,
DOI 10.1007/978-1-4419-0050-0_18,

high-water marks to document the severity of the floods in areas with no stream gauges, which helps emergency managers to estimate damages and to validate floodplain maps.

Take Indiana as an example, its Department of Natural Resources (DNR) in partnership with FEMA is completing and issuing county-wide preliminary Digital Flood Insurance Rate Maps (DFIRMs) and Flood Insurance Study reports throughout Indiana. The new digital floodplain maps will be based on updated topographic and orthophoto data. The mapping initiative will depict new or revised Base Flood Elevations (BFEs), which the flooding water may reach during the 1 percent annual chance flood (also known as the 100-Year Frequency Flood). The initial phase should be completed by 2011. In addition to DFIRMs, the State launched the Indiana Floodplain Mapping Initiative in 2004 to revise outdated maps for all 92 Indiana counties (http://www.in.gov/dnr/water/). These updated maps will help the communities see the revised floodplain boundaries for their future planning. However, until then, they will continue to use current Flood Insurance Rate Maps for permitting development, which are made available through the "Indiana Map", a public GIS data resource (http://inmap.indiana.edu/).

Despite the organized federal and state level activities for flood mapping described above, they are mostly focused on precaution or prediction. The produced maps are based on models and simulation, which may not fit to a specific flood event. Therefore, such maps are not sufficient for coping with a realistic flooding event. Quick assessment of the actual flood extent over a vast area by field visit is unrealistic, whereas satellite remote sensing can meet this need due to its large coverage and relatively short revisit time. It is one of the most effective and accurate technologies for mapping such natural hazards as floods over broad areas (Campbell 2002). The timely provision of such disaster related information and products for the local and state governments, the residents, and the disaster rescue and relief teams can help them quickly make response strategies, take rescue actions, and develop mitigation plans. It can help the authorities in rescue and relief efforts, damage assessment, mitigation and future planning to take remedial measures and to safe-guard such events effectively. The timely acquisition of satellite data over the flooded areas for pre-, during, and post-events is extremely valuable for assessing flood damages at a large scale. Evaluation of the collected satellite images, along with other ancillary data, can help us understand the flood situation and its dynamic development, identify designated floodplain boundaries, and further estimate the flood damages to the major standing crops, residents, roads and streets, and other infrastructure. The mapping results of the flood extent can then be published immediately on the Internet through a Web mapping tool or service, such as the Google Earth Application Programming Interface (API). They can also be regularly updated with the most-recent flood development, analysis and assessment results, relief and rescue activities.

This chapter utilizes remote sensing images to map the floods that occurred in southern Indiana in June 2008. First, the flood development is described and the importance and availability of remote sensing data is discussed. Second, damages to crops, roads, and residential areas are estimated, followed by the changes to the existing estimated floodplains. The final results are published on the Web through the Google Earth plug-in as a resource for damage assessment, planning and executing of disaster related rescue, relief and rehabilitation activities.

18.2 The Southern Indiana Floods

Southern Indiana experienced heavy rainfalls from the end of May to early June, 2008. A rainfall of 1 to 3 inches on the night of May 30 caused flooding to portions of the Wabash River (see Figure 18.4) in western Indiana. Additional rainfalls of 1 to 5 inches on June 4 and 5 developed into flooding in central and southern parts of Indiana. On June 6, central Indiana and eastern Illinois reccived another cycle of rainfall which created flooding in the creeks, streams and rivers in western central Indiana. On June 7, the highest official rainfall of 9.5 inches was recorded from the city of Center Point in Clay County. As a result of this week-long rainfall, both the Wabash and White rivers were near historic flood levels on June 8 as documented by USGS operated gauge stations distributed over the State. The rainfall cycle that started early in the first week and intermittently continued until the second week of June, made and kept all the streams and rivers in the flooding situation. The average rainfall for this month was 8.00 inches, 3.87 inches above the normal conditions. Figure 18.1 is the precipitation map for June 7 and for the first week of June (June 1-7), which demonstrates the severity of this rainfall in southern Indiana.

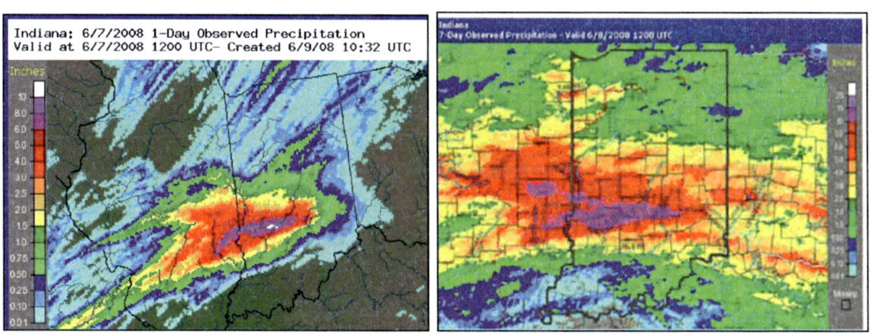

Fig. 18.1 Precipitation maps for June 7, 2008 (left) and June 1- 7, 2008 (right) in western central Indiana (courtesy National Weather Service)

These high rainfalls caused severe flooding in the rivers flowing through southern central and southern Indiana and caused heavy destructions to both urban areas and agricultural lands. River water levels were almost twice as high as the flood stage level as recorded by the National Weather Service/USGS operated gauging stations. These water levels were much higher than many previous years.

Although such data did help USGS predict floods downstream, the flow of excessive rainfall over the plain areas added more severity than predicted. Flowing through these areas are the Wabash, White and East Fork White rivers and a number of streams. The heavy flooding situation in these water channels remained for about 6 days, from June 8 to 13. In addition to the rivers and streams, vast areas away from them were also flooded due to heavy surface runoff, which added more severity than predicted from the gauge data. The situation of this heavy flooding can be seen from the comparison of the water levels recorded at various locations for the Wabash and White rivers for year 2007, 2008 and the preset flood stage levels, given below in Figure 18.2 and 18.3.

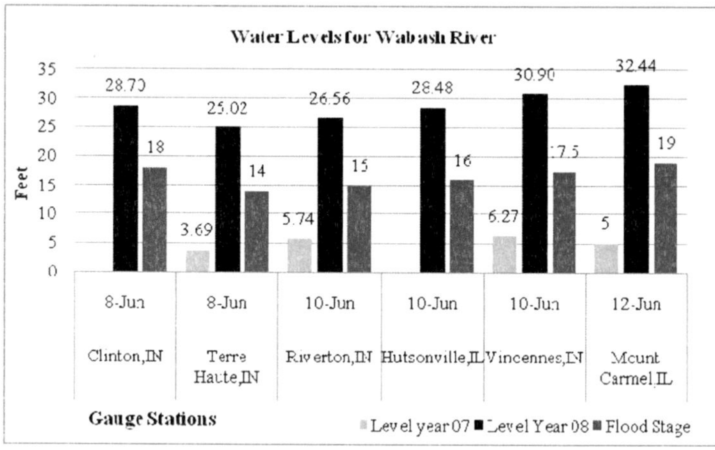

Fig. 18.2 USGS water gauge station data for the Wabash River of Indiana

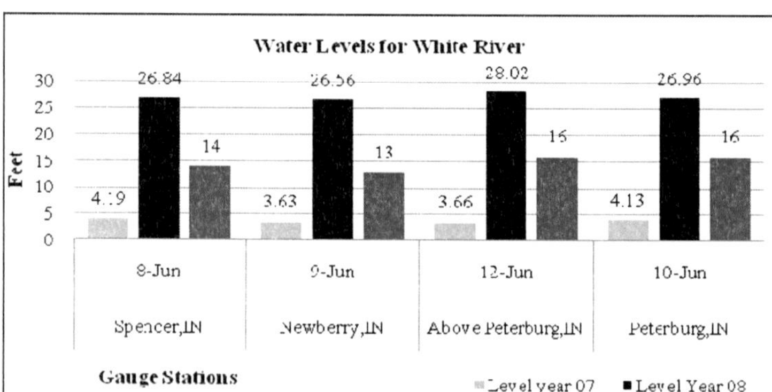

Fig. 18.3 USGS water gauge station data for the White River of Indiana

18.3 Remote Sensing Data Portals and Images

18.3.1 Data Portals

There are a number of portals and repositories, from which we can acquire satellite images along with other geographic data for this study. Indiana View (http://www.indianaview.org/) is a state-wide consortium of a total of 14 universities and institutions in Indiana. The vision for Indiana View is to facilitate and promote the sharing and use of public domain remote sensing image data by Indiana universities, four-year colleges, community colleges, K-12 (from kindergarten to high school) institutions, libraries, museums, government agencies, and private sectors. Indiana View was accepted to the national America View (http://www.americaview.org/) consortium in 2004 with Purdue University as the lead institution. The America View consortium is administered through the USGS Land Remote Sensing Program. The objectives of Indiana View are to:

- Create and maintain an online portal to provide easy access to remote sensing data holdings of the member institutions.
- Promote the use of remote sensing data from kindergarten to college education through tutorials and training.
- Facilitate the use of remote sensing data to monitor statewide issues, such as urban development, forestry, water quality, air pollution, flooding and other natural disasters, crop development, resource management, and homeland security.
- Disseminate real-time satellite data utilizing the Purdue Terrestrial Observatory (PTO).

An online portal (http://www.indianaview.org/glovis/) has been created using the USGS provided GeoVis software. The portal currently provides access to more than 250 Landsat TM (Thematic Mapper) scenes (1984 to 2008), 350 ASTER (Advanced Spaceborne Thermal Emission and Reflection Radiometer) scenes (2000 to 2007), 25 Landsat MSS (Multispectral Scanner) scenes (1972 to 1992), 2 Corona scenes (1965), and the National Agricultural Statistics Service (NASS) Indiana Cropland Data. A link is also provided to the Indiana University Spatial Data Portal that contains over 14 terabytes of 6-inch to 1-meter ortho image data for Indiana (1998 to 2008). Indiana View received the Landsat data for pre-and post-event scenes of the flooded areas from the USGS Earth Resources Observation and Science (EROS) Data Center and other commercial satellite images as a result of the activation of the International Charter: Space and Major Disasters (http://www.disastercharter.org)[1]. Indiana View made the Landsat scenes avail-

---

[1] All Landsat images have been available for free since December 2008.

able for Indiana state government agencies and universities for their analysis and monitoring of the flooding.

Another portal for access to the near real-time data, Purdue Real-time Satellite Information Gateway (PRESTIGE) (http://www.purdue.teragrid.org/prestige), allows users to subscribe to basic MODIS (Moderate Resolution Imaging Spectroradiometer) and GOES (Geostationary Operational Environmental Satellites) sensor data products. The basic products include MODIS Level 1B calibrated and geo-referenced (MOD02) data, reflectance, true color, NDVI (Normalized Difference Vegetation Index) and EVI (Enhanced Vegetation Index) vegetation indices, NDVI composites, sea surface temperature, and fire detection image files. Users can request the time period that they wish to obtain the data for, the data file format and coverage areas. Users will receive an email message, if desired, to notify them once the data products are ready. Daily MODIS image data were made available to the researchers and users around the State one (1) hour after the data were collected for a period of 7 days after the flood event.

18.3.2 Satellite Images

Through Indiana View we can obtain medium resolution satellite images, including MODIS and Landsat, at no cost. However, higher resolution satellite images, such as SPOT, IKONOS, and QuickBird, are only available from the International Charter program through USGS for this specific flood study. The selection of a particular data source depends mainly on its timely coverage, availability, spatial, spectral and temporal resolutions, cloud cover, and costs.

June 11, 2008 had clear sky in southern Indiana, which provided a good opportunity to record the flood extent with Landsat satellite imagery. The large coverage of Landsat images provides the primary data source for studying the extent and damages caused by the flood. The Landsat imagery of June 11, 2008 acquired between 09:30 -10:00 AM was the best data for mapping the flood extent as the peak flooding in these areas was observed on June 10 and 11. By this time, the water from northern (upper stream) parts and the adjoining areas subsequently surged along central and western areas leading to the southern areas. The real time data reported by the NWS hydrologic prediction services and the readings from the gauge stations indicated that the Wabash and White Rivers along with smaller rivers and streams experienced heavy surges. Visual observation of the imagery revealed that nine (9) counties in southern Indiana were severely affected, and thus were selected as the study area. The county boundaries with gauge station locations, rivers, and the available Landsat images are shown in Figure 18.4.

For comparison purposes, the Landsat image collected between 09:30 -10:00 AM on June 9, 2007 over the same area is also acquired from USGS as a result of the declaration of International Charter through the Indiana View archive. This image is shown in Figure 18.4. This data covers most of the southern areas of Illinois and Indiana States. The availability of the temporal datasets helps greatly in observing the changes caused by the flooding. The June 2007 image clearly shows the normal situation of water bodies like rivers, streams, lakes and ponds in this area. The image also helps visualize the extent of the flood plains along the river courses.

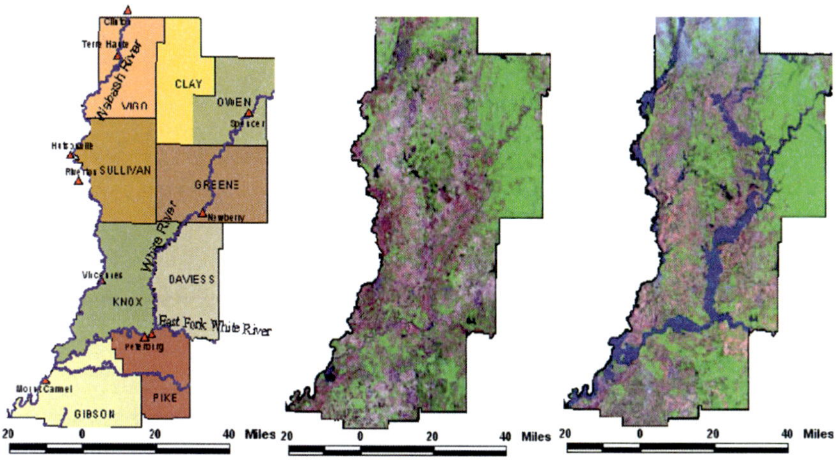

Fig. 18.4 County map with rivers and gauge stations (left), Landsat images of year 2007 (middle) and 2008 (right) for the study area. Landsat bands 5, 4, 3 were displayed as red, green and blue channels, respectively.

In total, we used two (2) scenes of Landsat TM images, one from 2007 and one from 2008, at 30 m resolution with six (6) bands. The thermal band was excluded from image classification due to its coarser spatial resolution. Each scene consists of 6,167 x 6,167 pixels and covers an area of about 34,225 square kilometers.

MODIS data are available for free on a daily basis. MODIS imagery collected at 15:00 on June 11, 2008 over the flood area was acquired through Indiana View from the Purdue Terrestrial Observatory. The data consists of seven (7) bands at the resolution of 500 m. Each MODIS scene has 950 x 620 pixels and covers an area of about 147,250 square kilometers. We used only one MODIS scene in this study.

18.4 Image Classification

For the assessment of the flood extent and damages of the heavily flooded nine (9) counties, the Landsat images are classified using an object-based classification method (Blaschke and Lang 2005). This method uses fuzzy rules-based techniques to classify image objects or segments rather than individual pixels. The object-based image analysis is often carried out in two steps: image segmentation, and classification of the segmented objects. The first step, image segmentation, is the division of an image into contiguous, disjoint, and homogeneous regions. It is the prerequisite for the object-based classification, in which an object or segment is used for classification. The aim of the segmentation step is to create meaningful objects by using appropriate methods and using scale, shape and color parameters of the objects to maintain the geometry of the segmented objects as closely as possible to actual objects (Blaschke and Strobl 2001). With a careful selection of segmentation parameters, object primitives being very similar to the original object's geometry can be achieved (Baatz and Schape 2000). At the second step, these objects are classified using their spectral, contextual, and textural features. The selection of these features is based on the object's physical characteristics and their spectral responses. A proper selection of features with well-defined fuzzy membership functions for every class would allow for a good classification and produce a homogenous thematic map (Benz *et al.* 2004). The classification results depend on these input features and the degree of membership value of a class. The closer the membership value of a class to 1, with less or no alternative assignment membership value to other classes, the more stable the classification results (Definiens 2000).

For this study, the Landsat images are first segmented to create image segments and then classified into five (5) major classes, i.e., Water, Wet Area, Vegetation, Open Area, and Urban Area. The Water class includes the areas with sufficient visible flowing or standing water. Because the flood waters are turbid, its reflection increases especially in the NIR (Near Infrared) and MIR (Mid Infrared) region (Campbell 2002). The means respectively from the NIR and MIR bands within a segment are used for classification. Means of the blue band within a segment are used for the Wet Area class, which appears in light blue color showing the high moisture soil. The Vegetation class includes all types of crops, shrubs and forests. The Open Area 1 class shows no plausible land cover, however, some areas with light cloud patches have been classified as Open Area 2, because the area underneath the clouds consists of open land and light vegetation. For Water classification, only spectral object features, such as NDVI and band means, are used. The detailed object features used for classification are listed in Table 18.1.

Table 18.1 Object features used in Landsat image classification

Class	Object features
Water	Mean NIR (Band 4) and Mean MIR (Band 5)
Wet Area	Mean Blue (Band 1)
Vegetation	NDVI and Mean NIR (Band 4)
Urban Area	Mean Blue (Band 1), Mean MIR(Band 7)
Open Area 1 & 2	Mean MIR (Band 5)

The resultant quality metrics in Table 18.2 show the mean, standard deviation, minimum and maximum membership values for image classification. The urban areas are mostly misclassified with open area and vegetation area due to the presence of intense tree canopies and grassy areas within the urban areas. A patch of light cloud in the upper part of the image is classified as urban area, but some of urban area is misclassified as open area. The final classification map is shown in Figure 18.5. The higher mean, lower standard deviation and the membership values closer to 1 indicate the high classification accuracy of the classes (Syed *et al.* 2005). These results indicate that the Water and Wet Area classes are very accurately determined; the other four (4) classes have some degrees of misclassification among each other.

Table 18.2 Object-based fuzzy classification quality metrics

Class	Mean	Std. Dev.	Min	Max
Water	1	0	1	1
Vegetation	0.88	0.21	0.10	1
Open Area1	0.71	0.29	0.10	1
Open Area2	0.83	0.15	0.15	0.98
Wet Area	1	0	1	1
Urban Area	0.73	0.29	0.10	1

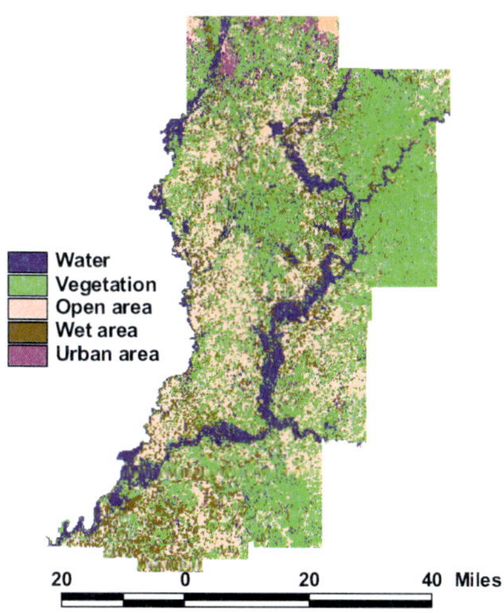

Fig. 18.5 Classification results of the Landsat image of June 11, 2008

Since the objective of this study is to map the flood extent, assess damages, and map floodplain boundaries, further analysis will be based on only two composite classes, Water and No Water, which is a combination all but water classes. We first repeat the above classification process to obtain the water distribution in year 2007. By comparing the pre- (2007) and post- (2008) flooding classification results, a flood map is produced. Figure 18.6 shows the classification results for both years (left and middle) and the flood map (right). It shows the pre- and post-flood water body extent and the most flood prone areas. The red (darker) color in the flood map is the water area exceeding that of year 2007. Visual analysis of the flood image reveals that the most serious flooding is along the river courses. This flood map helps quantify the areas subject to flooding. The classification results are converted to vector data for further analysis and assessment through a GIS (Geographic Information System). The comparison of the Water class for the two years and related statistical results show an increase of about 93,369 hectares or, about 9% of the area was flooded in June 2008. The most affected counties are Knox, Daviess, Greene, and Vigo, mainly because of the overflowing rivers. Knox County is mostly affected by the White river as it flows all along the County's eastern and southern boundary. The Wabash River on its western border did not affect Knox County as badly as it did near the Illinois counties to the west. Daviess County experienced flooding mainly from the White River and the East Fork White River. The summary and comparison of each county's flooded area are shown in Table 18.3. The change detection to assess flooding in urban areas is limited by the relatively low resolution (30 m) of the images.

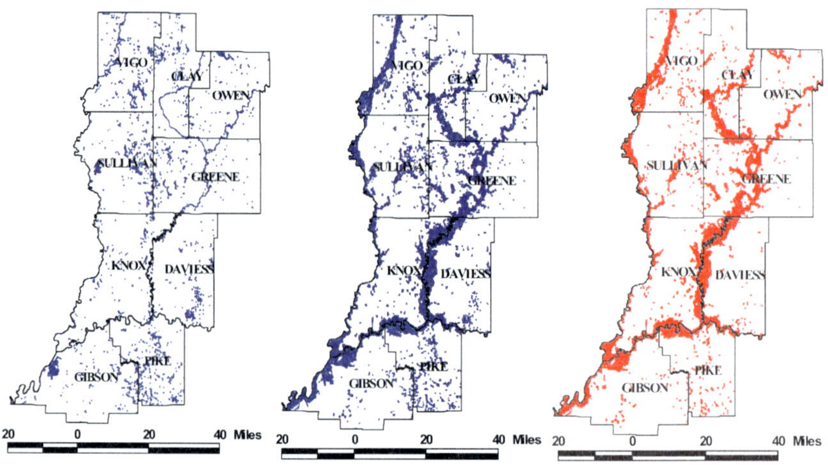

Fig. 18.6 Water of year 2007 (left), 2008 (middle) and the flooded area (right)

In addition to the Landsat images, MODIS data are also used to map the extent of the flood. For this purpose, the MODIS image is classified to water and no-water classes based on the same process applied to the Landsat images. To estimate the MODIS flood water extent, Landsat 2007 water class from the Landsat image is used as the reference of normal water body extent. The flood extent derived from MODIS is then compared with the one from the Landsat 2008 image to analyze the effect of resolution variation on the mapping .The resultant MODIS classification yields 97,507 hectares as flood area, about 4% more as compared to 93,369 hectares from the Landsat 2008 image. This inconsistence is partially due to the large pixel size, which resolves the areas adjacent to rivers as the Water class (Corina *et al.* 2006). The MODIS large pixel size resolves the high moisture wetlands as water, which occurs in close proximity along waterways and at the edges of flood extents. Besides, in the southern part of the study area where the three rivers join, the gauge station data showed increasing flood water levels on the night (at 23:00) of June 11, 2008, close to the time (15:00) the MODIS image was collected. The MODIS image and its flood extent with reference to the Landsat 2007 image are shown in Figure 18.7.

Table 18.3 Overall county areas and the flooded areas

County	Area (Hectare)	Flooded area (Hectare)	% Flooded area
Vigo	106224	10209	10
Clay	93237	8272	9
Owen	100317	4360	4
Sullivan	117465	6715	6
Greene	141291	14534	10
Knox	135653	20638	15
Daviess	113004	12709	11
Pike	88302	5994	7
Gibson	129159	9940	8
Total	1024652	93371	9

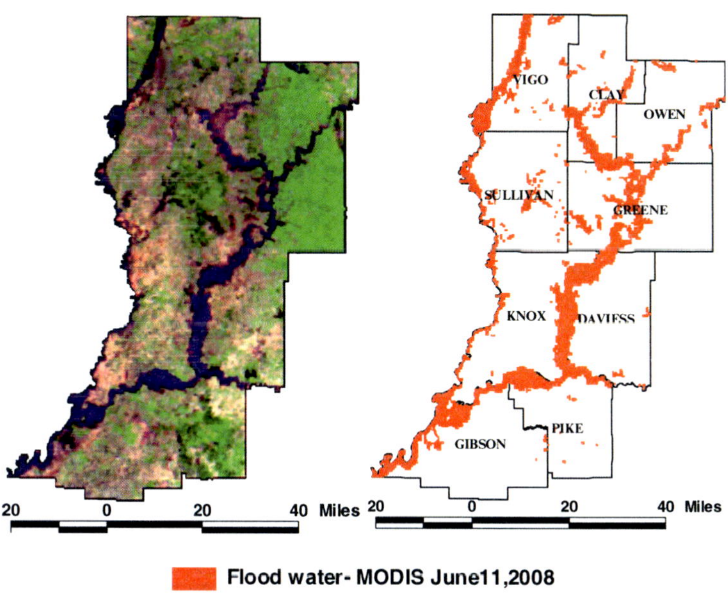

Fig. 18.7 MODIS June 11, 2008 image (left) and its flood extent (right)

18.5 Damage Assessment

The above classification results are used to estimate the damages to the standing crops, the roads submerged by flood waters, and the areas out of the designated floodplains. The details of the analysis process and results are described below.

18.5.1 Crop Damages

Every year the National Agricultural Statistics Service of the United States Department of Agriculture, along with the Farm Service Agency (FSA) and the participating State governments, record and produce a digital data layer named "Cropland Data Layer" (CDL) for major crops (http://www.nass.usda.gov). The CDL program annually focuses on corn, soybean, and cotton agricultural regions in the participating states to produce digitally categorized, geo-referenced output products for crop acreage estimation. To produce such CDL data, the Resource-Sat-1 (AWiFS, 56 m resolution) imagery and FSA registered grower's reported field data are used. This data layer is suitable for use in GIS applications for analysis purpose.

For this study, the latest available crop data layer for the year 2008 released by USDA in the first week of March 2009 is used. Corn and soybeans had been 100% planted before the occurrence of the flood in June 2008. Therefore, these data were used to estimate the crop areas and types affected by the June 2008 floods. The authors realized that the flood affected areas based on the 2008 crops layer may not be totally accurate as some of the flood affected croplands may have been replanted after the floods. However, these results show the crops as damaged or flood affected within the flood water extent. The CDL data consists of a quite detailed number of crops, however, for this assessment only the major crops grown in the study area are used. It includes corn, soybean, wheat (spring and winter wheat), pasture (including grass, grassland herbaceous, and hay), and woods (including woodland; deciduous, evergreen and mixed forests; shrub lands; woody wetlands). All the crops damage estimation is made by comparing each crop area with the flood extent layer and summarized by county. As most of the agricultural areas in these counties are under corn and soybean other than wheat, the resultant high percentage damages are also to these crops. Mostly the crop areas along the river courses and floodplains are affected. The CDL 2008 data (left) and the 2008 flood extent overlay (middle) are shown in Figure 18.8.

The crops damage statistics show that an average of 6% of corn, 22% of soybeans, and 12.5% of wheat are damaged in these counties. The crops in Greene County were severely hit with the highest percentage damages of 11.7% of corn, 35.4% of soybean, and 28% of wheat. Other highest percentage damages to corn and soybean were in Daviess followed by Knox, Pike and Vigo counties. The least crop damages were in Sullivan County because the flood water in the Wabash

River flowing along Sullivan County's western boundary caused more inundation in Illinois rather than in Indiana. It should be noted that the crop damage estimates only show the crop areas affected by the floods rather than a measure of complete damage since some of these affected crops may eventually survive. The degree of final damage would depend on factors such as ponding time, soil erosion, partial or total submerge, saturation time period of fields, formation of dense surface crust, and deposition of mud over the plants. Figure 18.8 (right) illustrates the spatial distribution of damaged crop areas for each county in the study area. Table 18.4 and Figure 18.9 show the crop areas under the flood and the summary of percentage crop damages for each county, respectively.

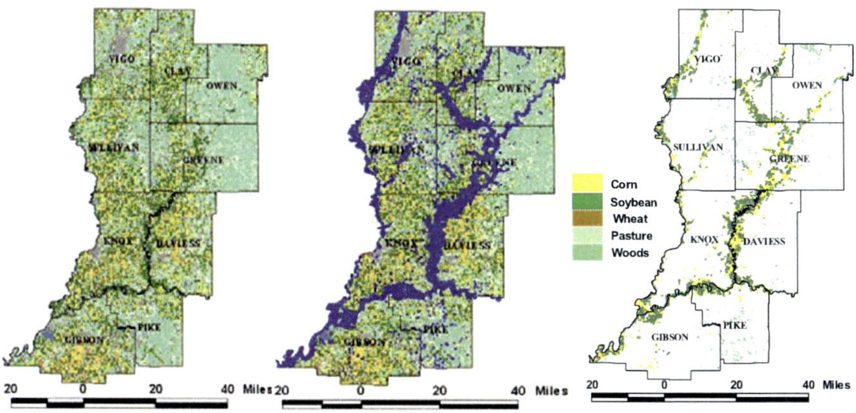

Fig. 18.8 2008 crop data layer without (left) and with (middle) flood extent overlay, and damaged crops (right)

Table 18.4 Areas of crops under flood (hectare)

County	Corn	Soybean	Wheat	Pasture, Hay, Grass	Forest, Shrub
Vigo	1823	3911	49	680	1675
Clay	1586	4058	23	208	1025
Owen	393	1336	31	510	1539
Sullivan	752	1994	18	647	1633
Greene	1607	7315	118	652	2061
Knox	1852	10496	63	526	2604
Daviess	1682	6544	38	297	1359
Pike	252	3265	2	249	1071
Gibson	780	2543	5	293	1362
Total Area	10728	41462	347	4062	14329

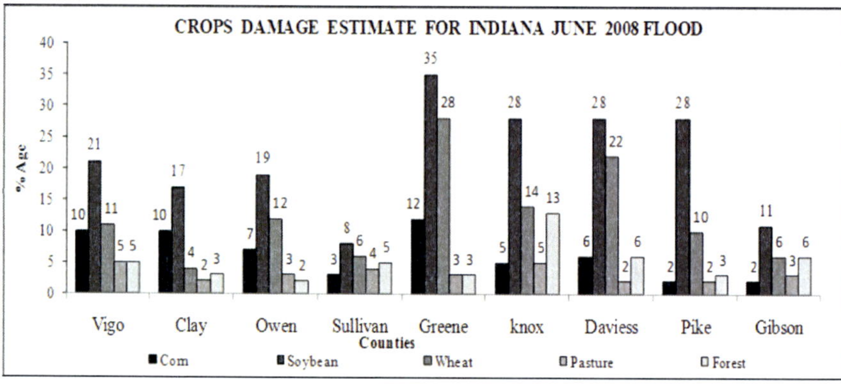

Fig. 18.9 Crop damage (in percentage) in the nine counties of Indiana

18.5.2 Road Damages

To assess the road damages, the Indiana Department of Transportation (INDOT) roads and streets 2005 data layer is used. First, the length of the roads within each county is calculated from the INDOT layer and then these roads are compared and checked with the flood water classification layer. Some of the roads observed under the flood waters may not be accurate as their width may not commensurate with the image pixel size. Even if a road is higher than the adjacent land, it is counted as flood category since its width is small and its adjacent land area is under the flood. Figure 18.10 shows the INDOT 2005 roads (left) and the flooded roads (right).

The evaluation of these results reveals that the roads passing through the floodplains were mostly affected. A large number of streets and small roads were also affected. In addition, a number of State Highways and US Highways passing through these counties were subject to the flood. State Road 57 and State Road 157 were most affected, with a total 13 km of road segments being under water at different locations, followed by the State Road 59, which had 6.8 km under water. Overall, about 37 km of different State Roads were affected. Of the US Highways, US-231 was the most affected with a total of about 12 km road segments being subject to the flood. This evaluation shows that the most affected were the county roads. A total of about 1,082 km of road segments were affected by the flood waters in these nine counties. The percentage of the roads in Knox, Greene and Daviess counties that were under flood waters of the White River were 10%, 6% and 8%, respectively. Gibson County had about 7% of the roads under flood water due to flooding along both the Wabash and White Rivers. Figure 18.11 and Table 18.5 show the statistics of these flood affected roads in each of these nine counties.

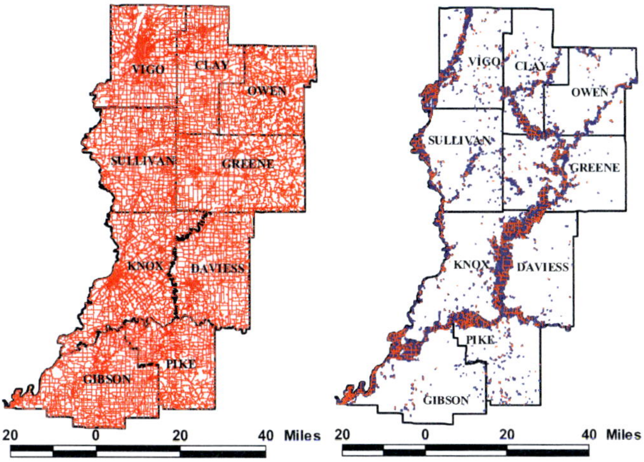

Fig. 18.10 INDOT 2005 roads and streets (left) and the flooded ones (right)

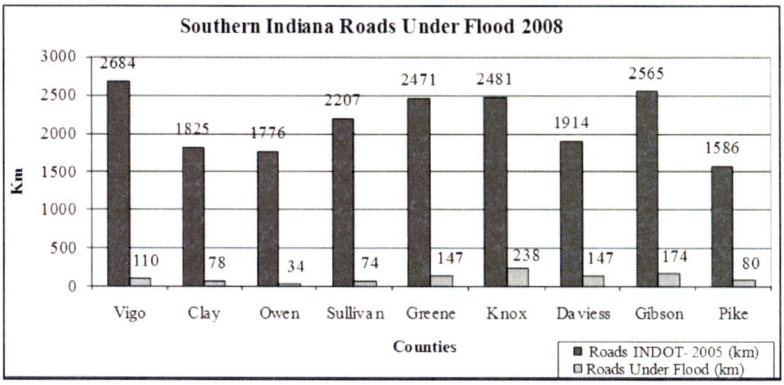

Fig. 18.11 Roads in counties and roads submerged under the flood water

Table 18.5 Statistics of the INDOT roads and flooded roads

County	Roads 2005 (km)	Roads flooded (km)	% Flooded
Vigo	2684	110	4.10
Clay	1825	78	4.32
Owen	1776	34	1.91
Sullivan	2207	74	3.35
Greene	2471	147	5.95
Knox	2481	238	9.59
Daviess	1914	147	7.68
Gibson	2565	174	6.78
Pike	1586	80	5.04
Total	19509	1082	5.5

18.5.3 Floodplain Changes

Floodplains are the land areas situated adjacent to rivers and streams that are subject to recurring inundation. The extent of these floodplains are marked and designated based on the analysis of occurrence of floods into the rivers and streams and the overflow of the water to the adjacent land over a long period of time. Usually, the water on the floodplain drains back to the water course as its flow recedes. However, for places where the water course has higher banks as compared to floodplain levels, the water may not drain back and cause inundation or ponding over a long period of time, thus making the area unusable for any activity. The analysis of every flood through remote sensing data acquired during the peak time can greatly help examine, inventory, and incorporate changes to the extents of these floodplains and other flood prone areas. The analysis and comparison of the 2008 flood extent to the designated floodplains show that most of the actual flood water remains within the floodplains; however, it did overflow the boundaries of the floodplains at a few places (Figure 18.12, dark red areas) due to high water levels in the rivers. It mostly happened in the White River floodplains in Greene and Daviess counties and in the Wabash River floodplains in Knox County. These general pre-designated floodplains and the 2008 flood water overlay are shown in Figure 18.12. The light red color shows the designated floodplain extents, blue the actual 2008 flood extents, and the dark red the actual flood extent beyond the designated floodplains. We found in this study that 30% of the area of the designated floodplains is subject to floods and about 6,000 hectares (2%) of lands are beyond the designated floodplain areas, which are mostly in Greene and Daviess counties. In these counties, flood water did flow beyond the floodplains by about 1-3 km as shown in Figure 18.13.

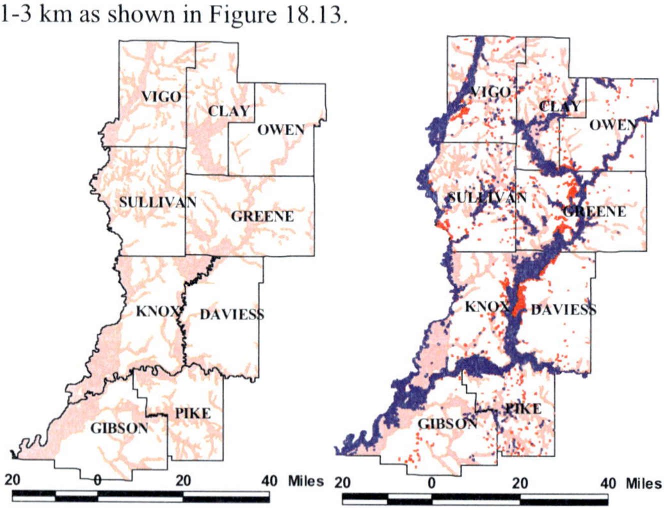

Fig. 18.12 General floodplains (left, light red) and the actual flood extents (right, blue for flood water within floodplains and dark red for flood water outside floodplains)

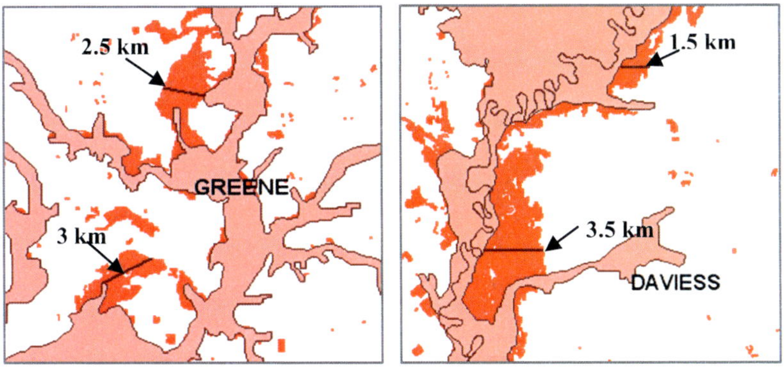

Fig. 18.13 Extent of flood water beyond designated floodplains (dark red)

18.6 Web-based Map Service

18.6.1 Web GIS

Web GIS is currently an active research field in the information technology industry. Governments, universities, companies, and individuals are producing a large amount of information, which is delivered to the public through the Internet. Much the same is true in the Earth observation technology. Many leading vendors have developed Web GIS tools to effectively share and publish valuable geospatial information. The main advantage of Web GIS is that it can visualize data more effectively with the help of digital maps such as remote sensing images and application-specific data layers. Among the various software tools, Google Earth and Google Earth plug-in are widely used to share and publish geospatial information because of their powerful visualization functionalities, abundant available data, and easiness to use.

18.6.2 Google Earth Plug-in

Google released its Google Earth plug-in in June, 2008. The Google Earth plug-in is the Web version of Google Earth desktop and enables us to embed Google Earth functionalities into one's own websites. Using this plug-in, one does not need to install Google Earth to use the virtual globe and can easily publish user specific geographic information onto the Internet along with other basic geospatial data, such as background images, roads, and terrain provided by Google Earth. To customize the Google Earth plug-in, Google Earth provides a java script library, called Google Earth API (Application Programming Interface), which allows users to add Google Earth plug-in objects into one's own sites. Google Earth API also enables us to access Google Earth plug-in objects and their features. Using Google Earth API, we can publish geographic information into a website, draw geometric objects (point, line and polygon), and add realistic 3D models. The current version of Google Earth plug-in operates on Microsoft Windows 2000, XP, Vista, and Mac operating systems. To access a website developed based on the Google Earth plug-in, the Google Earth plug-in program should be installed on the local client computer. Using Google Earth API, one can incorporate many functionalities and data layers of Google Earth. The following is a brief list:

- Display background satellite imagery and aerial photos provided by Google Earth
- Display predefined layers (terrain, roads, boundaries, and 3D buildings)
- Create and manipulate geometry objects (place marker, line and polygon)
- Display KML (Keyhole Markup Language) or KMZ objects (parsing KML)
- Manage mouse and keyboard events for user interface

18.6.3 File Format and Conversion

The basic file format of Google Earth plug-in is KML or KMZ format. KMZ is a zipped KML file with *.kmz extension and includes any icons and images referenced in the KML file. KML is an XML-based language schema for displaying and visualizing geographic data. KML consists of many tags and nested structures to define various features (marker, icon, objects). In the Google Earth API, an interface whose name begins with ‘kml’ represents KML-related objects. The following is a simple kml example to display the place marker on the given cordinates. The internal coordinate system of Google Earth is geographic coordinates (latitude/longitude) based on the WGS 84 datum. More detailed information about KML can be found at the reference site http://code.google.com/apis/kml/documentation/kmlreference.html.

```
<?xml version="1.0" encoding="UTF-8"?>
<kml xmlns="http://www.opengis.net/kml/2.2">
        <Placemark>
                <name>Simple placemark</name>
                <description>Attached to the ground. Intelligently places itself
                 at the height of the underlying terrain.</description>
        <Point>
                <coordinates>-
122.0822035425683,37.42228990140251,0</coordinates>
        </Point>
        </Placemark>
</kml>
```

To display the existing vector GIS data such as ESRI shapefile (*.shp) in Google Earth, one needs to convert these GIS data into KML format. The output

KML file from conversion software also follows the standard KML format, and one can modify the contents of the KML file by using text editor, if needed.

18.6.4 Data Display in Google Earth

A *networklink* in Google Earth API can be used to load and display geographic files in Google Earth. A *networklink* contains a link property that specifies files to be loaded in server or network. Using *networklink*, one can add the following data into Google Earth

- Generic image files used by icons in icon styles, ground overlays and screen overlays
- Model files used in the model object
- KML or KMZ files converted from a shapefile

Below is a sample code that adds 2008 flood map into Google Earth by the networkLink tag:

```
var networkLink = ge.createNetworkLink("");
networkLink.setDescription("Flood June 2008 data");
networkLink.setName("Open NetworkLink");
networkLink.setFlyToView(true);
var link = ge.createLink("");
link.setHref("https://engineering.purdue.edu/CE/floodmaps/Flood 2008.kmz");
networkLink.setLink(link);
ge.getFeatures().appendChild(networkLink);
```

Similarly, a raster image can also be overlaid into Google Earth using ground overlay in Google Earth API. The following example shows how these functions are used to add 2008 crop data files into Google Earth plug-in by using the groundOverlay tag.

```
var groundOverlay = ge.createGroundOverlay('');
groundOverlay.setIcon(ge.createIcon(''))
groundOverlay.getIcon(). setH-
ref("https://engineering.purdue.edu/CE/floodmaps/cdl_9c_in_2007.jpg");
groundOverlay.setLatLonBox(ge.createLatLonBox(''));
var center =
ge.getView().copyAsLookAt(ge.ALTITUDE_RELATIVE_TO_GROUND);
var north = 39.61;
var south = 38.162;
var east = -86.63
var west = -87.997;
var rotation = 0;
var latLonBox = groundOverlay.getLatLonBox();
latLonBox.setBox(north, south, east, west, rotation);
ge.getFeatures().appendChild(groundOverlay);
```

18.6.5 The Flood Map Website

Using the features and functions of the Google Earth plug-in and data provided by Google Earth, a Web-mapping service has been designed. One can access this site through Web browsers such as Internet Explorer or Firefox at https://engineering.purdue.edu/CE/floodmaps/2008_flood_map.htm. The main interface is shown in Figure 18.14. The left region is an embedded Google Earth plug-in and some controls are placed on the right side of the page to control the Google Earth API.

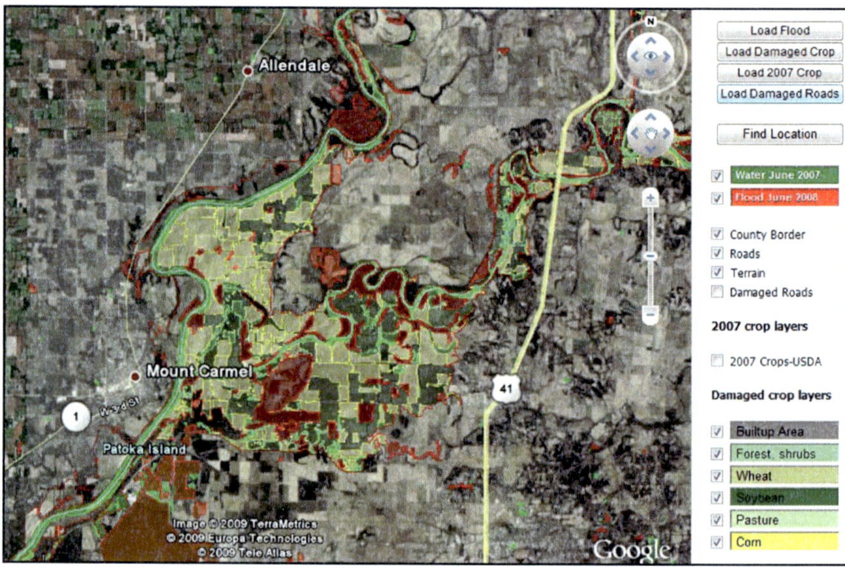

Fig. 18.14 Main interface of the flood map web page

The developed Web service can display different group layers (water, crops, flood, damaged crops and roads data). All of these data are in KML/KMZ format converted from original ESRI shapefile except the crops data, which is a generic image data format. By selecting each load data button, one can load and display each layer in the Google Earth plug-in. The visibility property of each layer can also be turned on and off. The predefined Google Earth layers such as boundaries, roads and terrain can be displayed in the same way. As shown in Figure 18.15, one can easily check the influence of 2008 flood compared with 2007 water data using this website.

Furthermore, Figure 18.16 shows a number of zoomed-in images to assess the local damages. Figure 18.16(a) shows the flooded area around Evansville. One can check which residential areas were affected by the flood. Figures 18.16(b) and 18.16(c) show the areas with different damaged crop types. Figure 18.16(d) illustrates how one can identify the specific damaged roads by overlaying the road layer with the flood layer. In summary, with the help of the abundant imagery data in Google Earth combined with the produced flood data, the developed Web mapping service provides an easy and effective way for the public and agencies to assess flood damages.

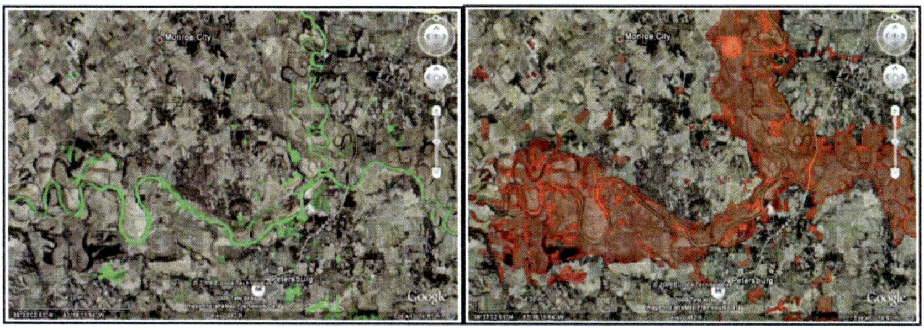

(a) 2007 Water (b) 2008 Flood

Fig. 18.15 Comparison between 2007 water and 2008 flood

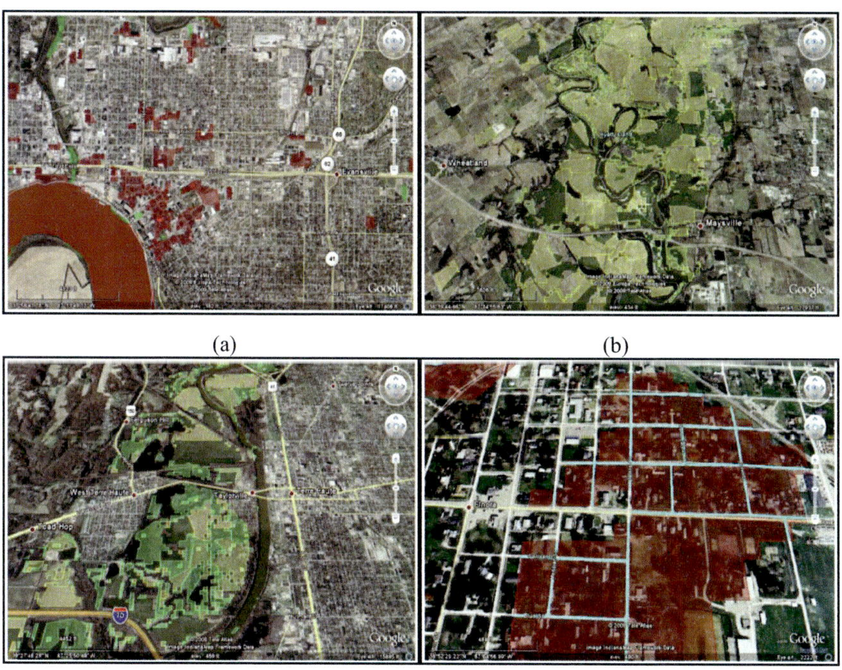

(a) (b) (c) (d)

Fig. 18.16 Zoom-in views to assess the local damages caused by the 2008 flood

18.7 Conclusions

Rainfall data for the first week of June 2008 and the river gauge station readings clearly showed the severity of this flood in southern Indiana. There was an average increase of roughly 12 feet in the Wabash and White River levels for almost five (5) days as compared to the flood stage. State of emergency was declared in 23 counties by the State Governor and 39 counties were declared as major disaster by the President. A total of 51 counties were affected by the flood causing damages of approximately $ 1 billion. This study is focused on the evaluation of nine counties in this region where an area of about 93,369 hectares was affected by floods. Southern Indiana had a variety of summer crops in the area, but corns and soybeans were the most affected, resulting in about 6-22% losses as compared to the 2008 crops data. Greene, Knox, Daviess and Vigo counties were among the severely affected ones by the overflowing Wabash, White, and East Fork White rivers, both for crops and road damages. A total of about 1,082 km roads were submerged by the flood, spreading in almost every county, including State highways, US roads, County roads and many other small roads, with the most affected being the county roads. The flood water did spill over the designated floodplains in a number of places along the river courses. A total of 6,000 hectares of land outside the floodplain areas, mostly in Greene and Daviess counties, was flooded, which suggests an underestimation in the flood modeling and simulation studies. The medium spatial resolution (30m) Landsat multispectral images proved to be a useful and suitable data source for mapping flood extent, evaluating the designated floodplains, and estimating damages to major crops and roads. The flood extent has been very effectively mapped using timely satellite image data collected over the flood peek period. The geospatial distributions of estimated damages show a high concentration of crop losses along the rivers and streams.

The primary objective of using remote sensing imagery for mapping widespread disasters is to provide the planners and disaster management institutions with a quick view and assessment of the ground conditions and disaster impact. The timely availability of the remote sensing imagery and its derived products to the public and disaster management authorities helps to quickly understand the disaster impact and support disaster response activities during and after the flood. Web-mapping techniques have played an essential role in populating, visualizing, and evaluating the flooding situation and damages during and after the flood. This study demonstrates that the Google Earth plug-in can be quickly adopted to build specific applications and is easy to use and access through Web browsers for the public. It can help the agencies make a quick response plan and move swiftly to take appropriate remedial measures. It can also help people in or concerned with the flood affected areas to view and assess the amount of flooding and losses.

Although the use of satellite remote sensing image data proved very helpful for rapid flood mapping, the medium resolution Landsat imagery is unable to detect damages at a finer level. The damage estimate was therefore mainly limited to the broader categories. Higher resolution data are needed, along with other GIS data, to further estimate the damages in individual properties, farms, and owners. Such work can be enhanced in urban areas if city property data are made available. Combined with high resolution digital elevation models, the flood extent map would be a valuable data source for estimating the flood depth and calibrating flood models.

References

Baatz, M. and Schape, A., 2000. Multiresolution Segmentation – an optimization approach for high quality multi-scale image segmentation. In: STROBL, J. *et al.* (Hrsg.): Angewandte Geographische Informations verarbeitung XII. Beiträge zum AGIT-Symposium Salzburg 2000, Karlsruhe, Herbert Wichmann Verlag: 12–23.

Benz, U.C., Hofmann, P., Willhauck, G., Lingenfelder, I. and Heynen, M., 2004. Multiresolution, object-oriented fuzzy analysis of remote sensing data for GIS ready information. Journal of Photogrammetry & Remote Sensing, 58, pp. 239–258

Blaschke, T. and Strobl, J., 2001. What's wrong with pixels? Some recent developments interfacing remote sensing and GIS. GeoBIT/GIS, 6, 12– 17.

Blaschke, T. Lang, S., 2006. Object based image analysis for automated information extraction-a synthesis. Measuring the Earth II ASPRS Fall Conference, 2006, San Antonio, Texas, CD-ROM

Campbell, J.B., 2002. Introduction to remote sensing, 3rd Ed., Guilford Press, New York

Corina, A., Simona, O., Emily, B., 2006. MODIS versus ASTER water classification, Spectral Imaging: Eighth International Symposium on Multispectral Color Science, Proc. of SPIE-IS&T Electronic Imaging, SPIE Vol. 6062

Definiens, 2000. Developer User Guide. http:\\ www.definiens-iamging.com

Syed, S., Dare, P., and Jones, S., 2005. Automatic classification of land cover features with high resolution imagery and lidar data: an object oriented approach, Proceedings of SSC2005 Spatial Intelligence, Innovation and Praxis: The National Biennial Conference of the Spatial Sciences Institute

Chapter 19

DECISION MAKING BASED ON EARTH OBSERVATION TECHNOLOGY

Jixian Zhang, Yu Zeng, Wenhan Xie, Tao Wang

19.1 Geo-Spatial Decision Support Systems

Spatial data derived from earth observation are applied in domains as varied as agriculture, environmental resource management, territorial planning, etc. They are used to learn about the environment and to assist in making management decisions. Earth observation has been widely used in China (Chen 1999). Disaster monitoring and evaluation based on remote sensing has played an increasingly important role in the government's emergency management. In response of the Wenchuan Earthquake, high resolution imagery was used for geological hazard monitoring, influence evaluation and loss statistics. There are severe flood hazards in southern China each summer, leading to the establishment of the data exchange center for monitoring and evaluating the flood hazards in related governmental sectors. The remote sensing data and the monitoring results can be automatically uploaded to the information system for decision makers. Large scale management and dynamic monitoring of natural resources and ecosystems have been implemented in China under the support of quantitative remote sensing and GIS techniques. This chapter presents the working experiences of China in response to unexpected disaster events, evaluation of seasonal hazards and dynamic monitoring of natural resources.

In 1992, Spatial Decision Support System of China was established and put into operation as part of the e-Government. It provides quantitative, visual and model-based support for decision makers in central government. This section introduces the structure and components of GIS platform which incorporates various remote sensing data of SDSS. Acquisition and application of earth observation data are discussed further in scenarios of monitoring and evaluating of water resources, desertification and western development of China. Three typical information systems running in some government agencies are described.

D. Li et al. (eds.), *Geospatial Technology for Earth Observation*, DOI 10.1007/978-1-4419-0050-0_19,

19.1.1 E-Government

E-Government is an important application domain of spatial support decision-making. It bases on government intranet networks and uses a variety of information as static resource for government management purposes. The information resource consists of natural geography (mountains, rivers, climate, land cover, mineral resources), land-use conditions, distribution of enterprises, public facilities (roads, airports, docks, squares, parks, power facilities, gas facilities, water supply and drainage facilities), distribution of residential areas, and so on. Based on the management needs of governments, spatial and non-spatial (statistical) information which are real-time or quasi-real-time data are integrated, then various methods of visualization and spatial analysis are provided to policy makers and assist government decision-making. The feedback information in the executive process should be collected timely so that the government can adjust the plan and procedure promptly to reach the most reasonable and effective results.

E-Government spatial assistant decision-making application is characterized by the following:

(1) Wide range of needs (various branches and industries), multi-source (spatial, non-spatial,) and heterogeneous data sources (different formats of spatial and non-spatial data).

(2) The government has a large number of communication facilities and information resources which can coordinate and collect all information resources from society, industries, and enterprises.

(3) E-government applications include both long-term municipal planning work and real-time tasks such as disasters and emergency management. Also they need both an extensive and long-lasting accumulation of information resources and a capacity of fast integrating of multi-source, heterogeneous information.

(4) Information resources in e-Government are mainly based on department management; and departments take charge of the data resources. At the same time, the information resource is required to be sharing between different departments. So it has the feature of information resources sharing and self-government.

(5) Most information in e-Government applications is confidential. So both access control and usage record have great significance. Furthermore, it is also important to identify the user's identity, manage access privileges and record the usage log of information

19.1.2 Earth Observation Data

Earth observation data and fundamental geo-spatial data are important for the framework of spatial decision making. The latter has been described in Section 19.2, the acquisition of remote sensing data is necessary for monitoring and evaluation of land cover, water resource and desertification etc. Different thematic applications need different data sources of remote sensing for the optimal results of information extraction.

As a small satellite for disaster monitoring, Beijing-1 can obtain both panchromatic image with the resolution of 4m, swath of 24km, and multispectral image with the resolution of 32m and swath of 6000 km. The multispectral data, which include three bands of 0.52 - 0.62 μ m, 0.63 - 0.69 μ m, 0.76 - 0.90 μ m, can be applied for dynamic information extraction of desertification information. It can form temporal sequence dataset, and create national desertification remote sensing image and multi-scale vector database. Land desertification and land salinization are due to the changes of weather, land cover and land use that result in the material (i.e. soil and alkali) composition enrichment and content increase in land surface layer, which in turn causes the decrease of land biomass. In the monitoring cases, the remote sensing data such as NOAA-AVHRR and Terra-MODIS can achieve the annual and seasonal analysis with the precision of 1 km.

Beijing-1 and Landsat TM data can be applied for the monitoring of water resources, which need to acquire basic geographic properties of river, lake and reservoir, such as the range and area etc. For example, from the former type of data one can extract water distribution vector data around Beijing city. Aiming to the areas, ranges, reserves of the important water sources and these daily, seasonal or annual changes, the national remote sensing water level monitoring database is created.

In the 1980s, under the organization of the National Agricultural District Committee, State Bureau of Surveying and Mapping, Ministry of Forestry and related units took national land using status investigation for 15 land categories by using MSS satellite images, and then had the mapping at the scale of 1:500,000. That investigation macroscopically reflected the basic conditions of land resources in China. At present the main information updating sources are Landsat TM, Terra-ASTER with the resolution of 20-30m and land use data at the scale of 1:100,000, where the update cycle is about 5 years. At the same time the field survey and other type of results are also necessary.

In the west development application system, monitoring work using remote sensing was concentrated on land biomass. Using satellite image sequence with high temporal resolution (MODIS, NOAA and SPOT Vegetation etc), time series analysis and extraction of multiple cropping indices can be developed for monitoring national agricultural multiple cropping index condition and its changes. Grassland biomass is an important index for the quality of the grassland ecosystem, which affects the development and degradation of grassland ecosystem. Combining normalized NDVI with the sampling of biomass data, the correlation mode between land biomass and NDVI can be created to estimate grassland biomass. The work provides support for national development strategies.

The wetland ecosystem is closely related to weather and biodiversity. Remote sensing data of Landsat ETM+/ TM and ASTER have been widely used to the resource monitoring of wetland ecosystem change. In the typical wetland research areas of rivers and lakes, like the Northeast, Tibetan Plateau, and Yangtze and

Yellow Rivers, several remote sensing models of wetland category and wetland ecosystem vegetation retrieval have been formed with field surveying, and then the national geo-spatial database for wetland resources is created for management and research of wetland ecological evolution and its change.

During the process of decision making, remote sensing monitoring information should be integrated or fused mutually with other types of information (such as field survey and statistic information etc) for achieving scientific and accurate results.

19.1.3 Data Model and Organization

E-Government in spatial support decision-making adopts a common spatial data model, (1) different scales (2) on the horizontal sub-regional and sheeting, (3) in structure, there are vector and raster. The spatial data model diagram is shown in Figure 19.1.

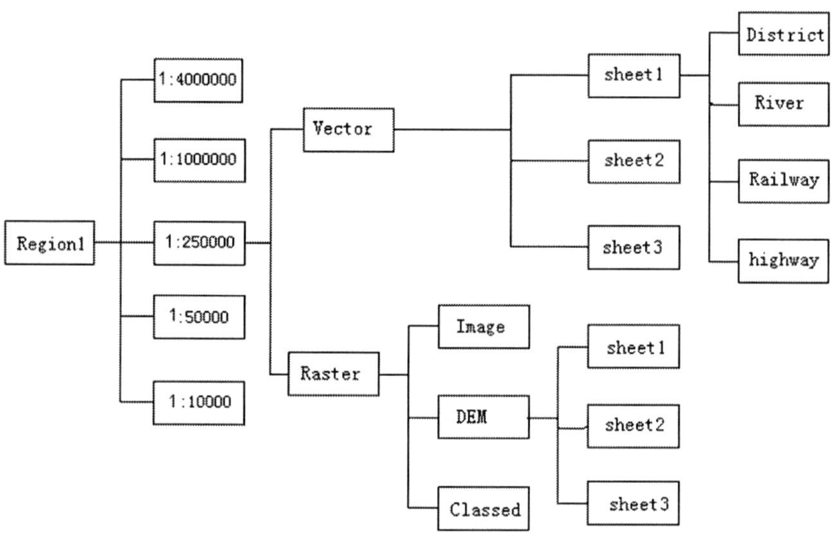

Fig. 19.1 Data organization

(1) Hierarchical on the vertical grade: It is divided into different map libraries by the original resolution map (scale). According to the current Chinese National standard, the spatial databases are divided into 3 basic libraries which are 1:1000000, 1:250000, and 1:50000. There exists some regions with 1:10000 or even larger-scale spatial data library. In the library, spatial data are divided into sub-vertical layers. The map layers can be divided into fundamental layers and thematic layers. Generally speaking, the foundation layers in each scale level have covered the whole region, and the theme layer can only be found in a layer scale,

covering a local area. Despite the thematic layers only exist in a particular scale, they should be displayed in all scale levels.

(2) Horizontal on the sub-area and sheet. That is, in the horizontal spatial data, it is divided into different regions which can be connected or disconnected. The map is further divided into small tile map sheets. The sheet dividing adopts the national standard GB/T13989-92. In order to improve the management and the efficiency of data scheduling, image, DEM, raster data are often classified in the sub-standard rate on the basis of another block; generally they are divided into 1024x1024, 512x512 or 128x128.

In e-Government applications, one important feature of the horizontal partition is the district administrative regions. District administrative regions of the system are independent of the standard topographic map of the sheeting, which is a typical structure as a tree. Administrative divisions of the region have regional codes of national unity, "the code of administrative divisions" to the provincial, county, and then the next level administrative divisions (townships and villages, small cities, plant, work area, etc.), should be expanded by national standards.

(3) It can be classified into vector data and raster data according to the geographical structure of the data. Vector data include four basic types: point, line, surface and annotation. Vector data express features as four-type elements of point, line, surface and annotation, presented by the sampling coordinate. A point feature is expressed by a coordinates (x, y); line by line elements of the elements on the string of data points $(x_1, y_1), (x_2, y_2)\ldots (x_n, y_n)$. The border of area element do not use coordinate string to express directly, but use the indirect coordinates, that is, recorded a number of elements of the code from the arc line of code related to the concentration of the corresponding elements of the arc. Vector data are generally layered management. For example, a regional of vector data can be divided into layers of district, rivers, railway and highway.

Raster data is recorded in all geographic grids on the value of the geographical elements, such as integrated optical properties of value - the value of image, the value of the elevation - DEM, elements of the classification of value - such as land-use classification.

Raster data can be classified into image, DEM and cover class. Image data are saved as a standard-general of the image files, such as BMP, GIF, JPG, and so on, with additional information as a plain text file format. DEM is a terrain elevation matrix. Coverage class is a classification for terrain coverage, such as soil type, land-use type or land covered type.

Management of spatial data employs the commercial object-relational database, which supports Oracle and SQL Server. Vector data consists of map sheets. All area, line, point and annotation objects are organized as a binary large object existing in the image field. Similarly, images, DEM, and surface grid map data are stored in the block image field.

19.1.4 Software Architecture

The spatial decision-making system used in e-Government applications is designed as a multi-layer architecture which includes a system service layer, GIS and DSS-based functional layer, application service layer, and user service layer.

(1) System service layer: to provide spatial data, statistical data, multimedia data, metadata and other information management and services.

(2) GIS and DSS-based function layer: to provide various spatial analysis and operations, such as buffer analysis , overlay analysis, network analysis, and as the domain polyglot analysis, as well as some assistant decision-making analysis method, such as predictive analysis, Evaluation of comparison, the analysis of consumer behavior, population control and simulation analysis.

(3) Application service layer: to provide application services such as application structure, query service, analyze service, display services and Web services.

(4) User service layer: to provide some functions such as individually tailored, response to user's operation, Graphic presentation, plotting, and printing

19.1.5 Data Management and User Rights

In e-Government applications, user rights and data management are very essential. They are based on two important principles: information sharing and management of user data. Government spatial support decision-making needs information sharing in different departments. At the same time, each department has the right to decide the authority of access and also should update and maintain their own data.

(1) User management is based on Agent-Staff-Tree, so that users have the different access authority according to their positions.

(2) For public information resource, data access authority is distributed by different departments and positions.

(3) Various departments manage their own data. Each department can decide what kind of data can be accessed by which department and the people at which level. The information resources that could be accessed by other departments are registered in a catalog center and marked with access right.

All departments are required to provide shared information resource from other departments as much as possible. Meanwhile, the access log should be kept once any confidential information is accessed.

19.2 Monitoring and Management of Earthquake Hazards

Earthquake is a recurrent natural disaster that has wide and destructive effects. Before people find an effective way for earthquake prediction, it is of great importance to obtain information on the earthquake and its secondary disasters, as well as to effectively monitor and evaluate the situation of the disaster promptly after the earthquake, which can reduce the loss of the lives and properties.

On May 12, 2008, a large earthquake occurred with the Richter magnitude scale of 8.0 in Wenchuan County, Sichuan Province, China. Because the hazard-affected areas are hilly and mountainous areas, landslides, which are the major secondary disasters of earthquakes, caused road blocking and burying of residential areas, and brought severe impacts on hazard rescue. This section firstly presents the method for spatial distribution information extraction of landslide using high-resolution imagery, and then discusses the method for hazard assessment and loss statistics.

19.2.1 Landslide Determination Using High-Resolution Imagery

19.2.1.1 Characteristics of Landslide in Image

After image processing, landslides can be interpreted and their location, type, boundary, size, activity model and stability can be determined. Based on this, their influence on projects can be further predicted. The interpretation characters of landslides include shape, tone, shadow, texture, etc. Landslides usually have typical geomorphic features, as illustrated in Figure 19.2.

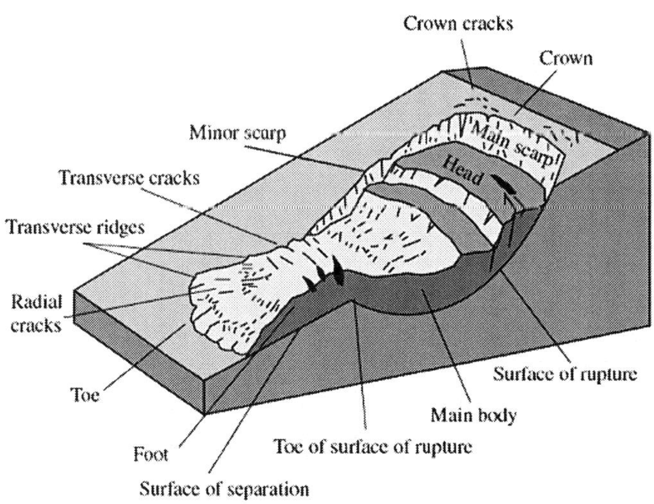

Fig. 19.2 Sketch map of landslide structure (Adopted from: About.com: Geology)

Landslide hazard interpretation based on remote sensing is one of the most complicated information extractions of slope deformation. Slope has various shapes in the nature, particularly in those long-term deformations, which are usually syntheses of different deformation. It is difficult to interpret these kinds of landslides, especially for ancient landslides whose typical shapes are almost removed, which in turn increases the difficulty of landslide interpretation. Therefore, landslide elements, sliding characteristics, and image characteristics should be studied carefully before interpretation. In an aerial-photo, landslides usually have plane shapes of dust pans, landslide walls, landslide steps, landslide toes, and landslide cracks. Besides, the characteristics of landslides in a remote sensing image include (Niu 2002, Xie 2007, Ma and Tian 2007):

(1) Average slope of landslide is flatter than surrounding mountainous areas, whereas some of landslide could be plane or depression. Because of different lithology, geologic structure, groundwater activity, and landslide volume, the landslide usually presents different shapes, such as tongue, pear, arc, etc. In addition, new landslide has similar color to rocks in true color images, and has a visible contrast with surrounding vegetation and water bodies. These characteristics of landslides are clear in remote sensing images, which help us recognize them.

(2) The crown, toe, two sides and middle of a landslide could generate cracks before and during sliding. These cracks will grow to gullies with different sizes under the effect of surface water and other stresses, which cause strip shadows and tone differences in remote sensing images.

(3) Drunken forests or yataghan-like trees, as well as destroyed vegetation are all good signs for landslide interpretation, which differ from other plants in the image.

(4) In the place where landslides occur, usually, the river becomes dry or changes its route, roads get blocked or cracked, and there are displacements and fractures in mountains.

(5) In the place where there is water sharp downcut, lateral erosion, and ground water activity, landslides will occur if there is no protection of stable layer.

Usually a landslide occurs along rivers or roads, thus road rupture or accumulation can be used as a clue for landslide interpretation. Moreover, characteristics of landslides, such as toe and contrast of tone with the surroundings will help in landslide recognition.

19.2.1.2 Landslide Interpretation in Beichuan County

The Wenchuan earthquake, occurred on May 12, 2008 in Sichuan province of China, was destructive to Beichuan County (Figure 19.3). After the earthquake, relevant people were organized and sent to the disaster area for rescue by the Chinese government. At the same time, the interpretation for geologic hazards was launched. The number of interpreted landslides in Beichuan was 435 with a total area of 27,199,327.6m^2. For reconstruction purposes, most of the landslides distributed along the Minjiang River and close to built-up areas were interpreted. The workflow for landslide hazard interpretation is shown in Figure 19.4.

Fig. 19.3 Part of Beichuan after the earthquake (aerial-photo)

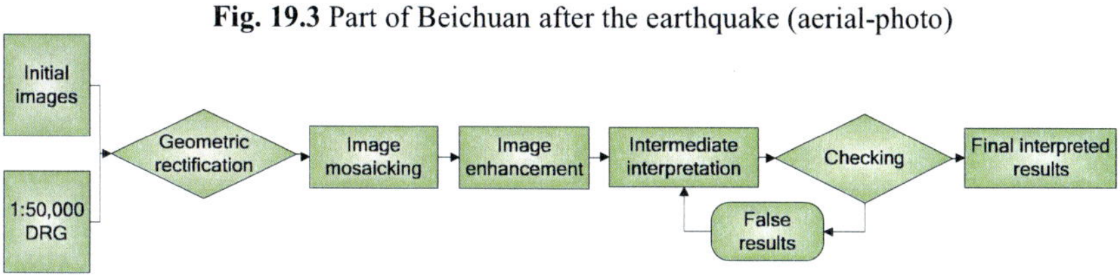

Fig. 19.4 Interpretation flow chart of landslide hazard

Interpretation Preparation

Remote sensing images used for interpretation are aerial-photos with 0.3m ground spacing distance (GSD) taken on May 21, 2008. In areas where aerial-photos can not cover, ALOS multispectral images (10 m GSD) are also used. The aerial-photos are clear and have good quality (Figure 19.5), whereas the cloud coverage of the ALOS image is about 20% (Figure 19.6). Beijing-1 satellite images (Figure 19.7) are used for interpretation assistance and result verification.

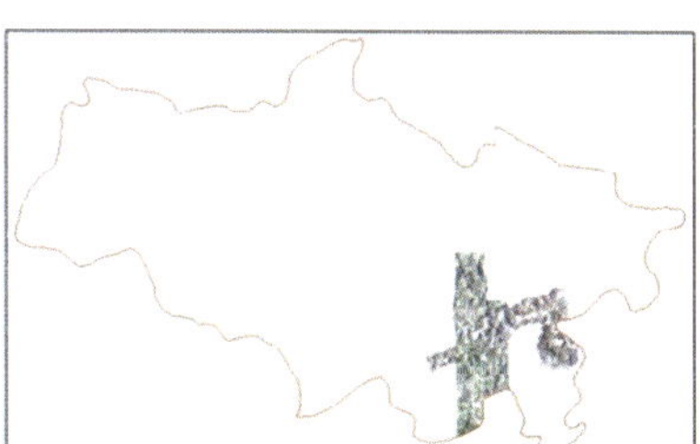

Fig. 19.5 Coverage of aerial photos

Fig. 19.6 Coverage of ALOS images

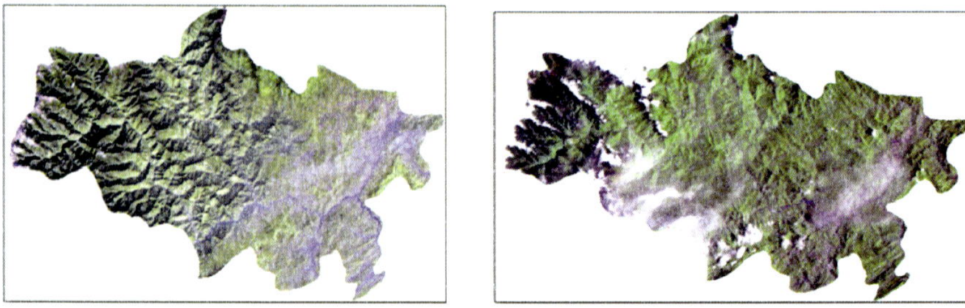

(a) 2006 (b) May, 2008

Fig. 19.7 Color composition image of Beijing-1 satellite

Interpretation Requirements:

(1) ArcMap or ArcView is used in the interpretation. Transverse Mercator projection, Krasovsky ellipsoid, and the Xi'an 1980 Geodetic Coordinate System are adopted;

(2) Attributes include: number of parcel (TBBH), code of hazards (CODE), coordinates X, coordinates Y, area of parcel (TBMJ), level of hazard, monitoring time (JCSJ), and annotation (BZ) (Figure 19.8).

FID	XH	Shape *	TBBH	TBMJ	CODE	x	y	JCSJ
0	1	Polygon	1	62841. 7	306	964420. 2	3558934	20080521
1	2	Polygon	2	51820. 3	306	965233. 9	3558536	20080521
2	3	Polygon	3	229159. 4	306	974477. 1	3554951. 1	20080521
3	4	Polygon	4	169202. 4	306	971005. 7	3554230. 9	20080521
4	5	Polygon	5	177357. 3	306	1034017. 7	3556848. 9	20080521
5	7	Polygon	7	210668. 1	306	1033304. 5	3555745. 4	20080521
6	8	Polygon	8	35290. 5	306	1033225. 5	3555360. 6	20080521
7	10	Polygon	10	112375. 7	306	1032045. 1	3554363. 1	20080521
8	12	Polygon	12	135615. 3	306	1033490. 5	3553369. 3	20080521
9	13	Polygon	13	330346. 1	306	1034017. 4	3552831. 3	20080521
10	14	Polygon	14	208061	306	1030605. 1	3552527. 3	20080521
11	15	Polygon	15	636223. 6	306	978253	3547220	20080521
12	17	Polygon	17	8565	306	1029088. 4	3549669. 3	20080521

Fig. 19.8 Attributes of interpreted landslides

Landslides should be interpreted properly. Some landslides may be covered by vegetation, and some new landslides may have the similar color to rocks. More attentions should be paid to interpretation on these kinds of landslides (Figure 19.9.).

Interpretation of landslides

There were many landslides in Beichuan after the earthquake, which threatened human lives and properties. These landslides can be identified using high-resolution images. In the images, there are color differences between landslides and their surroundings, and this is noticeable at the plane and edge of the landslides. Figure 9 and 10 demonstrate a number of landslides identified in aerial and satellite images, respectively.

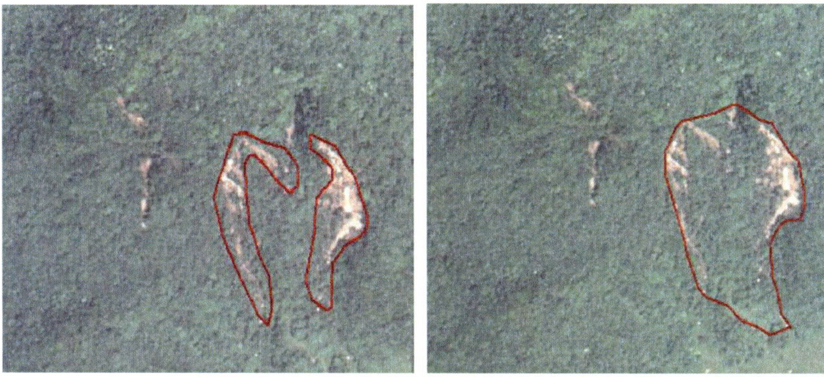

(a) one example

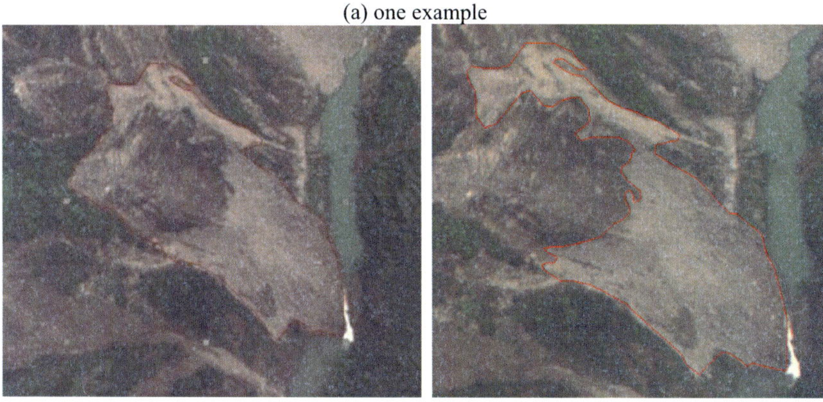

(b) another example

Fig. 19.9 Landslide interpretation in images. (left map is wrong interpretation; right one is correctly interpretation)

Fig. 19.10 Landslides in aerial-photos

Fig. 19.11 Landslides in ALOS image

Related image signs like scarp, toe, crack and main body can be clearly identified from the image after an earthquake. The track of sliding can be seen easily, which looks like a broom, a sector or a pear with the small area at the beginning and larger area at the end of sliding. Boundary shape and texture characteristics of different land covers, such as boundary shape between water and bank, boundary shape between soil and vegetation, strip shadows vertical to water can be used for interpretation. Moreover, landslide accumulations, which lead to blocking of rivers and roads, can be seen clearly in high-resolution aerial images (Figure 19.12).

Fig. 19.12 Landslides and their accumulations

In high-resolution images, if there are large volume landslides, a 2D image can be used for landslide interpretation, otherwise, 3D simulation using remote sensing images can be used, especially for ancient landslide recognition (Figure 19.13). Digital elevation model (DEM) presents terrain relief. Based on textures of remote sensing images, the environment of studied area and conditions of hazard can be analyzed by overlapping textures on DEM, which is derived by reconstructing 3D scenes of the studied area. For the landslides which are difficult to interpret directly in a 2D image, 3D simulation can be used for landslide interpretation. Field checking is necessary in order to verify the recognized landslides (Nichol *et al.* 2006).

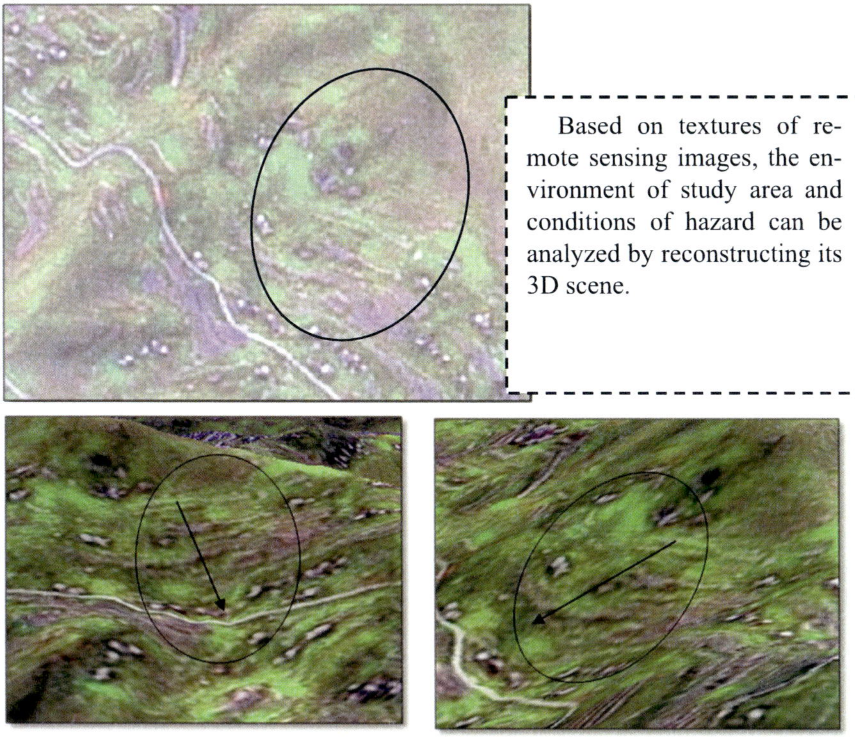

Fig. 19.13 3D image simulation (SPOT5 image false color composition)

19.2.2 Hazard Assessment and Loss Estimation

Hazard assessment method includes analytical model of the hazard-affected body in geospatial distribution, analysis model of the hazards environment, hazard factors and vulnerability of hazard-affected body, etc. Using distribution model based on geospatial grid information, the quantitative spatial distribution of the hazard-affected body can be calculated, which is one of important factors for hazard analysis and assessment. By analyzing the spatial and interactive relationship among the environment of developing hazards, hazard factors and the hazard-affected body at a certain time, and by analyzing the hazard vulnerability, the hazard alarm and real-time emergency response can be implemented. Through the research on hazard assessment model, the common characteristics of hazard assessment methods can be found.

In earthquake hazard assessment and loss estimation, GIS can be used to extract the extent and level of earthquake damage. Using spatial analysis and statistical analysis, the damages on natural condition and socio-economic condition can be quickly assessed based on the earthquake emergency database. The assessment results can provide scientific supports for governmental hazard relief management and decision-making.

19.2.2.1 Spatial Analysis Methods

GIS' advantage in contrast with computer-aided mapping system is providing advanced tools for spatial analysis. Using spatial analysis technique, through the observation and experiments on the raw data, new experience and knowledge can be attained, which can be used as the supports for decision-making. Typical spatial analysis techniques include overlay analysis, network analysis, buffering, and so on.

Overlay analysis is an operation that overlaps features in two or more maps and produces a new layer. It divides original elements into several new elements, and the new elements integrate all attributes of the original two or more layers. In other words, the overlay analysis not only generates new spatial relationships, but also generates new attributes and spatial relationships. Overlay analysis uses mathematical models to calculate attributes of new elements.

(1)Polygons overlay: The procedure overlays polygons in two layers, which results in a new polygon element of the output layer. At the same time, the new element links to their attributes to meet the needs of the establishment of model. For example, when census map is overlapped with campus map, the results for each school district and its corresponding survey can be acquired.

(2)Polygon overlay with point: Polygon overlay with point can be transformed to point-in-polygon test. It produces a new property for each point. For example, when the position of wells is overlaid with a planning area, a region that contains each of the wells can be found.

(3)Polygon overlay with line: The polygon where the line is partly or wholly located can be determined.

19.2.2.2 Hazard Assessment Based on GIS

(1) Selection of the assessment indexes

The background data of natural and socio-economic conditions is the foundation for hazard assessment. Background information mainly includes terrain, water body, population, transportation, power, communication, urban and rural settlements, economic output and important facilities. The population, settlement data and economic output are organized using the existing administrative division system. The background data of natural and socio-economic conditions can be partly utilized by analyzing disasters, loss vulnerability and risks.

(2) Determination of assessment content

According to the workflow of national emergency management, three-stage (before, during and after hazard) evaluation is adopted. The population in hazard affected area, population that needs transfer and resettlement, the loss of crop, residential building and infrastructure should be assessed.

(3)Hazard assessment analysis

Based on the geographical characteristics and statistics data of the hazard affected area, the spatial analysis model on area of hazard influence can be determined by the integrated remote sensing image database, GIS database and the hazard image library. Then, the hazard assessment and analysis system can be established to achieve rapid hazard assessment, and obtain the extent and level of population loss, crop loss, housing loss, etc., which will provide supports for planning and reconstruction after the hazard.

(4) Loss estimation

Using the statistical analysis method to analyze the hazard loss data, users can mine the information behind the statistical data, and discover the distribution rule of the hazard.

19.2.2.3 Visualization of the Hazard Affected Region

Based on fundamental geographic information data, remote sensing images, and hazard assessment results, 2D or 3D maps and statistical charts can be produced. It reflects the spatial distribution characteristics and statistical characteristics of the hazard statistical data and provides comprehensive hazard information for decision makers.

(a)Damage estimation (b) Visualization of disaster situation

Fig. 19.14 Geographic Information Service System of the Wenchuan earthquake

For efficiency of data management, statistics and analysis, using remote sensing images from different sensors, the interpreted information of the disaster situation, and the integrated fundamental geospatial data, we can establish the comprehensive geographic information database of the disaster situation. Based on the self-developed 2D/3D geospatial information service platform, we established the Geographic Information Service System of the Wenchuan Earthquake. This system not only supports enquiry and statistics of disaster situations, but also provides 3D visualization, measurement and analysis based on parcels of hazard-affected counties. It can provide support to decision making for reconstruction after the earthquake (Figure 19.14).

19.3 Monitoring and Assessment of Floods

Each summer, floods and storms always severely threaten lives and properties of the populace in southern and coastal regions of China. Earth observation provides necessary data and powerful techniques for forecasting, monitoring and evaluating of flood hazards. An information system with user-friendly GUI and powerful functions can support the government in making decisions and reducing damages (Goodchild 2003). The following section discusses data content, data exchange center, data integration, and functions of the China Flood Disaster Monitoring and Evaluation System.

19.3.1 Data Sources

19.3.1.1 Fundamental Geospatial Database

Fundamental geospatial data provides the analysis for different hazard thematic data. It is the product from all levels of surveying and mapping governmental organizations, including multi-scale geospatial databases covering the whole country. The massive data has a complex data structure and multi-resolutions under different taxonomy. The design of database structure is of great importance to effectively integrate the fundamental geospatial data and the hazard thematic data (Andreas 2001). It is crucial for the efficiency and quality of the hazard monitoring and assessment system.

The large scale relational database is used in managing fundamental geospatial data, where data manipulation language (DML) is extended to encapsulate spatial operations of geospatial data. At logic level, geospatial data is categorized according to scales. For the sake of positioning efficiency and visualization quality, only one spatial reference framework is applied in each sub-geospatial database. Vector data and raster data (including image data) are stored in separate tables. Vector data and thematic raster data are further classified into different coverages according to semantics. Vector layers consist of the administrative boundary, transportation system, residential area, and hydrological features according to elements of basic topographic maps, while raster layers consist of soil and vegetation types. In order to improve the efficiency of data operation, each feature layer is stored by map sheets where spatial objects are managed. Therefore, there are four levels in the logical model which include the spatial database, coverage, sheet and spatial objects (as indicated in Figure 19.15). Relational tables are employed to manipulate spatial objects and relations among them.

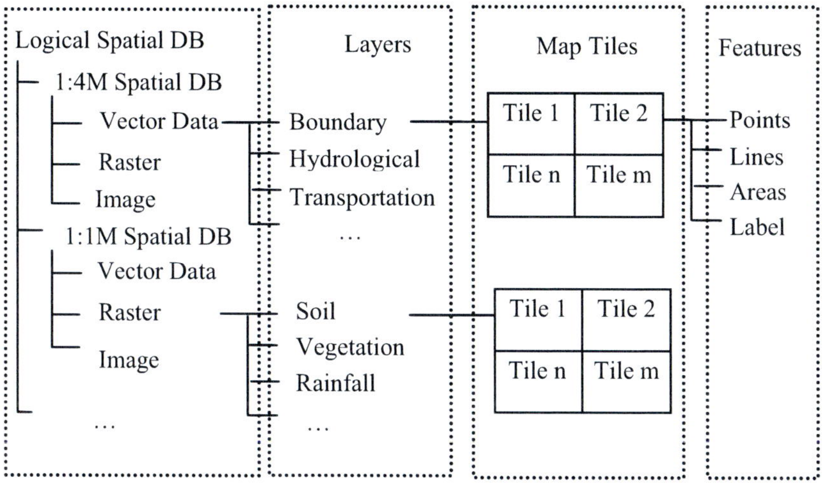

Fig. 19.15 Hierarchical Storage Structure of Spatial Data

The fundamental geospatial database covers all features of the national topographic map series. The types of data are DLG, DEM, DRG, DOM, Geographic Name and metadata. The data volume in the database is about 7 GB.

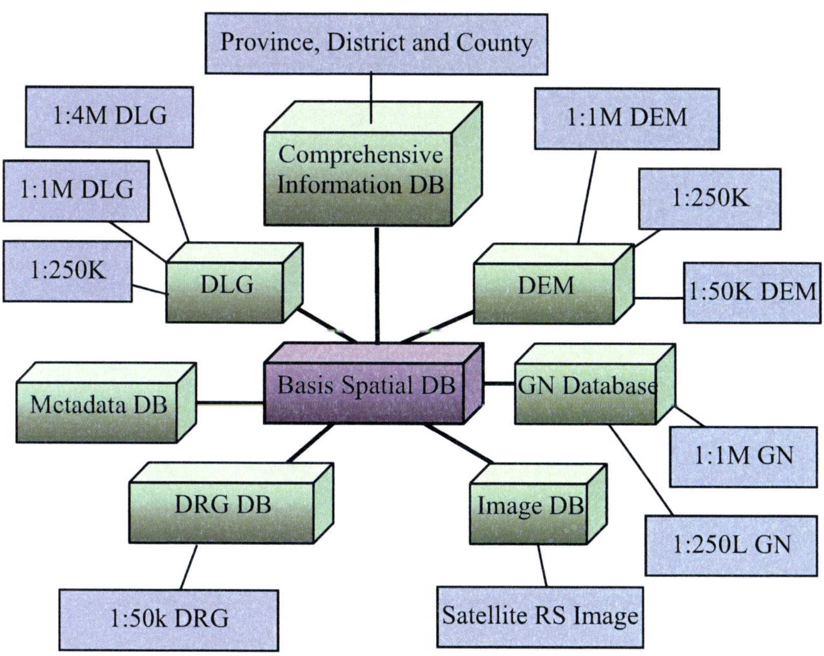

Fig. 19.16 Content of Fundamental Spatial Database

- Topographic database. Three scales, which are 1:4 million, 1:1 million and 1:250k are included in this database. 1:1 million data is updated in 2000 and 1:250k data is updated from 2000 to 2002.
- DEM database. Three scales, which are 1:1 million, 1:250k and 1:50k with the resolution 500 meters, 100 meters and 25 meters respectively, are included in this database.
- Geographic name database. Two scales, which are 1:1 million and 1:250k are included in this database. The production date of the former scale is 2001.
- Orthorectified image database. This database consists of TM color images acquired at 2000.
- Map product database. 1:50k scale DRGs of some important regions are stored in this database.

19.3.1.2 Thematic Databases

In order to effectively monitor and evaluate the flood hazard, not only information of reservoir, weather forecast, satellite cloud map, and precipitation forecast and statistics, but also information related to flood hazard, such as geological hazard, meteorological hazard, and earthquake hazard, should be considered and used. By relating all thematic data in one information system, decision makers can consider all aspect information of the flood and determine the best strategy.

The thematic databases include:

(1) Flood database, which includes situation of reservoir, hydrological information, historical cases of flood, river records and statistical data.

(2) Meteorologic database, which includes important weather forecast data, satellite cloud map, precipitation information and typhoon information.

(3) Earthquake database, which includes information of different levels of earthquakes.

(4) Geological disaster database, which includes information of landslide, land cracks, collapse, and debris flow.

(5) Remote sensing disaster monitoring database, which includes analysis report and disaster images.

(6) Historical statistics of disasters, which counted by year and region.

19.3.2 Data Exchange

The information system of flood monitoring and assessment is established based on a distributed network. The data exchange center of GIS is the central module of the system, whilst effectively integrating multi-level data sharing and data exchanging is the basis for data exchange centre. Generally there are four types of data sharing and exchanging (Henry 1997, Jia 2004, Longley 2005):

1) File-based data exchanging is the most commonly used pattern in exchanging centre. GIS data is composed not only of spatial data, but also of attribute, symbol, topology and rules. Therefore, information must be kept during data exchanging. There are two solutions for this work. The first one is to use the converting tool in GIS platform which can import and export different data formats. The other one is to use industrial standards such as OGC GML or Chinese National Spatial Data Transforming Format (CNSDTF).

2) Database DML interface can also be used for data exchanging by defining the public data operation interface. Client application can access GIS data under interface specifications. Traditional commercial databases extend the abstract data type to accommodate complex spatial objects, such as DB2, Informix and Oracle. In order to represent various spatial data, most database management systems extend their data type based on OGC simple feature (examples are DB2 and informix's ST_Geometry and Oracle's MDSYS.SDO_GEOMETRY) and develop data management plugins (examples are Spatial Extender, DataBlade). The other type is independent spatial data middle ware developed by GIS vendors based on ordinary relational database management system.

3) Data exchanging based on function level can ensure openness of an information system. The client application system must be developed based on some GIS platform in order to access GIS data in the central database. The development platform of application systems should be compatible with the one of the central database, otherwise the platform should provide interface for data manipulation like ODBC. The system model can be Client/Server or Browser/Server. Most GIS vendors provide both types. The C/S model is fast and efficient, however it may need additional work such as deployment and update of client software. In B/S model, users only need an internet browser and most processing is finished on the server side, however it has the disadvantage of low efficiency.

4) Data sharing based on Web Services can hide most technical details in a distributed architecture, which is popular nowadays. By Web Services, functions of GIS data sharing can be integrated in various applications. Open Geospatial Consortium (OGC) has launched Web Services Initiative which proposes an evolutional, interoperable-specification-based, and integrated seamless framework (OGC Web Services, OWS). The OWS is a loosely coupling deployment based on OGC abstract and implementation specifications of related GIS data sharing which include the Web Map Service (WMS) and Web Feature Service (WFS).

In the National Flood Monitoring and Assessment System, the third type data sharing and exchanging is adopted. Thematic data from various governmental departments have different taxonomy of their own data. Visualization, query and analysis of all data must be achieved in one application environment in order to provide an integrated service to the central government. So each governmental department should embed the data provision interface or convert data to predefined format. A special data interface and data exchange platform is required for those specialized departments which have wide applications. Aiming at the diversity of different thematic data, data exchange and integration are handled by foreground and background, which are used for static data and dynamic data, respectively (Figure 19.17).

Data exchanging centre is automatically connected with other related data nodes. It examines if there are new data available at any time and then converts them to the format needed by disaster monitoring and assessment system. Data exchanging system can adjust its checking frequency, data formats and configuration of sub data nodes. Data to be transformed are satellite cloud maps, precipitation forecast data and in-situ data, typhoon forecast data, etc., as indicated above.

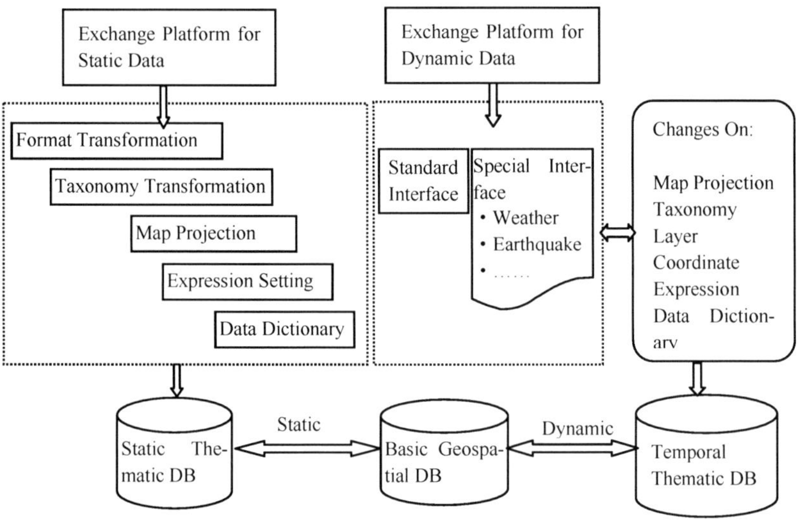

Fig. 19.17 Schema of Data Exchange

19.3.3 Integration of Multi-Source Information

Flood disaster data with the feature of multi-source, complex structure cover the departments of the Ministry of Civil Affairs, the Ministry of Land and Resources, the Ministry of Water Resources, the Ministry of Agriculture, the Ministry of Environment protection, etc. These data from different industries and departments have the characteristics of multi-type, multi-scale, multi-resolution, multi-temporal, multi-coordinate system, and so on, which result in inconsistent and discontinuous data, accordingly it is difficult to apply the data directly in disaster monitoring and assessment in emergency response (Abel *et al.* 1998).

Emergency response and spatial decision supporting system are based on geospatial data, so it needs to integrate various flood disaster information from different departments. Through dynamic data exchange, integrated display, query, statistics and analysis, the system provides the integrated information services such as the distribution of flood disaster, disaster, disaster relief and post-disaster reconstruction to the leaders of the State Council and the staffs of the General Office of the State Council.

Integration of multi-source information can be realized by the methods of metadata, exclusive interface and XML, so as to assure that data matches the spatial position; data organization is consistent with data storage. Multi-source data are integrated by thematic data conversion, theme-oriented data organization and spatial information link. The integrated data can be displayed, queried and analyzed under the unified application software environment. At present, the spatial data are integrated with water information, hydrology information, water quality data, weather cloud data, rainfall data, rainfall forecast data, remote sensing disaster data, groundwater data, socio-economic data, and so on. Some of the integrated results are illustrated here:

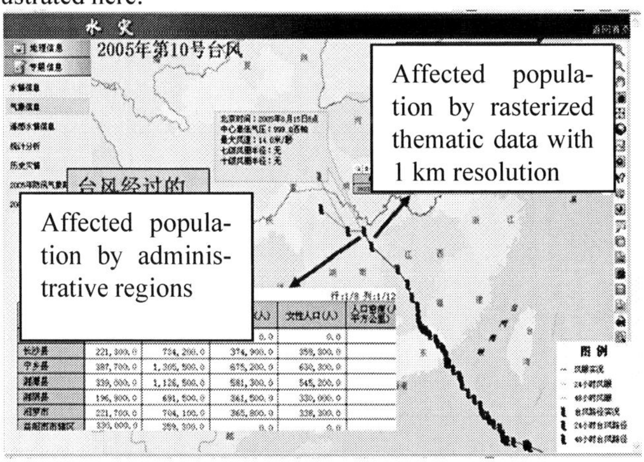

Fig. 19.18 Integration of weather information and population information. (Population statistics affected by Typhoon)

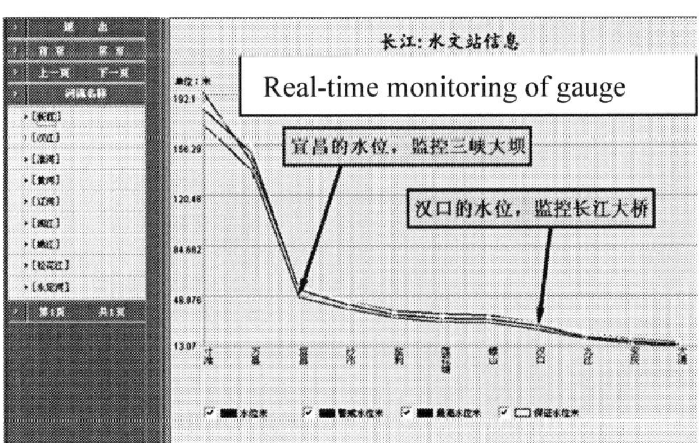

Fig. 19.19 Integration of geographic information and hydrology information. (Dynamic monitoring on gauge station of Yangtze River

19.3.4 Services

China is one of the countries that often suffered from flood hazards in the world. Defending and reducing the damage and loss caused by floods is extremely important and urgent to China. Along with a wide application of the non-engineering measures in all aspects of flood prevention and disaster reduction, GIS technology can play a role in flood forecast, on-site command and post-disaster evaluation. Multi-scale spatial data are not only the basis of all kinds of thematic data, but can also provide data support for retrieval and analysis.

The services and contents on flood disaster monitoring and assessment are mainly used to solve the spatial data management (including spatial data input, output, conversion, transmission, storage, maintenance, etc.), data querying, flood information integration, flood information publication, and so on.

(1) Conversion of thematic data on-line. The new generated data will be dynamically (timing or on a regular basis) converted to the data format required by flood prevention and weather information services system. Its main functions include time setting, data format conversion, and so on. The converted data include weather cloud data, rainfall data, rainfall data, rainfall forecast data, and typhoon forecast data.

(2) Scheduling and display of multi-scale basic geospatial data. Through the scheduling of multi-scale basic geospatial data, the corresponding thematic contents can be displayed and queried according to different needs from the users, so as to meet the business needs of different users and different levels.

(3) Loading, query and display of water information. It mainly includes the reservoir inquiries, the reservoir alarm inquiries, the reservoir flood inquiries, hydrological alarm inquiry, hydrological flood inquiries, river flood inquiries, reservoir search, the hydrological station search, and so on. The spatial distribution of flood facilities (reservoirs, hydrological stations) can be showed based on the geospatial data. The reservoir cross-section map with real-time water level information can be shown by inquiring indicators of the reservoir. It can query the historical water information of the reservoir and water information of the hydrological stations to get the cross-section map of the major river's water level change.

(4) Query, display, and output of weather information. Query and display of the status and description of rainfall in terms of time periods (for example, from 8 am to 8 pm) are available. Searching and displaying the distribution of rainfall forecast according to the time span (such as 24 hours or 48 hours) can be realized. Also, it is necessary to search and generate the cloud animations and weather cloud map description in terms of time. Moreover, it is possible that one can search and display the path of typhoon movement and its main technical parameters. Flood weather information can be printed and outputted as PDF file. Meanwhile, it can be browsed and transmitted through the internet.

(5) Flood simulation and analysis. According to the comparison of the DEM data and the elevation data of the water level, flooded regions can be ac-

quired. A fitted surface can be calculated based on the hydrological measuring stations records. Thus, the grid data sets lower than the water level can be obtained by the comparison of the DEM data and the tilted surface. It can calculate the submerged area and flood storage capacity. Furthermore, it can categorize the important facilities such as residential areas and traffic lines in the submerged areas.

(6) Web publishing of the flood weather information. This publishes the flood weather information on web browser daily, including meteorological information (for example, rainfall status from 8:00am to 8:00pm, the 24-hour or 48-hour rainfall forecasting, the weather cloud map at 8 o'clock, the path of the typhoon movement, the weather bulletin, important weather report, etc.) distributed at a specific electronic publication format (PDF format) , water level information (such as the chart of water level), remote sensing information (such as images, video information commentary), flood information (such as flood and drought information, the reservoir water level, etc.), and basic spatial information.

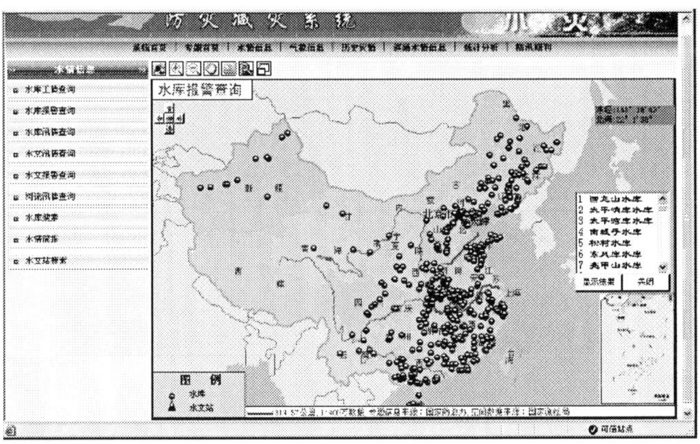

Fig. 19.20 Query of reservoir alarms

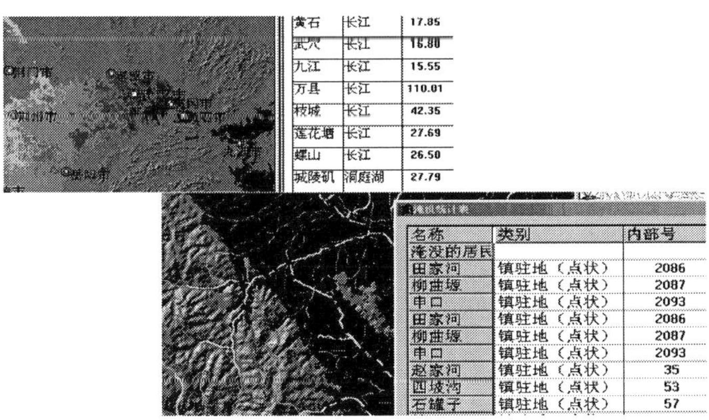

Fig. 19.21 Damage estimation of the flood

19.4 Water Resources and Desertification Monitoring System

19.4.1 Background and Necessity

China is a serious drought and water shortage country. Per capita water resource is only 2300 cubic meters, only 1/4 of the world average. Canada is 48 times larger than China on this index, Brazil is 16 times, Indonesia is 9 times, the former Soviet Union is 7 Times, the United States is 5 times, and even Japan and Mexico have higher water per capita. Therefore, China is one of the poorest countries in water per capita. But on the other hand, China is the largest water consumption country in the world. Only in 2002, the fresh water consumption of the country reaches 549.7 billion cubic meters, which is about 13% of the world. It is about 1.2 times of the United States in 1995, whose fresh water supply was 470 billion cubic meters. From the 1970s, the water shortage gradually spread from the locals to the whole country and became a more serious issue. China's agriculture and the national economy are seriously affected. The area under drought-hitting is 2 million to 2.6 million square kilometers each year, which affect food production by 15 billion to 20 billion kg and industrial output value more than 2,000 billion Yuan. Moreover, 70 million people have difficulty in water drinking. The distributing of water resources is unequal severely from the North to the South in China. In Yangtze River watershed and its southern area, the population accounts for 54% and water resources accounts for 81%. There is 46% of the population in the north, but only 19% of water resources. Due to the natural environment as well as the high intensity of human activities, the shortage of water resources is becoming more and more serious in the north of China.

At the same time, China's desert area is larger and becoming even wider. China is one of the countries which suffer from serious desertification. At the northwest region of China, there are 12 deserts which stretch into miles of sand line in the north. The area of the desert and desertification land in the country is 1,533,000 square kilometers. It accounts for 15.9% of the land area. About 50 million people are affected directly by desertification. In the region of northwest, northern parts of north-China, and western parts of northeast ("Three North"), there are about 2 million ha. of farmland affected sandstorm disasters each year. The food production is low and instable. There are 1.5 billion ha. of grassland that are in serious degradation. As a result of sandstorms, thousands of water conservancy facilities are broken. Desertification and drought have a serious impact on industry and agriculture in some area of China. 60% of poor counties in China are here. In the most serious region, problems of food and clothing have not yet been resolved.

Over the years, the Chinese Government has been committed to the work of water resources protection, rational development and utilization, and land desertification prevention and control.

Through using scientific methods, such as remote sensing to monitor and analyze national water resources and desertification land and establish water resources and desertification monitoring system, it can provide scientific basis and decision-making information for the government about water resource protection and desertification prevention and control.

Combined with remote sensing (RS), geographic information systems (GIS) and global navigation satellite system (GNSS) have become common technologies of observing resources and environment. The resources and environment monitoring supported by GIS can provide not only dynamic and friendly interfaces for displaying the status and historical change of resources and environment, but also simulate the future process of the development about resources and environment. As a result, based on spatial fundamental geographic data integrated with satellite remote sensing data, Chinese Academy Surveying and Mapping established the Water Resources and Desertification Monitoring Analysis System which provides related information services for the management and decision-making of government.

19.4.2 Data Content

The data content of the Water Resources and Desertification Monitoring Analysis System includes:

Multi-scale fundamental spatial geographic data, including 1:4000000, 1:1000000, 1:250000 and 1:50000 maps with many features, such as administrative divisions, water, rivers, roads and ancillary facilities, residents places, geographical names, terrain, peak, and so on. They are all basic scale geographic information data of China;

Multi-temporal and multi-resolution satellite remote sensing monitoring data, including the Landsat TM remote sensing image data with 30 meter resolution, "Beijing-1" small satellite remote sensing monitoring image data with 32 meter resolution and the national desertification image data;

Remote sensing interpretative information, including multi-scale national land desertification distribution vector data based on analysis and interpretation of remote sensing image, the main water-body distribution vector data in Beijing and its surrounding.

Statistical data, including various types of water bodies' area in Beijing and its surrounding for many years, the statistical data of desertification land area of major provinces and typical cities, the statistical data of multi-year change about the area of desertification land.

19.4.3 Service Themes

The Water Resources and Desertification Monitoring Analysis System consists of two parts: Beijing and its surrounding water resources, the status of national land desertification.

(1) Water resources. The change of water bodies in Beijing and its surrounding region are monitored by using "Beijing-1" satellite multi-spectral data. In particular, monitoring results of important water sources of Beijing's reservoirs are made monthly, the change of water area is analyzed based on the monitoring image.

- The distribution of water resources of Beijing and surrounding region (Figure 19.22 left). This part mainly includes the surface water resources of the five major river basins, the surface water resources of the Chaobai River Basin, Yongding River basin, North Canal basin, Daqing River basin, and Jiyun River basin. Various visualization methods are provided to show the spatial distribution of the various water bodies in different river basins.

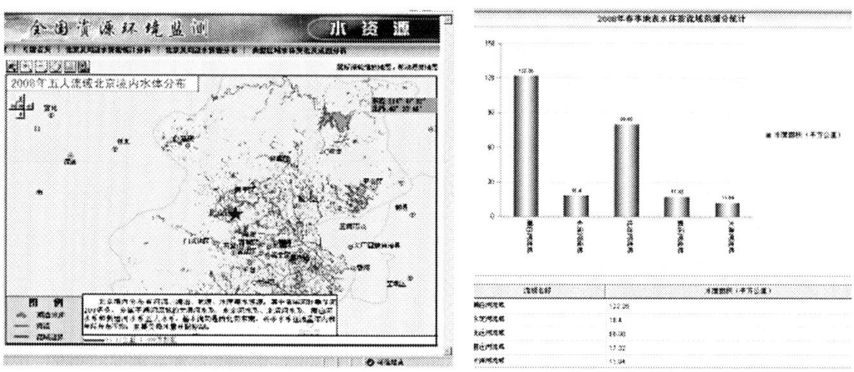

Fig. 19.22 Distribution and statistics of water resources in Beijing

- Statistics of Water resources in Beijing and its surrounding (Figure 19.22 right). Using statistical maps and statistical charts, we achieve the water resources statistics of the river basins, the water resources statistics of the administrative regions and the water resources statistics of all types of water bodies.
- Analysis of the water bodies change and causes in typical region (Figure 19.23). It includes all the important water bodies in Beijing and its surrounding, such as all the reservoirs of Beijing, Miyun reservoir and Guanting reservoir, the water sources of the South-North Water Transfer and Baiyangdian. Based on the fundamental geographic information, we achieved that the dynamic changes show of the water area and the scope of the important water bodies for many years, the comparison of water scope at different times and three-dimensional visualization. At the same time, the system achieves statistics of the water area changes and analysis of the change causes.

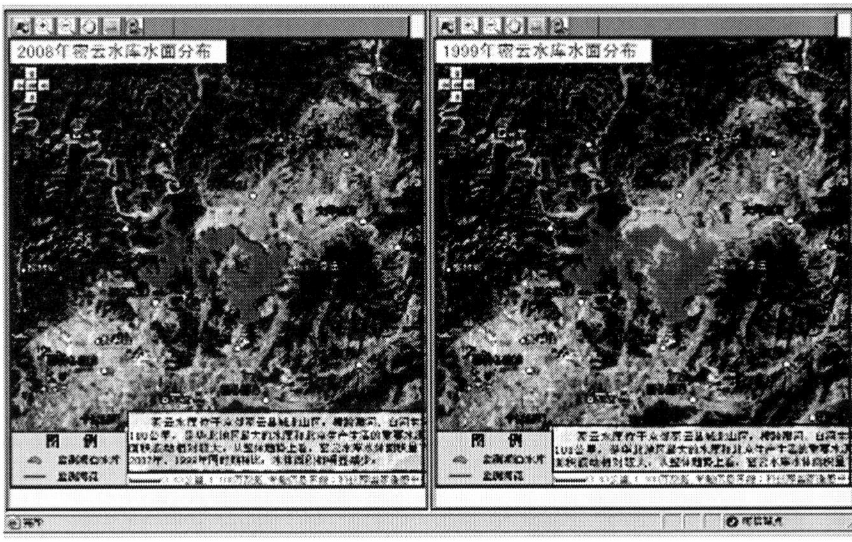

Fig. 19.23 Water distribution in 2008 (left) and 1999 (right) of Beijing Miyun reservoir

(2) Desertification. The condition of the national desertification is monitored by using the "Beijing-1" satellite multi-spectral data annually. Based on monitoring image data, the vector boundaries of the eight main deserts are extracted for state analysis of the national desertification land. Taking the counties and cities of serious desertification for example, the analysis about the changes trend and causes of desertification is performed. On the basis of the above work, we established the topic of desertification application based on spatial fundamental geography information.

Fig. 19.24 Distribution of deserts in China

- Distribution of the national desertification land. It includes the overall distribution of the National desertification, the provincial (autonomous regions) distribution of desertification land and area statistics of desertification land by county (Figure 19.24).
- Analysis on the status of national desertification land. It includes analysis of the provincial desertification extent in the Northern China and the distribution of desertification in Beijing and its surrounding (Figure 19.25). This processing includes extraction and statistical analysis of the desertification land on different radius of the buffer zone in Beijing and its surrounding, the extent analysis of land desertification in provinces with severe desertification.

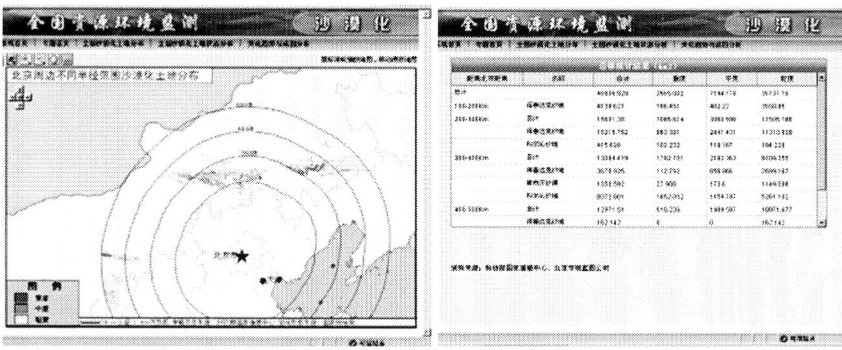

Fig. 19.25 Statistics of desert around Beijing

- Analysis of change trend and causes. Taking Yulin City and the Blue-Flag County in Inner Mongolia Autonomous Region with serious desertification as case areas, the analysis includes change trend and causes of land desertification at micro-scale. For example, Yulin City, the main functions include: the distribution of land desertification in 2002, the distribution of land desertification in 2006, the changes of desertification land from 2002 to 2006, the transfer matrix of desertification land and analysis of land changes trend from 2002 to 2006.

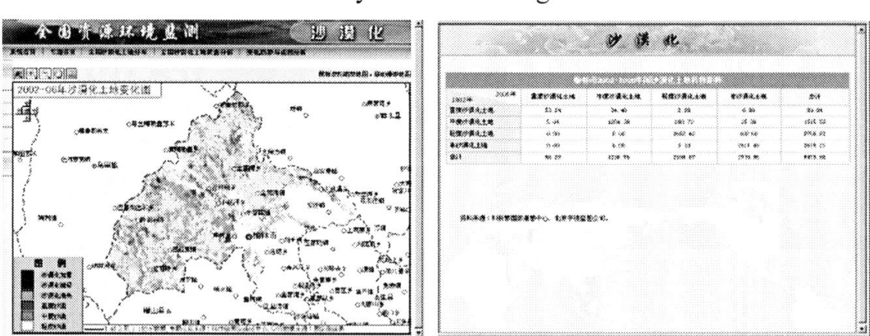

Fig. 19.26 Analysis of trend and driving forces of Desertification

19.4.4 Benefits

The Water Resources and Desertification Monitoring Analysis System has been deployed at government agencies and management department. It provides the real-time and historical change information of water resources and desertification for manager and decision-maker. It played an important role in analysis and decision-making for the rational development and utilization of water resources and land desertification control.

19.5 West Development Information Service System

19.5.1 Background and Necessity

Western China is abundant in natural resources and has vast potential market with its important strategic location. Owing to some reasons of nature, history and society, the western region has comparatively laggard economic standard. Per capita GDP is only two thirds of average standard in China and less than 40 percents compared with the eastern region. Therefore, it is urgent for the western region to step forth and mend its pace to carry out reasonable exploitation.

West development is very complicated and requires vast system engineering. Scientific decision-making is required to make the engineering run well. Moreover, scientific foundation is the basis of scientific decision-making, which must depend on application system construction. The West Development Information Service System is established by related departments of the State Council. The system can provide some important managing information on western economic development including strategic reports, planning and a great deal of west information such as different levels of districts, societal and economic development, infrastructure, related contrast among Western, Eastern and Middle Regions, etc. It can be an effective aided tool for different levels of governments to carry out West development after incessant applications. Moreover, sub-information systems, which can be connected and interoperated over the computer network, have also established in the western region including Chongqing, Yunnan, Guizhou, Gansu, and other western provinces and municipalities.

19.5.2 Data Contents

Compared with the eastern region, the greatest competitiveness of the western region is its rich resources. In a certain sense, the western region of China is far from developed on natural resources, social and economic resources, information and resources, etc. Information (data) is a carrier of resources. A comprehensive understanding and grasp of the resources in the western region development can be finished through information collection, collation and analysis to these resources. Under the condition of the correct analysis of the characteristics and the potential advantages of the western region, scientific development programming, implementation plans and objectives of action measures can be worked out.

With the characteristics of being authoritative and timely, the data of the West Development Information Service System roots in more than 10 authorized business sectors such as the contents of national economy, social development, environment, infrastructure and so on. These data can be broadly divided into two categories: spatial data (vector data, image data, DEM data, and raster data) and non-spatial data (statistical data, text data, and multimedia data).

And all the data are stored and managed uniformly in the large-scale database of Oracle9i database or SQL Server 2000.Different kinds of data can be made to match by property keywords and achieve effective integration.

19.5.3 Service Themes

The West Development Information Service System and its provincial application sub-systems are established to be a dynamic multi-purpose information system under the condition of uniform technical regulations and platform support according to the needs of governmental departments of all levels. The latter is divided into four components, i.e. Resources and Ecological Environment, Economic and Social Development, Returning Farmland to Forests and Grassland, Free Trade Area. These sub-systems are mainly reflected in natural resources, national economic and social situation and sustainable development and can provide the leadership and departments aided decision-making information from macrostructure, mesostructure to microstructure levels.

Among them, Gansu Province has finished the deployment of Gansu Province Returning Farmland to Forest and Grassland Application System. Chongqing City has deployed the Chongqing Economic and Social Development Aided Decision-making System. Guizhou Province has established the Guizhou Province Resources and Entironment System. Meanwhile, Yunnan Province has carried out the deployment of Yunnan - the ASEAN Free Trade Area Application System.

Gansu Province Returning Farmland to Forest and Grassland Application System is established based on the Zhuanglang County as a pilot according to the demands of 'detailed to the village to monitor the land, corresponds to the farmers'. By comparing remote sensing images with different time stamp the effectiveness of farmland can be checked and confirmed without too much field work. Gradually, combined with the application of global positioning system (GPS), the work can be done to achieve real-time access the data of check and acceptance provided by field investigators, supervision and inspection work in the indoor field.

Chongqing Economic and Social Development Aided Decision-making System is constructed based on the requirement of Chongqing's current economic and social development of e-Government to speed up the pace of development and finish the overall implementation put forward by the central government. In this system, population database, corporation database and basic spatial database can be made to realize effective integration. And the establishment of the Chongqing Municipality level of urbanization evaluation index system and evaluation model can be reflected in the development of Chongqing's urbanization. Some good attempts have been made in the aspect of internal city and town evaluation and model application in order to provide application services to the government in the field of aided decision-making about national economic and social development.

Guizhou Province Resources and Environment System consists of the two parts of provincial and country levels and is becoming an information network covering country, provincial and county levels combining with e-government network of State Council. The purpose of establishing this system is used to carry out rational development and utilization of resources, environmental protection and promote sustainable economic and social development services for the leadership of the department in order to provide reasonable and scientific decision-making services and ultimately realize the goal of scientific use, rational development, effective protection and the use of scientific and rational development, effective protection so as to achieve sustainable development.

The ASEAN Free Trade Area Application System provides macro-scale geospatial data framework of Yunnan Province and ASEAN. This system can integrate social, economic information resources based on B/S structure and reflect the geographical distribution of ASEAN countries, Yunnan's regional advantages in China-ASEAN Free Trade Area as well as construction information of international channels to connect Southeast Asia.

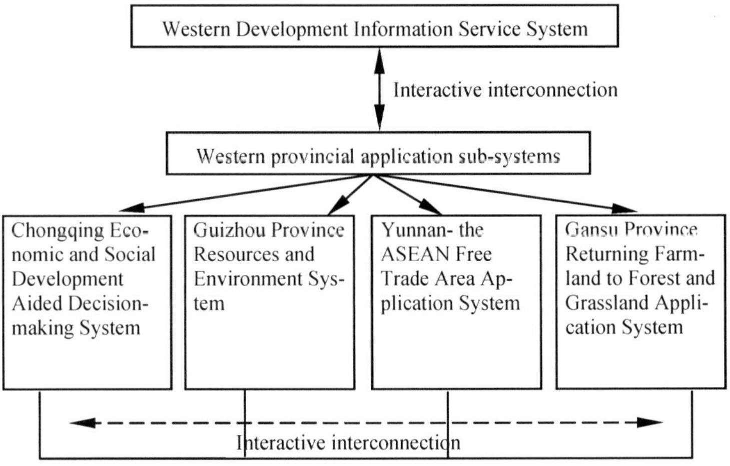

Fig. 19.27 Western Development Information Service System and its provincial application sub-systems

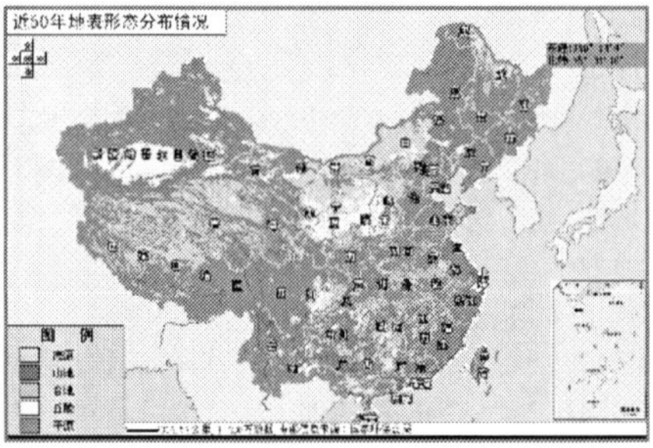

The distribution of the surface in the western provinces and cities

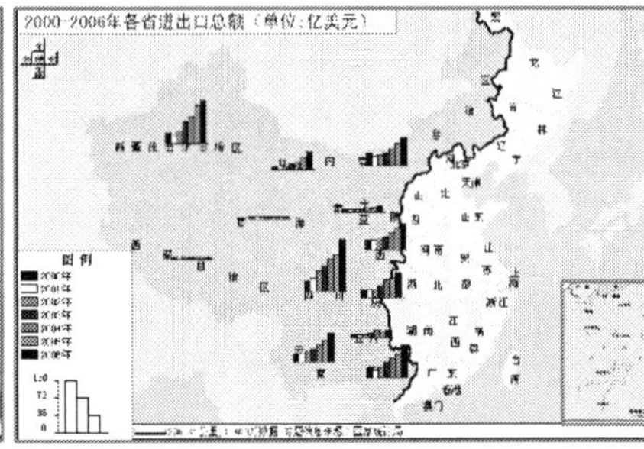

Western provinces foreign trade in 2002-2006

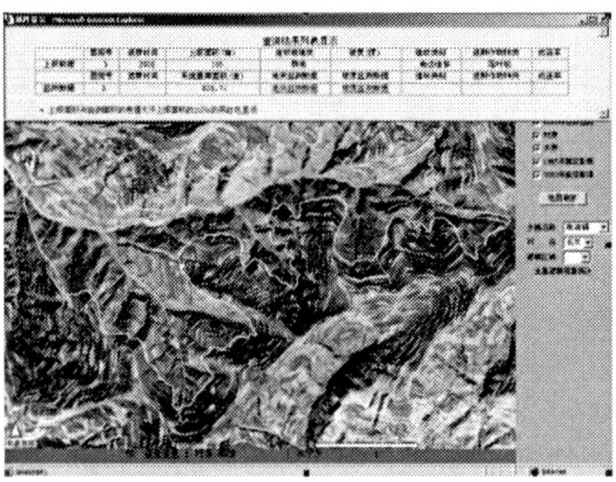

Gansu Province Returning Farmland to Forest and Grassland Application System

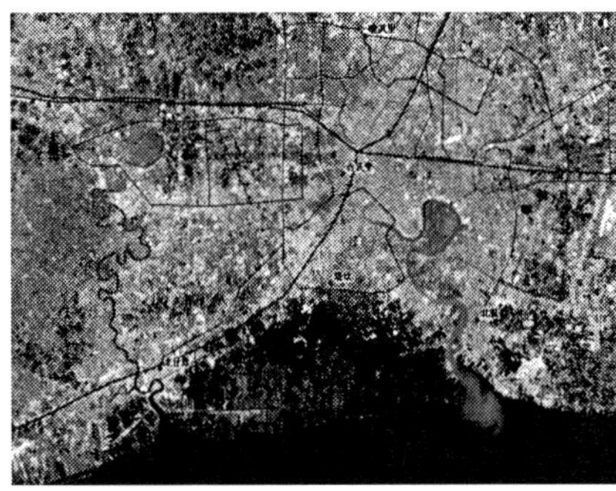

Yunnan- the ASEAN Free Trade Area Application System

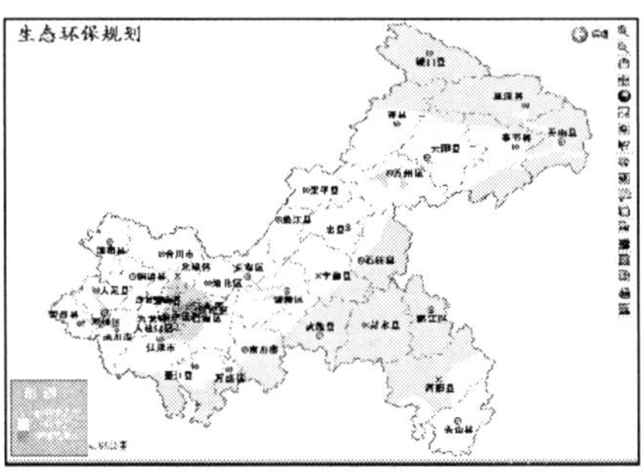

Chongqing Economic and Social Development

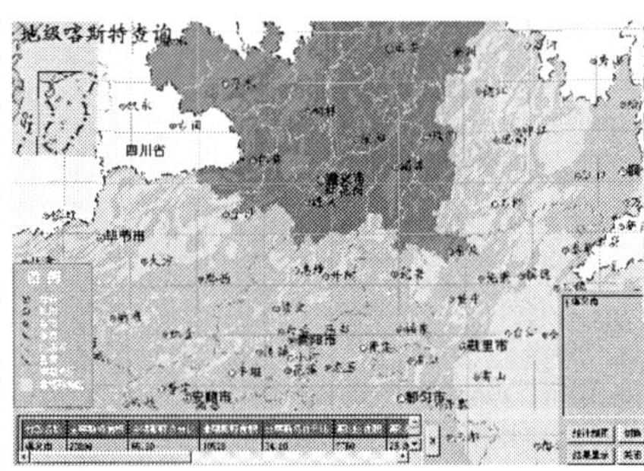

Guizhou Province Resources and Environment System

Fig. 19.28 West development information service system

- Supports integration of GIS and DSS. On the technical goals, it is the core for West Development Information Service System to adopt the technology of geo-spatial information and decision-making support. And this system supports integration of GIS and DSS and is applied to construct the universal geo-spatial aided decision-making application platform oriented to e-Government.
- Supports constructing unified, integrated resource database with coordinated services and adopt large-scale database to storage and manage uniformly spatial and non-spatial information of e-Government.
- Supports integrating massive non-spatial thematic data based on the framework of geo-spatial data and establish the e-government integrated resources database in order to build application system oriented to different government departments and make it run well to deal with a variety of information resources development and applications, etc.
- Supports industrial standards by using XML/GML as the standard and norm of data transmission and conversion, which can improve the level of standardization and regularization effectively.

19.5.4 Benefits

West Development Information Service System has been used in national planning of western China and helped policy makers on improving efficiency and avoiding bias. The development of the system also boosted the application of earth observation data in e-Government and facilitating information of the western region of China.

(1) Integration and sharing of information can be achieved among different departments so as to provide services for decision-making to government and hence government decision-making can be more scientific and efficient.

(2) To promote the technical development of domestic copyright software and enhance their application level.

(3) The process of information sharing and standardization is promoted by integrating e-Government database of information resources.

(4) The fields of spatial data applications have been widened to a new height of distributed government network application and it can be extended to construct e-Government in the government agencies.

(5) This system can be applied not only for the policy makers in central government and the four western provinces, but also for all over the sub-provincial local government departments.

19.6 Conclusions

This chapter summarized the type of Earth observation data, data model and management (spatial databases), system architecture, key functional modules and roles in e-Government of spatial decision support systems. Furthermore, we described application systems running in some Chinese governmental agencies, which concentrated on fast response of the disaster.

During the Wenchuan earthquake, the rescue work has benefited greatly from Earth observation data and information system. Earth observation technology is becoming an indispensable means for spatial decision making. The acquisition and processing of Earth observation data and fundamental geospatial data is necessary for monitoring and management of earthquake hazards, monitoring and evaluation of land cover, water resource and desertification etc.

With the fast development of economy and society, problems of environment and natural resource management pose new challenges to the government, such as deterioration of water, desertification and typical gradual disasters. This demands the exploiting of new Earth observation technology for decision making.

References

Abel J *et al.* (1998). Towards Integrated Geographical Information Processing. International Journal of Geographic Information System, 12(4): 353-371.

Andreas I (2001). Multi-scale GIS: from generalization to data integration. Proceedings of the 20th International Cartographic Conference, Beijing.

Chen S (1987). Geographical Data Handling and GIS In China. International Journal of Geographical Information Science, 1(3): 219-228.

Goodchild F (2003). Geospatial Data in Emergencies. In: Cutter S *et al* (ed) The Geographical Dimensions of Terrorism. Routledge, New York.

Henry T (1997). The Global Spatial Data Infrastructure. Proceedings of IEAS 97 and IWGIS 97, Beijing.

Janet E.Nichol *et al.*(2006). Application of High-Resolution Stereo Satellite Images to Detailed Landslide Hazard Assessment. Geomorphology, 10.

Jia WJ (2004). Web Service Based Web Feature Service. International Archives of the Photogrammetry. Remote Sensing and Spatial Information Sciences XXXV, (B4) :163-166.

Longley A *et al.* (2005). Geographic Information Systems and Science. 2nd edn. Wiley, New York.

Liu JY *et al.* (2003). Digital Simulation of Population Density in China. Acta Geographica Sinica, 58(1): 17-24.

Ma Y, Tian W X (2007). The High Resolution Remote-Sensing Images in Landslide Interpretation of Loess Platea. Resources Environment& Engineerining, 21(2).

Niu B R (2002). Remote Sensing Investigation and Analysis of Landslide Hazard. Highway, 10.

Xie T (2007). Sensitive Appraisal of Landslide Base on RS and GIS in Reservoir of Hydroelectric Power in Dashixia. Master's degree thesis, Chengdu Polytechnic University.

List of Contributors

The corresponding or leading contributors of a chapter are denoted with an asterisk (*), followed by their brief biographies. The list is in the order of chapters and authors.

***Deren Li, Chapter 1**

State Key Laboratory of Information Engineering in Surveying, Mapping, and Remote Sensing, Wuhan University, Wuhan 430079, P.R. China

Prof. Deren Li received his Dr.-Ing degree from the University of Stuttgart, Germany in 1985. He is currently the Chair of the Academic Committees of Wuhan University and the State Key Laboratory of Information Engineering in Surveying, Mapping and Remote Sensing, an academician of both the Chinese Academy of Sciences and the Chinese Academy of Engineering, and holds an honorary doctorate from ETH, Zurich. Prof. Li was the President of ISPRS Comm. III and Comm. VI in 1988-1992 and 1992-1996, respectively. Since 1989, Prof. Li has supervised 108 Ph.D students and nine post-doctoral researchers. He has published over 615 papers and nine books. He received the Samuel G. Gamble Award at the ISPRS XXI Congress in 2008. He also has received numerous academic honors, including the National Great Invention Award, the National Science and Technology Progress Award, the National Excellent Textbook Award, and the Excellent Educational Achievements Award.

***Lei Yan, Chapter 2**

Beijing Key Laboratory for Spatial Information Integration and its Applications, School of Earth and Space Sciences, Peking University, Beijing, 100871, China

Prof. Yan received his Ph.D degree in Instrument Science and Technology from Tsinghua University, Beijing, China. Since then he has worked as a faculty member at Peking University, Beijing and is currently a full professor in the Institute of Remote Sensing and GIS at the School of Earth Science and Space Technology of Peking University. His research interests include remote sensing, digital imaging, UAV remote sensing, polarization remote sensing, location-based service, and space-time closed loop cybernetics. He has led and participated in many research projects and has authored and co-authored numerous academic books and articles.

Zhiyang Gou, Chapter 2

Beijing Key Laboratory for Spatial Information Integration and its Applications School of Earth and Space Sciences, Peking University, Beijing, 100871, China

Yini Duan, Chapter 2

Beijing Key Laboratory for Spatial Information Integration and its Applications School of Earth and Space Sciences, Peking University, Beijing, 100871, China

***Qingquan Li, Chapter 3**

State Key Laboratory of Information Engineering in Surveying, Mapping, and Remote Sensing, Wuhan University, Wuhan 430079, P.R. China

Transportation Research Center, Wuhan University, Wuhan, 430079, China

Professor Qingquan Li is the director of the Transportation Research Center, Wuhan University, as well as vice president of the University. His research interests are GIS-T, three-dimensional and dynamic data modeling in GIS, location-based service, integration of GIS, GPS, and remote sensing, intelligent transportation system, and surveying engineering. Prof. Li has published four books and over 100 papers in his research field and serves on the editorial boards of a number of open journals. He also has completed over 20 research projects and has received various national and professional prizes in recognition of his research achievements.

Yong Liu, Chapter 3

State Key Laboratory of Information Engineering in Surveying, Mapping, and Remote Sensing, Wuhan University, Wuhan 430079, P.R. China

School of Electronic Information, Wuhan University, Wuhan, 430079, China

Qingzhou Mao, Chapter 3

State Key Laboratory of Information Engineering in Surveying, Mapping, and Remote Sensing, Wuhan University, Wuhan 430079, P.R. China

Transportation Research Center, Wuhan University, Wuhan, 430079, China

***Mattia Crespi, Chapter 4**

Dipartimento Idraulica Trasporti Strade, Area di Geodesia e Geomatica, Sapienza Università di Roma, via Eudossiana 18, 00184 Rome, Italy

Mattia Crespi received his M.Sc. in Civil Engineering from the Polytechnic of Milan, Italy in 1987 with a dissertation pertainng to numerical problems for solving large systems of equations. In 1992, he received his PhD in Geodesy and Surveying at the Polytechnic of Turin with a dissertation pertaining to GPS data processing and in particular, a new strategy for ambiguity fixing. He achieved the rank of full professor in surveying and geomatics in 2005 at the University of Rome “La Sapienza,” Rome, Italy. He has supervised about 80 B.Sc. and M.Sc. students and 10 PhD students. Professor Crespi is an author of more than 130 papers concerning different topics in geomatics. His current research interests are focused on high resolution satellite imagery orientation and quality assessment, digital terrain and surface models generation and quality investigation, GNSS permanent arrays for post-processing and real time positioning services, and application of geomatics to archaeology.

Francesca Fratarcangeli, Chapter 4

Dipartimento Idraulica Trasporti Strade, Area di Geodesia e Geomatica, Sapienza Università di Roma, via Eudossiana 18, 00184 Rome, Italy

Francesca Giannone, Chapter 4

Dipartimento Idraulica Trasporti Strade, Area di Geodesia e Geomatica, Sapienza Università di Roma, via Eudossiana 18, 00184 Rome, Italy

Francesca Pieralice, Chapter 4

Dipartimento Idraulica Trasporti Strade, Area di Geodesia e Geomatica, Sapienza Università di Roma, via Eudossiana 18, 00184 Rome, Italy

***Xiuxiao Yuan, Chapter 5**

School of Remote Sensing and Information Engineering, Wuhan University, 129 Luoyu Road, Wuhan 430079, P.R. China

Dr. Yuan focuses on research and education in remote sensing and global positioning systems and their integration. He has made unique and original contribution to the areas of theories and methods for high precision photogrammetric positioning, GPS/IMU-supported aerotriangulation, geometric processing of high-resolution satellite imagery, etc. He published five monographs and more than 100 papers.

***Shunlin Liang, Chapter 6**

Department of Geography, University of Maryland, College Park, MD 20742, USA

Shunlin Liang received his Ph.D. degree in remote sensing and GIS from Boston University, Boston, Massachusetts, USA. He was a post-doctoral research associate with Boston University from 1992 to 1993 and a validation scientist with the NOAA/NASA Pathfinder AVHRR Land Project from 1993 to 1994. He is currently a professor with the University of Maryland, College Park. His main research interests focus on spatio temporal analysis of remotely sensed data, integration of data from different sources and numerical models, and linkage of remote sensing with carbon and water cycles. He authored the book "Quantitative Remote Sensing of Land Surfaces" (Wiley, 2004) and edited the book "Advances in Land Remote Sensing: System, Modeling, Inversion and Application" (Springer, 2008). Dr. Liang is a member of NASA's ASTER and MODIS science teams and a co-chairman of the International Society for Photogrammetry and Remote Sensing Commission VII/I Working Group on Fundamental Physics and Modeling. He is an Associate Editor of the IEEE Transactions on Geoscience and Remote Sensing.

Kaicun Wang, Chapter 6

Department of Geography, University of Maryland, College Park, MD 20742, USA

Wenhui Wang, Chapter 6

NOAA/NESDIS/STAR and I. M. System Groups, Inc., MD, USA

Dongdong Wang, Chapter 6

Department of Geography, University of Maryland, College Park, MD 20742, USA

Sheng Gui, Chapter 6

Department of Geography, University of Maryland, College Park, MD 20742, USA

School of Resource and Environmental Science, Wuhan University, Wuhan 430079, China

Xiaotong Zhang, Chapter 6

Department of Geography, University of Maryland, College Park, MD 20742, USA

School of Resource and Environmental Science, Wuhan University, Wuhan 430079, China

Jeremy Mirmelstein, Chapter 6

Department of Geography, University of Maryland, College Park, MD 20742, USA

Xiufang Zhu, Chapter 6

Department of Geography, University of Maryland, College Park, MD 20742, USA

Hye-yun Kim, Chapter 6

NOAA/NESDIS/STAR and I. M. System Groups, Inc., MD, USA

Juan Du, Chapter 6

College of Resources Science and Technology, Beijing Normal University, Beijing, P.R. China

Steven Running, Chapter 6

School of Forestry, University of Montana, Missoula, MT 59812, USA

John Townshend, Chapter 6

Department of Geography, University of Maryland, College Park, MD 20742, USA

Si-Chee Tsay, Chapter 6

NASA Goddard Space Flight Center, Greenbelt, MD 20771, USA

Robert Wolf, Chapter 6

Code 614.5, NASA Goddard Space Flight Center, Greenbelt, MD 20771, USA

Crystal Schaaf, Chapter 6

Department of Geography, Boston University, Boston, MA 02215, USA

Alan Strahler, Chapter 6

Department of Geography, Boston University, Boston, MA 02215, USA

***Rudiger Gens, Chapter 7**

Alaska Satellite Facility, Geophysical Institute, University of Alaska Fairbanks, 903 Koyukuk Dr., P.O. Box 757320, Fairbanks, AK 99775-7320, USA

Dr. Rudiger Gens is a remote sensing scientist at the Alaska Satellite Facility (ASF) at the Geophysical Institute, University of Alaska Fairbanks (UAF). His Ph.D. dissertation in Engineering from the University of Hannover, Germany, was entitled "Quality Assessment of SAR Interferometric Data." His research specialization is in the processing and applications of Synthetic Aperture Radar (SAR) and Interferometric SAR (InSAR) data. At ASF Dr. Gens has developed a variety of software tools for making SAR data more accessible and usable, provided technical support to a large remote sensing user community, and used satellite data to improve understanding of the changing Arctic landscapes and the processes guiding these changes. Dr. Gens is also a cooperating faculty member in the Department of Geology and Geophysics at UAF, where he teaches courses in the principles and applications of SAR and InSAR. He is an associate editor for the International Journal of Remote Sensing.

***Björn Waske, Chapter 8**

Faculty of Electrical and Computer Engineering, University of Iceland, 107 Reykjavik, Iceland

Björn Waske received his degree in Applied Environmental Sciences with a major in Remote Sensing from Trier University, Germany, in 2002. Until mid-2004 he was a research assistant in the Department of Geosciences at Munich University, Germany and conducted research on the use of remote sensing data for flood forecast modeling. From 2004 to 2007 he pursued a PhD in Geography at the Center for Remote Sensing of Land Surfaces (ZFL) at the University of Bonn, Germany. In 2006, he was a visiting faculty member in the Department of Electrical and Computer Engineering at the University of Iceland for three months; and since early 2008 he has been a post-doctoral researcher there. His research activities concentrate on advanced concepts for image classification and data fusion, with a focus on SAR and multispectral and hyperspectral images.

Mingmin Chi, Chapter 8

School of Computer Science, Fudan University, Shanghai, P.R. China

Jón Atli Benediktsson, Chapter 8

Faculty of Electrical and Computer Engineering, University of Iceland, 107 Reykjavik, Iceland

Sebastian van der Linden, Chapter 8

Geomatics Laboratory, Humboldt-Universität zu Berlin, 10099 Berlin, Germany

Benjamin Koetz, Chapter 8

Remote Sensing Laboratories, Universität Zürich, 8057 Zurich, Switzerland

***Liangpei Zhang, Chapter 9**

State Key Laboratory of Information Engineering in Surveying, Mapping, and Remote Sensing, Wuhan University, Wuhan 430079, P.R. China

Liangpei Zhang received his B.Sc. degree in physics from Hunan Normal University in 1982, his M.Sc. degree in optics from the Xi'an Institute of Optics and Precision Mechanics of Chinese Academy of Sciences in 1988, and his Ph.D. degree in photogrammetry and remote sensing from Wuhan University in 1998. From 1997 to 2000, he was a professor at the School of the Land Sciences in Wuhan University. In 2000, he joined the State Key Laboratory of Information Engineering in Surveying, Mapping and Remote Sensing as a professor and the head of the Remote Sensing Division. Since 2007 he has been a "ChangJiang Scholar" professor (chair professor), which is selected by the Education Ministry of China. He has published over 160 papers. His research interests include hyperspectral remote sensing, high resolution remote sensing, image processing, and artificial intelligence. Dr. Zhang has served as co-chair of the SPIE Series Conferences on Multispectral Image Processing and Pattern Recognition (MIPPR), the Conference on Asia Remote Sensing in 1999, editor of the MIPPR01, MIPPR05, Geoinformatics Symposiums, associate editor of Geo-spatial Information Science Journal, a member of the Chinese National Committee for the International Geosphere-Biosphere Program, and an executive member for the China Society of Image and Graphics.

Yanfei Zhong, Chapter 9

State Key Laboratory of Information Engineering in Surveying, Mapping, and Remote Sensing, Wuhan University, Wuhan 430079, P.R. China

***Peng Han, Chapter 10**

State Key Laboratory of Information Engineering in Surveying, Mapping, and Remote Sensing, Wuhan University, Wuhan 430079, P.R. China

Han Peng received his Ph.D. degree from the State Key Lab for Information Engineering in Surveying, Mapping and Remote Sensing, Wuhan University, Wuhan, China, in 2008. He is the author or coauthor of three books and more than 20 journal and conference papers. His current research interests include the scale effect in image processing and pattern recognition.

Zhilin Li, Chapter 10

Department of Land Surveying and Geo-Informatics, Hong Kong Polytechnic University, Hong Kong

State Key Laboratory of Information Engineering in Surveying, Mapping, and Remote Sensing, Wuhan University, Wuhan 430079, P.R. China

Jianya Gong, Chapter 10

See the author list for Chapter 13.

***John Trinder, Chapter 11**

School of Surveying and Spatial Information Systems, The University of New South Wales, Sydney NSW 2052 Australia

John Trinder graduated from the University of New South Wales (UNSW) with a BSurv and PhD and also a MSc degree at ITC in The Netherlands. He was employed at the School of Surveying and SIS at UNSW from 1965 to 1999, progressing to the position of professor in 1991 and head of the school from 1990-1999. He is currently an emeritus professor at UNSW. Dr. Trinder has taught and conducted research at UNSW for about 40 years, specializing in photogrammetry and remote sensing, and has published more than 150 scientific papers in journals and conference proceedings. He has held a number of executive positions in the Council of the International Society for Photogrammetry and Remote Sensing (ISPRS), including treasurer from 1992-1996, secretary general from 1996-2000, president from 2000-2004, first vice president from 2004-2008, and is an honorary member.

Arcot Sowmya, Chapter 11

School of Computer Science and Engineering, The University of New South Wales, Sydney NSW 2052 Australia

Gianni Lisini, Chapter 12

Center for Risk and Security, IUSS_Institute for Advanced Studies, V.le Lungo Ticino Sforza, 56, 27100 Pavia, Italy

Fabio Dell'Acqua, Chapter 12

Dipartimento di Elettronica, Università di Pavia,Via A. Ferrata, 1, 27100 Pavia, Italy

Paolo Gamba, Chapter 12

Dipartimento di Elettronica, Università di Pavia,Via A. Ferrata, 1, 27100 Pavia, Italy

Paolo Gamba is currently an associate professor of Telecommunications at the University of Pavia, Italy. He received the Laurea degree in Electronic Engineering, "cum laude," from the University of Pavia, Italy, in 1989, and the Ph.D. degree in Electronic Engineering from the same University in 1993. He was the organizer and technical chair of the biennial GRSS/ISPRS Joint Workshops on "Remote Sensing and Data Fusion over Urban Areas" from 2001 to 2009. He has been the chair of Technical Committee 7 "Pattern Recognition in Remote Sensing" of the International Association for Pattern Recognition (IAPR) from October 2002 to October 2004 and chair of the Data Fusion Committee of the IEEE Geoscience and Remote Sensing Society. He is a Senior Member of IEEE, and since 2009 he has served as editor-in-chief of the IEEE Geoscience and Remote Sensing Letters. He has been a guest editor of special issues for a number of remote sensing journals. He has published more than 60 papers in international peer-review journals and presented more than 150 papers in workshops and conferences.

***Jianya Gong, Chapter 13**

State Key Laboratory of Information Engineering in Surveying, Mapping, and Remote Sensing, Wuhan University, Wuhan 430079, P.R. China

Prof. Jianya Gong is currently the director of the State Key Laboratory of Information Engineering in Surveying, Mapping and Remote Sensing. He received his PhD from Wuhan Technical University of Surveying and Mapping in 1992. He has been a visiting professor in the Department of Geography, University of Massachusetts, Boston, USA, a Changjiang Chair professor at the State Key Laboratory of Information Engineering in Surveying, Mapping and Remote Sensing, a visiting professor at the Department of Surveying and Land Information, Hong Kong Polytechnic University, and a visiting professor in the Department of Geography and GeoInformation Science at George Mason University, Fairfax, Virginia, USA. Prof. Gong has served multiple times as ISPRS Commission Secretary and Working Group Chair since 1988. His research interests include geospatial data structure and data model, geospatial data integration and management, geographical information systems software, geospatial data sharing and interoperability, photogrammetry, and GIS and remote sensing applications.

Huayi Wu, Chapter 13

State Key Laboratory of Information Engineering in Surveying, Mapping, and Remote Sensing, Wuhan University, Wuhan 430079, P.R. China

Wenxiu Gao, Chapter 13

State Key Laboratory of Information Engineering in Surveying, Mapping, and Remote Sensing, Wuhan University, Wuhan 430079, P.R. China

Peng Yue, Chapter 13

State Key Laboratory of Information Engineering in Surveying, Mapping, and Remote Sensing, Wuhan University, Wuhan 430079, P.R. China

Xinyan Zhu, Chapter 13

State Key Laboratory of Information Engineering in Surveying, Mapping, and Remote Sensing, Wuhan University, Wuhan 430079, P.R. China

***Qing Zhu, Chapter 14**

State Key Laboratory of Information Engineering in Surveying, Mapping, and Remote Sensing, Wuhan University, Wuhan 430079, P.R. China

School of Communication and Information Engineering , Shanghai University, Shanghai 200072, P.R. China

Dr. Zhu is a full professor in photogrammetry and GIS. He received his MSc in photogrammetry and remote sensing in 1989 from the Southwestern Jiaotong University and his PhD in 1995 from the Northern Jiaotong University Since 1995, he worked at the Wuhan Technical University of Surveying and Mapping as a post-doctoral associate at the the Hong Kong Polytechnic University as a visiting research scientist, the Chinese University of Hong Kong as a Croucher visiting scholar, and the Nanyang Technological University of Singapore as a Tan Chin Tuan Exchange Fellow. Since 1997, professor Zhu has been working at the State Key Laboratory of Information Engineering in Surveying, Mapping and Remote Sensing (Wuhan University), and is the co-chair of ISPRS WG V/4. His main research themes include digital photogrammetry, 3D GIS, and virtual geographic environments.

Xiaoxia Yang, Chapter 14

State Key Laboratory of Information Engineering in Surveying, Mapping, and Remote Sensing, Wuhan University, Wuhan 430079, P.R. China

Haifeng Li, Chapter 14

State Key Laboratory of Information Engineering in Surveying, Mapping, and Remote Sensing, Wuhan University, Wuhan 430079, P.R. China

Gottfried Konecny, Chapter 15

Institute for Photogrammetry and Geoinformation, Leibniz University Hannover, Nienburger Strasse 1, 30167 Hannover, Germany

Emeritus Prof. Konecny received his Dr.-Ing. degree from the Technical University of Munich, Germany. He taught at the University of New Brunswick, Canada, and then led the Institute for Photogrammetry and Engineering Surveys at the University of Hannover, Germany from 1971 to 1989. He has honorary doctorates from four different countries, is an honorary professor of Wuhan University, and a member of four academies. He has been involved in a number of space programs in the USA and Europe. He was the president of ISPRS from 1984 to 1988, and chairman of the European Association of Remote Sensing Laboratories from 1993 to 1997. He is actively involved in a number of geoinformatics and Earth observation consulting activities around the world. He has authored and co-authored six books and over 230 papers.

***Qiming Zhou, Chapter 16**

Department of Geography, Hong Kong Baptist University, Kowloon Tong, Kowloon, Hong Kong

Qiming Zhou is a professor of geography in the field of geographical information science, specialized in GIS and remote sensing. He has been actively engaged in and published in the fields of digital terrain analysis, landuse and land cover change detection, monitoring and modeling, spatio-temporal pattern analysis in multi-temporal remote sensing, and GIS and remote sensing applications to environment and natural resource management. Prof Zhou is currently the chairman of ISPRS Working Group VII/5 "Methods of Change Detection and Process Modeling," vice-president of the Hong Kong GIS Association, vice-president of the Hong Kong Society of Photogrammetry and Remote Sensing, council member of the China GIS Association, a member of the board of directors of the International Association of Chinese Professionals in GIS and an editor of numerous international journals in geospatial science and technology.

***Xiaoling Chen, Chapter 17**

State Key Laboratory of Information Engineering in Surveying, Mapping and Remote Sensing, Wuhan University, Wuhan 430079, P.R. China

The Key Laboratory of Poyang Lake Ecological Environment and Resource Development, Jiangxi Normal University, Nanchang 330022, P.R. China

Prof. Xiaoling Chen received her PhD from the Department of Geography, Nanjing University, China in 1997. She was then employed by the State Key Laboratory of Information Engineering in Surveying, Mapping and Remote Sensing, Wuhan University, as an associate professor from 1998 to 2000, a professor since 2000, and director of the Ocean Observation and Digital

Engineering Center, and general secretary of Sirindhorn International Center for Geoinformatics which is named by the Thailand Princess. In 2008, she was named the dean of the School of Geography and Environment at Jiangxi Normal University and director of the Key Laboratory of Poyang Lake Wetland and Watershed Research, Ministry of Education of China. Her research interests are Quantitative Remote Sensing especially Ocean Color Remote Sensing, Application of RS/GIS on Ecology, Environment and Disaster Mitigation. She was a visiting scholar with NASA, the Institute for Crisis, Disaster and Risk Management of the United States, the United Nations, the International Ocean Institute Canada, IHDP, START, IOC, and the National Center for Geocomputation of Ireland, the Hong Kong Polytechnic University, and GISTDA Thailand. She has published 127 papers and eight books.

Zhifeng Yu, Chapter 17

State Key Laboratory of Information Engineering in Surveying, Mapping and Remote Sensing, Wuhan University, Wuhan 430079, P.R. China

***Jie Shan, Chapter 18**

School of Civil Engineering, Purdue University, IN 47907, USA

Prof. Shan received his PhD in photogrammetry and remote sensing from Wuhan University, China. Since then he has held faculty positions at universities in China, Sweden, and the United States, and has been a research fellow in Germany. Currently, he is an associate professor in geomatics engineering in the School of Civil Engineering of Purdue University, where he teaches remote sensing and geospatial information science and technology. His research interests are digital mapping, geospatial data mining, and urban modeling. He is the recipient of multiple academic awards, among which are the Talbert Abrams Grand Award, the ESRI Award for Best Scientific Paper in GIS (First Place), and the Excellence in GIS Award. He has been a co-chair of the ISPRS Data Fusion Working Group and the Pattern Recognition for Remote Sensing Working Group. He is an associate editor for IEEE Geoscience and Remote Sensing, the assistant editor for Photogrammetric Engineering and Remote Sensing, and has served as a reviewer for more than 20 technical journals.

Ejaz Hussain, Chapter 18

School of Civil Engineering, Purdue University, IN 47907, USA

KyoHyouk Kim, Chapter 18

School of Civil Engineering, Purdue University, IN 47907, USA

Larry Biehl, Chapter 18

Purdue Terrestrial Observatory, Purdue University, IN 47907, USA

***Jixian Zhang, Chapter 19**

Chinese Academy of Surveying and Mapping, Beijing, China

Dr. Jixian Zhang earned his Ph.D. from the Remote Sensing and Information Engineering Department of Wuhan University, China in 1994, and then worked in the State Key Laboratory of Image Information Processing and Artificial Control, Huazhong University of Science and Technology as a post-doctoral researcher. He is currently the president of the Chinese Academy of Surveying and Mapping.. Dr. Zhang has published three books and 140 publications, won 16 prestigious national awards, and has worked on 40 research projects as PI or Co-PI. He has been an editorial board member of the International Journal of Geoinformatics, the International Journal of Applied Earth Observation and Geoinformation, and the Journal of Surveying and Mapping; the editor-in-chief of the International Journal of Image and Data Fusion. He served as the chair of ISPRS Working Group VII/6: Remote Sensing Data Fusion, since 2004 and committed to continue in the 2008-2012 term. Dr. Zhang is currently a professor at Wuhan University, Capital Normal University, Shandong University of Science and Technology, and China University of Mining and Technology, and has instructed over 50 M.Sc and PhD students.

Yu Zeng, Chapter 19

Chinese Academy of Surveying and Mapping, Beijing, China

Wenhan Xie, Chapter 19

Chinese Academy of Surveying and Mapping, Beijing, China

Tao Wang, Chapter 19

Chinese Academy of Surveying and Mapping, Beijing, China

Index

G

H

M

N

O

S

T

X

Y

Z